C·A·B INTERNATIONAL

BRITISH CROP PROTEC'

GW00630724

The UK
Pesticide Guide

Editor – G.W. Ivens M.A., D.Phil.

CAB International (CABI), formerly the Commonwealth Agricultural Bureaux, is a not-for-profit intergovernmental organization registered with the United Nations which is owned and administered by its 28 member governments. Established in 1928, CABI exists to provide information, scientific and development services for agriculture and allied disciplines throughout the world. Further details available from CABI Centre, Wallingford, Oxon, OX10 8DE.

The British Crop Protection Council (BCPC), is a self-supporting limited company with charity status which was formed in 1968 to promote the knowledge and understanding of the science and practice of crop protection. The corporate members include government departments and research councils; advisory services; associations concerned with the farming industry, agrochemical manufacturers, agricultural engineering, agricultural contracting and distribution services; universities; scientific societies; organizations concerned with the environment; and some experienced independent members. Further details available from BCPC, 49 Downing Street, Farnham, GU9 7PH.

British Library Cataloguing in Publication Data
U.K. pesticide guide 1991.
  1. Great Britain. Pesticides
  I. Ivens, G. W. (Giles William) *1923–*
632.95

ISBN 0-85198-704-4

Cover design by David Pratt
Printed and bound in Great Britain by Redwood Press Ltd., Melksham

PREFACE

The MAFF annual publication *Approved Products for Farmers and Growers* (the 'Orange Book') was discontinued after 1985 and there was then no readily available guide to approved uses for pesticides in the UK. The need for such a reference book became more urgent with the introduction of the Food and Environment Protection Act 1985 and the regulations introduced under the Act from October 1986 onwards. The British Crop Protection Council and CAB International have collaborated in a joint enterprise to fill this gap by producing *The UK Pesticide Guide*, an annual publication now in its fourth edition.

Information on the chemicals which are registered in the UK is obtained from the annual MAFF/HSE publication, *Pesticides* and the monthly *Pesticides Register*. Information about what products are on the market, their recommended uses and other details summarized in the pesticide profiles are drawn from the product labels or manuals provided by the suppliers of agrochemicals. This book is an independent publication the aim of which is to present reliable and impartial guidance in the selection of pesticides for specific uses. However, notwithstanding anything that appears in the book, before using any pesticide product the user must still **read the label** to ensure that it is suitable for the purpose intended.

In addition to information on pesticides the book includes a list of approved adjuvants, which are not themselves pesticides but are used in association with them. Another particular feature is the inclusion of the official 'off-label' approvals. These are mostly related to minor crops which may be of major significance to a small number of people but are not of sufficiently widespread importance to be included on firms' labels.

If pesticide chemicals are to be used safely and effectively it is important that the facts needed for making informed choices should be readily available. The encouraging reception of the first three editions of *The UK Pesticide Guide* suggests that it is already assuming a leading role in providing what is needed.

The editor is indebted to all the companies who supplied information on their products, and to members of the British Crop Protection Council who gave valuable advice on content and layout.

The Ministry of Agriculture, Fisheries and Food provided valuable assistance through its Pesticide Registration Department at Harpenden and staff of the Pesticides Safety Division in London. To Dr D.M. Smith of the Health and Safety Executive's Employment Medical Advisory Service and Miss D. Collings of C5 Division, Home Office, grateful thanks are extended for contributing the sections on Poisoning by Pesticides and Chemicals Subject to the Poisons Law respectively.

G.W. Ivens
Editor

DISCLAIMER

Every effort has been made to ensure that the information in this book is complete and correct at the time of going to press but the Editor and the publishers do not accept liability for any error or omission in the content, or for any loss, damage or other accident arising from the use of products listed herein. Omission of a product does not necessarily mean that it is not approved and available for use.

The information has been selected from official sources and from suppliers' labels and product manuals of pesticides approved under the Control of Pesticides Regulations 1986 and available for pest, disease and weed control applications in the United Kingdom.

It is essential to follow the instructions on the approved label before handling, storing or using any crop-protection product. Approved 'off-label' uses are made entirely at the risk of the user.

The contents of this publication are based on information received up to September 1990.

CONTENTS

ABOUT THE UK PESTICIDE GUIDE

This book provides a guide to the bewildering range of products available in the UK to farmers and growers for the control of pests, weeds and diseases or the regulation of plant growth. The aim is to help the user in the selection of pesticides, by providing details of all the products marketed and their approved uses, both those recommended on labels and those which have been officially approved but are not recommended on labels and for which the suppliers will not accept responsibility (off-label approvals). The book is revised annually.

Sources of information
The information for this edition has been drawn from these authoritative sources :
— approved labels and product manuals received from the suppliers of pesticide formulations up until September 1990.
— the **MAFF/HSE** publication *Pesticides 1990*, which gives approval numbers but not the uses of the products.
— entries in *The Pesticides Register*, the official monthly publication on approvals, up to and including Issue No. 8 (September 1990).
— Notices of Approval relating to off-label uses published in *The Pesticides Register*.

Coverage
As well as identifying the products available, the book provides guidance on how to use them safely and effectively, but without giving details of dose rates, volumes, spray schedules, or approved tank mixes. While we have tried to cover all other important factors, it is not possible to give full details of the recommendations for use of all products in a book of this size. **Before using any pesticide product it is essential that the user should read the label carefully and comply strictly with the instructions it contains.**

This edition covers those products approved for use in agriculture, horticulture (including amenity horticulture), forestry and areas near water. Similar formulations may have home garden, domestic, food storage, public health or animal health uses which are not covered here. Within the chosen fields of use all types of pesticide (acaricide, algicide, fungicide, herbicide, insecticide, lumbricide, molluscicide, nematicide and rodenticide) are included, together with the plant growth regulators covered by the Control of Pesticides Regulations which are employed as straw shorteners, sprout inhibitors and various other horticultural purposes.

Criteria for inclusion
To be included in the Guide, a product must meet the following conditions :
— It must have **MAFF/HSE** Approval under the 1986 Regulations.
— Information on the approved uses must have been provided by the supplier.
— The product must be expected to be on the UK market in 1991.
Where chemicals have recently (since 1987) been suspended or withdrawn the active ingredient is listed but no further details are given.

The primary aim of the book is to provide a practical guide to what products the farmer, grower or forester can legally and realistically obtain and to indicate the purposes for which they may be used.

LAYOUT AND CONTENT OF THE BOOK

Crop Guide
The Crop Guide serves as an index to the more complete information in the Pesticide Profiles. It is preceded by a Crop Index showing how the crops are grouped. Under the main headings individual crops are listed in alphabetical order. For larger crop groups such as cereals, the chemicals recommended in the Crop Guide must be confirmed by reference to the Pesticide Profiles to check whether they apply to all types or to some only, such as winter cereals. Because of differences in the wording of product labels it may sometimes be necessary to refer to broader categories of organism in addition to specific organisms (e.g. annual grasses and annual weeds in addition to annual meadow grass).

Pesticide Profiles
These are arranged in alphabetical order of the active ingredient using the common names approved by the British Standards Institution. Individual products can be found by looking up the Index of Proprietary Names at the back of the book. Within the profile a table covers all products approved and available on the market, in the following style :

Product name	Main supplier	Active ingredient content	Formulation type	MAFF* Approval No.
9 Roundup	Schering	360 g/l	SL	03947
10 Roundup Four 80	Monsanto	480 g/l	SL	03176

For mixtures of active ingredients, the respective contents are given in the same order as the ingredients appear in the profile heading. For a list of the formulation types in full, see the key to abbreviations on p. 4.

The Uses section lists all those uses Approved or Provisionally Approved at the date of going to press (October 1990), giving both the principal target organisms and the recommended crops or situations. Where there is an important condition of use or the approval is 'off-label' this is given in parentheses. Numbers in square brackets refer to the numbered products in the table above. Thus a typical use approved for Roundup but not for Roundup Four 80 appears as :

Weed beet in SUGAR BEET (*wiper application*) [8,9]

Below the Uses paragraph, notes are presented under the following headings. Any reference to dose rate is made in terms of product rather than active ingredient.

---

* In a few cases where products registered with the Health and Safety Executive are used in agriculture HSE numbers are quoted here, e.g. H0462

Efficacy	Factors important in making the most effective use of the treatment. Certain factors, such as the need to apply chemicals uniformly, the need to spray at appropriate volume rates and the need to prevent settling out of active ingredient in spray tanks, are assumed to be important for all treatments and are not emphasized in individual profiles.
Crop safety/ restrictions	Factors important in minimizing the risk of crop damage and, where applicable, maximum permitted number of applications. With many chemicals the stage of growth of the crop at the time of application is of critical importance and this is given both in descriptive terms and, where a standardized growth stage key is available (i.e. for cereals, oilseed rape and peas), in the form of the numerical codes published jointly by AAB, BCPC and NAC (GS numbers) – See Appendix 3.
Special precautions/ environmental safety	Inclusion under The Poisons Rules (1982); any warnings about organophosphate and other anticholinesterase compounds; the hazard classification as indicated on the label. The hazard classification is given with products classified as Very toxic, Toxic, Harmful, Irritant or Corrosive but, where no hazard classification is required, no mention of this is made. Any specific restrictions relating to use and any environmental hazards are given here, including any potential dangers to livestock, wildlife, bees and fish. The need to avoid drift onto neighbouring crops and to wash out equipment thoroughly after use are important with all pesticide treatments, but may be mentioned here if of special importance.
Protective clothing/ Label precautions	The COSHH regulations require that wherever there is a label recommendation for use of protective clothing it should be preceded by the phrase *'Engineering control of operator exposure must be used where reasonably practicable in addition to the following personal protective equipment:'* and followed by *'However, engineering controls may replace personal protective equipment if a COSHH assessment shows they provide an equal or higher standard of protection.'*

These phrases are not repeated in the profiles but an indication is given of the items of protective clothing which may be required in using the various products. Without a greater degree of uniformity in the wording of labels, it is not possible to show the items required for different operations e.g. handling concentrate, spraying with hand lance, cleaning equipment etc. However, the items which may be needed for one or other of the operations involved in using a product are given as a series of letters on the first line of this section and refer to the items of protective clothing listed in Appendix 2A.

Other label precautions not concerned with choice of protective clothing are given on the second line as a list of numbers referring to the key to standard phrases in Appendix 2B. **The letters referring to protective clothing and the numbered precautions are given for information only and should not be used for the purpose of making a COSHH assessment without reference to the current product label.**

Withholding period	Any requirements relating to pesticides with potentially harmful effects on livestock are given here.
Harvest interval	The period which must elapse after the last application of a product before harvesting for human or animal consumption is given in this section.
Approval	Includes notes on approval for aerial or ULV application and 'Off-label' approvals, giving references to the official approval document numbers (OLA numbers), copies of which can be obtained from ADAS or NFU and which must be consulted before the treatment is used.
Stop-press	Announcements of any major recent changes.

KEY TO ABBREVIATIONS

Formulation type

AE	Aerosol generator
AL	Other liquids to be applied undiluted
BB	Block bait
CB	Bait concentrate
CG	Encapsulated granule
CR	Crystals
CS	Capsule suspension
DP	Dustable powder
DS	Powder for seed treatment
EC	Emulsifiable concentrate
EW	Oil in water emulsion
FP	Smoke cartridge
FS	Flowable concentrate for seed treatment
FT	Smoke tablet
FU	Smoke generator
FW	Smoke pellets
GA	Gas
GB	Granular bait
GE	Gas-generating product
GG	Macrogranule
GP	Flo-dust
GR	Granules
GS	Grease
HN	Hot fogging concentrate
KK	Combi-pack (solid/liquid)
KL	Combi-pack (liquid/liquid)
KN	Cold-fogging concentrate
KP	Combi-pack (solid/solid)
LA	Lacquer
LI	Liquid, unspecified
LS	Solution for seed treatment
MG	Microgranules
OL	Oil miscible liquid
PA	Paste

PC	Gel or paste concentrate
PO	Pour-on
PS	Seed coated with a pesticide
PT	Pellet
RB	Ready-to-use bait
RH	Ready-to-use spray in hand-operated sprayer
SA	Sand
SC	Suspension concentrate (= flowable)
SG	Water soluble granules
SL	Soluble concentrate
SP	Water soluble powder
SS	Water soluble powder for seed treatment
SU	Ultra low-volume suspension
TB	Tablets
TC	Technical material
TP	Tracking powder
UL	Ultra low-volume liquid
VP	Vapour releasing product
WB	Water soluble bags
WG	Water dispersible granules
WP	Wettable powder
WS	Water dispersible powder for slurry treatment of seed
XX	Other formulations

Other abbreviations

a.i.	active ingredient
CDA	controlled droplet application
cm	centimetre(s)
d	day(s)
EBDC	ethylene-bis-dithiocarbamate fungicide
g	gram(s)
GS	growth stage (unless in formulation column)
h	hour(s)
ha	hectare(s)
kg	kilogram(s)
l	litre(s)
m	metre(s)
min	minute(s)
mm	millimetre(s)
MRL	Maximum Residue Limit
mth	month(s)
NA	notice of approval
OLA	off-label approval
PR	Pesticides Register
ULV	ultra-low volume
wk	week(s)
w/v	weight/volume
w/w	weight/weight
yr	year(s)

FEPA AND COSHH

The Food and Environment Protection Act 1985 (FEPA)

FEPA introduced statutory powers to control pesticides with the aims of protecting human beings, animals and plants, safeguarding the environment, ensuring safe, effective and humane methods of controlling pests and making pesticide information available to the public. Control of pesticides is achieved by the Control of Pesticides Regulations 1986 (COPR), which lay down the Approvals required before any pesticide may be sold, stored, supplied or advertised, and allow for the general requirements set out in various Consents, which specify the conditions subject to which approval is given. Consent A relates to advertisement, Consent B to sale, supply and storage, Consent C (i) to use and Consent C (ii) to aerial application of pesticides. The conditions of the Consents may be changed from time to time. Details are given in the MAFF/HSE reference book *Pesticides 1990* (revised annually) and are updated in the monthly periodical *The Pesticides Register.*

The controls currently in force include the following :

- Only approved products may be supplied, stored, advertised or used.
- Only products specifically approved for the purpose may be applied from the air.
- A recognised Storeman's Certificate of Competence is required by anyone who stores pesticides for sale or supply.
- A recognised Certificate of Competence is required by anyone who gives advice when selling or supplying pesticides.
- Users of pesticides must comply with the Conditions of Approval relating to use. The conditions may cover :
  - field of use
  - crops/situations where treatment may be applied
  - protective clothing requirements
  - maximum dose rate and number of applications
  - minimum dilution rate
  - minimum harvest interval
  - need to keep humans and animals out of treated areas
  - measures for protection of bees, wildlife and the environment

- A recognised Certificate of Competence is required by all persons applying pesticides as contractors and by those born after 31 December 1964 (unless working under direct supervision of a certificate holder).
- Only those adjuvants approved by MAFF may be used (see Adjuvants pp. 48–60).
- Regarding tank-mixes, 'Until 31 December 1991 no person shall combine or mix for use two or more pesticides which are anti-cholinesterase compounds unless the approved label of at least one of the pesticide products states that the mixture may be made; and no person shall combine or mix for use two or more pesticides if all the conditions of the approval relating to this use cannot be complied with'. A listing of permitted tank-mixtures is beyond the scope of this Guide. For information concerning farm crops the reader is referred to the annual Farmers Weekly Tank Mix Supplement (*Farmers Weekly*, 9 February 1990 and subsequent revisions).

The Control of Substances Hazardous to Health Regulations 1988 (COSHH)

The COSHH regulations, which came into force on 1 October 1989, were made under the Health and Safety at Work Act 1974 and are also important as a means of regulating the use of pesticides. The regulations cover virtually all subtances hazardous to health, including those pesticides classed as Very toxic, Toxic, Harmful, Irritant or Corrosive, other chemicals used in farming or industry and substances with occupational exposure limits. They also cover harmful micro-organisms, dusts and any other material, mixture or compound used at work which can harm people's health.

The basic principle underlying the COSHH regulations is that the risks associated with the use of any substance hazardous to health must be assessed before it is used and the appropriate measures taken to control the risk. The emphasis is changed from that pertaining under the Poisonous Substances in Agriculture Regulations 1984 (now repealed), whereby the principal method of ensuring safety was the use of protective clothing, to the prevention or control of exposure to hazardous substances by a combination of measures. In order of preference the measures should be :

a. substitution with a less hazardous chemical or product :
b technical or engineering controls (e.g. the use of closed handling systems etc.) :
c. operational controls (e.g. operators located in cabs fitted with air-filtration systems etc.) :
d. Use of personal protective equipment (PPE), which includes protective clothing.

Consideration must be given as to whether it is necessary to use a pesticide at all in a given situation and, if so, the product posing the least risk to humans, animals and the environment must be selected. Where other measures do not provide adequate control of exposure and the use of PPE is necessary, the items stipulated on the product label must be used as a minimum. It is essential that equipment is properly maintained and the correct procedures adopted. Where necessary, the exposure of workers must be monitored, health checks carried out and employees must be instructed and trained in precautionary techniques. Adequate records of all operations involving pesticide application must be made and retained for at least 3 years.

PESTICIDES: CODE OF PRACTICE

Detailed guidance on how to comply with the regulations is provided by the joint MAFF/HSE publication :

Pesticides: Code of practice for the safe use of pesticides on farms and holdings, 1990, ISBN 0 11 242892 4.

The Code of Practice is available from HMSO in a package, which includes additional up-to-date guidance on the consents to pesticide use, user certification, protection of honey bees, management of field margins and avoiding problems with pesticide wastes, together with an agricultural safety chart.

The Code of Practice gives comprehensive guidance on how to comply with COPR and the COSHH regulations under the headings User training and certification, Planning and preparation, Working with pesticides, Disposal of pesticide waste and Keeping records. In particular it details what is involved in making a COSHH Assessment. The principal source of information for making such an assessment is the approved product label. In most cases the label provides all the necessary information but in certain circumstances other sources must be consulted and these are listed in the Code of Practice.

The information given in the *UK Pesticide Guide* provides some of the answers needed to assess health risks, including the hazard classification and the level of operator protection required. However, the Guide cannot provide all the details needed for a complete hazard assessment, which must be based on the product label itself and, where necessary, by the Health and Safety Data Sheet and other official literature.

Further guidance and practical advice is available from the following publications :

Health and Safety Commission Approved Codes of Practice
 — COP 29, *Control of substances hazardous to health (General A CoP) and Control of Carcinogenic substances (Carcinogens A CoP)*, ISBN 0 11 885468 2, HMSO
 — COP 30, *Control of substances hazardous to health in fumigation operations*, ISBN 0 11 885469 0, HMSO

Other HSE publications

 — *COSHH Assessments*, ISBN 0 11 885470 4, HMSO
 — AS 6, *Crop spraying: agricultural safety*, 1988
 — AS 12, *Prevention of accidents*, 1984
 — AS 18, *Storage of pesticides on farms*, 1984
 — AS 25, *Training in the use of pesticides*, 1990
 — AS 26, *Protective clothing for use with pesticides*, 1986
 — AS 27, *Pesticides*, 1988
 — AS 28, *COSHH in agriculture*, 1989
 — Guidance Note CS 19, *Storage of approved pesticides: guidance for farmers and other professional users*, 1988, ISBN 0 11 885406 2, HMSO
 — Guidance Note EH 40/90, *Occupational exposure limits*, 1990, ISBN 0 11 885420 8, HMSO
 — Guidance Note HS(G) 51, *Storage of flammable liquids in containers*, 1990, ISBN 0 11 885533 6, HMSO

- Guidance Note MS 17 (revised), *Biological monitoring of workers exposed to organophosphorus pesticides*, 1987, ISBN 0 11 883951 9, HMSO
- Guidance Note EH 42, *Monitoring strategies for toxic substances*, 1989, ISBN 0 11 885412 7, HMSO
- IA C(L) 29, *Protective clothing in agriculture and allied industries*, 1989
- IND(G) 3(L), *First aid provision in small workplaces*
- IND(G) 64(L), *Introducing assessment*, 1989
- IND(G) 65(L), *Introducing COSHH*, 1989
- IND(G) 67(L), *Hazard and risk explained*, 1989

MAFF publications

- Booklet PB 0091, *Code of practice for suppliers of pesticides to agriculture, horticulture and forestry*, 1990
- Booklet 2078, *Guidelines for the use of herbicides on weeds in or near watercourses and lakes*, 1985
- Leaflet UL 79, *Pesticides: guide to the new controls*, 1987
- Reference Book 500, *Pesticides*, 1990, ISBN 0 11 242884 3 (and later editions), HMSO

Other useful publications

British Agrochemical Association
- *The plain man's guide to pesticides and the COSHH assessment*
- *How to avoid contaminating water with pesticides*, 1990

British Crop Protection Council
- *Nozzle selection handbook*, revised 1990, ISBN 0 948404 13 2
- *Hand-operated sprayers handbook*, 1989, ISBN 0 948404 30 2
- *Boom sprayers handbook*, 1990
- *Pesticide related law*, 1989, ISBN 0 948404 33 7
- *The management of cereal field margins*, 1989 (published jointly with Schering Agriculture)

Schering Agriculture
- *COSHH assessments* – an action plan
- *Working with pesticides guide: the regulations and your responsibilities*, revised May 1990

APPROVAL UNDER THE FEPA REGULATIONS

The FEPA regulations have replaced the earlier non-statutory Pesticides Safety Precautions Scheme (PSPS) and Agricultural Chemicals Approval Scheme (ACAS). Approval is now a legal requirement and it is an offence to use non-approved products or to use approved products in a manner which does not comply with the specific conditions of approval.

Full and Provisional Approval

The MAFF/HSE publication *Pesticides 1990* lists those products approved under the Control of Pesticides Regulations, together with the names of the marketing companies and the Registration numbers. The products listed may be approved either fully or provisionally (until a specified date) but no distinction is made between the two in the above book and the uses for which approval has been granted are not given.

The main source of information on the crops or situations in which the use of a particular pesticide is approved is the approved product label. From 1 January 1988 it became mandatory for operators to follow the conditions for use of a product laid down at the time of approval and the conditions must be clearly stated on the label.

Off-label arrangements

The interim arrangements for off-label use of pesticides announced in October 1987 expired in 1990 and new arrangements were announced in *The Pesticides Register* Issue No. 11, December 1989. The Pesticides Register announcement should be consulted for full details but the following summary gives an outline of the principal requirements :

1. Restrictions for extension to off-label use
   Safety precautions and statutory conditions relating to use as specified on the label must be observed. Application must be as stated on the label and in accordance with relevant codes of practice and requirements under COSHH.
   Pesticides must only be used on protected crops where specifically permitted (approval for use on tomatoes, cucumbers, lettuce, chrysanthemums and mushrooms apply to protected crops unless otherwise stated).
   Off-label use in or near water is not permitted under these arrangements.
   Off-label use by aerial application is not permitted under these arrangements.
   Rodenticides are not included in these arrangements.
   Use on land not intended for cropping is not included in these arrangements.
   Pesticides classed as harmful, dangerous or extremely dangerous to bees must not be used on any flowering crop (where use is recommended on flowering peas, cereals or oilseed-rape this use relates only to these crops).

2. Non-edible crops
   Subject to the above restrictions, pesticides approved for use on any growing crop may be used on hardy ornamental nursery stock, ornamental plants, bulbs and flowers, seed crops (not including seed potatoes, cereals, oilseeds, peas and beans), forest nursery crops prior to final planting out and flax grown for fibre.

3. Nursery fruit crops
   Subject to the above restrictions, pesticides, approved for use on any crop for human or animal consumption may be used on nursery fruit trees, bushes and non-fruiting strawberries provided that any fruit harvested within one year is destroyed.

4. Crops partly or wholly for consumption by humans or livestock
Subject to the above restrictions pesticides already approved for certain crops may also
be used at the owner's risk on minor crops as detailed below :

ADDITIONAL CROPS PERMITTED	CROPS ON WHICH ALREADY APPROVED
Arable crops	
Poppy, mustard, linseed, evening primrose, borage (for oilseed), gold of pleasure	Oilseed rape (products hazardous to bees must not be applied during flowering)
Grass seed crops	Wheat, barley, oats, rye, triticale, grass for grazing or fodder
Oats, rye, triticale, durum wheat	Wheat, barley, (treatments applied before stage GS 32 only)
Fruit crops	
Almond (orchard floor only)	Apple, cherry, plum (herbicides on orchard floor only)
Chestnut, hazelnut, walnut (orchard floor only)	Apple (herbicides on orchard floor only)
Quince, crab apple	Apple, pear
Blackberry, hybrid Rubus (loganberry etc.)	Raspberry
Red currant	Black/white currant
White currant	Black/red currant
Bilberry	Black/white/red currant
Vegetable crops	
Beetroot	Carrot, radish
Celeriac	Carrot
Horseradish	Potato, carrot, radish
Jerusalem artichoke	Potato, carrot, radish, turnip, swede
Parsnip	Carrot
Salsify	Potato, carrot, radish
Swede	Turnip
Turnip	Swede
Garlic, shallot	Bulb onion, salad onion
Pepper, aubergine	Tomato
Squash, pumpkin	Cucumber, gherkin, courgette, melon
Broccoli	Cauliflower, calabrese
Calabrese	Cauliflower, broccoli
Roscoff cauliflower	Cauliflower, broccoli, calabrese
Endive, (frisee, radicchio etc.), leaf herbs	Lettuce
Beet leaves, sorrel	Spinach
Peas, edible podded (mange-tout, sugar snap)	Dwarf french or runner beans

11

ADDITIONAL CROPS PERMITTED	CROPS ON WHICH ALREADY APPROVED
Peas, non-edible podded (harvested green)	Vining peas
Peas, harvested dry	Vining peas, dwarf french, runner or broad beans
Broad beans	Vining peas, dwarf french beans (products hazardous to bees not during flowering)
Runner beans	Dwarf french beans, edible-podded peas (products hazardous to bees not during flowering)
Beans, harvested dry	Peas, dwarf french, runner or broad beans (products hazardous to bees not during flowering)
Fennel (as a vegetable)	Celery, leek
Kohl rabi	Celery, leek, cabbage
Edible fungi (oyster mushroom etc.)	Mushroom
Edible flowers (e.g. Nasturtium, but not hops or globe artichokes)	Lettuce (products hazardous to bees not during flowering)

Additional information is provided by specific Off-label Notices of Approval. These are uses for which approval has been sought by individuals or organizations other than the manufacturers. An index of Off-label Approvals (and to off-label uses which have not been approved) is published at intervals by MAFF, together with the Notices of Approval themselves, and these are widely available from ADAS or NFU offices. An index to the Off-label Approval numbers is given on pp. 13–19. Operators making use of such treatments must obtain a copy of the relevant Notice of Approval and comply strictly with the conditions laid down therein. Although approved, it must be understood that such treatments are not endorsed by manufacturers and are made entirely at the risk of the user.

Details of new approvals, both full and provisional, of amendments to existing approvals and of off-label approvals are published in *The Pesticides Register.*

INDEX TO OFF-LABEL APPROVAL NUMBERS

CROP	APPROVED USES AND OLA NUMBERS
Apples	Collar rot 0117, 0118, 0170/88. Crop setting 0171/88. Phytophthora 0183, 0185/88
Apricots	Weeds 0062, 0063/88; 0754/90
Artichoke	American serpentine leaf miner 0271/88; 0372, 0373, 0403, 0406, 0407, 0422, 0425, 1058/89. Leaf miners 1058/89
Asparagus	Fungal diseases 0221/88. Western flower thrips 0096, 0409, 0410, 0657, 0658/89. Stemphylium 1209/88. American serpentine leaf miner 0271/88; 0219, 0373, 0403, 0406, 0407, 0422, 0425, 1058/89. Weeds 0531, 0537/89
Aubergine	American serpentine leaf miner 0970/88; 0147, 0199, 0219, 0372, 0373, 0403, 0406, 0407, 0422, 1058/89; 0089/90. Western flower thrips 0096, 0409, 0410, 0657, 0658/89. Fungal diseases 0755/89. Leaf miners 1058/89. Botrytis 0068/90
Beans, broad	Western flower thrips 0096, 0409, 0410, 0657, 0658/89. American serpentine leaf miner 0219, 0372, 0373, 0397, 0403, 0406, 0407, 0422, 1058/89; 0089/90. Fungal diseases 0570/89. Insect pests 0628, 0629, 0632/89
Beans, dwarf	American serpentine leaf miner 0291, 0372, 0373, 0397, 0403, 0406, 0407, 0422, 0425, 1058/89. Western flower thrips 0096, 0409, 0410, 0657, 0658/89. Insect pests 0628, 0629, 0632/89. Leaf miners 1058/89
Beans, french	Caterpillars, white flies, other insects 0064–0070, 0121, 0193, 0194, 0200/88. Western flower thrips 0096, 0409, 0410, 0657, 0658/89. Botrytis, Sclerotinia 1204/88. American serpentine leaf miner 0219, 0373, 0397, 0403, 0406, 0407, 0422, 0425, 1058/89. Leaf miners 1058/89. Insect pests 0628, 0629, 0632/89
Beans, navy	Western flower thrips 0096, 0409, 0410, 0657, 0658/89. Weed control, desiccation 0246/88. American serpentine leaf miner 0219, 0372, 0373, 0397, 0403, 0406, 0407, 0422, 0425, 1058/89. General weed control 0534/89. Insect pests 0628, 0629, 0632/89. Leaf miners 1058/89
Beans, runner	Western flower thrips 0096, 0409, 0410, 0657, 0658/89. American serpentine leaf miner 0219, 0372, 0373, 0397, 0403, 0406, 0407, 0422, 0425, 1058/89. Insect pests 0628, 0629, 0632/89. Leaf miners 1058/89
Beans, soya	Western flower thrips 0096, 0409, 0410, 0657, 0658/89. American serpentine leaf miner 0219, 0372, 0373, 0397, 0403, 0406, 0407, 0422, 0425, 1058/89; 0089/90. General weed control 0550, 0551/89; 0059, 0061/90. Desiccation 0538/89. Insect pests 0628, 0629, 0632/89. Leaf miners 1058/89
Beetroot	American serpentine leaf miner 0271/88; 0372, 0373, 0398, 0406, 0407, 0422, 1058/89; 0089/90. Western flower thrips 0437/89. Insect pests 0628, 0629, 0632/89; 0156/90. Leaf miners 1058/89

CROP	APPROVED USES AND OLA NUMBERS

Beetroot (seed) — Aphanomyces cochlioides 0178/88
Blackberry — Weed control 0754/89
Blackcurrants — Red spider mite 0158/89
Blueberry — Weed control 0754/89
Borage — Weeds 0359/89; 0054, 0055, 0058, 0070/90. Desiccation 0469/89. Carrot willow aphid 0076, 0077/90. Fungal diseases 0075/90
Brassicas — Downy mildew, damping-off 0170/88. Insect pests 0628, 0629, 0632/89
Broccoli — American serpentine leaf miner 0373, 0398, 0406, 0407, 0422, 1058/89. Fungal diseases 0232–0243/88. Downy mildew 0263/88. White blister 0183, 0185/88. Insect & nematode pests 0156/89. Thrips 0218/89. Western flower thrips 0437/89. Caterpillars 0218, 0607/89. Aphids 0607/89. Leaf miners 1058/89
Brussels sprouts — Fungal diseases 0232–0243/88. Downy mildew 0263/88. White blister 0183, 0185/88. American serpentine leaf miner 0373, 0398, 0406, 0407, 0422, 1058/89. Western flower thrips 0437/89. Leaf miners 1058/89
Cabbage — American serpentine leaf miner 0373, 0398, 0406, 0407, 0422, 1058/89. Fungal diseases 0232–0243/88. Downy mildew 0263/88. White blister 0183, 0185/88 Western flower thrips 0437/89. Leaf miners 1058/89
Calabrese — Broad-leaved weeds 0119/88. Fungal diseases 0232–0243/88. American serpentine leaf miner 0373, 0398, 0406, 0407, 0422, 1058/89. Spear rot 1382/88. White blister 0183, 0185/88. Western flower thrips 0437/89. Leaf miners 1058/90. Insect pests 0200/90
Cane fruit (Rubus species & hybrids) — Insect pests 0339, 0340/89
Carrots — Grass weeds 0167, 0168/88. American serpentine leaf miner 0271/88; 0373, 0398, 0406, 0407, 0422/89. Western flower thrips 0437/89
Cauliflower — Fungal diseases 0232–0243/88. Downy mildew 0263/88. White blister 0183, 0185/88. American serpentine leaf miner 0373, 0398, 0406, 0407, 0422, 1058/89. Western flower thrips 0437/89. Leaf miners 1058/89. Insect pests 0200/90
Cauliflower, Roscoff — Insect pests 0200/90
Causeways, harbour steps and landings — Seaweed, algae 0149–0152/89
Celeriac — Septoria leaf-spot 0103–0116/88. American serpentine leaf miner 0271/88; 0373, 0398, 0403, 0406, 0407, 0422/89. Carrot fly 0116/89. Celery fly 0116/89. Western flower thrips 0437/89. Weeds 0567/89; 0053, 0054, 0056, 0070/90. Insect pests 0628, 0629, 0632/89; 0156/90
Celery — Aphids 0201/88. American serpentine leaf miner 0271/88; 0219, 0373, 0406, 0407, 0422, 0425, 1058/89. Powdery mildew 0184/88. Thrips 0060, 0124–0133, 0135–0144/88. Western flower thrips 0096, 0409, 0410, 0657, 0658/89. Leaf miners 1058/89. Weeds 0053/90
Cherries — General weed control 0062, 0063/88

CROP	APPROVED USES AND OLA NUMBERS
Chicory	Western flower thrips 0096, 0409, 0410, 0657, 0658/89. American serpentine leaf miner 0271/88; 0219, 0373, 0403, 0406, 0407, 0422, 1058/89. Aphids, leaf miners & sciarid flies 0202/89. Leaf miners 1058/89
Chinese cabbage	American serpentine leaf miner 0219, 0372, 0373, 0391, 0397, 0398, 0403, 0406, 0407, 0422, 1058/89. Fungal diseases 0232–0236, 0238–0243/88. Western flower thrips 0096, 0409, 0410, 0437, 0657, 0658, 1058/89. Weeds 0344/90
Chinese cabbage, seedlings	Cabbage root fly 0748/89
Chrysanthemums	White rust 0501–0503/88
Clover	Blackgrass, wild oats, volunteer cereals 0623/88. Weeds 0392, 0751/89; 0246/90
Collards	Aphids, caterpillars 0607/89. Cabbage root fly, soil and seedling pests 1006/89
Coniferous tree seedlings	Root feeding nematodes 0157/89 Weeds 0460/89
Coniferous & broad-leaved trees	Weeds 0449/89
Courgettes	Bean seed fly 0229/88. Aphids 0208–0210/88. Western flower thrips 0096, 0409, 0410, 0437, 0657, 0658/89. American serpentine leaf miner 0271, 0970/88; 0147, 0199, 0219, 0373, 0398, 0403, 0406, 0407, 0422, 0425, 1058/89; 0089/90. Insect pests 0628, 0629, 0632/89. Powdery mildew, grey mould 0755/89. Leaf miners 1058/89. Botrytis 0068/90
Cranberry	Weeds 0754/89
Cucumbers	Downy mildew 0861, 0862/88. Caterpillars 0064–0070/89. Powdery mildew 0176, 0195, 0217/88; 0755/89. Root rots and damping-off 0214/88. Western flower thrips 0096, 0409, 0410, 0657, 0658/89. American serpentine leaf miner 0219, 0398, 0403, 0406, 0407, 0422, 0425/89; 0089/90
Damsons	Weeds 0754/89
Edible and non-edible crops	Slugs and snails 0120/89
Elm trees	Dutch elm disease 0297/90
Endive (leafy)	Western flower thrips 0096, 0409, 0410, 0657, 0658/89. American serpentine leaf miner 0271/88; 0219, 0373, 0403, 0406, 0407, 0422, 1058/89. Leaf miners 1058/89
Estuaries	Weeds 0394, 0395/90
Evening primrose	Weeds 0359, 0378, 0467, 0470, 0485, 0486, 0513, 0514, 0549, 0553/89; 0060, 0062, 0073, 0099/90. Fungal diseases 0568, 0572/89; 0073/90. Pollen beetles, flea beetles 0465/89
Fennel	Aphids 0201/88, 0663, 0939/89. Western flower thrips 0096, 0409, 0410, 0657, 0658/89. American serpentine leaf miner 0271/88; 0219, 0372, 0373, 0398, 0403, 0406, 0407, 0422, 0425, 1058/89. Carrot fly, celery fly 0116/89. Weeds 0054, 0056, 0058, 0070/90. Caterpillars 0663, 0939/89. Insect pests 0628, 0629, 0632/89; 0156/90. Leaf miners 1058/89. Carrot willow aphid 0076, 0077/90

Fenugreek	Weeds 0054, 0055, 0056, 0058, 0059, 0070/90. Carrot willow aphid 0076, 0077/90. Leaf miners 0272/90
Forestry areas	Bracken 0750/89
Forestry transplants	Hylobius, Hylastes 0126, 0127/90
Fruit	Insect pests 0628, 0629, 0632/89; 0156/90
Garlic	Thrips 0887/88; 0616/89. Cutworm 0887/88. Western flower thrips 0096, 0409, 0410, 0657, 0658/89. American serpentine leaf miner 0271/88; 0219, 0373, 0403, 0406, 0407, 0422, 0425/89; 0089/90. Stem nematodes 0197/89. Insect pests 0628, 0629, 0632/89; 0156/90. Weeds 0054, 0056, 0070/90
Gherkins	Aphids 0208–0210/88. Bean seed fly 0229/88. Western flower thrips 0096, 0409, 0410, 0437, 0657, 0658/89. American serpentine leaf miner 0271/88; 0219, 0373, 0398, 0403, 0406, 0407, 0422, 0425, 1058/89; 0089/90. Fungal diseases 0755/89. Leaf miners 1058/89. Botrytis 0068/90
Gold of pleasure	Weeds 0058/90
Gooseberry	Red spider mite 0158/89. Weeds 0754/89
Grapevines	Aphids, capsids, thrips 0120, 0134, 0145, 0146, 0148–0163, 0244, 1210/88. Red spider mite 0134, 0145, 0146, 0148–0163, 0244, 1210/88. Downy mildew 0164, 0182/88. Weeds 0173, 0177, 0215, 0216, 0798–0807, 0814–0817/88; 0535, 0536/89. Fungal diseases 0222, 1434/88. Insect pests 0628, 0629, 0632/89; 0156/90
Grass, grazing	Blackgrass, wild oats, volunteer cereals 0623/88. Cirsium spp. 0391/89
Grass seed	Seed treatment 0126/89
Grass seed crops	Growth regulation, volunteer ryegrass 0888–0892/88; 0166–0168/89. Aphids 0341/89. Weeds 0393, 0451–0453, 0457/89. Frit Fly 0725/89. Insect pests 0157/90
Greens (spring)	Broad-leaved weeds 0119/88
Hazel nuts	Gloeosporium 0245/88
Herbs	General weed control 0342, 0450, 0474–0477, 0487–0491, 0539, 0550, 0551, 0554, 0564, 0680, 0681, 0915/89; 0054, 0055, 0056, 0058, 0070, 0072/90. Insect pests 0628, 0629, 0632/89; 0156/90. Carrot willow aphid 0076, 0077/90. Leaf miners 0272/90
Hops	Canker 0072, 0075–0088/88. Alternaria 0075–0088/88. Quality promotion 0171/88. Cleavers 0197/88. Early season defoliation 0264, 0265/88. Weeds 0280/88. Aphids 0220/89
Horseradish	Weeds 0342, 0376, 0377, 0435, 0455/89; 0369/90
Kale	Fungal diseases 0232–0243/88. American serpentine leaf miner 0372, 0373, 0398, 0406, 0407, 0422, 1058/89. Western flower thrips 0096, 0409, 0410, 0437/89. White blister 0185/88. Caterpillars, aphids 0607/89. Leaf miners 1058/89. Cabbage root fly, soil and seedling pests 1006/89
Kohlrabi	Fungal diseases 0232, 0234–0243/88. American serpentine leaf miner 0372, 0373, 0398, 0406, 0407, 0422, 1058/89. Western flower thrips 0096, 0409, 0410, 0437/89. Insect pests 0405, 0408/89. Caterpillars, aphids 0607/89. Leaf miners 1058/89. Cabbage root fly, soil and seedling pests 1006/89; 0381/90

CROP	APPROVED USES AND OLA NUMBERS
Leeks	Thrips, cutworm 0887/88. Fungal diseases 0232–0243/88. Weeds 0212/88; 0054, 0056, 0070/90. Thrips 0060, 0124, 0133, 0135, 0144, 0166, 0887/88; 0616/89. White tip 0183, 0185/88. Western flower thrips 0096, 0409, 0410, 0657, 0658/89. American serpentine leaf miner 0271/88; 0219, 0373, 0406, 0407, 0422, 0425, 1058/89. Stem nematodes 0155/89. Leaf miners 1058/89. Insect pests 0156/90
Leek seedlings	Fusarium culmorum 0367/90
Lettuce	Annual weeds 0518, 1356, 1357/88. Broad leaved weeds 0196/88. Western flower thrips 0096, 0409, 0410, 0657, 0658/89. American serpentine leaf miner 0970/88; 0199, 0219, 0373, 0403, 0407, 1058/89. Downy mildew 0982/88. Root aphid 0436, 0744/89. Sclerotinia, botrytis 0571/89. Fungal diseases 0755/89. Leaf miners 1058/89. Insect pests 0156/90. Ringspot 0203, 0204/90
Linseed	Weeds 0363, 0456, 0581/89; 0052/90. Fungal diseases 0569/89
Loganberry	Weeds 0754/89
Lucerne	Blackgrass, wild oats, volunteer cereals, grass weeds 0623/88; 0246/90. Insect pests 0628, 0629, 0632/89; 0155/90
Marjoram	Leafhoppers 0664/89. Weeds 0054–0056, 0058, 0070/90. Carrot root aphid 0076, 0077/90. Leaf miners 0272/90
Marrow	Aphids 0208–0210/88; 0078/90. Fungal diseases 0232–0243/88. Powdery mildew 0184/88; 0755/89. Bean seed fly 0229/88. Thrips 0060, 0124–0133, 0135–0144/88. Western flower thrips 0096, 0409, 0410, 0437, 0657, 0658/89. American serpentine leaf miner 0271/88; 0219, 0373, 0398, 0403, 0406, 0407, 0422, 0425, 1058/89; 0089/90. Insect pests 0628, 0629, 0632/89. Botrytis 0755/89; 0068/90. Leaf miners 1058/89
Melon	Aphids 0208–0210/88. Bean seed fly 0229/88. Western flower thrips 0096, 0409, 0410, 0437, 0657, 0658/89. American serpentine leaf miner 0271/88; 0219, 0373, 0398, 0403, 0406, 0407, 0422, 0425, 1058/89; 0089/90. Fungal diseases 0755/89. Leaf miners 1058/89
Mint	Weeds 0319, 0320, 0357/90
Mooli	Cabbage root fly 0228/88. Insect pests 0405, 0408/89
Mushroom houses	Flies 1211/88; 0202/89
Mustard (for condiment production)	Seed weevil, pod midge 0227/89 Weeds 0466/89. Desiccation 0461/89
Narcissus (outdoors)	Large bulb fly 0749/89
Nectarine	Weeds 0754/89
Non-crop areas	Weeds 0614, 0615/89
Non-edible ornamentals (glasshouse)	Western flower thrips 0096, 0409, 0410, 0657, 0658/89. American serpentine leaf miner 0127, 0219, 1058, 1065/89. Insect pests 0007/89. Seed weevil, pod midge 0227/89. Vine weevil, sciarid fly 0374/89. Eelworm 0198/89. Fungal diseases 0360–0362/89. Powdery mildew 0351/89. Leaf miners 1058, 1065/89. Weeds 0055, 0057/90. Botrytis 0067, 0068/90. Growth regulation 0069/90

CROP	APPROVED USES AND OLA NUMBERS

Non-edible ornamentals (outdoors)
: Western flower thrips 0438/89. American serpentine leaf miner 0398, 0425, 1058, 1065/89; 0089/90. Insect pests 0007, 0205, 0337, 0338, 0424/89; 0156/90. Vine weevil, sciarid fly 0374. Leaf miners 0228, 1058, 1065/89. Codling moth 0293/89. Mites 0292, 0293/89. Fungal diseases 0360–0362, 06133, 0755/89; 0071, 0074/90. Powdery mildew 0351/89. Eelworm 0198/89. Weeds 0343, 0454, 0463, 0473, 0524–0529, 0532, 0552, 0581, 0684/89; 0055, 0058/90. Docks 0753/89. Botrytis 0067/90. Growth regulation 0069/90

Nursery fruit trees and soft fruit bushes
: Insect pests 0204, 0375, 0424/89. Vine weevil, sciarid fly 0374/89. Mites, codling moth 0294/89. Fungal diseases 0360–0362, 0613/89; 0071, 0074/90. Weeds 0343, 0454, 0524–0529, 0533, 0684/89; 0057/90. Docks 0753/89

Nuts
: Weeds 0174, 0175/88. Botrytis 1209/88

Onions
: Cutworm 0887/88. Thrips, 0887/88; 0616/89. Broad leaved weeds 0198/88. Downy mildew 0183, 0185/88. Western flower thrips 0096, 0409, 0410, 0657/89. American serpentine leaf miner 0271/88; 0219, 0373, 0403, 0406, 0407, 0422, 0425/89. Stem nematodes 0197/89. Weeds 0504/89; 0054, 0056, 0070/90. Insect pests 0628, 0629, 0632/89; 0156/90

Onions (salad)
: Fungal diseases 0232–0243/88. Thrips 0060, 0124–0133, 0135–0144, 0166, 0887/88. Cutworm 0887/88. Weeds 0275/89

Ornamental bulbs
: Soilborne pests, nematodes 0154/89. Fungal diseases 0458, 0459, 0472/89. Weeds 0301/89; 0051/90. Botrytis, fusarium 0755/89

Parsley
: Crown rot 0990/89; 0058/90. Carrot rot aphid 0076, 0077/90. Leaf miners 0272/90. Weeds 0304/90

Parsnips
: Grass weeds 0167, 0168, 0220/88. American serpentine leaf miner 0271/88; 0373, 0398, 0406, 0407, 0422/89. Carrot fly 0226/89. Western flower thrips 0437/89. Weeds 0515–0517/89

Peaches
: Weeds 0062, 0063/88; 0754/89

Peas (mange-tout)
: Broad leaved weeds 0186–0189, 0224, 0225/88. Damping-off, downy mildew 0190–0192/88. Wild oats, blackgrass, meadow grass 0226, 0227/88. Western flower thrips 0096, 0409, 0410, 0657, 0658/89. Botrytis, sclerotinia 1209/88. American serpentine leaf miner 0271/88; 0219, 0372, 0373, 0406, 0407, 0422, 0425, 1058/89. Thrips 0246/89. Weeds 0530/89. Insect pests 0628, 0629, 0632/89. Leaf miners 1058/89

Peas, combining and vining (where pod is not for consumption)
: Ascochyta, 0089–0102, 0169/88. Leaf spot 0089–0102/88. Botrytis, Sclerotinia 1209/88. Downy mildew, 0169/88. American serpentine leaf miner 1058/89. Botrytis, Sclerotinia 0231/88. American serpentine leaf miner 1058/89. Weeds 0567/89; 0057/90. Leaf miners 1058/89

Peas, seed
: Seed treatment 0336, 0344, 0345, 0352/89

Peppers
: Western flower thrips 0096, 0409, 0410, 0657, 0658/89. American serpentine leaf miner 0219, 0372, 0373, 0406, 0407, 0422, 1058/89; 0089/90. Fungal diseases 0755/89. Leaf miners 1058/89. Botrytis 0068/90

Plums
: Rust and gall mite 0211/88. Weeds 0062, 0063

CROP	APPROVED USES AND OLA NUMBERS

Potato
American serpentine leaf miner 0271/88; 0373, 0397, 0403, 0406, 0407, 0422, 1058/89; 0089/90. Colorado beetle 1366, 1367, 1383–1391, 1472/88. Leaf miners 1058/89

Pumpkin
Bean seed fly 0229/88. Aphids 0208–0210/88. American serpentine leaf miner 0271/88; 0147, 0373, 0398, 0403, 0406, 0407, 0422, 0425, 1058/89; 0089/90. Western flower thrips 0096, 0409, 0410, 0437/89. Leaf miners 1058/89

Quince
Weeds 0062, 0063/88; 0754/89

Radicchio (red leafy endive)
Western flower thrips 0096, 0409, 0410, 0657, 0658/89. American serpentine leaf miner 0219/89. Sclerotinia, Botrytis 0571/89. Insect pests 0628, 0629, 0632/89; 0156/90. Aphids, caterpillars 0663, 0939/89. Weeds 0680, 0681/89

Radish
Aphids 0202–0204/88. Cabbage root fly 0228/88. American serpentine leaf miner 0271/88; 0373, 0406, 0407, 0422/89. Insect pests 0405, 0408/89. Downy mildew 0349/90

Radish seed crops
Weeds 0058/90

Raspberries
Bud and leaf mite 0230, 1433/88. Spawn control 0266–0268/88. Root rot 0743/89. Weeds 0754/89

Red beet (see also Beetroot)
Insect pests 0628, 0629, 0632/89; 0156/90

Redcurrants
Red spider mite 0158/89. Weeds 0754/89

Red fescue seed crops
Broad leaved weeds, meadow grass, black grass 0622/88. Black grass, meadow grass, rye-grasses, wild oats, volunteer cereals 0246/90

Rhubarb
American serpentine leaf miner 0271/88; 0373, 0406, 0407, 0422, 0425, 1058/89. Leaf miners 1058/89

Rooting beds
Western flower thrips 0219, 0658/89

Sage
Weeds 0304, 0339/90

Sainfoin
Blackgrass, wild oats, volunteer cereals 0623/88

Sainfoin seed crops
Blackgrass, meadow grass, rye-grasses, wild oats, volunteer cereals 0246/90

Salsify
American serpentine leaf miner 0271/88; 0373, 0398, 0406, 0407, 0422/89. Western flower thrips 0437/89

Savoy
Aphids, caterpillars 0607/89. Cabbage root fly, soil and seedling pests 1006/89

Shallots
Western flower thrips 0096, 0409, 0410, 0657, 0658/89. American serpentine leaf miner 0271/88; 0219, 0373, 0403, 0406, 0407, 0422, 0425/89; 0089/90

Sorrel
Flea beetle 0617/89. Weeds 0054–0056, 0058, 0070/90. Carrot root fly 0076, 0077/90. Leaf miners 0272/90

Spinach, spinach beet
Aphids 0201/88. Leaf miner 0223/88; 1058/89. Weeds 0179–0181/88. Western flower thrips 0096, 0409, 0410, 0657, 0658/89. American serpentine leaf miner 0271/88; 0219, 0373, 0403, 0406, 0407, 0422, 1058/89. Insect pests 0628, 0629, 0632/89; 0156/90

Squash
Aphids 0208–0210/88. Bean seed fly 0229/88. Western flower thrips 0096, 0409, 0410, 0657, 0658/89. American serpentine leaf miner 0271/88; 0219, 0373, 0398, 0403, 0406, 0407, 0422, 0425, 1058/89; 0089/90. Fungal diseases 0755/89. Leaf miners 1058/89. Botrytis 0068/90

Strawberries	Red core 0170/88. Docks 0752/89
Sugar beet	American serpentine leaf miner 0271, 0970/88; 0199, 0373, 0406, 0407, 0422, 0425, 1058/89. Leaf miners 1058/89; 0057/90
Sunflower	Desiccation 0462/89. Weeds 0567/89
Swede	American serpentine leaf miner 0271/88; 0373, 0398, 0406, 0407, 0422/89. Western flower thrips 0437/89. Weeds 0468/89
Sweetcorn	Aphids 0205–0207/88. Thrips 0060, 0124–0133, 0135–0144, 0166/88. Frit fly 0153/89. Insect pests 0628, 0629, 0632/89; 0156/90
Tomato houses	Leaf miner, caterpillars 1211/88. Powdery mildew 0039, 0087/90
Tomatoes	American serpentine leaf miner 0219, 0407/89. Caterpillars 0064–0070, 0223/88. Root diseases, damping off 0165, 0213/88. Powdery mildew 0969/88; 0305/90. Western flower thrips 0096, 0409, 0410, 0657, 0658/89. Didymella stem rot 0249–0262/88. General insect control 0218, 0219/88. Leaf miners 0223/88. Fungal diseases 0755/89
Trefoil	Blackgrass, wild oats, volunteer cereals 0623/88
Trefoil seed crops	Blackgrass, meadow grass, rye-grasses, wild oats, volunteer cereals 0246/90
Turnip	American serpentine leaf miner 0271/88; 0373, 0398, 0406, 0407, 0422/89. Western flower thrips 0437/89
Watercourses	Common reed 0745/89
Watercress	Growth regulation 0171/88. Damping off 0604/89. Aphids 0790/89. General insect control 0747, 0790/89
Watermelons	Fungal diseases 0755/89. Botrytis 0068/90
White currants	Weeds 0754/89
White turnips	Alternaria 0674/8
Winter oilseed rape	Weeds 0471/89
Winter vetches	Weeds 0682, 0683/89

MAXIMUM RESIDUE LEVELS (MRLs)

Statutory limits have been established for pesticide residues in food crops and animal products. Approval for pesticide products is granted on the basis that, with the relatively small number of treatments which are liable to result in residues in foodstuffs, such residues will be below internationally agreed levels when the treatment is applied in accordance with the approved conditions of use. Thus, as long as products are used according to the label instructions and following good agricultural practice, the maximum residue levels should not be exceeded. The responsibility of the farmer or grower is discharged by ensuring compliance with the label instructions.

Details of the maximum residue levels (which apply to imported as well as to home-produced foodstuffs) are contained in *The Pesticides (Maximum Residue Levels in Food) Regulations,* Statutory Instrument No. 1378/1988, available from HMSO. A summary of the maximum residue levels applicable to the chemicals included in this Guide is given below. For the full details and definitions, however, and for information on chemicals not currently marketed in Britain it is essential to consult the Regulations. It should be noted that the existence of an MRL in a particular foodstuff does not necessarily indicate that use of the chemical has been approved on that crop.

The Pesticides Register, Issue No. 7, 1989 should also be consulted for details of an Advisory MRL for tecnazene in potatoes and advice on procedures which will ensure that the requirements can be met.

MAXIMUM RESIDUE LEVELS (mg residue/kg food) PERMITTED IN FOODSTUFFS

2-aminobutane	Potatoes 50; citrus 5
amitrole	Fruits, vegetables, mushrooms 0.05
azinphos-methyl	Peaches, nectarines 4; citrus, grapes, celery 2; apples, pears, plums, strawberries, raspberries, blackcurrants, bananas, Brussels sprouts 1; carrots, turnips, swedes, onions, tomatoes, cucumbers, cabbage, cauliflowers, beans, peas, leeks, lettuce 0.5; potatoes 0.2
captan	Apples, pears, grapes, strawberries, raspberries, blackcurrants, tomatoes 3; peaches, nectarines, plums, beans, peas, leeks, lettuce 2; citrus, bananas, potatoes, carrots, turnips, swedes, onions, cucumbers, brassicas, celery, mushrooms 0.1
carbaryl	Peaches, nectarines, plums, raspberries, blackcurrants, lettuce 10; citrus, strawberries 7; apples, pears, grapes, bananas, tomatoes, cabbage, beans, peas 5; cucumbers, celery 3; carrots, swedes 2; rice, turnips, onions, cauliflowers, Brussels sprouts, leeks, mushrooms 1; cereals (other than rice) 0.5
carbendazim	Citrus, peaches, nectarines, grapes 10; apples, pears, strawberries, raspberries, blackcurrants, tomatoes, lettuce 5; potatoes 3; plums, onions, celery 2; bananas, mushrooms 1; wheat, rye, barley, oats, cucumbers, Brussels sprouts 0.5; milk, eggs 0.1

MAXIMUM RESIDUE LEVELS (mg residue/kg food) PERMITTED IN FOODSTUFFS

chlordane	Meat products, dairy products (other than milk) 0.05; eggs, fruits, vegetables, mushrooms, 0.02; milk 0.002
chlorfenvinphos	Citrus 1; potatoes, carrots, turnips, swedes, onions, celery 0.5; meat products 0.2; tomatoes, cucumbers, brassicas, beans, peas, leeks, lettuce 0.1; apples, pears, peaches, nectarines, plums, grapes, strawberries, raspberries, blackcurrants, bananas, mushrooms 0.05; milk 0.008
chlorpyrifos-methyl	Cereals 10; meat products, eggs 0.05; milk 0.01
diazinon	Meat products 0.7; fruits, vegetables, mushrooms 0.5; cereals 0.05; milk 0.02
dichlofluanid	Grapes, raspberries, blackcurrants 15; strawberries, lettuce 10; citrus, apples, pears, peaches, nectarines, plums, bananas, carrots, turnips, swedes, onions, tomatoes, cucumbers, brassicas, beans, peas, leeks 5; potatoes 0.01
dichlorvos	Cereals 2; lettuce 1; vegetables (other than lettuce), mushrooms 0.5; fruits 0.1; meat products, eggs 0.05; milk 0.02
dicofol	Fruits, vegetables (other than tomatoes and cucumbers), mushrooms 5; cucumbers 2; tomatoes 1
diflubenzuron	Citrus, apples, pears, plums, tomatoes, cabbage, Brussels sprouts 1; mushrooms 0.1; meat products, milk, eggs 0.05
dimethoate	Citrus, peaches, nectarines, plums, blackcurrants, cucumbers, brassicas, beans 2; apples, pears, grapes, strawberries, raspberries, bananas, carrots, turnips, swedes, onions, tomatoes, peas, celery, leeks, mushrooms 1; potatoes 0.05
dithiocarbamates	Grapes, raspberries, blackcurrants 5; apples, pears, peaches, nectarines, strawberries, tomatoes 3; plums, bananas 1; carrots, cucumbers, beans 0.5; potatoes 0.1
endosulfan	Fruits, turnips, swedes, tomatoes, cucumbers, brassicas, beans, peas, celery, leeks, lettuce 2; onions 1; potatoes, carrots 0.2; cereals 0.1
etrimfos	Cereals 10
fenitrothion	Cereals 10; citrus 2; fruits (other than citrus), vegetables (other than potatoes), mushrooms 0.5; potatoes 0.05
gamma-HCH	Strawberries, raspberries, blackcurrants 3; sheepmeat, tomatoes, brassicas, lettuce 2; meat (other than sheepmeat), citrus, apples, pears, peaches, nectarines, plums, bananas, turnips, swedes, onions, cucumbers, beans, celery, leeks, mushrooms 1; grapes 0.5; dairy produce (other than milk), carrots 0.2; cereals, eggs, peas 0.1; potatoes 0.05; milk 0.008
imazalil	Citrus (whole fruit) 5; citrus (fruit without peel) 0.1
ioxynil	Onions 0.1
iprodione	Apples, pears, peaches, nectarines, plums, grapes, strawberries 10; raspberries, blackcurrants, tomatoes, cucumbers 5; onions 0.1
malathion	Cereals 8; onions, tomatoes, cucumbers, brassicas, beans, peas, celery, leeks, lettuce, mushrooms 3; citrus 2, fruits (other than citrus), potatoes, carrots, turnips, swedes 0.5
mercury compounds	Cereals, apples, pears, vegetables, mushrooms 0.02

MAXIMUM RESIDUE LEVELS (mg residue/kg food) PERMITTED IN FOODSTUFFS

methacrifos	Cereals 10
methyl bromide	Cereals 0.1
omethoate	Swedes, leeks 2; fruits (other than bananas), tomatoes 1; bananas, carrots, turnips, cucumbers, brassicas, beans, peas, celery, lettuce, mushrooms 0.2; onions 0.1; potatoes 0.05
phosalone	Apples, pears, peaches, nectarines 2; citrus, plums, grapes, strawberries, raspberries, blackcurrants, bananas, onions, tomatoes, cucumbers, brassicas, beans, peas, celery, leeks, lettuce, mushrooms 1; potatoes, carrots, turnips, swedes 0.1
pirimiphos-methyl	Cereals 10; citrus 0.5
pyrethrins	Cereals 3
quintozene	Lettuce 3; bananas 1; potatoes 0.2; tomatoes 0.1; cabbage, cauliflowers 0.02; beans 0.01
tecnazene	Potatoes (Advisory MRL, see *Pesticides Register* Issue No. 7 1989) 5; lettuce 2
thiabendazole	Potatoes 5
triazophos	Bananas 1; carrots, cabbage, Brussels sprouts 0.1; potatoes, onions 0.05
trichlorphon	Cereals 0.1
vinclozolin	Strawberries 10; peaches, nectarines, grapes, raspberries, blackcurrants, celery, lettuce 5; tomatoes, 3; beans 2; apples, pears, onions, cucumbers, cabbage, cauliflower, peas 1; potatoes 0.1

CHEMICALS SUBJECT TO THE POISONS LAW

Certain products in this book are subject to the provisions of the Poisons Act 1972, the Poisons List order 1982 and the Poisons Rules 1982 (copies of all these are obtainable from HMSO). These Rules include general and specific provisions for the storage and sale and supply of listed non-medicine poisons. Although in general poisons controlled by the poisons law should be packaged and labelled in accordance with the provisions of the Classification, Packaging and Labelling of Dangerous Substances Regulations 1984, the Poisons Rules do contain two requirements as to packaging; however these relate to sales.

The chemicals approved for use in the UK and included in this book are specified under Parts I and II of the Poisons List as follows :

Part I Poisons (sale restricted to registered retail pharmacists and to non-pharmacy registered businesses providing sales do not take place on retail premises)

aluminium phosphide
chloropicrin
methyl bromide
sodium cyanide

Part II Poisons (sale restricted to registered retail pharmacists and listed sellers registered with local authority)

aldicarb (a)	fentin hydroxide
alphachloralose	fonofos (a)
azinphos-methyl	mephosfolan
chlorfenvinphos (a,b)	nicotine (c)
demeton-S-methyl	omethoate
demeton-S-methyl sulphone	oxydemeton-methyl
dichlorvos	paraquat (d)
DNOC*	pirimiphos-ethyl
drazoxolon (b)	quinalphos
endosulfan	thiometon
fentin acetate	triazophos

(a) Granular formulations containing up to 12% a.i. are exempt
(b) Treatments on seeds are exempt
(c) Formulations containing not more than 7.5% of nicotine are exempt
(d) Pellets containing not more than 5% paraquat ion are exempt

---

* Approvals for use of DNOC products revoked 31 December 1989

OCCUPATIONAL EXPOSURE STANDARDS (OES)

Various substances hazardous to health have been assigned 'Occupational Exposure Standards' by the Health and Safety Commission in relation to specified reference periods. For such substances exposure by inhalation should not exceed the standard or, if exceeded, must be reduced as soon as practicable. Full details are given in HSE Guidance Note EA 40/90 *Occupational exposure limits*, 1990. Chemicals listed which are currently marketed as pesticides are given below with the approved standards (values in brackets are under review).

	Long term (8 hr) OES		Short term (10 min) OES	
	ppm	mg/m^3	ppm	mg/m^3
aluminium salts	–	2	–	–
atrazine	–	(10)	–	–
azinphos-methyl	–	0.2	–	0.6
benomyl	–	10	–	15
borates	–	1	–	1
bromacil	1	10	2	20
captan	–	5	–	15
carbaryl	–	5	–	10
carbofuran	–	0.1	–	–
chlordane	–	(0.5)	–	(2)
chloropicrin	0.1	0.7	0.3	2
chlorpyrifos	–	0.2	–	0.6
copper (as Cu)	–	1	–	2
cresols	5	22	–	–
2,4-D	–	10	–	20
diazinon	–	0.1	–	0.3
1,3-dichloropropene	(1)	(5)	(10)	(50)
dichlorvos	0.1	1	0.3	3
diquat	–	0.5	–	1
disulfoton	–	0.1	–	0.3
diuron	–	10	–	–
endosulfan	–	0.1	–	0.3
gamma-HCH	–	0.5	–	1.5
iron salts (as Fe)	–	1	–	2
malathion	–	10	–	–
methyl bromide	5	20	15	60
nicotine	–	0.5	–	1.5
paraquat (dust)	–	0.1	–	–
pentachlorophenol	–	0.5	–	1.5
picloram	–	10	–	20
pyrethrins	–	5	–	10
rotenone	–	5	–	10
TCA	(1)	(5)	–	–
thiram	–	5	–	10
warfarin	–	0.1	–	0.3

MAXIMUM EXPOSURE LIMITS (MEL)

Under the COSHH Regulations certain substances have been assigned maximum exposure limits in relation to specified reference periods. Chemicals listed in the Schedule* and currently marketed as pesticides are :

	Long term (8 hr) MEL		Short term (10 min) MEL	
	ppm	mg/m^3	ppm	mg/m^3
formaldehyde	2	2.5	2	2.5
hydrogen cyanide	–	–	10	10
(released by sodium cyanide)				

SUBSTANCES MOST DANGEROUS TO THE AQUATIC ENVIRONMENT – THE RED LIST

A Consent to Discharge issued by the National Rivers Authority is required for the disposal of all wastes into controlled waters and discharges containing the substances most damaging to the aquatic environment, The Red List, will also require approval from the Secretary of State for the Environment, advised by HM Inspectorate of Pollution. The pesticides currently marketed in the UK which are included in the initial priority Red List are the following :

atrazine
azinphos-methyl
dichlorvos
endosulfan
fenitrothion
gamma-HCH

malathion
mercury compounds
pentachlorophenol
simazine
trifluralin

substances on the list which are not marketed as pesticides in the UK are :

aldrin
cadmium and its compounds
DDT
1,2-dichloroethane
dieldrin
endrin

hexachlorobenzene
hexachlorobutadiene
polychlorinated biphenyls
tributyltin compounds
trichlorobenzene
triphenyltin compounds

Further substances may be added to the list from time to time.

* For further information see *COSHH Regulations 1988*, Statutory Instrument 1988 No. 1657

USE OF HERBICIDES IN OR NEAR WATER

Herbicides which could be used to control aquatic weeds or weeds growing along the banks of watercourses were granted special clearance under PSPS and will continue to receive special attention under the new regulations. The table below lists those products, the labels of which recommend or refer to clearance for this type of use.

Booklet 2078 *Guidelines for the Use of Herbicides on Weeds in or near Watercourses and Lakes* published by MAFF must be followed in applying herbicides in these situations and the appropriate River Authority Catchment Board, River Purification Authority or Waterworks Undertaking should be consulted before the spraying of such areas is undertaken.

CHEMICAL	PRODUCT	WEEDS CONTROLLED	SAFETY INTERVAL BEFORE IRRIGATION
asulam	Asulox	docks, bracken	nil
2,4-D	Atlas 2,4-D Dormone	water weeds, dicotyledon weeds on banks	3 wk
dalapon	BH Dalapon	bulrushes, reeds, grass weeds on banks	5 wk
dalapon + dichlobenil	Fydulan	weeds in or near water	–
dichlobenil	Casoron G Casoron G-SR	aquatic weeds	2 wk
diquat	Midstream Reglone	aquatic weeds	10 d
fosamine-ammonium	Krenite	woody weeds	nil
glyphosate	Mascot Sonic Roundup Roundup Pro Spasor	reeds, rushes, sedges, waterlilies, grass weeds on banks	nil
maleic hydrazide	Bos MH 180 Regulox K	suppression of grass growth on banks	3 wk
terbutryn	Clarosan 1FG	aquatic weeds	7 d

USE OF PESTICIDES IN FORESTRY

The table below lists those products, the labels of which refer to approval for use in forestry.

CHEMICAL	PRODUCT	USE
aluminium phosphide	Phostoxin Power Phosphine Pellets Talunex	Mole and rabbit control
ammonium sulphamate	Amcide Root-Out	Rhododendron and other woody weed control
asulam	Asulox	Bracken control
atrazine	Ashlade 4% At Gran Ashlade Atrazine 50FL Atlas Atrazine Atraflow Gesaprim 500FW MSS Atrazine 50FL MSS Atrazine 4G MSS Atrazine 80WP Unicrop Flowable Atrazine	Weed control in conifers
atrazine + cyanazine	Holtox	Weed control in conifer plantations
atrazine + dalapon	Atlas Lignum Granules	Weed control in plantations and hardwood nurseries
atrazine + terbuthylazine	Gardoprim A 500FW Transformer Gardoprim A 500FW	Weed control in conifers and broad-leaved trees
chlorpyrifos	Dursban 4 Spannit	Weevil and beetle control in transplant lines and cut logs
clopyralid	Dow Shield	Weed control in conifers and hardwoods (off-label)
clopyralid + cyanazine	Coupler SC	Weed control in conifers and hardwoods (off-label)
clopyralid + triclopyr	Grazon 90	Weed control in conifer plantations (off-label)

CHEMICAL	PRODUCT	USE
2,4-D	BASF 2,4-D Ester 480 BH 2,4-D Ester 50 Silvapron D	Weed, heather and willow control in plantations
2,4-D + dicamba + triclopyr	Broadshot	Perennial and woody weed control
dalapon	Atlas Dalapon	Grass weed control
dalapon + dichlobenil	Fydulan	Grass and perennial weed control
dalapon + di-1-p-menthene	Volunteered	Grass weed control in conifer plantations
dicamba	Tracker	Bracken control
dichlobenil	Casoron G Casoron G4 Prefix D	Annual and perennial weed control in plantations of certain tree species established for at least 2 years
diflubenzuron	Dimilin WP	Caterpillar control on forest trees
diphenamid	Enide 50W	Annual weed control in forest nurseries
diquat + paraquat	Farmon PDQ Parable	Pre-planting weed control, pre-emergence control in nurseries, directed sprays or ring weeding in plantations. Firebreak maintenance
fosamine-ammonium	Krenite	Woody weed control in forestry land (not conifer plantations)
gamma-HCH	Gamma-Col Gamma-HCH Dust Lindane Flowable	Various insect pests in nursery beds, transplant lines, plantations and cut timber
glyphosate	FAL Glyphosate Roundup Stirrup	Control of annual, perennial and woody weeds (directed sprays and wiper application). Chemical thinning
glyphosate + simazine	Rival	Weeds in conifer plantations

CHEMICAL	PRODUCT	USE
hexazinone	Velpar Liquid	Selective annual, perennial and woody weed control in conifers
paraquat	Dextrone X Gramoxone 100	Pre-planting weed control, pre-emergence control in nurseries, directed sprays or ring-weeding in plantations. Firebreak maintenance
propyzamide	Kerb Flo Kerb Flowable Kerb Granules Kerb 50W	Grass weed control in forest trees
resmethrin	Turbair Resmethrin Extra	Pine looper moth control in pines
simazine	Ashlade Simazine 50FL Gesatop 500FW Gesatop 50WP MSS Simazine 50FL MSS Simazine 80WP Simazine 500FL Truchem Simazine 500L Truchem Simazine WP Unicrop Flowable Simazine	Annual weed control in nursery beds and transplant lines
triclopyr	Garlon 4 Timbrel	Woody weed control
ziram	AAprotect	Bird and animal repellent

PESTICIDES USED AS SEED TREATMENTS
(including treatments on seed potatoes)

CHEMICAL	PRODUCT	FORMULATION	CROP(S)
aluminium ammonium sulphate	Curb	DP	Crops, general
2-aminobutane	Butafume	FU	Potatoes
bendiocarb	Seedox SC	FS	Maize, sweetcorn
benomyl	Benlate	DP	Field beans, onions, peas
benomyl + iodofenphos + metalaxyl	Polycote Pedigree	DS, FS	Onions
bromophos + captan + thiabendazole	Bromotex T	DS	Beans, brassicas, carrots, celery, leeks, onions, vegetables
captan + fosetyl-aluminium + thiabendazole	Aliette Extra	WS	Peas
captan + gamma-HCH	Gammalex	DS	Brassicas, linseed, oilseed rape
carbendazim + tecnazene	Various products, see carbendazim + tecnazene entry	DS	Potatoes
carboxin + gamma-HCH + thiram	Vitavax RS Flowable	FS	Fodder rape, oilseed rape
carboxin + imazalil + thiabendazole	Cerevax Extra	FS	Barley, oats
carboxin + thiabendazole	Cerevax	FS	Rye, wheat
chlorfenvinphos	Birlane LST	LS	Wheat

CHEMICAL	PRODUCT	FORMULATION	CROP(S)
drazoxolon	Mil-Col 30	FS	Beans, peas
ethirimol + flutriafol + thiabendazole	Ferrax	FS	Barley
fenpropimorph + gamma-HCH + thiram	Lindex Plus FS	FS	Oilseed rape
fonofos	Fonofos Seed Treatment	FS	Barley, wheat, ryegrass
fuberidazole + imazalil + triadimenol	Baytan IM	DS	Barley
fuberidazole + triadimenol	Baytan Baytan Flowable	DS FS	Cereals Cereals
gamma-HCH	Gammasan 30 New Kotol Wireworm FS	DS FS FS	Cereals Cereals Cereals
gamma-HCH + phenylmercury acetate	Mergamma 30	DS	Cereals
gamma-HCH + thiabendazole + thiram	New Hysede FL	FS	Broccoli, Brussels sprouts, cabbages, oilseed rape, swedes, turnips
gamma-HCH + thiram	New Hydraguard	LS	Brassicas, oilseed rape, swedes
guazatine	Rappor	LS	Wheat
guazatine + imazalil	Rappor Plus	LS	Barley, oats
hymexazol	Tachigaren 70WP	WP	Sugar beet
iodofenphos	Elocril 50WP	WP	Onions
iprodione	Rovral WP	WP	Brassicas, oilseed rape, potatoes, flowers
iprodione + metalaxyl + thiabendazole	Polycote Prime	DS	Carrots

CHEMICAL	PRODUCT	FORMULATION	CROP(S)
metalaxyl	Polycote Universal	FS	Flowers, vegetables
metalaxyl + thiabendazole + thiram	Apron Combi 453FS	FS	Beans, peas
2-methoxyethyl-mercury acetate	Panogen M	LS	Cereals
pencycuron	Monceren DS	DP	Potatoes
phenylmercury acetate	Agrosan D	DS	Cereals
	Ceresol	LS	Cereals
	Ceresol 50	LS	Cereals
	Single Purpose	LS	Cereals
tecnazene	Various products, see tecnazene entry	Various	Potatoes
tecnazene + thiabendazole	Various products, see tecnazene + thiabendazole entry	DP, SS	Potatoes
thiabendazole	Storite Clear Liquid	LS	Potatoes
	Storite Flowable	FS	Potatoes
	Tecto Dust	DS	Potatoes
thiabendazole + thiram	Ascot 480 FS	FS	Field beans, peas
	Hy-TL	FS	Peas
	Hy-Vic	FS	Various field crops and vegetables
thiram	Agrichem Flowable Thiram	FS	Beans, peas, maize, vegetables, flowers
tolclofos-methyl	Rizolex	DS	Potatoes
	Rizolex Flowable	FS	Potatoes

33

AERIAL APPLICATION OF PESTICIDES

Only those products specifically approved for aerial application may be so applied and they may only be applied to specific crops or for specified uses. A list of products approved for application from the air was published as a supplement to *The Pesticides Register* Issue 3, April 1989 and updated in July 1989, Issue 6 and August 1990, Issue 7, 1990. The list given below has been prepared from these sources.

It is emphasized that the list is for guidance only — reference must be made to the product labels for detailed conditions of use which must be complied with. The list does not include those products which have been granted restricted aerial approval limiting the area which may be treated.

Detailed rules are imposed on aerial application regarding prior notification of the Nature Conservancy, water authorities, bee keepers, Environmental Health Officers, neighbours, hospitals, schools etc. and the conditions under which application may be made. The full conditions are available from MAFF and must be consulted before any aerial application is made.

CHEMICAL	PRODUCT	CROPS/USES
asulam	Asulox	Bracken control
azinphos-methyl + demeton-S-methyl sulphone	Gusathion MS	Peas, brassicas, brassica seed crops
benodanil	Calirus	Barley, wheat
benomyl	Benlate Fungicide	Wheat, barley, dwarf and field beans
carbaryl	Murvin 85	Peas, brassicas, apples, pears, lettuce, strawberries, blackcurrants, raspberries, gooseberries
carbendazim	Battal FL	Barley, winter wheat, field and dwarf beans, oilseed rape
	Bavistin	Cereals, field and dwarf beans, onions, oilseed rape
	Bavistin DF	Winter wheat, winter barley, winter rye, field and dwarf beans, oilseed rape, onions
	Bavistin FL	Cereals, field and dwarf beans, onions, oilseed rape

CHEMICAL	PRODUCT	CROPS/USES
	Carbate Flowable	Wheat, barley, rye, field beans, onions, oilseed rape
	Delsene 50DF	Winter wheat, barley, field beans, oilseed rape
	Derosal WDG	Wheat, barley, field, broad and dwarf beans
	Stempor DG	Cereals, field and dwarf beans
carbendazim + chlorothalonil	Bravocarb	Wheat, barley, peas, beans
carbendazim + mancozeb	Kombat Kombat WDG	Cereals Winter wheat, winter barley, winter oilseed rape
	Septal WDG	Winter wheat, winter barley, winter oilseed rape
carbendazim + maneb	Ashlade M	Wheat, barley, oats, rye, triticale
	Ashlade Mancarb FL	Cereals
	Delsene M Flowable	Wheat, barley
	Squadron	Cereals
	Tripart Legion	Winter wheat, winter barley
carbendazim + maneb and sulphur	Bolda FL	Winter and spring wheat, winter barley, oilseed rape
carbendazim + maneb and tridemorph	Ashlade Cosmic FL Cosmic FL	Winter wheat, barley Winter wheat, barley
carbendazim + prochloraz	Sportak Alpha	Cereals
carbendazim + propiconazole	Hispor 45WP	Wheat, barley
chlorfenvinfos	Birlane 24 Sapecron 240EC	Wheat (Jan and Feb only) Winter cereals (Jan and Feb only)

CHEMICAL	PRODUCT	CROPS/USES
chlormequat	Atlas Chlormequat 700	Wheat, oats
	CCC 460	Wheat, oats
	CCC 700	Wheat, oats, barley
	Cleanacres PDR	Wheat, oats
	Clifton Chlormequat 46	Wheat, oats
	Farmacel 645	Wheat, oats
	Hyquat	Wheat, oats
	Hyquat 70	Wheat, oats, barley
	Hyquat 75	Wheat, oats, barley
	Mandops Barleyquat B	Barley
	Mandops Bettaquat B	Wheat, oats
	Mandops Chlormequat 700	Wheat, oats, barley
	Portman Chlormequat 400	Wheat, oats
	Portman Chlormequat 460	Wheat, oats
	Portman Chlormequat 700	Wheat, oats
	Power Chlormequat 700	Winter and spring wheat
	Quadrangle Chlormequat 700	Wheat, oats
	Titan	Cereals
	Tripart Brevis	Wheat, oats
	Tripart Chlormequat 460	Wheat, oats
chlormequat + choline chloride	Arotex Extra	Wheat, spring barley, oats, rye, triticale

CHEMICAL	PRODUCT	CROPS/USES
	Ashlade 5C	Wheat, oats
	New 5C Cycocel	Winter oilseed rape, wheat, oats, triticale, barley
	Tripart 5C	Wheat, oats
2-chloroethyl phosphonic acid	Cerone	Winter barley
chlorothalonil	BB Chlorothalonil	Wheat, barley
	Bombardier Bravo 500	Wheat, barley, potatoes, field beans Wheat, potatoes, peas, field beans
	Chiltern Ole	Wheat, barley, potatoes
	Jupital	Peas, beans
	Repulse	Peas, beans
	Sipcam UK Rover 500	Wheat, barley, potatoes, field beans
chlorotoluron	Chiltern Chlortoluron	Winter wheat, winter barley, durum wheat, triticale
	Dicurane 500 FW	Winter wheat, winter barley, durum wheat, triticale
	Toro	Winter wheat, winter barley, durum wheat, triticale
	Tripart Ludorum	Winter wheat, winter barley
	Tripart Ludorum 700	Winter wheat, winter barley, durum wheat, triticale
chlorpropham + linuron	Profalon	Daffodils, narcissi, tulips
chlorpyrifos	Dursban 4	Wheat, barley, oats
copper hydroxide	Chiltern Kocide 101	Barley, wheat, beans, potatoes
copper oxychloride	Cuprokylt	Potatoes

CHEMICAL	PRODUCT	CROPS/USES
cymoxanil + mancozeb	Ashlade Blight Fungicide	Potatoes
	Curzate M	Potatoes
	Fytospore	Potatoes
	Systol M	Potatoes
dalapon	BH Dalapon	Estuaries, watercourses and surrounding banks
demeton-S-methyl	Campbell's DSM	Cereals, peas, potatoes, sugar beet, carrots, brassicas, field beans
	Chafer Azotox 580	Cereals, peas, potatoes, sugar beet
	Chiltern Demeton-S-Methyl	Cereals, peas, beans, sugar beet
	Metasystox 55	Cereals, beet crops, potatoes, carrots, parsnips, celery, brassicas, top fruit, soft fruit, herbage seed crops, maize, mangolds
	Mifatox	Cereals, peas, potatoes, sugar beet, carrots, brassicas, field beans
	Power DSM 580	Cereals, peas, beans, beet crops, carrots, parsnips, celery, brassicas, top fruit, soft fruit, herbage seed crops, maize, mangolds, potatoes
	Quadrangle DSM Systemic Insecticide	Cereals, peas, potatoes, sugar beet, beans
	Tripart Systemic Insecticide	Cereals, peas, potatoes, sugar beet, beans
	Vassgro DSM	Cereals, grass seed crops, peas, beans, potatoes, sugar beet
desmetryn	Semeron 25WP	Brassicas

CHEMICAL	PRODUCT	CROPS/USES
difenzoquat	Avenge 2	Wheat, rye, barley, maize, herbage seed crops
dimethoate	Ashlade Dimethoate	Wheat, barley, peas, potatoes, sugar beet, brassicas, carrots
	Atlas Dimethoate 40	Cereals, peas, potatoes, sugar beet, carrots, brassicas
	BASF Dimethoate 40	Cereals, peas, potatoes, beet crops, brassicas, carrots, beans, herbage seed crops
	Brabant Dimethoate	Cereals, peas, potatoes, beet crops, soft fruit, carrots, tree nurseries
	Campbell's Dimethoate 40	Cereals, peas, potatoes, sugar beet
	Chiltern Dimethoate 40	Cereals, peas, potatoes, sugar beet, beans
	Clifton Dimethoate 40	Cereals, herbage seed crops, beans, peas, ware and leaf brassicas, potatoes, carrots, beet crops
	Portman Dimethoate EC	Cereals, peas, potatoes, sugar beet, beans
	Power Dimethoate	Sugar beet, mangolds, fodder beet, red beet, carrots, peas, beans, potatoes
	Quad Dimethoate 40	Cereals, peas, potatoes, sugar beet, beans
	Tripart Dimethoate 40	Cereals, peas, potatoes, sugar beet, beans
	Unicrop Dimethoate 40	Cereals, peas, potatoes, sugar beet, herbage seed crops
disulfoton	Campbell's Disulfoton P-10	Brassicas, broad beans
	Disyston P-10	Brassicas, beans, sugar beet, carrots

CHEMICAL	PRODUCT	CROPS/USES
fenitrothion	Dicofen	Cereals, peas
fentin acetate + maneb	Brestan 60	Potatoes
	Hytin	Potatoes
	Trimastan	Potatoes
fentin hydroxide	Ashlade Flotin	Potatoes
	Chiltern Super-Tin 4L	Potatoes
	Du-ter 50	Potatoes
	Farmatin 50	Potatoes
	Farmatin 560	Potatoes
	Quadrangle Super-Tin 4L	Potatoes
fentin hydroxide + maneb and zinc	Chiltern Tinman	Potatoes
ferbam + maneb and zineb	Trimanzone	Cereals, oilseed rape, Brussels sprouts, potatoes
flamprop-M-isopropyl	Commando	Wheat, barley, rye, triticale, durum wheat
	Gunner	Wheat, barley, durum wheat
heptenophos	Hostaquick	Cereals, peas, brassicas, beans
iprodione	Rovral Flo	Oilseed rape, peas
	Rovral WP	Bulb onions, salad onions
iprodione + thiophanate-methyl	Compass	Wheat, barley, oilseed rape, field beans
isoproturon	Arelon	Wheat, winter barley, rye, triticale
	Arelon WDG	Wheat, winter barley, winter rye, triticale
	Chiltern IPU	Wheat, barley, rye, triticale

CHEMICAL	PRODUCT	CROPS/USES
	Hytane 500FW	Autumn or spring sown cereals
	Sabre	Winter wheat, winter barley, rye, triticale
	Sabre WDG	Wheat, winter barley, winter rye, triticale
	Tolkan	Winter wheat, barley
linuron	Afalon	Potatoes, carrots, parsnips, celery, parsley
	Liquid Linuron	Cereals
malathion	Malathion 60	Arable field crops, except oilseed rape
mancozeb	Ashlade Mancozeb FL	Potatoes
	Dithane 945	Potatoes
	Dithane Dry Flowable	Potatoes
	Manzate 200	Potatoes
	Penncozeb	Potatoes
	Portman Mancozeb 80	Cereals
	Unicrop Mancozeb	Potatoes
mancozeb + metalaxyl	Fubol 58WP	Potatoes
	Fubol 75WP	Potatoes
mancozeb + oxadixyl	Recoil	Early and maincrop potatoes
maneb	Campbell's X-Spor SC	Potatoes, winter cereals, spring barley
	Clifton Maneb	Winter wheat, potatoes
	Headland Spirit	Barley, winter wheat, potatoes
	Manzate	Winter cereals, potatoes
	Manzate Flowable	Barley, wheat, potatoes

CHEMICAL	PRODUCT	CROPS/USES
	Trimangol 80	Potatoes, tomatoes, flower bulbs
	Unicrop Maneb	Potatoes, winter wheat
maneb + zinc	Chiltern Manex	Potatoes, cereals
	Quadrangle Manex	Potatoes, cereals
	Vassgro Manex	Potatoes, cereals
maneb + zinc oxide	Mazin	Potatoes
manganese zinc ethylenebisdithio-carbamate + ofurace	Patafol Plus	Potatoes, oilseed rape
metaldehyde	Chiltern Metaldehyde Slug Killer Mini Pellets	All crops
	Doff Metaldehyde Slug Killer Mini Pellets	All crops
	Farmon Mini Slug Pellets	All crops
	Mifaslug	All crops
	Optimol	All crops
	PBI Slug Pellets	All crops
	Quad Mini Slug Pellets	All agricultural and horticultural crops
	Slug Destroyer	All crops
	Tripart Mini Slug Pellets	All crops
	Unicrop 6% Mini Slug Pellets	All crops
	Vassgro Mini Slug Pellets	All crops
methabenzthiazuron	Tribunil	Barley, winter wheat, autumn sown spring wheat, winter oats, winter rye, durum wheat, triticale, perennial ryegrass

CHEMICAL	PRODUCT	CROPS/USES
methiocarb	Club	All crops
	Draza	All crops
metoxuron	Dosaflo	Winter wheat, winter barley
monolinuron	Arresin	Potatoes, dwarf beans, leeks
omethoate	Folimat	Winter wheat
oxydemeton-methyl	Metasystox R	Cereals, beet crops, potatoes, carrots, parsnips, celery, beans, peas, top fruit, soft fruit, herbage seed crops, maize, mangolds
phenmedipham	Betanal E	Beet crops, strawberries
phorate (granules only)	BASF Phorate	Field and broad beans, beet crops, carrots
	Campbell's Phorate	Field and broad beans
	Terrathion Granules	Field and broad beans
phosalone	Zolone Liquid	Cereals, brassica seed crops
pirimicarb	Aphox	Cereals, peas, potatoes, sugar beet, beans, brassicas, maize, swedes, oilseed rape, turnips, carrots
	Pirimicarb 50DG	Cereals, field beans, potatoes, sugar beet, peas
potassium sorbate + sodium metabisulphite + sodium propionate	Brimstone Plus	Wheat, barley, oilseed rape, potatoes, sugar beet, swedes
prochloraz	Sportak	Barley, wheat, rye, oilseed rape
prometryn	Gesagard 50WP	Peas, potatoes, carrots, celery, leeks
propiconazole	Radar	Wheat, barley
	Tilt 250EC	Wheat, barley

CHEMICAL	PRODUCT	CROPS/USES
propiconazole + tridemorph	Tilt Turbo 475EC	Wheat, barley
simazine	Gesatop 500WP	Beans
	Gesatop 50FW	Beans
sulphur	Clifton Sulphur	Oilseed rape, wheat, barley, sugar beet
	Intrachem Sulfur W80	Cereals, sugar beet, oilseed rape, swedes, turnips, hops
	Kumulus S	Sugar beet, wheat, barley, oilseed rape
	Magnetic 6	Sugar beet, swedes, turnips, cereals, oilseed rape
	Thiovit	Cereals, sugar beet
terbutryn	Prebane 500FW	Cereals, rye, triticale
triadimefon	Bayleton	Sugar beet, cereals, brassicas (including turnips and swedes)
	Bayleton 5	Blackcurrants
triadimenol + tridemorph	Dorin	Winter wheat, barley, oats
tri-allate	Avadex BW Granular	Wheat, barley, winter beans, peas
triazophos	Hostathion	Peas
trichlorphon	Dipterex 80	Beet crops, brassicas, spinach
tridemorph	Calixin	Barley, oats, winter wheat, swedes, turnips
triforine	Saprol	Spring barley
vinclozolin	Ronilan	Oilseed rape, dwarf and field beans, ornamentals
	Ronilan FL	Oilseed rape, dwarf and french beans, ornamentals
zineb	Unicrop Zineb	Potatoes

ULV APPLICATION

Certain pesticide products are specifically formulated for use with ULV applicators of various types, such as mistblowers, aerosol projectors, CDA sprayers and fogging machines. Others may be used without dilution as ULV treatments or may be diluted for application at higher volume rates. Products of both types are included in the table below.

CHEMICAL	PRODUCT	CROP
amitrole + atrazine	Atlazin CDA	non-crop
amitrole + atrazine + 2,4-D	CDA Python	non-crop
	Snapper CDA	non-crop
	Terramix CDA	non-crop
	Torapron	non-crop
amitrole + atrazine + dicamba	Sarapron	non-crop
amitrole + atrazine + diuron	Bullseye CDA	non-crop
	CDA Viper	non-crop
	Duranox	non-crop
	Rassamix CDA	non-crop
	Rassapron	non-crop
amitrole + atrazine + MCPA	Meganox Plus	non-crop
	Serramix CDA	non-crop
amitrole + simazine	CDA Clearaway	non-crop
	CDA Simflow Plus	non-crop
	Mascot Highway Liquid	non-crop
	Primatol SE 500FW	non-crop
	Transformer Primatol SE 500FW	non-crop
asulam	Asulox	bracken in hill land, forestry and non-crop land
atrazine	Gesaprim 500 FW	non-crop
	Mascot Gauntlet Liquid	non-crop
atrazine + imazapyr	Arsenal XL CDA	non-crop
	Moderator	non-crop
atrazine + terbuthylazine	Transformer Gardoprim A 500 FW	forestry
carbaryl + pyrethrins	Microcarb-T	poultry houses

CHEMICAL	PRODUCT	CROP
chloridazon	Trojan SC	beet crops
chloridazon + ethofumesate	Spectron	beet crops
chlorotoluron	Chlortoluron 500 Dicurane 500 FW Tripart Ludorum	cereals cereals cereals
chlorpropham	Mirvale 500 HN Warefog 25	potatoes potatoes
chlorpropham + propham	Atlas Indigo Pommetrol M Power Gro-Stop	potatoes potatoes potatoes
2,4-D	CDA Dicotox Extra Silvapron D	turf forestry
2,4-D + mecoprop	CDA Supertox 30 Verdone CDA	turf turf
dichlorvos	Darmycel Dichlorvos Nuvan 500 EC	mushroom houses poultry houses
dimethoate	Turbair Systemic Insecticide	cereals, vegetables, fruit crops, ornamentals
fenitrothion + permethrin + resmethrin	Turbair Grain Store Insecticide	grain stores
formaldehyde	Dyna-Form Formaldehyde	glasshouses mushroom houses
gamma-HCH + resmethrin + tetramethrin	Dairy Fly Spray	livestock houses
glyphosate	Barclay Gallup CDA Spasor Roundup Roundup Four 80 Roundup Pro Spasor Stirrup	amenity, forestry and non-crop situations
glyphosate + simazine	Mascot Ultra-sonic Rival	non-crop non-crop
imazapyr	Arsenal Arsenal 50	non-crop non-crop

CHEMICAL	PRODUCT	CROP
ioxynil + isoproturon + mecoprop	Post-Kite	cereals
iprodione	CDA Rovral	turf
	Turbair Rovral	lettuce, tomatoes, chrysanthemums
isoproturon	Hytane 500FW	cereals
	Tolkan 500	cereals
	Tolkan Liquid	cereals
linuron + trietazine + trifluralin	Pre-Empt	winter cereals
MCPA	BASF MCPA Amine 50	cereals, grassland
methabenzthiazuron	Tribunil	cereals, ryegrass
nonylphenoxypoly (ethyleneoxy) ethanol-iodine + thiabendazole	Tubazole Tubazole M	potatoes
oxycarboxin	Plantvax 75	glasshouse ornamentals
permethrin	Turbair Permethrin	vegetables
phenothrin	Sumicidin	livestock houses
pirimiphos-methyl	Actellifog	glasshouse crops
propyzamide	Kerb Flo	forestry
	Kerb Flowable	forestry
	Kerb 50W	
resmethrin	Turbair Resmethrin Extra	vegetables, soft fruit, ornamentals, mushrooms
simazine + trietazine	Remtal SC	peas, beans, lupins
tecnazene	Nebulin	potatoes
terbuthylazine + terbutryne	Opogard 500 FW	peas, beans, lupins
terbutryn	Prebane 500 FW	cereals
thiabendazole	Storite Flowable	potatoes
thiophanate-methyl	CDA Mildothane	turf
triadimefon	Bayleton	cereals, grassland, field crops

ADJUVANTS

Pesticide adjuvants are not themselves classed as pesticides but under the Control of Pesticides Regulations 'No person shall use a pesticide in conjunction with an adjuvant except in accordance with the conditions of the approval given originally in relation to that pesticide, or as varied subsequently by lists of authorised adjuvants published by the Ministers'.

Lists of authorised adjuvants are published in *The Pesticides Register* at intervals, the first appearing in Issue 3 1989 with subsequent additions in Issues 9 and 12 1989 and 1, 2, 5, 6 and 7 1990. The products listed below are included, with the names of suppliers and authorised adjuvant numbers, and are divided into three categories as follows :

Category 1	Extenders
Category 2	Sticking agents
Category 3	Wetting agents, including Adjuvant oils.

Adjuvant product labels must be consulted for details of compatible chemicals or products, rates etc. but the table below provides a summary of the label information to indicate the area of use of the adjuvant. Protective clothing requirements and label precautions refer to the keys given in Appendix 2 and may include warnings about products harmful or dangerous to fish.

	PRODUCT	SUPPLIER	ADJ. NO.	CATEGORY
	Actipron	Bayer	0013	3
	Actipron	BP	0013	3
Type	Adjuvant oil containing 97% highly refined mineral oil			
Use with	Asulox		Dow Shield + Goltix WG	
	Basagran		Fusilade 5 + Goltix WG	
	Benlate		Gesaprim 500 FW	
	Betanal E		Goltix WG	
	Betanal E + Goltix WG		Pilot	
	Checkmate		Roundup	
	Checkmate + Goltix WG			
Label precautions	21, 28, 29, 54, 63, 65			
	Adder	Embetec	0019	3
Type	Adjuvant oil containing 97% refined mineral oil			
Use with	Ambush C + Checkmate		Checkmate + Goltix WG	
	Asulox*		Checkmate + Metasystox 55	
	Basagran		Checkmate + Thiovit	
	Butisan S + Checkmate		Checkmate + Toppel 10	
	Checkmate		Dow Shield + Goltix WG	

	PRODUCT	SUPPLIER	ADJ. NO.	CATEGORY
	Checkmate + Decis		Gesaprim 500 FW	
	Checkmate + Dow Shield		Goltix WG	
	Checkmate + Fubol 58 WP		phenmedipham	
	Checkmate + gamma-HCH			
	(SC formulations)			
	*Not in forestry			
Label precautions	21, 28, 29, 54, 63, 65, 70			

	PRODUCT	SUPPLIER	ADJ. NO.	CATEGORY
	Addwet	Agstock	0056	3
	No details available			

	Agral	ICI	0033	3
Type	Non-ionic wetter/spreader containing 900 g/l alkyl phenol ethylene oxide			
Use with	Any spray for which additional wetter is approved and recommended			
Label precautions	A, C/ 6a, 6b, 12c, 18, 21, 29, 36, 37, 54, 63, 65, 70			

	Agrisorb	ABM	0057	2, 3
Type	Wetter/spreader containing 850 g/l tallow amine ethoxylate			
Use with	glyphosate			
Label precautions	A, C/ 5c, 6a, 6b, 12a, 18, 21, 25, 29, 36, 37, 53, 67, 78			

	Agriwet	ABM	0057	3
Type	Non-ionic wetter containing 25% alkyl phenol ethoxylate			
Use with	Numerous fungicides, insecticides and nutrients. See label for details			
Label precautions	A, C/ 6a, 6b, 18, 21, 36, 37, 53, 63, 70			

	Agropen	Ideal	0110	3
Type	Adjuvant oil containing 95% emulsifiable vegetable oil			
Use with	Compatible with most pesticide formulations. See label for details			
Label precautions	A, C/ 6a, 6b, 16, 21, 29, 36, 37, 54, 63, 66, 70, 78			

	PRODUCT	SUPPLIER	ADJ. NO.	CATEGORY
	Anphix	ANP Developments	0103	3

No details available

	Ashlade Adjuvant Oil	Ashlade	0067	3
Type	Adjuvant oil containing 99% highly refined mineral oil			
Use with	Goltix WG		phenmedipham	
Label precautions	6a, 6b, 6c, 18, 21, 28, 29, 36, 37, 53, 63, 66			

	Atlas Adherbe	Atlas	0023	3
Type	Adjuvant oil containing 83% highly refined mineral oil			
Use with	Checkmate Fusilade 5 Goltix WG		Goltix WG + Protrum K Pilot Protrum K	
Label precautions	6a, 6b, 6c, 18, 21, 28, 29, 36, 37, 53, 60, 63, 66			

	Atlas Adjuvant Oil	Atlas	0021	3
Type	Adjuvant oil containing 95% highly refined mineral oil			
Use with	Goltix WG Goltix WG + Protrum K		phenmedipham Protrum K	
Label precautions	6a, 6b, 6c, 18, 21, 28, 29, 36, 37, 53, 60, 63, 66			

	Atlas Libsorb	Atlas	0050	3
Type	Non-ionic wetter/spreader containing 900 g/l alkyl alcohol ethoxylate			
Use with	Any spray for which additional wetter is approved and recommended			
Label precautions	A, C/ 6a, 6b, 18, 21, 29, 36, 37, 53, 63, 65			

	PRODUCT	SUPPLIER	ADJ. NO.	CATEGORY
	Chiltern Cropspray 11E	Chiltern	0005	3
Type	Adjuvant oil containing 99% highly refined mineral oil			
Use with	Asulox		Dosaflo	
	atrazine		glyphosate	
	bromoxynil + ioxynil		Goltix WG	
	Checkmate		Goltix + phenmedipham	
	Clout		isoproturon	
	Commando		phenmedipham	
	dichlorprop, MCPA		propiconazole	
	mecoprop and mixtures		triadimenol	
Label precautions	28, 29, 54, 60, 63, 66			

	Citowett	BASF	0047	3
Type	Non-ionic wetter/spreader containing 99–100% alkyarylpolyglycol ether			
Use with	Any spray for which additional wetter is approved and recommended			
Label precautions	A, C/ 6a, 6b, 18, 21, 29, 36, 37, 53, 63, 66			

	Clifton Alkyl 90	Clifton	0118	3
	No details available			

	Clifton Glyphosate Additive	Clifton	0024	2, 3
Type	Wetter/spreader containing 850 g/l tallow amine ethoxylate			
Use with	glyphosate			
Label precautions	A, C/ 5c, 6a, 6b, 12a, 18, 21, 25, 29, 36, 37, 53, 63, 66, 70, 78			

	Clifton Wetter	Clifton	0028	3
Type	Non-ionic wetter/spreader containing 250 g/l alkyl phenol ethoxylate			
Use with	Numerous fungicides, herbicides, insecticides and nutrients. See label for details			
Label precautions	A, C/ 6a, 6b, 18, 21, 36, 37, 53, 63			

	PRODUCT	SUPPLIER	ADJ. NO.	CATEGORY
	Codacide Oil	Microcide	0011	1, 2, 3
Type	Adjuvant oil containing 95% emulsifiable vegetable oil			
Use with	Numerous fungicides, herbicides and insecticides. See label for details			
Label precautions	28, 29, 54, 63, 66			
	Cutonol	Midkem	0066	3
	No details available			
	Du Pont Adjuvant	Du Pont	0019	3
Type	Non-ionic wetter/spreader containing 900 g/l ethylene oxide condensate			
Use with	Any spray for which additional wetter is approved and recommended			
Label precautions	A, C/ 6a, 6b, 12c, 18, 21, 30, 36, 37, 53, 63, 65			
	Emerald	Intacrop	0031	3
Type	Extender and antitranspirant			
	No further information available			
	Enhance	Midkem	0060	3
Type	Non-ionic wetter/spreader containing 90% alkyl phenol ethoxylate			
Use with	Any spray for which additional wetter is approved and recommended			
Label precautions	A, C/ 6a, 6b, 18, 21, 29, 36, 37, 53, 63, 65			
	Enhance Low Foam	Midkem	0068	3
	No details available			

	PRODUCT	SUPPLIER	ADJ. NO.	CATEGORY
	Ethokem	Midkem	0052	3
Type	Cationic surfactant containing 870 g/l polyethanoxy alkyl amine			
Use with	glyphosate metamitron phenmedipham Asulox Checkmate		Fusilade 5 Garlon 4 Gesaprim 500 FW Gramoxone 100 Reglone	
Label precautions	A, C/ 5c, 6a, 6b, 12c, 18, 21, 25, 28, 29, 36, 37, 53, 63, 65			
	Ethokem C/12	Midkem	0049	3
Type	Cationic surfactant containing bis-2 hydroxyethyl coco-amine			
Use with	glyphosate		ammonium sulphate	
Label precautions	A, C/ 5c, 6a, 6b, 12c, 18, 21, 25, 28, 29, 36, 37, 53, 63, 65			
	Event	Ideal	0018	2, 3
Type	Compatibility agent containing 90% alkyl and aryl ether phosphates			
Use with	Many pesticide/liquid fertilizer mixtures. See label for details			
Label precautions	A, C/ 16, 21, 29, 35, 36, 54, 63, 66, 70, 78			
	Exell	Truchem	0017	2, 3
Type	Wetter/spreader containing 64% polyethoxylated tallow amine			
Use with	diquat		glyphosate	
Label precautions	A, C/ 6a, 6b, 18, 21, 28, 29, 36, 37, 53, 63, 65, 70, 78			
	Farmon Blue	Farm Protections	0003	3
Type	Non-ionic wetter/spreader containing 900 g/l alkyl phenol ethylene oxide			
Use with	Any spray for which additional wetter is approved and recommended			
Label precautions	A, C/ 6a, 6b, 12c, 18, 21, 29, 36, 37, 54, 63, 65, 70, 78			

	PRODUCT	SUPPLIER	ADJ. NO.	CATEGORY
	Farmon Wetter	Farm Protection	0120	3
	No details available			

	Frigate	Fermenta	0044	3
Type	Wetter/spreader containing 800 g/l tallow amine ethoxylate			
Use with	glyphosate			
Label precautions	A, C/ 5c, 6a, 6b, 6c, 12c, 18, 21, 25, 29, 36, 37, 53, 63, 65, 78			

	Fyzol 11E	Schering	0072	3
Type	Adjuvant oil containing 99% highly refined mineral oil			
Use with	Betanal E Pilot		Betanal E + Goltix WG	
Label precautions	25, 29, 53, 63, 66, 70			

	Galion	Intracrop	0001	3
	Galion	Service	0001	3
Type	Non-ionic wetter/spreader containing ethylene oxide condensates			
Use with	diclofop-methyl		mecoprop	
Label precautions	No information			

	Headland Guard	WBC Technology	0073	1, 2
Type	Sticker/spreader containing 550 g/l organic co-polymer and surfactants			
Use with	A wide range of agricultural chemicals and trace elements			
Label precautions	18, 29, 36, 37, 63			

	Headland Intake	WBC Technology	0074	3
Type	Penetrant containing 450 g/l organic acids and surfactants			
Use with	A wide range of fungicides, herbicides, desiccants and growth regulators			
Label precautions	A, C/ 6a, 6b, 18, 21, 28, 29, 36, 37, 53, 63, 65			

	PRODUCT	SUPPLIER	ADJ. NO.	CATEGORY
	High Trees Mixture B	Service	0004	3
Type	Non-ionic wetter/spreader			
Use with	Roundup			
Label precautions	No information			
	HT Non-ionic Wetter	Service	0009	3
Type	Wetter/spreader containing 90% alkyl phenol ethylene oxide condensate			
Use with	Any spray for which additional wetter is approved and recommended			
Label precautions	No information			
	Hyspray	Fine	0020	2, 3
Type	Cationic surfactant containing 800 g/l polyethoxylated tallow amine			
Use with	glyphosate			
label precautions	A, C/ 5c, 6a, 6b, 18, 21, 25, 28, 29, 35, 36, 53, 63, 65, 70			
	Jogral	Ideal	0109	3
Type	Cationic surfactant containing 800 g/l tallow amine ethoxylate			
Use with	glyphosate			
Label precautions	A, C/ 5c, 6a, 6b, 12c, 18, 21, 28, 29, 36, 37, 53, 63, 65, 70, 78			
	Lo-Dose	Quadrangle	0059	3
Type	Cationic surfactant containing 800 g/l tallow amine ethoxylate			
Use with	Roundup		Roundup Four 80	
Label precautions	A, C/ 5c, 6a, 6b, 12c, 18, 21, 28, 29, 36, 37, 53, 63, 65, 70, 78			
	Mangard	Mandops	0075	1, 2
	No details available			

	PRODUCT	SUPPLIER	ADJ. NO.	CATEGORY
	Minder	Stoller	0087	3
Type	Rape oil adjuvant containing 94% vegetable oil			
Use with	glyphosate			
Label precautions	29, 37, 54, 63, 67			
	Nu Film P	Intracrop	0039	2, 3
	No details available			
	PBI spreader	PBI	0034	3
Type	Non-ionic wetter/spreader			
Use with	Any spray for which additional wetter is approved and recommended			
Label precautions	30, 54, 63, 65, 70			
	Planet	Ideal	0025	3
Type	Non-ionic wetter/spreader containing 85% alkyl polyglycol ether and fatty acid			
Use with	Any spray for which additional wetter is recommended			
Label precautions	5c, 6a, 6b, 12c, 21, 29, 36, 37, 63, 65, 78			
	Power Non-ionic Wetter	Power	0081	3
Type	Non-ionic wetter/spreader containing 900 g/l alkyl phenol ethylene oxide concentrate			
Use with	Any spray for which additional wetter is approved and recommended			
Label precautions	A, C/ 6a, 6b, 12c, 18, 21, 30, 36, 37, 54, 63, 66, 70			
	Power Spray Save	Power	0061	3
Type	Cationic surfactant containing 800 g/l tallow amine ethoxylate			
Use with	glyphosate			
Label precautions	A, C/ 5c, 6a, 6b, 12c, 14, 18, 21, 28, 29, 36, 37, 53, 63, 65, 70, 78			

	PRODUCT	SUPPLIER	ADJ. NO.	CATEGORY
	Quad-Fast	Quadrangle	0054	2
Type	Coating agent/surfactant containing di-1-p-menthene			
Use with	Glyphosate and other pesticides. Growth regulators and nutrients for which a coating agent is approved and recommended			
Label precautions	No information			

	PRODUCT	SUPPLIER	ADJ. NO.	CATEGORY
	Quadrangle Cropspray 11E	Quadrangle	0010	3
Type	Adjuvant oil containing 99% highly refined mineral oil			

Use with

Asulox	Dosaflo
atrazine	glyphosate
bromoxynil + ioxynil	Goltix WG
Checkmate	Goltix WG + phenmedipham
Clout	isoproturon
Commando	phenmedipham
2,4-DP, MCPA	propiconazole
mecoprop and mixtures	triadimenol

Label precautions 28, 29, 54, 63, 66

	PRODUCT	SUPPLIER	ADJ. NO.	CATEGORY
	Quadrangle Q 900	Quadrangle	0012	3

No details available

	PRODUCT	SUPPLIER	ADJ. NO.	CATEGORY
	Rapide	Intracrop	0016	3

No details available

	PRODUCT	SUPPLIER	ADJ. NO.	CATEGORY
	Solar	Ideal	0110	3
Type	Non-ionic spreader/activator containing 75% polypropoxy propanol and 15% alkyl polyglycol ether			
Use with	Foliar applied plant growth regulators			
Label precautions	A, C/ 16, 21, 29, 36, 37, 54, 63, 66, 70, 78			

	Sprayfast	Mandops	0053	1, 2, 3
Type	Coating agent/surfactant containing di-1-p-menthene and nonyl phenol ethylene oxide condensate			
Use with	Glyphosate and other pesticides, growth regulators and nutrients for which a coating agent is approved and recommended			
Label precautions	30, 54, 63, 65			

	Spraymate Activator 90	Newman	0062	3
Type	Non-ionic wetter/spreader containing 750 g/l alkylphenyl hydroxypolyoxyethylene			
Use with	Any spray for which additional wetter is approved and recommended			
Label precautions	A, C/ 6a, 6b, 18, 21, 28, 29, 36, 37, 53, 63, 65			

	Spraymate Bond	Newman	0037	1, 2
Type	Sticker/extender containing 450 g/l synthetic latex			
Use with	A wide range of contact herbicides, fungicides and insecticides. Contact supplier for details			
Label precautions	A, C/ 6a, 6b, 18, 21, 28, 29, 36, 37, 53, 63, 65			

	Spraymate L1-700	Newman	0038	3
Type	Penetrating, acidifying surfactant containing 750 g/l soyal phospholids			
Use with	Chlormequat and a wide range of systemic pesticides and trace elements. See label for details			
Label precautions	6a, 6b			

	Sprayprover	Fine	0027	2, 3
Type	Adjuvant oil containing 800 g/l highly refined mineral oil			
Use with	atrazine Beetomax + metamitron		metamitron phenmedipham	
Label precautions	A, C/ 6a, 6b, 18, 21, 28, 29, 36, 37, 53, 63, 66			

	PRODUCT	SUPPLIER	ADJ. NO.	CATEGORY
	Stik-It	Quadrangle	0071	3
Type	Non-ionic wetter/spreader			
Use with	A wide range of fungicides and insecticides			
Label precautions	No information			
	Swirl	Shell	0085	3
Type	Adjuvant oil containing 590 g/l highly refined mineral oil			
Use with	Commando			
Label precautions	12c, 21, 29, 52, 63, 66, 70			
	Team	Monsanto	0079	3
Type	Cationic surfactant containing 800 g/l ethoxylated tallow amine			
Use with	Roundup			
Label precautions	A, C/ 5c, 6a, 6b, 12c, 14, 18, 21, 28, 29, 36, 37, 53, 63, 65, 70, 78			
	Tonic	Brown Butlin	0094	1,2
	No details available			
	Topup	FCC	0080	3
Type	Cationic surfactant containing 800 g/l ethoxylated tallow amine			
Use with	Roundup			
Label precautions	A, C/ 5c, 6a, 6b, 12c, 14, 18, 21, 28, 29, 36, 37, 53, 63, 65, 70, 78			
	Tradename A	Farm Protection	0094	1, 2
	No details available			
	Tripart Acer	Tripart	0097	3
	No details available			

PRODUCT	SUPPLIER	ADJ. NO.	CATEGORY
Tripart Cropspray 11E	Tripart	0006	3

Type	Adjuvant oil containing 99% highly refined mineral oil
Use with	Asulox / Dosaflo

Use with:

Asulox	Dosaflo
atrazine	Fusilade 5
bromoxynil + ioxynil	glyphosate
Checkmate	Goltix WG
Clout	Goltix WG + phenmedipham
Commando	isoproturon
2,4-DP, MCPA, mecoporp	phenmedipham
and mixtures	propiconazole
	triadimenol

Label precautions	28, 29, 54, 60, 63, 66

PRODUCT	SUPPLIER	ADJ. NO.	CATEGORY
Tripart Lentus	Tripart	0017	1, 2

No details available

PRODUCT	SUPPLIER	ADJ. NO.	CATEGORY
Tripart Minax	Tripart	0016	3

Type	Non-ionic wetter/spreader containing alkyl alcohol ethoxylate
Use with	Any spray for which additional wetter or spreader is approved and recommended
Label precautions	No information

PRODUCT	SUPPLIER	ADJ. NO.	CATEGORY
Vassgro Spreader	Vass	0035	3

Type	Non-ionic wetter/spreader containing nonyl phenol-ethylene oxide condensates
Use with	Any spray for which additional wetter or spreader is approved and recommended
Label precautions	29, 54, 63, 65, 70

PRODUCT	SUPPLIER	ADJ. NO.	CATEGORY
Wayfarer	Service	0045	3

Type	Cationic wetter/spreader containing 80% tallow amine ethoxylate
Use with	glyphosate
Label precautions	No information

POISONING BY PESTICIDES — FIRST AID MEASURES

If pesticides are handled in accordance with the required safety precautions, as given on the container label, poisoning should not occur. It is difficult however, to guard completely against the occasional accidental exposure. Thus, if a person handling or exposed to pesticides becomes ill it is a wise precaution to apply first aid measures appropriate to pesticide poisoning even though the cause of illness may eventually prove to have been quite different.

The first essential in case of suspected poisoning is for the person involved to stop work, to be moved away from any area of possible contamination and for a doctor to be called at once. If no doctor is available the patient should be taken to hospital as quickly as possible. In either event it is most important that the name of the chemical being used should be recorded and preferably the whole product label or leaflet should be shown to the doctor or hospital concerned.

Some pesticides which are unlikely to cause poisoning in normal use are extremely toxic if swallowed accidentally or deliberately. In such cases get the patient to hospital as quickly as possible, with all the information you have.

Regular pesticide users should consider appointing a trained first aider, even if numbers of employees are not large, since there is a specific hazard.

General measures
Measures appropriate in all cases of suspected poisoning include :

- Keep the patient at rest and under shelter.
- Remove any protective or other contaminated clothing (taking care to avoid personal contamination).
- Wash any contaminated areas carefully with water or with soap and water if available.
- Lay the patient down and cover with one clean blanket or coat etc. Avoid overheating.
- In cases of eye contamination, flush with plenty of clean water for at least 15 min.
- If breathing ceases or weakens, ensure that breathing passages are clear, remove false teeth or any other obstructions from the mouth, remove any constrictions around the neck, draw the tongue forwards and apply artificial respiration. Do not use mouth to mouth method if poisonous chemical has been swallowed.
- If consciousness is lost, place the casualty on his or her side with head down and tongue drawn forward to prevent inhalation of vomit.

Specific measures
In case of poisoning with particular chemical groups, the following measures may be taken before transfer to hospital :

Dinitro compounds* (dinoseb, DNOC, dinocap etc.)
Keep the patient at rest and cool.

---

* Approvals for use of dinoseb, dinoseb acetate, dinoseb amine, dinoterb and binapacryl were revoked on 22 Jan 1988. Approval for storage of these materials was withdrawn on 30 June 1988. Approvals for supply of DNOC were revoked in Dec 1988 and for its use in Dec 1989.

Organophosphorus and carbamate insecticides
(aldicarb, azinphos-methyl, benfuracarb, carbofuran, chlorfenvinphos, chlorpyrifos, dichlorvos, heptenophos, malathion, methiocarb, mevinphos, propoxur, quinalphos, thiometon, triazophos etc)
Keep the patient at rest. The patient may suddenly stop breathing so be ready to give artificial respiration.

Organochlorine compounds (aldrin*, dienochlor, endosulfan, gamma-HCH)
If convulsions occur, do not interfere unless the patient is in danger of injury; if so any restraint must be gentle. When convulsions cease, place the casualty on his or her side with head down, tongue forward (recovery position).

Paraquat, diquat
Irrigate skin and eye splashes copiously with water. If any chemical has been swallowed, take to hospital for tests.

Cyanide (including sodium cyanide)
Send for medical aid. Remove casualty to fresh air, if necessary using breathing apparatus and protective clothing. If the casualty is NOT BREATHING, start resuscitation using a manual method where practicable. If there is no heart beat or pulse, start cardiac massage. Remove casualty's contaminated clothing. Gently brush solid particles from the skin, making sure you protect your own skin from contamination. Wash the skin and eyes copiously with water. Transfer casualty to nearest accident and emergency hospital by the quickest possible means together with the first aid cyanide antidote, if held on the premises.

Reporting of pesticide poisoning
Any cases of poisoning by pesticides must be reported without delay to an HM Agricultural Inspector of the Health and Safety Executive. In addition any cases of poisoning by substances named in schedule 2 of The Reporting of Injuries, Diseases and Dangerous Occurrences Regulations 1985, must also be reported to HM Agricultural Inspectorate (this includes organophosphorus chemicals, mercury and some fumigants).

Cases of pesticide poisoning should also be reported to the manufacturer concerned.

Additional information
General advice on the safe use of pesticides is given in Health and Safety Executive free leaflets :

AS6	*Crop spraying*
AS12	*Prevention of accidents*
AS18	*Storage of pesticides on farms*
AS25	*Training in the use of pesticides*
AS26	*Protective clothing for use with pesticides, and wallchart*
AS27	*Pesticides*
MS(A)9	*Cyanide poisoning*
IND(G)3(L)	*First aid provision in small workplaces*

---

* Approvals for supply, storage and use of aldrin were revoked in May 1989.

all available from HSE local offices and Head Office, St Hugh's House, Stanley Precinct, Bootle, Merseyside, L20 3QY.

Also :
Guidance Note MS18 *Health Surveillance by routine procedures* obtainable from HMSO.

A useful booklet produced by GIFAP (International Group of National Associations of Manufacturers of Agrochemical Products) titled 'Guidelines for emergency measures in cases of pesticide poisoning' is available from GIFAP, Avenue Hamoir 12, 1180 Brussels, Belgium.

The major agrochemical companies are able to provide authoritative medical advice about their own pesticide products. Detailed advice is also available to doctors, from :

- The National Poisons Information Service, New Cross Hospital, London SE14 5ER (071 407 7600)

and from regional centres in :

- Belfast, Royal Victoria Hospital (0232 240503)
- Birmingham, Dudley Road Hospital (021 5543801)
- Cardiff, Ambulance Headquarters (0222 569200)
- Edinburgh, Royal Infirmary (031 229 2477)
- Leeds, Leeds General Infirmary (0532 432799)
- Newcastle, Royal Victoria Hospital (091 232 5131)

CROP INDEX

CROP GUIDE

Arable crops

General

Diseases	Soil-borne diseases	dazomet, methyl bromide with amyl acetate, methyl bromide with chloropicrin
Pests	Birds	aluminium ammonium sulphate, quassia, ziram
	Damaging mammals	aluminium ammonium sulphate, quassia, ziram
	Nematodes	dazomet, methyl bromide with amyl acetate, methyl bromide with chloropicrin
	Slugs and snails	aluminium sulphate + copper sulphate + sodium tetraborate, metaldehyde, methiocarb
	Soil pests	dazomet, methyl bromide with amyl acetate, methyl bromide with chloropicrin
Weeds	Annual dicotyledons	diquat + paraquat (autumn stubble, pre-planting/sowing, stubble burning, sward destruction/direct drilling), glyphosate (autumn stubble, pre-planting/sowing), glyphosate (sward destruction/direct drilling, minimum cultivation), paraquat (autumn stubble), paraquat (minimum cultivation), paraquat (pre-emergence)
	Annual grasses	diquat + paraquat (autumn stubble, pre-planting/sowing, stubble burning, sward destruction/direct drilling), glyphosate (autumn stubble, pre-planting/sowing), glyphosate (sward destruction/direct drilling, minimum cultivation), paraquat (autumn stubble), paraquat (minimum cultivation), paraquat (pre-emergence), paraquat (sward destruction/direct drilling)
	Annual weeds	amitrole, amitrole (pre-sowing, autumn stubble), glyphosate (autumn stubble, pre-planting/sowing)
	Barren brome	paraquat (autumn stubble), propyzamide
	Couch	amitrole, amitrole (pre-sowing, autumn stubble), glyphosate (autumn stubble, pre-planting/sowing), glyphosate (sward destruction/direct drilling, minimum cultivation), propyzamide
	Creeping bent	paraquat (autumn stubble), paraquat (minimum cultivation), paraquat (sward destruction/direct drilling)
	Docks	amitrole, amitrole (pre-sowing, autumn stubble)
	General weed control	dazomet, methyl bromide with amyl acetate, methyl bromide with chloropicrin
	Perennial dicotyledons	glyphosate (autumn stubble, pre-planting/sowing), glyphosate (sward destruction/direct drilling, minimum cultivation)
	Perennial grasses	diquat + paraquat (autumn stubble, pre-planting/sowing, stubble burning, sward destruction/direct drilling), glyphosate (autumn stubble, pre-planting/sowing), glyphosate (sward destruction/direct drilling, minimum cultivation)

Perennial ryegrass	glyphosate (sward destruction/direct drilling, minimum cultivation), paraquat (sward destruction/direct drilling)	
Perennial weeds	amitrole, amitrole (pre-sowing, autumn stubble)	
Rough meadow grass	glyphosate (sward destruction/direct drilling, minimum cultivation), paraquat (sward destruction/direct drilling)	
Total vegetation control	atrazine, glyphosate	
Volunteer cereals	diquat + paraquat (autumn stubble, pre-planting/sowing, stubble burning, sward destruction/direct drilling), glyphosate (autumn stubble, pre-planting/sowing), glyphosate (sward destruction/direct drilling, minimum cultivation), paraquat (autumn stubble), paraquat (minimum cultivation), paraquat (pre-emergence)	
Volunteer potatoes	glyphosate (autumn stubble)	
Wild oats	paraquat (autumn stubble)	

Beans, field/broad

Diseases	Ascochyta	benomyl (seed treatment), metalaxyl + thiabendazole + thiram, thiabendazole + thiram
	Chocolate spot	benomyl, carbendazim + chlorothalonil, carbendazim (mbc), chlorothalonil, iprodione, iprodione (off-label), iprodione + thiophanate-methyl, thiabendazole + thiram, thiophanate-methyl, vinclozolin
	Damping off	drazoxolon, metalaxyl + thiabendazole + thiram, thiabendazole + thiram, thiram (seed treatment)
	Downy mildew	chlorothalonil + metalaxyl, fosetyl-aluminium, metalaxyl + thiabendazole + thiram
	Fusarium diseases	drazoxolon
	Rust	fenpropimorph
	Seed-borne diseases	thiabendazole + thiram
Pests	Aphids	demeton-S-methyl, dimethoate, disulfoton, fatty acids, heptenophos, malathion, nicotine, oxydemeton-methyl, phorate, pirimicarb, resmethrin
	Bean beetle	triazophos
	Capsids	phorate
	General insect control	demeton-S-methyl (off-label)
	Leaf miners	heptenophos (off-label), nicotine (off-label), oxamyl (off-label), triazophos (off-label)
	Mealy bugs	fatty acids
	Nematodes	aldicarb
	Pea and bean weevils	aldicarb, cyfluthrin, cypermethrin, deltamethrin, phorate, triazophos
	Scale insects	fatty acids
	Soil pests	gamma-HCH

	Thrips	demeton-S-methyl, fatty acids
	Weevils	fenvalerate
	Whitefly	fatty acids
	Woolly aphid	fatty acids
Weeds	Annual dicotyledons	bentazone, carbetamide, chlorpropham + cresylic acid + fenuron, chlorpropham + diuron, chlorpropham + fenuron, cyanazine, cyanazine (Scotland only), prometryn + terbutryn, propyzamide, simazine, simazine + trietazine, terbuthylazine + terbutryn, terbutryn + trietazine, trifluralin
	Annual grasses	alloxydim-sodium, carbetamide, chlorpropham + cresylic acid + fenuron, chlorpropham + diuron, chlorpropham + fenuron, cyanazine, cyanazine (Scotland only), diclofop-methyl, prometryn + terbutryn, propyzamide, sethoxydim, simazine, TCA (delayed sowing), terbuthylazine + terbutryn, trifluralin
	Annual meadow grass	terbutryn + trietazine, tri-allate
	Annual weeds	glyphosate (pre-harvest)
	Blackgrass	cycloxydim, diclofop-methyl, sethoxydim, tri-allate
	Canary grass	diclofop-methyl
	Chickweed	chlorpropham + cresylic acid + fenuron, chlorpropham + fenuron
	Couch	cycloxydim, glyphosate (pre-harvest), sethoxydim, TCA (delayed sowing)
	Creeping bent	cycloxydim
	Perennial grasses	diclofop-methyl, propyzamide, sethoxydim, TCA (delayed sowing)
	Perennial weeds	glyphosate (pre-harvest)
	Rough meadow grass	diclofop-methyl
	Ryegrass	diclofop-methyl
	Volunteer cereals	alloxydim-sodium, carbetamide, cycloxydim, diclofop-methyl, propyzamide, sethoxydim
	Wild oats	alloxydim-sodium, cycloxydim, diclofop-methyl, propyzamide, sethoxydim, tri-allate
Crop control	Pre-harvest desiccation	diquat
Plant growth regulation	Increasing yield	chlormequat, chlormequat + di-1-p-menthene
	Lodging control	chlormequat + di-1-p-menthene

Beans, soya

Pests	Leaf miners	oxamyl (off-label)
Weeds	Annual dicotyledons	trifluralin (off-label)
	Annual grasses	trifluralin (off-label)

71

Crop control	Pre-harvest desiccation	diquat (off-label)

Beet crops

Diseases

	Black leg	hymexazol
	Cercospora leaf spot	copper hydroxide, maneb, maneb + zinc
	Damping off	hymexazol
	Phoma leaf spot	fentin hydroxide
	Powdery mildew	copper sulphate + sulphur, potassium sorbate + sodium metabisulphite + sodium propionate, propiconazole, sulphur, triadimefon, triadimenol
	Ramularia leaf spot	fentin hydroxide, propiconazole
	Rust	propiconazole, triadimenol

Pests

	Aphids	aldicarb, aldicarb + gamma-HCH, carbosulfan, demeton-S-methyl, dimethoate, disulfoton, oxamyl, oxydemeton-methyl, phorate, pirimicarb
	Capsids	phorate
	Cutworms	alphacypermethrin, cypermethrin, gamma-HCH, triazophos
	Docking disorder vectors	aldicarb, aldicarb + gamma-HCH, benfuracarb, oxamyl
	Flea beetles	benfuracarb, carbofuran, carbosulfan, deltamethrin, gamma-HCH, trichlorfon
	Leaf miners	aldicarb, aldicarb + gamma-HCH, benfuracarb, heptenophos (off-label), nicotine (off-label), pirimiphos-methyl, triazophos (off-label)
	Leatherjackets	chlorpyrifos, gamma-HCH
	Mangold fly	carbofuran, carbosulfan, demeton-S-methyl, dimethoate, oxamyl, trichlorfon
	Millipedes	aldicarb, aldicarb + gamma-HCH, bendiocarb, benfuracarb, carbofuran, carbosulfan, gamma-HCH, oxamyl
	Nematodes	aldicarb, carbofuran, carbosulfan
	Pygmy beetle	aldicarb, aldicarb + gamma-HCH, bendiocarb, benfuracarb, carbofuran, carbosulfan, chlorpyrifos, gamma-HCH, oxamyl
	Springtails	aldicarb + gamma-HCH, bendiocarb, benfuracarb, carbofuran, carbosulfan, gamma-HCH
	Symphylids	aldicarb + gamma-HCH, bendiocarb, benfuracarb, carbosulfan, gamma-HCH
	Tortrix moths	carbofuran
	Virus yellows vectors	deltamethrin + heptenophos
	Wireworms	aldicarb + gamma-HCH, bendiocarb, benfuracarb, carbofuran, carbosulfan, gamma-HCH

Weeds	Annual dicotyledons	carbetamide, chloridazon, chloridazon + chlorpropham + fenuron + propham, chloridazon + ethofumesate, chloridazon + fenuron + propham, chloridazon + lenacil, chlorpropham + fenuron + propham, clopyralid, ethofumesate, ethofumesate + phenmedipham, lenacil, lenacil + phenmedipham, metamitron, paraquat, phenmedipham, propham, propyzamide, trifluralin
	Annual grasses	alloxydim-sodium, carbetamide, chloridazon + chlorpropham + fenuron + propham, chloridazon + fenuron + propham, chlorpropham + fenuron + propham, dalapon, dalapon + di-1-p-menthene, diclofop-methyl, fluazifop-P-butyl, metamitron, paraquat, propham, propyzamide, quizalofop-ethyl, sethoxydim, TCA, TCA (delayed sowing), trifluralin
	Annual meadow grass	chloridazon, chloridazon + ethofumesate, chloridazon + lenacil, ethofumesate, lenacil, metamitron, tri-allate
	Blackgrass	chloridazon + ethofumesate, cycloxydim, diclofop-methyl, ethofumesate, sethoxydim, tri-allate
	Canary grass	diclofop-methyl
	Chickweed	propham
	Corn marigold	clopyralid
	Couch	cycloxydim, dalapon, quizalofop-ethyl, sethoxydim, TCA (delayed sowing)
	Creeping bent	cycloxydim
	Creeping thistle	clopyralid
	Fat-hen	metamitron
	Mayweeds	clopyralid
	Perennial dicotyledons	clopyralid
	Perennial grasses	dalapon, dalapon + di-1-p-menthene, diclofop-methyl, fluazifop-P-butyl, propyzamide, quizalofop-ethyl, sethoxydim, TCA (delayed sowing)
	Polygonums	propham
	Rough meadow grass	diclofop-methyl
	Ryegrass	diclofop-methyl
	Volunteer cereals	alloxydim-sodium, carbetamide, cycloxydim, dalapon, dalapon + di-1-p-menthene, diclofop-methyl, fluazifop-P-butyl, paraquat, quizalofop-ethyl, sethoxydim, TCA
	Weed beet	glyphosate (wiper application)
	Wild oats	alloxydim-sodium, cycloxydim, diclofop-methyl, fluazifop-P-butyl, propham, sethoxydim, TCA, tri-allate

Cereals

Diseases	Blue mould	fuberidazole + triadimenol
	Botrytis	carbendazim + chlorothalonil + maneb, iprodione + thiophanate-methyl, potassium sorbate + sodium metabisulphite + sodium propionate, thiophanate-methyl

Brown foot rot and ear blight

carbendazim + chlorothalonil + maneb, carbendazim + maneb + sulphur, carboxin + imazalil + thiabendazole, carboxin + thiabendazole, ethirimol + flutriafol + thiabendazole, fuberidazole + imazalil + triadimenol, fuberidazole + triadimenol, maneb + zinc, potassium sorbate + sodium metabisulphite + sodium propionate, thiophanate-methyl

Covered smut

2-methoxyethylmercury acetate, carboxin + imazalil + thiabendazole, ethirimol + flutriafol + thiabendazole, fuberidazole + imazalil + triadimenol, fuberidazole + triadimenol, gamma-HCH + phenylmercury acetate, phenylmercury acetate

Crown rust

triadimenol + tridemorph

Damping off

thiabendazole + thiram

Disease control/foliar feed

sulphur

Eyespot

benomyl, carbendazim + chlorothalonil, carbendazim + chlorothalonil + maneb, carbendazim + flusilazole, carbendazim + flutriafol, carbendazim + mancozeb, carbendazim + maneb, carbendazim + maneb + sulphur, carbendazim + maneb + tridemorph, carbendazim (mbc), carbendazim + prochloraz, carbendazim + propiconazole, flusilazole, maneb, prochloraz, propiconazole, thiophanate-methyl

Foot rot

guazatine + imazalil

Leaf stripe

2-methoxyethylmercury acetate, carboxin + imazalil + thiabendazole, ethirimol + flutriafol + thiabendazole, fuberidazole + imazalil + triadimenol, fuberidazole + triadimenol, gamma-HCH + phenylmercury acetate, guazatine + imazalil, phenylmercury acetate

Loose smut

2-methoxyethylmercury acetate, carboxin + imazalil + thiabendazole, carboxin + thiabendazole, ethirimol + flutriafol + thiabendazole, fuberidazole + imazalil + triadimenol, fuberidazole + triadimenol, gamma-HCH + phenylmercury acetate, phenylmercury acetate

Net blotch

2-methoxyethylmercury acetate, carbendazim + chlorothalonil + maneb, carbendazim + flusilazole, carbendazim + flutriafol, carbendazim + mancozeb, carbendazim + maneb, carbendazim + maneb + tridemorph, carbendazim + prochloraz, carbendazim + propiconazole, carboxin + imazalil + thiabendazole, ethirimol + flutriafol + thiabendazole, fenpropimorph + prochloraz, ferbam + maneb + zineb, flusilazole, fuberidazole + imazalil + triadimenol, gamma-HCH + phenylmercury acetate, guazatine + imazalil, iprodione, iprodione + thiophanate-methyl, mancozeb, maneb, nuarimol, phenylmercury acetate, potassium sorbate + sodium metabisulphite + sodium propionate, prochloraz, propiconazole, propiconazole + tridemorph, pyrazophos, triforine

Powdery mildew	carbendazim + chlorothalonil + maneb, carbendazim + flusilazole, carbendazim + flutriafol, carbendazim + mancozeb, carbendazim + maneb, carbendazim + maneb + sulphur, carbendazim + maneb + tridemorph, carbendazim + prochloraz, carbendazim + propiconazole, chlorothalonil + fenpropimorph, chlorothalonil + flutriafol, chlorothalonil + propiconazole, copper oxychloride + maneb + sulphur, ethirimol + flutriafol + thiabendazole, fenpropidin, fenpropimorph, fenpropimorph + prochloraz, ferbam + maneb + zineb, flusilazole, fuberidazole + imazalil + triadimenol, fuberidazole + triadimenol, mancozeb, maneb + zinc, manganese zinc ethylenebisdithiocarbamate complex, nuarimol, potassium sorbate + sodium metabisulphite + sodium propionate, prochloraz, propiconazole, propiconazole + tridemorph, pyrazophos, thiophanate-methyl, triadimefon, triadimenol, triadimenol + tridemorph, tridemorph, triforine
Pyrenophora leaf spot	carboxin + imazalil + thiabendazole, fuberidazole + triadimenol, gamma-HCH + phenylmercury acetate, guazatine + imazalil, phenylmercury acetate
Rhynchosporium	benomyl, carbendazim + chlorothalonil, carbendazim + chlorothalonil + maneb, carbendazim + flusilazole, carbendazim + flutriafol, carbendazim + mancozeb, carbendazim + maneb, carbendazim + maneb + sulphur, carbendazim + maneb + tridemorph, carbendazim (mbc), carbendazim + prochloraz, carbendazim + propiconazole, chlorothalonil, copper oxychloride + maneb + sulphur, ethirimol + flutriafol + thiabendazole, fenpropidin, fenpropimorph, fenpropimorph + prochloraz, ferbam + maneb + zineb, flusilazole, fuberidazole + imazalil + triadimenol, fuberidazole + triadimenol, iprodione + thiophanate-methyl, mancozeb, maneb, maneb + zinc, nuarimol, potassium sorbate + sodium metabisulphite + sodium propionate, prochloraz, propiconazole, propiconazole + tridemorph, pyrazophos, thiophanate-methyl, triadimefon, triadimenol, triadimenol + tridemorph
Rust	benodanil, carbendazim + chlorothalonil + maneb, carbendazim + flusilazole, carbendazim + flutriafol, carbendazim + mancozeb, carbendazim + maneb, carbendazim + maneb + sulphur, carbendazim + maneb + tridemorph, carbendazim + propiconazole, chlorothalonil + fenpropimorph, chlorothalonil + flutriafol, chlorothalonil + propiconazole, ethirimol + flutriafol + thiabendazole, fenpropidin, fenpropimorph, fenpropimorph + prochloraz, ferbam + maneb + zineb, flusilazole, fuberidazole + imazalil + triadimenol, fuberidazole + triadimenol, mancozeb, maneb, maneb + zinc, manganese zinc ethylenebisdithiocarbamate complex, potassium sorbate + sodium metabisulphite + sodium propionate, propiconazole, propiconazole + tridemorph, triadimefon, triadimenol, triadimenol + tridemorph
Seed-borne diseases	thiabendazole + thiram

	Septoria diseases	2-methoxyethylmercury acetate, carbendazim + chlorothalonil, carbendazim + chlorothalonil + maneb, carbendazim + flusilazole, carbendazim + flutriafol, carbendazim + mancozeb, carbendazim + maneb, carbendazim + maneb + sulphur, carbendazim + maneb + tridemorph, carbendazim + prochloraz, carbendazim + propiconazole, chlorothalonil, chlorothalonil + fenpropimorph, chlorothalonil + flutriafol, chlorothalonil + propiconazole, copper hydroxide, copper oxychloride + maneb + sulphur, fenpropidin, fenpropimorph + prochloraz, ferbam + maneb + zineb, flusilazole, fuberidazole + triadimenol, guazatine, iprodione, iprodione + thiophanate-methyl, mancozeb, maneb, maneb + zinc, manganese zinc ethylenebisdithiocarbamate complex, nuarimol, potassium sorbate + sodium metabisulphite + sodium propionate, prochloraz, propiconazole, propiconazole + tridemorph, triadimenol + tridemorph
	Sharp eyespot	prochloraz
	Snow rot	benodanil, triadimenol, triadimenol + tridemorph
	Sooty moulds	carbendazim + chlorothalonil + maneb, carbendazim + mancozeb, carbendazim + maneb, carbendazim + maneb + sulphur, carbendazim + maneb + tridemorph, chlorothalonil + flutriafol, iprodione + thiophanate-methyl, mancozeb, maneb, maneb + zinc, manganese zinc ethylenebisdithiocarbamate complex, potassium sorbate + sodium metabisulphite + sodium propionate, propiconazole, propiconazole + tridemorph, thiophanate-methyl
	Stinking smut	2-methoxyethylmercury acetate, carboxin + thiabendazole, fuberidazole + triadimenol, gamma-HCH + phenylmercury acetate, guazatine, phenylmercury acetate
	Stripe smut	2-methoxyethylmercury acetate, carboxin + thiabendazole, fuberidazole + triadimenol, gamma-HCH + phenylmercury acetate, phenylmercury acetate
	Yellow rust	benodanil, carbendazim + chlorothalonil + maneb, carbendazim + flusilazole, carbendazim + flutriafol, carbendazim + mancozeb, carbendazim + maneb, carbendazim + maneb + sulphur, carbendazim + maneb + tridemorph, carbendazim + propiconazole, chlorothalonil + fenpropimorph, chlorothalonil + flutriafol, chlorothalonil + propiconazole, ethirimol + flutriafol + thiabendazole, fenpropidin, fenpropimorph, fenpropimorph + prochloraz, flusilazole, fuberidazole + imazalil + triadimenol, fuberidazole + triadimenol, maneb, oxycarboxin, potassium sorbate + sodium metabisulphite + sodium propionate, propiconazole, propiconazole + tridemorph
Pests	Aphids	alphacypermethrin, chlorpyrifos, cypermethrin, deltamethrin, demeton-S-methyl, dimethoate, fenvalerate, heptenophos, omethoate, oxydemeton-methyl, phosalone, pirimicarb
	Barley yellow dwarf virus vectors	bifenthrin, cyfluthrin, cypermethrin, deltamethrin
	Birds	aluminium ammonium sulphate
	Cutworms	gamma-HCH

	Damaging mammals	aluminium ammonium sulphate, sulphonated cod liver oil
	Frit fly	chlorpyrifos, fonofos, omethoate, pirimiphos-methyl, triazophos
	Leatherjackets	chlorpyrifos, fenitrothion, gamma-HCH, methiocarb (reduction), quinalphos, triazophos
	Saddle gall midge	fenitrothion
	Thrips	chlorpyrifos, fenitrothion
	Wheat bulb fly	chlorfenvinphos, chlorpyrifos, dimethoate, fonofos, omethoate, pirimiphos-methyl
	Wheat-blossom midges	chlorpyrifos, fenitrothion, triazophos
	Wireworms	fonofos, gamma-HCH, gamma-HCH + phenylmercury acetate
	Yellow cereal fly	alphacypermethrin, bifenthrin, chlorfenvinphos, cypermethrin, deltamethrin, fonofos, omethoate
Weeds	Annual dicotyledons	2,4-D, 2,4-DB + MCPA, benazolin + bromoxynil + ioxynil, bifenox + chlorotoluron, bifenox + isoproturon, bifenox + isoproturon + mecoprop, bromoxynil + chlorsulfuron + ioxynil, bromoxynil + clopyralid, bromoxynil + dichlorprop + ioxynil, bromoxynil + fluroxypyr, bromoxynil + fluroxypyr + ioxynil, bromoxynil + ioxynil, bromoxynil + ioxynil + isoproturon + mecoprop, bromoxynil + ioxynil + mecoprop, chlorotoluron, clopyralid, clopyralid + cyanazine, clopyralid + dichlorprop + MCPA, clopyralid + fluroxypyr + ioxynil, clopyralid + mecoprop, cyanazine, cyanazine + fluroxypyr, cyanazine + isoproturon, cyanazine + mecoprop, dicamba + dichlorprop + MCPA, dicamba + MCPA + mecoprop, dicamba + mecoprop, dichlorprop, dichlorprop + MCPA, dichlorprop + mecoprop, diflufenican + isoproturon, diflufenican + trifluralin, fluroxypyr, imazamethabenz-methyl, imazamethabenz-methyl + isoproturon, ioxynil + isoproturon + mecoprop, isoproturon, isoproturon + isoxaben, isoproturon + metsulfuron-methyl, isoproturon + pendimethalin, isoproturon + trifluralin, isoxaben, isoxaben + methabenzthiazuron, linuron, linuron + trietazine + trifluralin, linuron + trifluralin, MCPA, MCPA + MCPB, mecoprop, mecoprop-P, methabenzthiazuron, metoxuron, metoxuron + simazine, metsulfuron-methyl, metsulfuron-methyl + thifensulfuron-methyl, pendimethalin, pyridate, terbutryn, terbutryn + trifluralin, trifluralin
	Annual grasses	bifenox + chlorotoluron, bifenox + isoproturon, bifenox + isoproturon + mecoprop, bromoxynil + ioxynil + isoproturon, chlorotoluron, cyanazine, cyanazine + isoproturon, dalapon (stubble treatment), diclofop-methyl, diflufenican + isoproturon, imazamethabenz-methyl + isoproturon, ioxynil + isoproturon + mecoprop, isoproturon, isoproturon + isoxaben, isoproturon + metsulfuron-methyl, isoproturon + pendimethalin, isoproturon + trifluralin, linuron + trietazine + trifluralin, linuron + trifluralin, metoxuron, metoxuron + simazine, pendimethalin, TCA (delayed sowing), terbutryn + trifluralin, trifluralin

Annual meadow grass	bromoxynil + ioxynil + isoproturon + mecoprop, diflufenican + isoproturon, diflufenican + trifluralin, ioxynil + isoproturon + mecoprop, isoxaben + methabenzthiazuron, linuron, linuron + trifluralin, methabenzthiazuron, pendimethalin, terbutryn, terbutryn + trifluralin, tri-allate
Annual weeds	glyphosate (pre-harvest)
Barren brome	cyanazine + isoproturon, metoxuron
Black bindweed	dichlorprop, dichlorprop + MCPA, fluroxypyr, linuron
Blackgrass	bifenox + chlorotoluron, bifenox + isoproturon, bifenox + isoproturon + mecoprop, chlorotoluron, diclofop-methyl, diflufenican + isoproturon, fenoxaprop-ethyl, imazamethabenz-methyl, imazamethabenz-methyl + isoproturon, ioxynil + isoproturon + mecoprop, isoproturon, isoproturon + isoxaben, isoproturon + metsulfuron-methyl, isoproturon + pendimethalin, isoproturon + trifluralin, methabenzthiazuron, metoxuron, metoxuron + simazine, pendimethalin, terbutryn + trifluralin, tri-allate
Canary grass	diclofop-methyl
Chickweed	bromoxynil + chlorsulfuron + ioxynil, bromoxynil + dichlorprop + ioxynil + MCPA, bromoxynil + fluroxypyr + ioxynil, clopyralid + dichlorprop + MCPA, clopyralid + fluroxypyr + ioxynil, cyanazine + fluroxypyr, dicamba + MCPA + mecoprop, dicamba + mecoprop, fluroxypyr, ioxynil + isoproturon + mecoprop, linuron, mecoprop, mecoprop-P, metsulfuron-methyl, terbutryn + trifluralin
Cleavers	bromoxynil + fluroxypyr + ioxynil, clopyralid + fluroxypyr + ioxynil, cyanazine + fluroxypyr, dicamba + MCPA + mecoprop, dicamba + mecoprop, fluroxypyr, ioxynil + isoproturon + mecoprop, mecoprop, mecoprop-P, metsulfuron-methyl + thifensulfuron-methyl, pendimethalin, pyridate
Corn marigold	clopyralid, clopyralid + mecoprop, linuron
Couch	amitrole (direct-drilled), dalapon (stubble treatment), glyphosate (pre-harvest), TCA (delayed sowing)
Creeping thistle	clopyralid
Docks	clopyralid + mecoprop, dicamba + MCPA + mecoprop, fluroxypyr
Fat-hen	linuron, MCPA
Field pansy	bifenox + isoproturon, metsulfuron-methyl + thifensulfuron-methyl
General weed control	paraquat (stubble burning)
Hemp-nettle	bromoxynil + chlorsulfuron + ioxynil, bromoxynil + fluroxypyr + ioxynil, clopyralid + dichlorprop + MCPA, clopyralid + fluroxypyr + ioxynil, dichlorprop + MCPA, fluroxypyr, MCPA
Loose silky bent	imazamethabenz-methyl

	Mayweeds	bromoxynil + chlorsulfuron + ioxynil, bromoxynil + dichlorprop + ioxynil, clopyralid, clopyralid + dichlorprop + MCPA, clopyralid + fluroxypyr + ioxynil, clopyralid + mecoprop, dicamba + dichlorprop + MCPA, dicamba + MCPA + mecoprop, dicamba + mecoprop, metsulfuron-methyl, terbutryn + trifluralin
	Perennial dicotyledons	2,4-D, 2,4-DB + MCPA, clopyralid, cyanazine + mecoprop, dicamba + dichlorprop + MCPA, dicamba + MCPA + mecoprop, dicamba + mecoprop, dichlorprop, dichlorprop + MCPA, dichlorprop + mecoprop, MCPA, MCPA + MCPB, mecoprop, mecoprop-P
	Perennial grasses	dalapon (stubble treatment), diclofop-methyl, imazamethabenz-methyl, TCA (delayed sowing)
	Perennial ryegrass	linuron + trifluralin
	Perennial weeds	glyphosate (pre-harvest)
	Polygonums	2,4-DB + MCPA, bromoxynil + chlorsulfuron + ioxynil, bromoxynil + dichlorprop + ioxynil, clopyralid + dichlorprop + MCPA, dicamba + dichlorprop + MCPA, dicamba + MCPA + mecoprop, dicamba + mecoprop, dichlorprop, dichlorprop + MCPA, linuron, metsulfuron-methyl + thifensulfuron-methyl
	Rough meadow grass	chlorotoluron, diclofop-methyl, fenoxaprop-ethyl, isoproturon, isoxaben + methabenzthiazuron, linuron + trifluralin, methabenzthiazuron, terbutryn
	Ryegrass	diclofop-methyl
	Speedwells	bromoxynil + fluroxypyr + ioxynil, clopyralid + fluroxypyr + ioxynil, cyanazine + fluroxypyr, ioxynil + isoproturon + mecoprop, metsulfuron-methyl + thifensulfuron-methyl, pendimethalin, pyridate, terbutryn + trifluralin
	Volunteer cereals	diclofop-methyl
	Volunteer oilseed rape	imazamethabenz-methyl
	Volunteer potatoes	amitrole (stubble), fluroxypyr
	Wild oats	bifenox + chlorotoluron, bifenox + isoproturon + mecoprop, chlorotoluron, diclofop-methyl, difenzoquat, diflufenican + isoproturon, fenoxaprop-ethyl, flamprop-M-isopropyl, glyphosate (wiper glove), imazamethabenz-methyl, imazamethabenz-methyl + isoproturon, isoproturon, isoproturon + pendimethalin, pendimethalin, tri-allate
Crop control	Pre-harvest desiccation	diquat, glyphosate
Plant growth regulation	Increasing yield	chlormequat, chlormequat + choline chloride
	Lodging control	2-chloroethylphosphonic acid, 2-chloroethylphosphonic acid + mepiquat chloride, chlormequat, chlormequat + 2-chloroethylphosphonic acid, chlormequat + choline chloride, chlormequat + di-1-p-menthene

Cereals undersown

Weeds	Annual dicotyledons	2,4-DB, 2,4-DB + linuron + MCPA, 2,4-DB + MCPA, 2,4-DB + mecoprop, benazolin + 2,4-DB + MCPA, bentazone + cyanazine + 2,4-DB, bentazone + MCPA + MCPB, bromoxynil + ioxynil, dicamba + dichlorprop + MCPA, MCPA + MCPB, MCPB
	Chickweed	2,4-DB + mecoprop, benazolin + 2,4-DB + MCPA, bentazone + cyanazine + 2,4-DB
	Cleavers	benazolin + 2,4-DB + MCPA
	Mayweeds	dicamba + dichlorprop + MCPA
	Perennial dicotyledons	2,4-DB, 2,4-DB + MCPA, 2,4-DB + mecoprop, benazolin + 2,4-DB + MCPA, dicamba + dichlorprop + MCPA, MCPA + MCPB, MCPB
	Polygonums	2,4-DB + MCPA, benazolin + 2,4-DB + MCPA, dicamba + dichlorprop + MCPA

Evening primrose

Diseases	Fungus diseases	iprodione (off-label)
Pests	Flea beetles	deltamethrin (off-label)
	Pollen beetles	deltamethrin (off-label)
Weeds	Annual dicotyledons	benazolin + clopyralid (off-label), clopyralid + propyzamide (off-label), pendimethalin (off-label), propyzamide (off-label), trifluralin (off-label)
	Annual grasses	clopyralid + propyzamide (off-label), pendimethalin (off-label), propyzamide (off-label), trifluralin (off-label)
	Annual weeds	linuron (off-label)
	Chickweed	benazolin + clopyralid (off-label)
	Cleavers	benazolin + clopyralid (off-label)
	Mayweeds	benazolin + clopyralid (off-label), clopyralid + propyzamide (off-label)
	Perennial grasses	propyzamide (off-label)
Crop control	Pre-harvest desiccation	diquat

Fodder brassicas

Diseases	Alternaria	iprodione
	Botrytis	iprodione
	Canker	carboxin + gamma-HCH + thiram
	Damping off	carboxin + gamma-HCH + thiram
	Fungus diseases	zineb (off-label)
	White blister	mancozeb + metalaxyl (off-label)
Pests	Aphids	chlorpyrifos + dimethoate, cypermethrin, malathion

	Cabbage root fly	carbofuran (off-label), chlorpyrifos + dimethoate
	Cabbage stem flea beetle	cypermethrin
	Caterpillars	cypermethrin
	Cutworms	chlorpyrifos + dimethoate
	Flea beetles	carboxin + gamma-HCH + thiram
	Leatherjackets	chlorpyrifos + dimethoate
	Pod midge	phosalone
	Seed weevil	phosalone
	Soil pests	carbofuran (off-label)
	Wireworms	chlorpyrifos + dimethoate
Weeds	Annual dicotyledons	benazolin + clopyralid (off-label), clopyralid, desmetryn, propachlor, sodium monochloroacetate, tebutam, trifluralin
	Annual grasses	dalapon, dalapon + di-1-p-menthene, fluazifop-P-butyl (stockfeed only), propachlor, TCA, TCA (delayed sowing), tebutam, trifluralin
	Annual meadow grass	tri-allate
	Blackgrass	tri-allate
	Chickweed	benazolin + clopyralid (off-label)
	Cleavers	benazolin + clopyralid (off-label)
	Corn marigold	clopyralid
	Couch	dalapon, TCA (delayed sowing)
	Creeping thistle	clopyralid
	Fat-hen	desmetryn
	Mayweeds	benazolin + clopyralid (off-label), clopyralid
	Perennial dicotyledons	clopyralid
	Perennial grasses	dalapon, dalapon + di-1-p-menthene, fluazifop-P-butyl (stockfeed only), TCA (delayed sowing)
	Volunteer cereals	dalapon + di-1-p-menthene, fluazifop-P-butyl (stockfeed only), TCA, tebutam
	Wild oats	fluazifop-P-butyl (stockfeed only), TCA, tri-allate

Gold-of-pleasure

Weeds	Annual dicotyledons	trifluralin (off-label)
	Annual grasses	trifluralin (off-label)

Grass seed crops

Diseases	Crown rust	propiconazole
	Damping off	drazoxolon, thiabendazole + thiram
	Drechslera leaf spot	propiconazole

81

	Fungus diseases	prochloraz
	Fusarium diseases	drazoxolon
	Powdery mildew	propiconazole
	Rhynchosporium	propiconazole
	Seed-borne diseases	benomyl (off-label), thiabendazole + thiram
Pests	Aphids	deltamethrin (off-label), demeton-S-methyl, dimethoate, oxydemeton-methyl
	Frit fly	omethoate (off-label)
	General insect control	oxydemeton-methyl (off-label)
Weeds	Annual dicotyledons	2,4-D, 2,4-D + mecoprop, 2,4-DB + linuron + MCPA (off-label), clopyralid + dichlorprop + MCPA (off-label), clopyralid + mecoprop (off-label), clopyralid (off-label), dicamba + MCPA + mecoprop, isoxaben, MCPA, mecoprop, mecoprop-P, methabenzthiazuron, methabenzthiazuron (off-label)
	Annual grasses	ethofumesate, fluazifop-P-butyl (off-label)
	Annual meadow grass	methabenzthiazuron, methabenzthiazuron (off-label)
	Blackgrass	ethofumesate, fluazifop-P-butyl (off-label), methabenzthiazuron (off-label), sethoxydim (off-label)
	Chickweed	dicamba + MCPA + mecoprop, ethofumesate, mecoprop, mecoprop-P
	Cleavers	dicamba + MCPA + mecoprop, ethofumesate, mecoprop, mecoprop-P
	Docks	dicamba + MCPA + mecoprop
	Mayweeds	dicamba + MCPA + mecoprop
	Perennial dicotyledons	2,4-D, 2,4-D + mecoprop, clopyralid + mecoprop (off-label), clopyralid (off-label), dicamba + MCPA + mecoprop, MCPA, mecoprop, mecoprop-P
	Polygonums	dicamba + MCPA + mecoprop
	Rough meadow grass	methabenzthiazuron
	Ryegrass	chlorpropham (off-label)
	Volunteer cereals	ethofumesate, fluazifop-P-butyl (off-label), sethoxydim (off-label)
	Wild oats	difenzoquat, fluazifop-P-butyl (off-label), sethoxydim (off-label)
Plant growth regulation	Growth retardation	chlorpropham (off-label)

Legumes

Diseases	Damping off	thiabendazole + thiram
	Seed-borne diseases	thiabendazole + thiram
Pests	General insect control	deltamethrin (off-label), demeton-S-methyl (off-label), oxydemeton-methyl (off-label)

82

Weeds	Annual dicotyledons	2,4-DB, 2,4-DB + linuron + MCPA, 2,4-DB + MCPA, benazolin + 2,4-DB + MCPA, carbetamide, chlorpropham, clopyralid (off-label), MCPA + MCPB, MCPB, paraquat, propyzamide, propyzamide (off-label)
	Annual grasses	carbetamide, chlorpropham, dalapon, diclofop-methyl, fluazifop-P-butyl (off-label), paraquat, propyzamide, propyzamide (off-label), TCA
	Annual meadow grass	tri-allate
	Barren brome	paraquat
	Blackgrass	diclofop-methyl, fluazifop-P-butyl (off-label), sethoxydim (off-label), tri-allate
	Canary grass	diclofop-methyl
	Chickweed	benazolin + 2,4-DB + MCPA, chlorpropham
	Cleavers	benazolin + 2,4-DB + MCPA
	Couch	dalapon, TCA
	Creeping bent	paraquat
	Docks	asulam (off-label)
	Perennial dicotyledons	2,4-DB, 2,4-DB + MCPA, benazolin + 2,4-DB + MCPA, clopyralid (off-label), MCPA + MCPB, MCPB
	Perennial grasses	dalapon, diclofop-methyl, propyzamide, propyzamide (off-label), TCA
	Polygonums	2,4-DB + MCPA, benazolin + 2,4-DB + MCPA, chlorpropham
	Rough meadow grass	diclofop-methyl
	Ryegrass	diclofop-methyl
	Volunteer cereals	carbetamide, diclofop-methyl, fluazifop-P-butyl (off-label), paraquat, sethoxydim (off-label)
	Wild oats	diclofop-methyl, fluazifop-P-butyl (off-label), paraquat, sethoxydim (off-label), tri-allate
Crop control	Pre-harvest desiccation	diquat

Linseed/flax

Diseases	Damping off	thiabendazole + thiram
	Fungus diseases	iprodione (off-label)
	Seed-borne diseases	captan + gamma-HCH, prochloraz, thiabendazole + thiram
Pests	Flea beetles	captan + gamma-HCH
Weeds	Annual dicotyledons	bentazone, bromoxynil + clopyralid, clopyralid, MCPA, trifluralin (off-label)
	Annual grasses	diclofop-methyl, sethoxydim, trifluralin (off-label)
	Annual meadow grass	tri-allate
	Annual weeds	glyphosate (pre-harvest), linuron (off-label), linuron + trifluralin (off-label)

	Blackgrass	diclofop-methyl, sethoxydim, tri-allate
	Canary grass	diclofop-methyl
	Corn marigold	clopyralid
	Couch	glyphosate (pre-harvest)
	Creeping thistle	clopyralid
	Mayweeds	clopyralid
	Perennial dicotyledons	clopyralid, MCPA
	Perennial grasses	diclofop-methyl
	Perennial weeds	glyphosate (pre-harvest)
	Rough meadow grass	diclofop-methyl
	Ryegrass	diclofop-methyl
	Volunteer cereals	diclofop-methyl, sethoxydim
	Wild oats	diclofop-methyl, sethoxydim, tri-allate
Crop control	Pre-harvest desiccation	diquat

Lupins

Diseases	Damping off	thiabendazole + thiram
	Seed-borne diseases	thiabendazole + thiram
Weeds	Annual dicotyledons	simazine + trietazine, terbuthylazine + terbutryn
	Annual grasses	diclofop-methyl, terbuthylazine + terbutryn
	Blackgrass	diclofop-methyl
	Canary grass	diclofop-methyl
	Perennial grasses	diclofop-methyl
	Rough meadow grass	diclofop-methyl
	Ryegrass	diclofop-methyl
	Volunteer cereals	diclofop-methyl
	Wild oats	diclofop-methyl
Crop control	Pre-harvest desiccation	diquat

Maize/sweetcorn

Diseases	Damping off	thiram (seed treatment)
Pests	Aphids	pirimicarb, pirimicarb (off-label)
	Frit fly	aldicarb, aldicarb (off-label), bendiocarb, carbofuran, chlorfenvinphos, chlorpyrifos, fenitrothion, phorate, triazophos
	General insect control	demeton-S-methyl (off-label), oxydemeton-methyl (off-label)
	Nematodes	aldicarb
	Thrips	dimethoate (off-label), fenitrothion (off-label)
	Wireworms	bendiocarb

Weeds	Annual dicotyledons	alachlor, atrazine, atrazine + cyanazine, clopyralid, cyanazine, pyridate, simazine
	Annual grasses	alachlor, atrazine, atrazine + cyanazine, cyanazine, simazine
	Annual meadow grass	tri-allate
	Blackgrass	tri-allate
	Cleavers	pyridate
	Corn marigold	clopyralid
	Couch	atrazine
	Creeping thistle	clopyralid
	Mayweeds	clopyralid
	Perennial dicotyledons	clopyralid
	Speedwells	pyridate
	Wild oats	difenzoquat, tri-allate

Oilseed rape

Diseases	Alternaria	carbendazim + chlorothalonil, carbendazim + maneb + sulphur, carbendazim + prochloraz, fenpropimorph + gamma-HCH + thiram, ferbam + maneb + zineb, iprodione, iprodione (seed treatment), iprodione + thiophanate-methyl, maneb, maneb + zinc, prochloraz, propiconazole, vinclozolin
	Botrytis	benomyl, carbendazim + chlorothalonil, carbendazim + maneb, carbendazim + maneb + sulphur, carbendazim + prochloraz, chlorothalonil, iprodione, iprodione + thiophanate-methyl, potassium sorbate + sodium metabisulphite + sodium propionate, prochloraz, thiophanate-methyl, vinclozolin
	Canker	carbendazim + prochloraz, carboxin + gamma-HCH + thiram, fenpropimorph + gamma-HCH + thiram, gamma-HCH + thiabendazole + thiram, prochloraz, thiabendazole (seed treatment)
	Damping off	carboxin + gamma-HCH + thiram, fenpropimorph + gamma-HCH + thiram, gamma-HCH + thiabendazole + thiram, gamma-HCH + thiram
	Disease control/foliar feed	sulphur
	Downy mildew	carbendazim + mancozeb, carbendazim + maneb, chlorothalonil, chlorothalonil + metalaxyl, mancozeb, maneb, maneb + zinc, manganese zinc ethylenebisdithiocarbamate + ofurace
	Light leaf spot	benomyl, carbendazim + chlorothalonil, carbendazim + mancozeb, carbendazim + maneb, carbendazim + maneb + sulphur, carbendazim (mbc), carbendazim + prochloraz, iprodione + thiophanate-methyl, potassium sorbate + sodium metabisulphite + sodium propionate, prochloraz, propiconazole, thiophanate-methyl, vinclozolin

85

	Sclerotinia stem rot	carbendazim + prochloraz, iprodione, iprodione + thiophanate-methyl, prochloraz, thiophanate-methyl, vinclozolin
	Seed-borne diseases	captan + gamma-HCH
	White leaf spot	carbendazim + prochloraz, prochloraz
Pests	Aphids	chlorpyrifos + dimethoate, deltamethrin, fenvalerate, pirimicarb
	Birds	aluminium ammonium sulphate
	Cabbage root fly	carbofuran, chlorpyrifos + dimethoate, fonofos, gamma-HCH + thiabendazole + thiram, phorate
	Cabbage seed weevil	azinphos-methyl + demeton-S-methyl sulphone, endosulfan, malathion
	Cabbage stem flea beetle	alphacypermethrin, bifenthrin, captan + gamma-HCH, carbofuran, cyfluthrin, cypermethrin, deltamethrin, fenvalerate, fonofos, gamma-HCH, phorate, pirimiphos-methyl
	Cabbage stem weevil	deltamethrin, gamma-HCH, malathion
	Cutworms	chlorpyrifos + dimethoate
	Damaging mammals	aluminium ammonium sulphate, sulphonated cod liver oil
	Flea beetles	captan + gamma-HCH, carbaryl, carboxin + gamma-HCH + thiram, fenpropimorph + gamma-HCH + thiram, gamma-HCH + thiabendazole + thiram, gamma-HCH + thiram
	Leatherjackets	chlorpyrifos + dimethoate
	Millipedes	gamma-HCH + thiabendazole + thiram
	Nematodes	aldicarb
	Pod midge	alphacypermethrin, cypermethrin (restricted area), endosulfan, fenvalerate, phosalone, triazophos
	Pollen beetles	alphacypermethrin, azinphos-methyl + demeton-S-methyl sulphone, cypermethrin (restricted area), deltamethrin, endosulfan, fenvalerate, gamma-HCH, malathion
	Rape winter stem weevil	bifenthrin, carbofuran, cyfluthrin, cypermethrin, deltamethrin, phorate
	Seed weevil	alphacypermethrin, cypermethrin (restricted area), fenvalerate, gamma-HCH, phosalone, triazophos
	Weevils	gamma-HCH
	Wireworms	chlorpyrifos + dimethoate
Weeds	Annual dicotyledons	alachlor, benazolin + clopyralid, carbetamide, clopyralid, clopyralid + propyzamide, cyanazine, metazachlor, napropamide, napropamide + trifluralin, propachlor, propyzamide, pyridate, tebutam, trifluralin
	Annual grasses	alachlor, carbetamide, clopyralid + propyzamide, cyanazine, dalapon, dalapon + di-1-p-menthene, diclofop-methyl, fluazifop-P-butyl, napropamide, napropamide + trifluralin, propachlor, propyzamide, quizalofop-ethyl, sethoxydim, TCA, tebutam, trifluralin
	Annual meadow grass	metazachlor, tri-allate

	Annual weeds	glyphosate (pre-harvest)
	Barren brome	clopyralid + propyzamide
	Blackgrass	cycloxydim, diclofop-methyl, metazachlor, sethoxydim, tri-allate
	Canary grass	diclofop-methyl
	Chickweed	benazolin + clopyralid
	Cleavers	benazolin + clopyralid, napropamide, pyridate
	Corn marigold	clopyralid
	Couch	cycloxydim, dalapon, glyphosate (pre-harvest), quizalofop-ethyl
	Creeping bent	cycloxydim
	Creeping thistle	clopyralid
	Groundsel	napropamide
	Groundsel, triazine resistant	metazachlor
	Mayweeds	benazolin + clopyralid, clopyralid, clopyralid + propyzamide
	Perennial dicotyledons	clopyralid
	Perennial grasses	dalapon, dalapon + di-1-p-menthene, diclofop-methyl, fluazifop-P-butyl, propyzamide, quizalofop-ethyl
	Perennial weeds	glyphosate (pre-harvest)
	Rough meadow grass	diclofop-methyl
	Ryegrass	diclofop-methyl
	Speedwells	pyridate
	Volunteer cereals	carbetamide, cycloxydim, dalapon, dalapon + di-1-p-menthene, diclofop-methyl, fluazifop-P-butyl, propyzamide, quizalofop-ethyl, sethoxydim, TCA, tebutam
	Wild oats	cycloxydim, diclofop-methyl, fluazifop-P-butyl, propyzamide, sethoxydim, TCA, tri-allate
Crop control	Pre-harvest desiccation	diquat
Plant growth regulation	Increasing yield	chlormequat, chlormequat + choline chloride, chlormequat + di-1-p-menthene
	Lodging control	chlormequat + di-1-p-menthene

Peas

Diseases	Ascochyta	benomyl (seed treatment), captan + fosetyl-aluminium + thiabendazole, carbendazim + chlorothalonil, chlorothalonil, chlorothalonil (off-label), fosetyl-aluminium (off-label), metalaxyl + thiabendazole + thiram, thiabendazole + thiram
	Botrytis	benomyl, carbendazim + chlorothalonil, chlorothalonil, iprodione, iprodione (off-label), vinclozolin

	Damping off	captan + fosetyl-aluminium + thiabendazole, drazoxolon, metalaxyl + thiabendazole + thiram, thiabendazole + thiram, thiram (seed treatment)
	Downy mildew	captan + fosetyl-aluminium + thiabendazole, carbendazim + chlorothalonil, chlorothalonil, fosetyl-aluminium (off-label), metalaxyl + thiabendazole + thiram
	Fusarium diseases	drazoxolon
	Sclerotinia	iprodione (off-label)
Pests	Aphids	aldicarb, azinphos-methyl + demeton-S-methyl sulphone, bifenthrin, cyfluthrin, cypermethrin, deltamethrin, deltamethrin + heptenophos, demeton-S-methyl, dimethoate, disulfoton, fatty acids, fenitrothion, fenvalerate, heptenophos, malathion, oxydemeton-methyl, phorate, pirimicarb, triazophos
	Birds	aluminium ammonium sulphate
	Capsids	phorate
	Damaging mammals	aluminium ammonium sulphate
	Leaf miners	aldicarb, heptenophos (off-label)
	Leatherjackets	chlorpyrifos
	Mealy bugs	fatty acids
	Nematodes	aldicarb
	Pea and bean weevils	aldicarb, azinphos-methyl + demeton-S-methyl sulphone, carbaryl, cyfluthrin, cypermethrin, deltamethrin, fenitrothion, fenvalerate, lambda-cyhalothrin, oxamyl, phorate, triazophos
	Pea cyst nematode	oxamyl
	Pea midge	azinphos-methyl + demeton-S-methyl sulphone, carbaryl, demeton-S-methyl, dimethoate, disulfoton, fenitrothion, oxydemeton-methyl, triazophos
	Pea moth	azinphos-methyl + demeton-S-methyl sulphone, bifenthrin, carbaryl, cyfluthrin, cypermethrin, deltamethrin, deltamethrin + heptenophos, fenitrothion, fenvalerate, lambda-cyhalothrin, triazophos
	Scale insects	fatty acids
	Soil pests	gamma-HCH
	Thrips	aldicarb, azinphos-methyl + demeton-S-methyl sulphone, carbaryl, dimethoate, disulfoton, fatty acids, fenitrothion
	Tortrix moths	azinphos-methyl + demeton-S-methyl sulphone
	Virus vectors	deltamethrin + heptenophos
	Whitefly	fatty acids
	Woolly aphid	fatty acids
Weeds	Annual dicotyledons	aziprotryne, bentazone, bentazone + MCPB, chlorpropham + cresylic acid + fenuron, chlorpropham + diuron, chlorpropham + fenuron, cyanazine, MCPA + MCPB, MCPB, pendimethalin, prometryn, prometryn + terbutryn, propham, simazine + trietazine, terbuthylazine + terbutryn, terbutryn + trietazine, trifluralin (off-label)

	Annual grasses	alloxydim-sodium, aziprotryne, chlorpropham + cresylic acid + fenuron, chlorpropham + diuron, chlorpropham + fenuron, cyanazine, diclofop-methyl, fluazifop-P-butyl, pendimethalin, prometryn, prometryn + terbutryn, propham, sethoxydim, TCA, TCA (delayed sowing), terbuthylazine + terbutryn, trifluralin (off-label)
	Annual meadow grass	pendimethalin, terbutryn + trietazine, tri-allate
	Annual weeds	glyphosate (pre-harvest)
	Blackgrass	cycloxydim, diclofop-methyl, pendimethalin, sethoxydim, tri-allate
	Canary grass	diclofop-methyl
	Chickweed	chlorpropham + cresylic acid + fenuron, chlorpropham + fenuron, propham
	Cleavers	pendimethalin
	Couch	cycloxydim, glyphosate (pre-harvest), TCA (delayed sowing)
	Creeping bent	cycloxydim
	Perennial dicotyledons	MCPA + MCPB, MCPB
	Perennial grasses	diclofop-methyl, fluazifop-P-butyl, TCA (delayed sowing)
	Perennial weeds	glyphosate (pre-harvest)
	Polygonums	propham
	Rough meadow grass	diclofop-methyl
	Ryegrass	diclofop-methyl
	Speedwells	pendimethalin
	Volunteer cereals	alloxydim-sodium, cycloxydim, diclofop-methyl, fluazifop-P-butyl, sethoxydim, TCA
	Wild oats	alloxydim-sodium, cycloxydim, diclofop-methyl, fluazifop-P-butyl, pendimethalin, propham, sethoxydim, TCA, tri-allate
Crop control	Pre-harvest desiccation	diquat
Plant growth regulation	Increasing yield	chlormequat, chlormequat + di-1-p-menthene
	Lodging control	chlormequat + di-1-p-menthene

Potatoes

Diseases	Black scurf and stem canker	imazalil + thiabendazole, iprodione, pencycuron, tolclofos-methyl
	Blight	benalaxyl + mancozeb, Bordeaux mixture, chlorothalonil, copper hydroxide, copper oxychloride, copper sulphate + cufraneb, copper sulphate + sulphur, cymoxanil + mancozeb, cymoxanil + mancozeb + oxadixyl, fentin acetate + maneb, fentin hydroxide, fentin hydroxide + maneb + zinc, fentin hydroxide + metoxuron, ferbam + maneb + zineb, mancozeb, mancozeb + metalaxyl, mancozeb + oxadixyl, maneb, maneb + zinc, maneb + zinc oxide, manganese zinc ethylenebisdithiocarbamate + ofurace, potassium sorbate + sodium metabisulphite + sodium propionate, zineb

	Dry rot	thiabendazole
	Gangrene	2-aminobutane, thiabendazole
	Powdery scab	maneb + zinc oxide
	Silver scurf	2-aminobutane, imazalil + thiabendazole, thiabendazole
	Skin spot	2-aminobutane, imazalil + thiabendazole, thiabendazole
Pests	Aphids	aldicarb, demeton-S-methyl, dimethoate, disulfoton, malathion, oxamyl, oxydemeton-methyl, phorate, pirimicarb
	Capsids	phorate
	Colorado beetle	azinphos-methyl + demeton-S-methyl sulphone (off-label), carbaryl (off-label), chlorfenvinphos (off-label), chlorpyrifos (off-label), cypermethrin (off-label), deltamethrin (off-label), endosulfan (off-label), triazophos (off-label)
	Cutworms	alphacypermethrin, chlorpyrifos, cypermethrin, triazophos
	Leaf miners	heptenophos (off-label), nicotine (off-label), triazophos (off-label)
	Leaf roll virus vectors	deltamethrin + heptenophos
	Leafhoppers	demeton-S-methyl, oxydemeton-methyl, phorate
	Nematodes	1,3-dichloropropene, aldicarb, oxamyl
	Potato cyst nematode	1,3-dichloropropene, aldicarb, carbofuran, ethoprophos, oxamyl
	Potato virus vectors	deltamethrin + heptenophos, nicotine
	Spraing vectors	oxamyl
	Wireworms	ethoprophos, phorate
Weeds	Annual dicotyledons	bentazone, cyanazine, diquat + paraquat, linuron, linuron + terbutryn, linuron + trietazine, metribuzin, monolinuron, monolinuron + paraquat, paraquat, pendimethalin, prometryn, prometryn + terbutryn, terbuthylazine + terbutryn, terbutryn + trietazine
	Annual grasses	alloxydim-sodium, cyanazine, dalapon, diclofop-methyl, diquat + paraquat, EPTC, linuron + trietazine, metribuzin, monolinuron, monolinuron + paraquat, paraquat, pendimethalin, prometryn, prometryn + terbutryn, sethoxydim, TCA (delayed sowing), terbuthylazine + terbutryn
	Annual meadow grass	linuron, linuron + terbutryn, monolinuron, monolinuron + paraquat, pendimethalin, terbutryn + trietazine
	Black bindweed	linuron
	Blackgrass	cycloxydim, diclofop-methyl, monolinuron + paraquat, pendimethalin, sethoxydim
	Canary grass	diclofop-methyl
	Chickweed	EPTC, linuron
	Cleavers	pendimethalin
	Corn marigold	linuron

	Couch	cycloxydim, dalapon, EPTC, sethoxydim, TCA (delayed sowing)
	Creeping bent	cycloxydim
	Fat-hen	EPTC, linuron, monolinuron
	Fumitory	linuron + terbutryn
	Perennial grasses	dalapon, diclofop-methyl, diquat + paraquat, sethoxydim, TCA (delayed sowing)
	Polygonums	linuron, monolinuron
	Rough meadow grass	diclofop-methyl
	Ryegrass	diclofop-methyl
	Speedwells	EPTC, pendimethalin
	Volunteer cereals	alloxydim-sodium, cycloxydim, dalapon, diclofop-methyl, diquat + paraquat, paraquat, sethoxydim
	Wild oats	alloxydim-sodium, cycloxydim, diclofop-methyl, monolinuron + paraquat, pendimethalin, sethoxydim
Crop control	Haulm destruction	fentin hydroxide + metoxuron
	Pre-harvest desiccation	diquat
Plant growth regulation	Volunteer suppression	maleic hydrazide

Seed brassicas/mustard

Diseases	Alternaria	iprodione
	Botrytis	iprodione
Pests	Cabbage seed weevil	azinphos-methyl + demeton-S-methyl sulphone, endosulfan, malathion
	Cabbage stem weevil	malathion
	Pod midge	endosulfan, phosalone, triazophos (off-label)
	Pollen beetles	azinphos-methyl + demeton-S-methyl sulphone, endosulfan, malathion, phosalone
	Seed weevil	phosalone, triazophos (off-label)
Weeds	Annual dicotyledons	benazolin + clopyralid (off-label), carbetamide, chlorthal-dimethyl + propachlor, propachlor, propyzamide, trifluralin
	Annual grasses	carbetamide, chlorthal-dimethyl + propachlor, diclofop-methyl, propachlor, propyzamide, quizalofop-ethyl, sethoxydim, TCA (delayed sowing), trifluralin
	Annual meadow grass	tri-allate
	Annual weeds	glyphosate (pre-harvest)
	Blackgrass	diclofop-methyl, sethoxydim, tri-allate
	Canary grass	diclofop-methyl
	Chickweed	benazolin + clopyralid (off-label)
	Cleavers	benazolin + clopyralid (off-label)

	Couch	glyphosate (pre-harvest), quizalofop-ethyl, TCA (delayed sowing)
	Mayweeds	benazolin + clopyralid (off-label)
	Perennial grasses	diclofop-methyl, propyzamide, quizalofop-ethyl, TCA (delayed sowing)
	Perennial weeds	glyphosate (pre-harvest)
	Rough meadow grass	diclofop-methyl
	Ryegrass	diclofop-methyl
	Volunteer cereals	carbetamide, diclofop-methyl, quizalofop-ethyl, sethoxydim
	Wild oats	diclofop-methyl, sethoxydim, tri-allate
Crop control	Pre-harvest desiccation	diquat (off-label)

Sunflowers

Weeds	Annual dicotyledons	trifluralin (off-label)
	Annual grasses	trifluralin (off-label)
Crop control	Pre-harvest desiccation	diquat (off-label)

Grain/crop store uses

Stored grain/rapeseed/linseed

Pests	Grain storage pests	chlorpyrifos-methyl, etrimfos, fenitrothion, fenitrothion + permethrin + resmethrin, gamma-HCH, iodofenphos, methacrifos, methyl bromide with chloropicrin, permethrin, pirimiphos-methyl
	Weevils	pyrethrins

Stored cabbages

Diseases	Alternaria	iprodione
	Botrytis	carbendazim + metalaxyl, iprodione
	Phytophthora	carbendazim + metalaxyl

Stored potatoes

Diseases	Black scurf and stem canker	nonylphenoxypoly(ethyleneoxy)ethanol-iodine complex + thiabendazole
	Dry rot	carbendazim + tecnazene, nonylphenoxypoly(ethyleneoxy)ethanol-iodine complex + tecnazene, tecnazene, tecnazene + thiabendazole
	Gangrene	nonylphenoxypoly(ethyleneoxy)ethanol-iodine complex + thiabendazole, tecnazene + thiabendazole

	Silver scurf	nonylphenoxypoly(ethyleneoxy)ethanol-iodine complex + thiabendazole, tecnazene + thiabendazole
	Skin spot	nonylphenoxypoly(ethyleneoxy)ethanol-iodine complex + tecnazene, tecnazene + thiabendazole
	Soft rot	nonylphenoxypoly(ethyleneoxy)ethanol-iodine complex + tecnazene
	Soil-borne diseases	nonylphenoxypoly(ethyleneoxy)ethanol-iodine complex + thiabendazole
Plant growth regulation	Sprout suppression	carbendazim + tecnazene, chlorpropham, chlorpropham + propham, maleic hydrazide, nonylphenoxypoly(ethyleneoxy)ethanol-iodine complex + tecnazene, tecnazene, tecnazene + thiabendazole

Grassland

Grassland

Diseases	Crown rust	propiconazole
	Disease control/foliar feed	sulphur
	Drechslera leaf spot	propiconazole
	Powdery mildew	propiconazole, triadimefon
	Rhynchosporium	propiconazole, triadimefon
	Rust	triadimefon
Pests	Aphids	pirimicarb
	Cutworms	gamma-HCH
	Frit fly	chlorpyrifos, fonofos, triazophos
	Leatherjackets	chlorpyrifos, gamma-HCH, triazophos
	Wireworms	gamma-HCH
Weeds	Annual dicotyledons	2,4-D, 2,4-D + dicamba + mecoprop, 2,4-D + dicamba + triclopyr, 2,4-D + mecoprop, 2,4-DB + linuron + MCPA, 2,4-DB + mecoprop, benazolin + 2,4-DB + MCPA, benazolin + bromoxynil + ioxynil, bentazone + cyanazine + 2,4-DB, bentazone + MCPA + MCPB, bromoxynil + ioxynil + mecoprop, clopyralid, clopyralid + mecoprop, dicamba + dichlorprop + MCPA, dicamba + MCPA + mecoprop, dicamba + mecoprop, dicamba + mecoprop + triclopyr, fluroxypyr, MCPA, MCPA + MCPB, mecoprop, mecoprop-P
	Annual grasses	ethofumesate
	Annual weeds	glyphosate, glyphosate (pre-cut/graze)
	Black bindweed	fluroxypyr
	Blackgrass	ethofumesate
	Bracken	asulam

Brambles	2,4-D + dicamba + mecoprop, 2,4-D + dicamba + triclopyr, clopyralid + triclopyr, dicamba + mecoprop + triclopyr, triclopyr
Broom	clopyralid + triclopyr, dicamba + mecoprop + triclopyr, triclopyr
Buttercups	MCPA
Chickweed	2,4-DB + mecoprop, benazolin + 2,4-DB + MCPA, bentazone + cyanazine + 2,4-DB, dicamba + MCPA + mecoprop, ethofumesate, fluroxypyr, mecoprop, mecoprop-P
Cleavers	benazolin + 2,4-DB + MCPA, dicamba + MCPA + mecoprop, ethofumesate, fluroxypyr, mecoprop, mecoprop-P
Corn marigold	clopyralid, clopyralid + mecoprop
Creeping thistle	clopyralid, clopyralid (off-label)
Destruction of short term leys	glyphosate
Docks	2,4-D + dicamba + triclopyr, asulam, clopyralid + mecoprop, clopyralid + triclopyr, dicamba + MCPA + mecoprop, dicamba + mecoprop, dicamba + mecoprop + triclopyr, fluroxypyr, MCPA, triclopyr
Gorse	2,4-D + dicamba + triclopyr, clopyralid + triclopyr, dicamba + mecoprop + triclopyr, triclopyr
Hemp-nettle	fluroxypyr
Japanese knotweed	2,4-D + dicamba + triclopyr
Mayweeds	clopyralid, clopyralid + mecoprop, dicamba + dichlorprop + MCPA, dicamba + MCPA + mecoprop
Perennial dicotyledons	2,4-D, 2,4-D + dicamba + mecoprop, 2,4-D + dicamba + triclopyr, 2,4-D + mecoprop, 2,4-DB + mecoprop, benazolin + 2,4-DB + MCPA, clopyralid, clopyralid + triclopyr, dicamba + dichlorprop + MCPA, dicamba + MCPA + mecoprop, dicamba + mecoprop, dicamba + mecoprop + triclopyr, dicamba (wiper application), glyphosate (wiper application), MCPA, MCPA + MCPB, mecoprop, mecoprop-P, triclopyr
Perennial weeds	glyphosate (pre-cut/graze)
Polygonums	benazolin + 2,4-DB + MCPA, dicamba + dichlorprop + MCPA, dicamba + MCPA + mecoprop
Rhododendrons	2,4-D + dicamba + triclopyr
Stinging nettle	2,4-D + dicamba + mecoprop, 2,4-D + dicamba + triclopyr, dicamba + mecoprop + triclopyr, triclopyr
· Volunteer cereals	ethofumesate
Volunteer potatoes	fluroxypyr
Woody weeds	2,4-D + dicamba + mecoprop, 2,4-D + dicamba + triclopyr, dicamba + mecoprop + triclopyr, triclopyr

Leys

Diseases	Fungus diseases	prochloraz
Pests	Frit fly	cypermethrin, omethoate
	Leatherjackets	cypermethrin
Weeds	Annual dicotyledons	2,4-DB + MCPA, benazolin + 2,4-DB + MCPA, bromoxynil + ethofumesate + ioxynil, cyanazine + mecoprop, dicamba + MCPA + mecoprop, dicamba + mecoprop, dicamba + mecoprop + triclopyr, isoxaben, linuron, MCPA, MCPA + MCPB, MCPB, mecoprop, mecoprop-P, methabenzthiazuron
	Annual grasses	ethofumesate
	Annual meadow grass	bromoxynil + ethofumesate + ioxynil, linuron, methabenzthiazuron
	Black bindweed	linuron
	Blackgrass	ethofumesate
	Brambles	dicamba + mecoprop + triclopyr
	Broom	dicamba + mecoprop + triclopyr
	Chickweed	benazolin + 2,4-DB + MCPA, dicamba + MCPA + mecoprop, ethofumesate, linuron, mecoprop, mecoprop-P
	Cleavers	benazolin + 2,4-DB + MCPA, dicamba + MCPA + mecoprop, ethofumesate, mecoprop, mecoprop-P
	Corn marigold	linuron
	Docks	dicamba + MCPA + mecoprop, dicamba + mecoprop, dicamba + mecoprop + triclopyr
	Fat-hen	linuron
	Gorse	dicamba + mecoprop + triclopyr
	Mayweeds	dicamba + MCPA + mecoprop
	Perennial dicotyledons	2,4-DB + MCPA, benazolin + 2,4-DB + MCPA, cyanazine + mecoprop, dicamba + MCPA + mecoprop, dicamba + mecoprop, dicamba + mecoprop + triclopyr, MCPA, MCPA + MCPB, MCPB, mecoprop, mecoprop-P
	Polygonums	2,4-DB + MCPA, benazolin + 2,4-DB + MCPA, dicamba + MCPA + mecoprop, linuron
	Rough meadow grass	methabenzthiazuron
	Stinging nettle	dicamba + mecoprop + triclopyr
	Volunteer cereals	ethofumesate
	Woody weeds	dicamba + mecoprop + triclopyr

Turf/amenity grass

Turf/amenity grass

Diseases	Anthracnose	chlorothalonil

	Brown patch	chlorothalonil, iprodione
	Dollar spot	benomyl, carbendazim (mbc), dichlorophen, iprodione, quintozene, thiabendazole, thiophanate-methyl, vinclozolin
	Fairy rings	benodanil, dichlorophen, oxycarboxin, triforine
	Fusarium patch	benomyl, carbendazim (mbc), chlorothalonil, iprodione, quintozene, thiabendazole, thiophanate-methyl, vinclozolin
	Grey snow mould	chlorothalonil, iprodione
	Melting out	chlorothalonil, iprodione, vinclozolin
	Red thread	benomyl, carbendazim (mbc), dichlorophen, iprodione, quintozene, thiabendazole, thiophanate-methyl, vinclozolin
	Take-all patch	chlorothalonil
Pests	Chafer grubs	gamma-HCH
	Earthworms	carbaryl, carbendazim (mbc), chlordane, gamma-HCH + thiophanate-methyl
	Frit fly	chlorpyrifos
	Leatherjackets	chlorpyrifos, gamma-HCH, gamma-HCH + thiophanate-methyl
	Millipedes	gamma-HCH
	Wireworms	gamma-HCH
Weeds	Algae	dichlorophen, ferrous sulphate
	Annual dicotyledons	2,4-D, 2,4-D + dicamba, 2,4-D + dicamba + ioxynil, 2,4-D + dicamba + mecoprop, 2,4-D + mecoprop, 2,4-D + picloram, dicamba + maleic hydrazide + MCPA, dicamba + MCPA + mecoprop, dicamba + paclobutrazol, dichlorprop + ferrous sulphate + MCPA, dichlorprop + MCPA, ioxynil, ioxynil + mecoprop, isoxaben, MCPA, MCPA + mecoprop, mecoprop, picloram
	Annual grasses	dalapon
	Annual weeds	glyphosate (wiper application)
	Bracken	dicamba, picloram
	Brambles	2,4-D + dicamba + mecoprop, 2,4-D + picloram, triclopyr
	Broom	triclopyr
	Buttercups	dichlorprop + MCPA
	Chickweed	mecoprop
	Cleavers	mecoprop
	Couch	dalapon
	Creeping thistle	2,4-D + picloram, MCPA
	Docks	2,4-D + picloram, dicamba + maleic hydrazide, MCPA, triclopyr
	Gorse	triclopyr
	Japanese knotweed	2,4-D + picloram, picloram

	Mosses	chloroxuron + ferrous sulphate, cresylic acid, dichlorophen, dichlorophen + ferrous sulphate, dichlorprop + ferrous sulphate + MCPA, fatty acids, ferrous sulphate
	Perennial dicotyledons	2,4-D, 2,4-D + dicamba, 2,4-D + dicamba + ioxynil, 2,4-D + dicamba + mecoprop, 2,4-D + mecoprop, 2,4-D + picloram, dicamba + maleic hydrazide + MCPA, dicamba + MCPA + mecoprop, dicamba + paclobutrazol, dicamba (wiper application), dichlorprop + ferrous sulphate + MCPA, dichlorprop + MCPA, ioxynil + mecoprop, MCPA, MCPA + mecoprop, mecoprop, picloram, triclopyr
	Perennial grasses	dalapon
	Perennial weeds	glyphosate (wiper application)
	Ragwort	2,4-D + picloram
	Slender speedwell	chlorthal-dimethyl
	Speedwells	2,4-D + dicamba + ioxynil, ioxynil + mecoprop
	Stinging nettle	2,4-D + dicamba + mecoprop, triclopyr
	Woody weeds	2,4-D + dicamba + mecoprop, 2,4-D + picloram, picloram, triclopyr
Plant growth regulation	Growth retardation	dicamba + maleic hydrazide + MCPA, dicamba + paclobutrazol
	Growth suppression	maleic hydrazide, mefluidide

Setaside

Setaside

Weeds	Annual weeds	glyphosate
	Perennial weeds	glyphosate

Field vegetables

General

Diseases	Damping off	etridiazole, propamocarb hydrochloride, thiabendazole + thiram, thiram (seed treatment)
	Seed-borne diseases	thiabendazole + thiram
	Soil-borne diseases	dazomet
Pests	Aphids	gamma-HCH, nicotine, rotenone
	Birds	aluminium ammonium sulphate
	Capsids	gamma-HCH, nicotine
	Cutworms	chlorpyrifos, trichlorfon
	Damaging mammals	aluminium ammonium sulphate

	Leaf miners	gamma-HCH, nicotine
	Leafhoppers	gamma-HCH, nicotine
	Nematodes	dazomet
	Sawflies	nicotine
	Slugs and snails	aluminium sulphate + copper sulphate + sodium tetraborate, metaldehyde, methiocarb
	Soil pests	dazomet
	Thrips	nicotine
	Woolly aphid	nicotine
Weeds	Annual dicotyledons	diquat, diquat + paraquat, paraquat, paraquat (pre-emergence)
	Annual grasses	diquat + paraquat, paraquat, paraquat (pre-emergence)
	Annual weeds	ammonium sulphamate (pre-planting)
	General weed control	dazomet
	Perennial grasses	diquat + paraquat
	Perennial weeds	ammonium sulphamate (pre-planting)
	Volunteer cereals	diquat + paraquat, paraquat, paraquat (pre-emergence)

Asparagus

Diseases	Fungus diseases	thiabendazole (off-label)
	Stemphylium	iprodione (off-label)
Pests	Asparagus beetle	cypermethrin (off-label)
	Leaf miners	cypermethrin (off-label), heptenophos (off-label), nicotine (off-label), triazophos (off-label)
Weeds	Annual dicotyledons	diquat (off-label), diuron, MCPA, simazine, terbacil
	Annual grasses	dalapon, diuron, simazine, terbacil
	Couch	dalapon, terbacil
	Perennial dicotyledons	MCPA
	Perennial grasses	dalapon, terbacil
	Volunteer cereals	dalapon

Beans, french/runner

Diseases	Anthracnose	carbendazim (mbc)
	Ascochyta	metalaxyl + thiabendazole + thiram
	Botrytis	benomyl, carbendazim (mbc), iprodione (off-label), vinclozolin
	Damping off	drazoxolon, metalaxyl + thiabendazole + thiram, thiram (seed treatment)
	Downy mildew	metalaxyl + thiabendazole + thiram
	Fusarium diseases	drazoxolon

	Halo blight	copper hydroxide
	Sclerotinia	iprodione (off-label)
Pests	Aphids	demeton-S-methyl, dimethoate, disulfoton, heptenophos, malathion, nicotine, oxydemeton-methyl, pirimicarb
	Caterpillars	Bacillus thuringiensis (off-label), cypermethrin
	General insect control	nicotine (off-label)
	Leaf miners	nicotine (off-label)
	Pea and bean weevils	cyfluthrin
	Red spider mites	tetradifon
	Thrips	demeton-S-methyl
	Whitefly	cypermethrin
Weeds	Annual dicotyledons	bentazone, chlorthal-dimethyl, diphenamid, diquat (off-label), monolinuron, monolinuron + paraquat, trifluralin
	Annual grasses	diclofop-methyl, diphenamid, monolinuron, monolinuron + paraquat, trifluralin
	Annual meadow grass	monolinuron, monolinuron + paraquat, tri-allate
	Blackgrass	diclofop-methyl, monolinuron + paraquat, tri-allate
	Canary grass	diclofop-methyl
	Fat-hen	monolinuron
	Perennial grasses	diclofop-methyl
	Polygonums	monolinuron
	Rough meadow grass	diclofop-methyl
	Ryegrass	diclofop-methyl
	Volunteer cereals	diclofop-methyl
	Wild oats	diclofop-methyl, monolinuron + paraquat, tri-allate
Crop control	Pre-harvest desiccation	diquat (off-label)

Brassicas

Diseases	Alternaria	carbendazim (mbc), chlorothalonil, fenpropimorph, iprodione, iprodione (seed treatment), maneb, maneb + zinc
	Botrytis	chlorothalonil, iprodione
	Canker	gamma-HCH + thiabendazole + thiram, thiabendazole + thiram
	Clubroot	cresylic acid, ferbam + maneb + zineb, mercurous chloride, thiophanate-methyl
	Damping off	fosetyl-aluminium (off-label), gamma-HCH + thiabendazole + thiram, gamma-HCH + thiram, propamocarb hydrochloride
	Damping off and wirestem	chlorothalonil, quintozene, thiabendazole + thiram, tolclofos-methyl

	Downy mildew	chlorothalonil, copper hydroxide, copper oxychloride + metalaxyl (off-label), dichlofluanid, fosetyl-aluminium (off-label), maneb, maneb + zinc, propamocarb hydrochloride
	Fungus diseases	zineb (off-label)
	Light leaf spot	benomyl, carbendazim (mbc), fenpropimorph
	Powdery mildew	copper sulphate + sulphur, pyrazophos, triadimefon, triadimenol
	Ring spot	benomyl, carbendazim (mbc), chlorothalonil, fenpropimorph
	Seed-borne diseases	captan + gamma-HCH
	Soil-borne diseases	nonylphenoxypoly(ethyleneoxy)ethanol-iodine complex + thiabendazole
	Spear rot	copper oxychloride (off-label)
	White blister	chlorothalonil + metalaxyl, mancozeb + metalaxyl (off-label)
Pests	Aphids	aldicarb, chlorpyrifos, chlorpyrifos + dimethoate, chlorpyrifos + disulfoton, cyfluthrin, cypermethrin, deltamethrin (off-label), demeton-S-methyl, dimethoate, disulfoton, disulfoton + quinalphos, fatty acids, fenvalerate, heptenophos, nicotine, permethrin, phorate, pirimicarb, quinalphos + thiometon, resmethrin
	Birds	aluminium ammonium sulphate
	Cabbage root fly	aldicarb, carbofuran, carbofuran (off-label), carbosulfan, chlorfenvinphos, chlorpyrifos, chlorpyrifos + dimethoate, chlorpyrifos + disulfoton, diazinon, disulfoton + quinalphos, fonofos, fonofos (off-label), gamma-HCH + thiabendazole + thiram, iodofenphos, phorate, triazophos, trichlorfon
	Cabbage stem flea beetle	cypermethrin, gamma-HCH
	Cabbage stem weevil	carbofuran, carbosulfan, gamma-HCH
	Caterpillars	azinphos-methyl + demeton-S-methyl sulphone (off-label), Bacillus thuringiensis, carbaryl, carbofuran, chlorpyrifos, cyfluthrin, cypermethrin, deltamethrin, deltamethrin (off-label), diflubenzuron, fenvalerate, iodofenphos, lambda-cyhalothrin, permethrin, pirimiphos-methyl, quinalphos, quinalphos + thiometon, resmethrin, triazophos, trichlorfon
	Cutworms	chlorpyrifos, chlorpyrifos + dimethoate, gamma-HCH, triazophos
	Damaging mammals	aluminium ammonium sulphate, sulphonated cod liver oil
	Flea beetles	aldicarb, captan + gamma-HCH, carbaryl, carbofuran, carbosulfan, cyfluthrin, deltamethrin, gamma-HCH, gamma HCH + thiabendazole + thiram, gamma-HCH + thiram
	General insect control	deltamethrin + heptenophos (off-label), demeton-S-methyl (off-label)
	Leaf miners	aldicarb, azinphos-methyl + demeton-S-methyl sulphone, heptenophos (off-label), nicotine (off-label), permethrin, pyrazophos (off-label), trichlorfon
	Leatherjackets	carbaryl, chlorpyrifos, chlorpyrifos + dimethoate, gamma-HCH

	Mealy bugs	fatty acids
	Millipedes	gamma-HCH + thiabendazole + thiram
	Pollen beetles	gamma-HCH
	Scale insects	fatty acids
	Soil pests	aldicarb (off-label), carbofuran (off-label)
	Thrips	azinphos-methyl + demeton-S-methyl sulphone (off-label), fatty acids
	Weevils	gamma-HCH
	Whitefly	fatty acids, permethrin, pirimiphos-methyl
	Wireworms	chlorpyrifos + dimethoate, gamma-HCH
	Woolly aphid	fatty acids
Weeds	Annual dicotyledons	aziprotryne, carbetamide, chlorthal-dimethyl, chlorthal-dimethyl + propachlor, clopyralid, cyanazine (off-label), desmetryn, metazachlor, propachlor, propachlor (off-label), propyzamide, sodium monochloroacetate, tebutam, trifluralin
	Annual grasses	aziprotryne, carbetamide, chlorthal-dimethyl + propachlor, cyanazine (off-label), diclofop-methyl, propachlor, propachlor (off-label), propyzamide, TCA (delayed sowing), tebutam, trifluralin
	Annual meadow grass	metazachlor, tri-allate
	Blackgrass	cycloxydim, diclofop-methyl, metazachlor, tri-allate
	Canary grass	diclofop-methyl
	Corn marigold	clopyralid
	Couch	cycloxydim, TCA (delayed sowing)
	Creeping bent	cycloxydim
	Creeping thistle	clopyralid
	Fat-hen	desmetryn
	Groundsel, triazine resistant	metazachlor
	Mayweeds	clopyralid
	Perennial dicotyledons	clopyralid
	Perennial grasses	diclofop-methyl, propyzamide, TCA (delayed sowing)
	Rough meadow grass	diclofop-methyl
	Ryegrass	diclofop-methyl
	Volunteer cereals	carbetamide, cycloxydim, diclofop-methyl, tebutam
	Wild oats	cycloxydim, diclofop-methyl, tri-allate

Brassica seed crops

Diseases	Alternaria	iprodione
	Botrytis	iprodione

Pests	Cabbage stem flea beetle	gamma-HCH
	Cabbage stem weevil	gamma-HCH
	Pod midge	phosalone
	Pollen beetles	gamma-HCH, phosalone
	Seed weevil	gamma-HCH, phosalone

Carrots/parsnips/parsley

Diseases	Alternaria blight	copper hydroxide, iprodione + metalaxyl + thiabendazole
	Black rot	benomyl (dip), iprodione + metalaxyl + thiabendazole
	Botrytis	benomyl (dip)
	Cavity spot	mancozeb + metalaxyl
	Damping off	iprodione + metalaxyl + thiabendazole, thiabendazole + thiram
	Liquorice rot	benomyl (dip)
	Powdery mildew	triadimefon
	Sclerotinia	benomyl (dip)
	Seed-borne diseases	thiabendazole + thiram
Pests	Aphids	aldicarb, carbofuran, carbosulfan, demeton-S-methyl, dimethoate, disulfoton, malathion, oxydemeton-methyl, phorate, pirimicarb
	Carrot fly	carbofuran, carbosulfan, chlorfenvinphos, diazinon, disulfoton, phorate, pirimiphos-methyl, quinalphos, triazophos, triazophos (off-label)
	Celery fly	malathion
	Cutworms	cypermethrin, quinalphos, triazophos
	Leaf miners	nicotine (off-label), pyrazophos (off-label)
	Nematodes	aldicarb, carbofuran, carbosulfan
Weeds	Annual dicotyledons	chlorpropham, chlorpropham + pentanochlor, linuron, metoxuron, metoxuron (off-label), pendimethalin (off-label), pentanochlor, prometryn, trifluralin
	Annual grasses	alloxydim-sodium, chlorpropham, dalapon, dalapon + di-1-p-menthene, diclofop-methyl, fluazifop-P-butyl, fluazifop-P-butyl (off-label), metoxuron, metoxuron (off-label), pendimethalin (off-label), prometryn, sethoxydim (off-label), TCA (delayed sowing), trifluralin
	Annual meadow grass	chlorpropham + pentanochlor, linuron, pentanochlor, tri-allate
	Black bindweed	linuron
	Blackgrass	diclofop-methyl, tri-allate
	Canary grass	diclofop-methyl
	Chickweed	chlorpropham, linuron
	Corn marigold	linuron

Couch	dalapon, TCA (delayed sowing)
Docks	asulam (off-label)
Fat-hen	linuron
Mayweeds	metoxuron, metoxuron (off-label)
Perennial grasses	dalapon, dalapon + di-1-p-menthene, diclofop-methyl, fluazifop-P-butyl, fluazifop-P-butyl (off-label), TCA (delayed sowing)
Polygonums	chlorpropham, linuron
Rough meadow grass	diclofop-methyl
Ryegrass	diclofop-methyl
Volunteer cereals	alloxydim-sodium, dalapon, dalapon + di-1-p-menthene, diclofop-methyl, fluazifop-P-butyl, sethoxydim (off-label)
Wild oats	alloxydim-sodium, diclofop-methyl, fluazifop-P-butyl, tri-allate

Celeriac

Diseases	Celery leaf spot	chlorothalonil (off-label)
Pests	Aphids	deltamethrin (off-label)
	Caterpillars	deltamethrin (off-label)
	Celery fly	quinalphos (off-label)
	General insect control	chlorfenvinphos (off-label), demeton-S-methyl (off-label), oxydemeton-methyl (off-label)
	Leaf miners	nicotine (off-label), pyrazophos (off-label)
Weeds	Annual dicotyledons	chlorpropham + pentanochlor, pentanochlor, trifluralin (off-label)
	Annual grasses	trifluralin (off-label)
	Annual meadow grass	chlorpropham + pentanochlor, pentanochlor
	Annual weeds	linuron (off-label)

Celery

Diseases	Bacterial blight	copper hydroxide
	Botrytis	benomyl, benomyl (dip), vinclozolin
	Celery leaf spot	Bordeaux mixture, carbendazim (mbc), chlorothalonil, copper hydroxide, copper oxychloride, cupric ammonium carbonate, zineb
	Damping off	propamocarb hydrochloride
	Liquorice rot	benomyl (dip)
	Powdery mildew	maneb (off-label)
	Sclerotinia	vinclozolin

Pests	Aphids	cypermethrin, deltamethrin (off-label), demeton-S-methyl, dimethoate, disulfoton, heptenophos, malathion, oxydemeton-methyl, pirimicarb
	Carrot fly	chlorfenvinphos, diazinon, diazinon (off-label), disulfoton, phorate
	Caterpillars	cypermethrin, deltamethrin (off-label)
	Celery fly	malathion, quinalphos, trichlorfon
	Cutworms	cypermethrin, triazophos
	Leaf miners	cypermethrin (off-label), heptenophos (off-label), nicotine, nicotine (off-label), triazophos (off-label)
	Thrips	dimethoate (off-label)
	Whitefly	permethrin
Weeds	Annual dicotyledons	chlorpropham, chlorpropham + pentanochlor, linuron, pentanochlor, prometryn
	Annual grasses	chlorpropham, diclofop-methyl, prometryn
	Annual meadow grass	chlorpropham + pentanochlor, linuron, pentanochlor
	Black bindweed	linuron
	Blackgrass	diclofop-methyl
	Canary grass	diclofop-methyl
	Chickweed	chlorpropham, linuron
	Corn marigold	linuron
	Fat-hen	linuron
	Perennial grasses	diclofop-methyl
	Polygonums	chlorpropham, linuron
	Rough meadow grass	diclofop-methyl
	Ryegrass	diclofop-methyl
	Volunteer cereals	diclofop-methyl
	Wild oats	diclofop-methyl
Plant growth regulation	Increasing germination	gibberellins
	Increasing yield	gibberellins

Chicory

Diseases	Sclerotinia	quintozene
Pests	Leaf miners	cypermethrin (off-label), heptenophos (off-label), nicotine (off-label)
Weeds	Annual dicotyledons	propyzamide
	Annual grasses	propyzamide
	Perennial grasses	propyzamide

Cucurbits

Diseases	Botrytis	iprodione (off-label)
	Damping off	propamocarb hydrochloride, thiabendazole + thiram
	Fungus diseases	zineb (off-label)
	Powdery mildew	bupirimate, imazalil, iprodione (off-label), maneb (off-label), quinomethionate
	Seed-borne diseases	thiabendazole + thiram
Pests	Aphids	disulfoton, heptenophos, pirimicarb (off-label)
	General insect control	demeton-S-methyl (off-label), oxydemeton-methyl (off-label)
	Leaf miners	cypermethrin (off-label), heptenophos (off-label), nicotine (off-label), oxamyl (off-label), pyrazophos (off-label), triazophos (off-label)
	Red spider mites	quinomethionate
	Thrips	dimethoate (off-label)
Weeds	Annual dicotyledons	diphenamid
	Annual grasses	diphenamid

Fennel

Pests	Aphids	deltamethrin + heptenophos (off-label), deltamethrin (off-label), pirimicarb (off-label)
	Caterpillars	deltamethrin + heptenophos (off-label), deltamethrin (off-label)
	Celery fly	quinalphos (off-label)
	General insect control	oxydemeton-methyl (off-label)
	Leaf miners	cypermethrin (off-label), heptenophos (off-label), nicotine (off-label), pyrazophos (off-label), triazophos (off-label)
Weeds	Annual dicotyledons	chlorpropham + pentanochlor, pentanochlor
	Annual meadow grass	chlorpropham + pentanochlor, pentanochlor

Herb crops

Diseases	Fungus diseases	iprodione (off-label)
	Rust	triadimefon (off-label)
Pests	Aphids	disulfoton (off-label)
	Flea beetles	deltamethrin (off-label)
	General insect control	demeton-S-methyl (off-label), oxydemeton-methyl (off-label)
	Leaf miners	deltamethrin (off-label)
	Leafhoppers	deltamethrin (off-label)

Weeds	Annual dicotyledons	chlorthal-dimethyl, clopyralid (off-label), diquat (off-label), lenacil, paraquat (off-label), pendimethalin (off-label), prometryn, propachlor, propyzamide (off-label), simazine (off-label), terbacil (off-label), trifluralin (off-label)
	Annual grasses	ethofumesate (off-label), paraquat (off-label), pendimethalin (off-label), prometryn, propachlor, propyzamide (off-label), simazine (off-label), terbacil (off-label), trifluralin (off-label)
	Annual meadow grass	lenacil
	Annual weeds	linuron (off-label)
	Chickweed	ethofumesate (off-label)
	Cleavers	ethofumesate (off-label)
	Couch	terbacil (off-label)
	Docks	asulam (off-label)
	Perennial dicotyledons	clopyralid (off-label)
	Perennial grasses	propyzamide (off-label), terbacil (off-label)
Crop control	Pre-harvest desiccation	diquat (off-label)

Lettuce, outdoor

Diseases	Big vein	carbendazim (mbc)
	Botrytis	benomyl, iprodione, iprodione (off-label), quintozene, thiram, vinclozolin
	Damping off	quintozene, tolclofos-methyl
	Downy mildew	fosetyl-aluminium (off-label), mancozeb, maneb, metalaxyl + thiram, zineb
	Sclerotinia	iprodione (off-label), quintozene
Pests	Aphids	cypermethrin, deltamethrin (off-label), demeton-S-methyl, dimethoate, fatty acids, gamma-HCH, heptenophos, malathion, nicotine, phorate, pirimicarb, resmethrin
	Caterpillars	carbaryl, cypermethrin, deltamethrin, deltamethrin (off-label)
	Cutworms	carbaryl, cypermethrin, gamma-HCH
	Damaging mammals	sulphonated cod liver oil
	General insect control	oxydemeton-methyl (off-label)
	Leaf miners	cypermethrin (off-label), diazinon, heptenophos (off-label), nicotine (off-label)
	Leatherjackets	carbaryl, gamma-HCH
	Lettuce root aphid	diazinon, phorate, phorate (off-label, higher rate)
	Mealy bugs	fatty acids
	Scale insects	diazinon, fatty acids
	Sciarid flies	resmethrin
	Thrips	fatty acids
	Whitefly	fatty acids, resmethrin

	Woolly aphid	fatty acids
Weeds	Annual dicotyledons	chlorpropham, chlorpropham + diuron + propham, propachlor (off-label), propyzamide, trifluralin
	Annual grasses	chlorpropham, chlorpropham + diuron + propham, diclofop-methyl, propachlor (off-label), propyzamide, trifluralin
	Blackgrass	diclofop-methyl
	Canary grass	diclofop-methyl
	Chickweed	chlorpropham
	Perennial grasses	diclofop-methyl, propyzamide
	Polygonums	chlorpropham
	Rough meadow grass	diclofop-methyl
	Ryegrass	diclofop-methyl
	Volunteer cereals	diclofop-methyl
	Wild oats	diclofop-methyl

Onions/leeks

Diseases	Allium leaf blotches	chlorothalonil, ferbam + maneb + zineb
	Botrytis	benomyl, benomyl + iodofenphos + metalaxyl, benomyl (seed treatment), carbendazim (mbc), chlorothalonil, iprodione, thiabendazole + thiram, vinclozolin
	Damping off	benomyl + iodofenphos + metalaxyl, thiabendazole + thiram
	Downy mildew	chlorothalonil + metalaxyl, copper hydroxide, ferbam + maneb + zineb, mancozeb + metalaxyl (off-label), propamocarb hydrochloride, zineb
	Fungus diseases	zineb (off-label)
	Rust	fenpropimorph, ferbam + maneb + zineb, triadimefon, triadimefon (off-label)
	Seed-borne diseases	thiabendazole + thiram
	White rot	mercurous chloride, vinclozolin
	White tip	chlorothalonil + metalaxyl, ferbam + maneb + zineb, mancozeb + metalaxyl (off-label)
Pests	Cutworms	cypermethrin, triazophos
	General insect control	deltamethrin (off-label), demeton-S-methyl (off-label), oxydemeton-methyl (off-label)
	Leaf miners	cypermethrin (off-label), heptenophos (off-label), nicotine (off-label), oxamyl (off-label), triazophos (off-label)
	Onion fly	benomyl + iodofenphos + metalaxyl, carbofuran, diazinon, iodofenphos
	Stem and bulb nematodes	oxamyl
	Stem nematodes	aldicarb, aldicarb (off-label), carbofuran
	Thrips	dimethoate (off-label), fenitrothion (off-label), malathion

Weeds	Annual dicotyledons	aziprotryne, chlorbufam + chloridazon, chlorbufam + chloridazon (off-label), chloridazon + propachlor, chlorpropham, chlorpropham + cresylic acid + fenuron, chlorpropham + fenuron, chlorthal-dimethyl, chlorthal-dimethyl + propachlor, clopyralid, cyanazine, ioxynil, monolinuron, pendimethalin, prometryn, prometryn (off-label), propachlor, sodium monochloroacetate
	Annual grasses	alloxydim-sodium, aziprotryne, chloridazon + propachlor, chlorpropham, chlorpropham + cresylic acid + fenuron, chlorpropham + fenuron, chlorthal-dimethyl + propachlor, cyanazine, diclofop-methyl, fluazifop-P-butyl, monolinuron, pendimethalin, prometryn, prometryn (off-label), propachlor, sethoxydim
	Annual meadow grass	monolinuron, tri-allate
	Annual weeds	bentazone (off-label), linuron (off-label)
	Blackgrass	diclofop-methyl, sethoxydim, tri-allate
	Canary grass	diclofop-methyl
	Chickweed	chlorpropham, chlorpropham + cresylic acid + fenuron, chlorpropham + fenuron
	Corn marigold	clopyralid
	Couch	sethoxydim
	Creeping thistle	clopyralid
	Fat-hen	monolinuron
	Mayweeds	bentazone (off-label), clopyralid
	Perennial dicotyledons	clopyralid
	Perennial grasses	diclofop-methyl, fluazifop-P-butyl, sethoxydim
	Polygonums	chlorpropham, monolinuron
	Rough meadow grass	diclofop-methyl
	Ryegrass	diclofop-methyl
	Volunteer cereals	alloxydim-sodium, diclofop-methyl, fluazifop-P-butyl, sethoxydim
	Wild oats	alloxydim-sodium, diclofop-methyl, fluazifop-P-butyl, sethoxydim, tri-allate
Plant growth regulation	Sprout suppression	maleic hydrazide

Peas, mange-tout

Diseases	Ascochyta	metalaxyl + thiabendazole + thiram, thiabendazole + thiram (off-label)
	Botrytis	iprodione (off-label)
	Damping off	metalaxyl + thiabendazole + thiram, thiabendazole + thiram (off-label)
	Downy mildew	metalaxyl + thiabendazole + thiram
	Root rot	drazoxolon (off-label)

	Sclerotinia	iprodione (off-label)
Pests	Aphids	demeton-S-methyl (off-label)
	General insect control	demeton-S-methyl (off-label)
	Leaf miners	nicotine (off-label), triazophos (off-label)
	Thrips	demeton-S-methyl (off-label)
Weeds	Annual dicotyledons	MCPB (off-label), simazine + trietazine
	Annual meadow grass	tri-allate (off-label)
	Blackgrass	tri-allate (off-label)
	Perennial dicotyledons	MCPB (off-label)
	Wild oats	tri-allate (off-label)

Radicchio

Diseases	Botrytis	iprodione (off-label)
	Sclerotinia	iprodione (off-label)
Pests	Aphids	deltamethrin + heptenophos (off-label), deltamethrin (off-label)
	Caterpillars	deltamethrin + heptenophos (off-label), deltamethrin (off-label)
	General insect control	demeton-S-methyl (off-label), oxydemeton-methyl (off-label)
Weeds	Annual dicotyledons	propyzamide (off-label)
	Annual grasses	propyzamide (off-label)
	Perennial grasses	propyzamide (off-label)

Radishes

Diseases	Downy mildew	propamocarb hydrochloride (off-label)
Weeds	Annual dicotyledons	trifluralin (off-label)
	Annual grasses	trifluralin (off-label)

Red beet

Pests	Aphids	malathion
	Flea beetles	trichlorfon
	General insect control	oxydemeton-methyl (off-label)
	Leaf miners	oxamyl (off-label), pyrazophos (off-label)
	Mangold fly	trichlorfon
Weeds	Annual dicotyledons	chlorpropham + fenuron + propham, clopyralid, ethofumesate, metamitron, phenmedipham
	Annual grasses	chlorpropham + fenuron + propham, diclofop-methyl, metamitron, quizalofop-ethyl
	Annual meadow grass	ethofumesate, metamitron

Blackgrass	diclofop-methyl, ethofumesate
Canary grass	diclofop-methyl
Corn marigold	clopyralid
Couch	quizalofop-ethyl
Creeping thistle	clopyralid
Fat-hen	metamitron
Mayweeds	clopyralid
Perennial dicotyledons	clopyralid
Perennial grasses	diclofop-methyl, quizalofop-ethyl
Rough meadow grass	diclofop-methyl
Ryegrass	diclofop-methyl
Volunteer cereals	diclofop-methyl, quizalofop-ethyl
Wild oats	diclofop-methyl

Rhubarb

Pests	Leaf miners	cypermethrin (off-label), heptenophos (off-label), nicotine (off-label), triazophos (off-label)
	Rosy rustic moth	chlorpyrifos
Weeds	Annual dicotyledons	chlorpropham + cresylic acid + fenuron, propyzamide, simazine
	Annual grasses	chlorpropham + cresylic acid + fenuron, dalapon, propyzamide, simazine, TCA
	Chickweed	chlorpropham + cresylic acid + fenuron
	Couch	dalapon, TCA
	Perennial grasses	dalapon, propyzamide, TCA
Plant growth regulation	Increasing yield	gibberellins

Root brassicas

Diseases	Alternaria	iprodione (off-label)
	Canker	gamma-HCH + thiabendazole + thiram
	Damping off	gamma-HCH + thiabendazole + thiram, gamma-HCH + thiram
	Fungus diseases	zineb (off-label)
	Powdery mildew	benomyl, carbendazim (mbc), copper sulphate + sulphur, potassium sorbate + sodium metabisulphite + sodium propionate, sulphur, triadimefon, triadimenol, tridemorph
Pests	Aphids	aldicarb, carbofuran, chlorpyrifos + dimethoate, cypermethrin, malathion, pirimicarb, pirimicarb (off-label)

110

	Cabbage root fly	aldicarb, carbofuran, carbofuran (off-label), carbosulfan, chlorfenvinphos, chlorfenvinphos (off-label), chlorpyrifos + dimethoate, diazinon, fonofos, gamma-HCH + thiabendazole + thiram, trichlorfon (off-label)
	Cabbage stem flea beetle	cypermethrin
	Cabbage stem weevil	carbofuran, carbosulfan
	Caterpillars	cypermethrin, deltamethrin
	Cutworms	chlorpyrifos + dimethoate
	Flea beetles	aldicarb, carbofuran, carbosulfan, deltamethrin, gamma-HCH + thiabendazole + thiram, gamma-HCH + thiram
	General insect control	chlorfenvinphos (off-label), deltamethrin + heptenophos (off-label)
	Leaf miners	nicotine (off-label), pyrazophos (off-label)
	Leatherjackets	chlorpyrifos + dimethoate
	Millipedes	gamma-HCH + thiabendazole + thiram
	Nematodes	aldicarb
	Soil pests	carbofuran (off-label)
	Turnip root fly	carbofuran
	Wireworms	chlorpyrifos + dimethoate
Weeds	Annual dicotyledons	clopyralid, metazachlor, propachlor, tebutam, trifluralin
	Annual grasses	alloxydim-sodium, dalapon + di-1-p-menthene, fluazifop-P-butyl (stockfeed only), propachlor, sethoxydim, TCA (delayed sowing), tebutam, trifluralin
	Annual meadow grass	metazachlor
	Blackgrass	cycloxydim, metazachlor, sethoxydim
	Corn marigold	clopyralid
	Couch	cycloxydim, sethoxydim, TCA (delayed sowing)
	Creeping bent	cycloxydim
	Creeping thistle	clopyralid
	Groundsel, triazine resistant	metazachlor
	Mayweeds	clopyralid
	Perennial dicotyledons	clopyralid
	Perennial grasses	dalapon + di-1-p-menthene, fluazifop-P-butyl (stockfeed only), sethoxydim, TCA (delayed sowing)
	Volunteer cereals	alloxydim-sodium, cycloxydim, dalapon + di-1-p-menthene, fluazifop-P-butyl (stockfeed only), sethoxydim, tebutam
	Wild oats	alloxydim-sodium, cycloxydim, fluazifop-P-butyl (stockfeed only), sethoxydim

111

Root crops

Pests	Cutworms	cypermethrin

Spinach

Diseases	Downy mildew	propamocarb hydrochloride, zineb
Pests	Aphids	pirimicarb (off-label)
	Flea beetles	trichlorfon
	General insect control	demeton-S-methyl (off-label), oxydemeton-methyl (off-label)
	Leaf miners	cypermethrin (off-label), heptenophos (off-label), nicotine (off-label), trichlorfon (off-label)
	Mangold fly	trichlorfon
Weeds	Annual dicotyledons	chlorpropham + fenuron, lenacil (off-label)
	Annual grasses	chlorpropham + fenuron, lenacil (off-label)
	Chickweed	chlorpropham + fenuron

Tomatoes, outdoor

Diseases	Blight	copper oxychloride
	Soil-borne diseases	metam-sodium (Jersey)
Pests	General weed control	metam-sodium (Jersey)
	Nematodes	metam-sodium (Jersey)
	Soil pests	metam-sodium (Jersey)

Watercress

Diseases	Damping off	mancozeb + metalaxyl (off-label)
	Rhizoctonia	benomyl (off-label)

Fruit and hops

General

Pests	Aphids	nicotine, resmethrin, rotenone
	Birds	aluminium ammonium sulphate
	Capsids	nicotine
	Damaging mammals	aluminium ammonium sulphate
	Leaf miners	nicotine
	Leafhoppers	nicotine
	Sawflies	nicotine

	Slugs and snails	aluminium sulphate + copper sulphate + sodium tetraborate, metaldehyde
	Thrips	nicotine
	Woolly aphid	nicotine
Weeds	Annual dicotyledons	diquat + paraquat
	Annual grasses	diquat + paraquat
	Perennial grasses	diquat + paraquat
	Volunteer cereals	diquat + paraquat

Fruit nursery stock

Diseases	Botrytis	iprodione (off-label)
	Fungus diseases	carbendazim (mbc) (off-label)
	Powdery mildew	dodemorph (off-label)
Pests	Capsids	triazophos (off-label)
	Caterpillars	triazophos (off-label)
	Codling moth	amitraz (off-label), triazophos (off-label)
	Cutworms	triazophos (off-label)
	General insect control	azinphos-methyl + demeton-S-methyl sulphone (off-label), deltamethrin (off-label)
	Mites	amitraz (off-label)
	Red spider mites	aldicarb (off-label), triazophos (off-label)
	Sciarid flies	fonofos (off-label)
	Tortrix moths	triazophos (off-label)
	Vine weevil	fonofos (off-label)
Weeds	Annual dicotyledons	clopyralid (off-label), linuron + trietazine (off-label), napropamide + trifluralin (off-label), propyzamide (off-label), trifluralin (off-label)
	Annual grasses	linuron + trietazine (off-label), napropamide + trifluralin (off-label), propyzamide (off-label), trifluralin (off-label)
	Docks	asulam (off-label)
	Perennial dicotyledons	clopyralid (off-label)
	Perennial grasses	propyzamide (off-label)

Apples/pears

Diseases	Blossom wilt	benomyl, tar oils, vinclozolin
	Canker	benomyl, Bordeaux mixture, carbendazim (mbc), copper hydroxide, copper oxychloride, copper sulphate + cufraneb, mercuric oxide, octhilinone, thiophanate-methyl
	Collar rot	copper hydroxide (drench), copper oxychloride + metalaxyl, copper oxychloride (off-label), fosetyl-aluminium
	Crown rot	fosetyl-aluminium

	Gloeosporium	captan
	Phytophthora fruit rot	captan, mancozeb + metalaxyl (off-label)
	Powdery mildew	benomyl, bupirimate, captan + nuarimol, captan + penconazole, carbendazim (mbc), dinocap, fenarimol, myclobutanil, nitrothal-isopropyl + zineb-ethylenethiuram disulphide complex, penconazole, pyrazophos, sulphur, thiophanate-methyl, triadimefon, triforine
	Pruning wounds	octhilinone
	Scab	benomyl, captan, captan + nuarimol, captan + penconazole, carbendazim (mbc), copper hydroxide, dithianon, dodine, fenarimol, mancozeb, maneb + zinc, myclobutanil, sulphur, thiophanate-methyl, thiram, triforine
	Silver leaf	octhilinone
	Storage rots	benomyl, carbendazim (mbc), carbendazim + metalaxyl, thiophanate-methyl, vinclozolin
Pests	Aphids	amitraz, azinphos-methyl + demeton-S-methyl sulphone, chlorpyrifos, cyfluthrin, cypermethrin, deltamethrin, demeton-S-methyl, dimethoate, fenitrothion, fenvalerate, heptenophos, lambda-cyhalothrin, malathion, nicotine, oxydemeton-methyl, phosalone, pirimicarb, pirimiphos-methyl, tar oils
	Apple blossom weevil	azinphos-methyl + demeton-S-methyl sulphone, chlorpyrifos, cyfluthrin, fenitrothion, gamma-HCH
	Bryobia mites	azinphos-methyl + demeton-S-methyl sulphone, demeton-S-methyl, dicofol + tetradifon, dimethoate, malathion, oxydemeton-methyl, phosalone
	Capsids	carbaryl, chlorpyrifos, cyfluthrin, cypermethrin, deltamethrin, dimethoate, fenitrothion, permethrin, pirimiphos-methyl, triazophos
	Caterpillars	chlorpyrifos, cypermethrin, deltamethrin, diflubenzuron, fenitrothion, fenpropathrin, fenvalerate, permethrin, pirimiphos-methyl
	Chafer grubs	carbaryl
	Cherry bark tortrix	trichlorfon
	Codling moth	amitraz, azinphos-methyl + demeton-S-methyl sulphone, carbaryl, chlorpyrifos, cyfluthrin, cypermethrin, deltamethrin, diflubenzuron, fenitrothion, fenvalerate, malathion, permethrin, pheromones, phosalone, pirimiphos-methyl, triazophos
	Earwigs	carbaryl, diflubenzuron
	Leaf midges	carbaryl, lambda-cyhalothrin
	Leaf miners	diflubenzuron
	Leafhoppers	carbaryl, demeton-S-methyl, malathion, oxydemeton-methyl
	Red spider mites	amitraz, azinphos-methyl + demeton-S-methyl sulphone, chlorpyrifos, clofentezine, demeton-S-methyl, dicofol, dicofol + tetradifon, dimethoate, fenpropathrin, malathion, oxydemeton-methyl, phosalone, tetradifon, triazophos

	Rhynchites	carbaryl
	Rust mite	diflubenzuron, pirimiphos-methyl
	Sawflies	azinphos-methyl + demeton-S-methyl sulphone, carbaryl, chlorpyrifos, cyfluthrin, cypermethrin, deltamethrin, demeton-S-methyl, diflubenzuron, dimethoate, fenitrothion, gamma-HCH, oxydemeton-methyl, permethrin, pirimiphos-methyl
	Scale insects	tar oils
	Slug sawflies	rotenone
	Suckers	amitraz, azinphos-methyl + demeton-S-methyl sulphone, carbaryl, chlorpyrifos, cyfluthrin, cypermethrin, deltamethrin, demeton-S-methyl, diflubenzuron, dimethoate, fenitrothion, fenvalerate, gamma-HCH, lambda-cyhalothrin, malathion, oxydemeton-methyl, permethrin, pirimiphos-methyl, tar oils
	Tortrix moths	azinphos-methyl + demeton-S-methyl sulphone, carbaryl, chlorpyrifos, cyfluthrin, cypermethrin, deltamethrin, diflubenzuron, fenitrothion, fenvalerate, permethrin, phosalone, pirimiphos-methyl, triazophos
	Weevils	carbaryl
	Winter moth	azinphos-methyl + demeton-S-methyl sulphone, carbaryl, chlorpyrifos, cyfluthrin, diflubenzuron, fenitrothion, phosalone, tar oils
	Woolly aphid	chlorpyrifos, demeton-S-methyl, heptenophos
Weeds	Annual dicotyledons	2,4-D + dichlorprop + MCPA + mecoprop, dicamba + MCPA + mecoprop, diuron, isoxaben, napropamide, oxadiazon, pendimethalin, pentanochlor, propyzamide, simazine, sodium monochloroacetate, terbacil
	Annual grasses	dalapon, diuron, napropamide, oxadiazon, pendimethalin, propyzamide, simazine, terbacil
	Annual meadow grass	pentanochlor
	Annual weeds	amitrole, amitrole + diuron, dichlobenil, glyphosate
	Bindweeds	oxadiazon
	Brambles	triclopyr
	Broom	triclopyr
	Chickweed	2,4-D + dichlorprop + MCPA + mecoprop
	Cleavers	2,4-D + dichlorprop + MCPA + mecoprop, napropamide, oxadiazon
	Couch	amitrole, dalapon, terbacil
	Creeping thistle	amitrole
	Docks	amitrole, asulam, dicamba + MCPA + mecoprop, triclopyr
	Gorse	triclopyr
	Groundsel	napropamide
	Perennial dicotyledons	2,4-D + dichlorprop + MCPA + mecoprop, dicamba + MCPA + mecoprop, dichlobenil, triclopyr

	Perennial grasses	dalapon, dichlobenil, propyzamide, terbacil
	Perennial weeds	amitrole, amitrole + diuron, glyphosate
	Polygonums	oxadiazon
	Stinging nettle	triclopyr
	Volunteer cereals	dalapon
	Woody weeds	triclopyr
Crop control	Sucker control	glyphosate
Plant growth regulation	Controlling vigour	paclobutrazol
	Fruit ripening	2-chloroethylphosphonic acid
	Fruit thinning	carbaryl
	Increasing fruit set	gibberellins, paclobutrazol
	Reducing fruit russeting	gibberellins
	Sucker inhibition	1-naphthylacetic acid

Apricots/peaches/nectarines

Diseases	Blossom wilt	vinclozolin
	Leaf curl	Bordeaux mixture, copper hydroxide, copper oxychloride, copper sulphate + cufraneb, cupric ammonium carbonate
Pests	Aphids	demeton-S-methyl, dimethoate, malathion, oxydemeton-methyl, tar oils
	Red spider mites	oxydemeton-methyl, tetradifon
	Scale insects	tar oils
	Winter moth	tar oils
Weeds	Annual weeds	amitrole (off-label)
	Docks	asulam (off-label)
	Perennial weeds	amitrole (off-label)

Blueberries/cranberries

Weeds	Docks	asulam (off-label)

Cane fruit

Diseases	Botrytis	benomyl, captan, carbendazim (mbc), chlorothalonil, dichlofluanid, iprodione, thiram, vinclozolin
	Cane blight	benomyl, dichlofluanid
	Cane spot	benomyl, carbendazim (mbc), chlorothalonil, copper hydroxide, copper oxychloride, copper sulphate + cufraneb, cupric ammonium carbonate, dichlofluanid, thiram

	Powdery mildew	benomyl, bupirimate, carbendazim (mbc), chlorothalonil, dichlofluanid, dinocap, fenarimol, triadimefon
	Purple blotch	benomyl
	Root rot	copper oxychloride + metalaxyl, mancozeb + metalaxyl (off-label)
	Rust	benodanil
	Spur blight	benomyl, carbendazim (mbc), copper hydroxide, thiram
Pests	Aphids	chlorpyrifos, demeton-S-methyl, dimethoate, heptenophos, malathion, oxydemeton-methyl, pirimicarb, tar oils
	Blackberry mite	endosulfan
	Bramble shoot moth	azinphos-methyl + demeton-S-methyl sulphone
	Capsids	dimethoate
	Caterpillars	Bacillus thuringiensis
	Leafhoppers	dimethoate, malathion
	Nematodes	1,3-dichloropropene
	Overwintering pests	cresylic acid
	Raspberry beetle	azinphos-methyl + demeton-S-methyl sulphone, carbaryl, chlorpyrifos, deltamethrin, fenitrothion, malathion, rotenone
	Raspberry cane midge	chlorpyrifos, fenitrothion, gamma-HCH
	Raspberry moth	carbaryl, tar oils
	Red spider mites	chlorpyrifos, demeton-S-methyl, dimethoate, oxydemeton-methyl, tetradifon
	Rhynchites	carbaryl
	Scale insects	tar oils
	Virus vectors	1,3-dichloropropene
	Winter moth	tar oils
Weeds	Annual dicotyledons	atrazine, atrazine + cyanazine, bromacil, chlorthal-dimethyl, isoxaben, lenacil, napropamide, paraquat, pendimethalin, propachlor, propyzamide, simazine, trifluralin
	Annual grasses	atrazine, atrazine + cyanazine, bromacil, dalapon, fluazifop-P-butyl, napropamide, paraquat, pendimethalin, propachlor, propyzamide, sethoxydim, simazine, trifluralin
	Annual meadow grass	lenacil
	Annual weeds	dichlobenil
	Blackgrass	sethoxydim
	Cleavers	napropamide
	Couch	bromacil, dalapon, sethoxydim
	Creeping bent	paraquat
	Docks	asulam (off-label)
	Groundsel	napropamide
	Perennial dicotyledons	dichlobenil

	Perennial grasses	bromacil, dalapon, dichlobenil, fluazifop-P-butyl, propyzamide, sethoxydim
	Perennial ryegrass	paraquat
	Rough meadow grass	paraquat
	Volunteer cereals	fluazifop-P-butyl, sethoxydim
	Wild oats	fluazifop-P-butyl, sethoxydim
Crop control	Sucker control	diquat + paraquat (off-label)
Plant growth regulation	Sucker inhibition	1-naphthylacetic acid

Currants

Diseases	Botrytis	benomyl, carbendazim (mbc), dichlofluanid, vinclozolin
	Currant leaf spot	benomyl, Bordeaux mixture, carbendazim (mbc), chlorothalonil, copper hydroxide, copper sulphate + cufraneb, cupric ammonium carbonate, dodine, mancozeb, maneb + zinc, quinomethionate, triforine, zineb
	Powdery mildew	benomyl, bupirimate, carbendazim (mbc), dinocap, fenarimol, quinomethionate, triadimefon, triforine
	Rust	Bordeaux mixture, copper oxychloride, thiram
Pests	Aphids	chlorpyrifos, demeton-S-methyl, dimethoate, heptenophos, malathion, oxydemeton-methyl, pirimicarb, tar oils
	Big-bud mite	endosulfan
	Blackcurrant leaf midge	azinphos-methyl + demeton-S-methyl sulphone, cyfluthrin, demeton-S-methyl, oxydemeton-methyl
	Capsids	carbaryl, chlorpyrifos, cyfluthrin, cypermethrin, fenitrothion
	Caterpillars	chlorpyrifos
	Clay-coloured weevil	carbaryl
	Earwigs	carbaryl, trichlorfon
	Gall mite	sulphur
	Leaf midges	carbaryl
	Overwintering pests	cresylic acid
	Red spider mites	chlorpyrifos, demeton-S-methyl, dicofol + tetradifon, dimethoate, oxydemeton-methyl, tar oils, tetradifon
	Sawflies	azinphos-methyl + demeton-S-methyl sulphone, carbaryl, cyfluthrin, cypermethrin, diflubenzuron, fenitrothion
	Scale insects	tar oils
	Slugs and snails	methiocarb
	Tortrix moths	carbaryl
	Winter moth	diflubenzuron, tar oils

Weeds	Annual dicotyledons	chlorpropham, chlorthal-dimethyl, diuron, isoxaben, lenacil, MCPB, napropamide, oxadiazon, paraquat, pendimethalin, pentanochlor, propachlor, propyzamide, simazine, sodium monochloroacetate
	Annual grasses	chlorpropham, dalapon, diuron, fluazifop-P-butyl, napropamide, oxadiazon, paraquat, pendimethalin, propachlor, propyzamide, simazine
	Annual meadow grass	lenacil, pentanochlor
	Annual weeds	dichlobenil
	Bindweeds	oxadiazon
	Chickweed	chlorpropham
	Cleavers	napropamide, oxadiazon
	Couch	dalapon
	Creeping bent	paraquat
	Docks	asulam, asulam (off-label)
	Groundsel	napropamide
	Perennial dicotyledons	dichlobenil, MCPB
	Perennial grasses	dalapon, dichlobenil, fluazifop-P-butyl, propyzamide
	Perennial ryegrass	paraquat
	Polygonums	chlorpropham, oxadiazon
	Rough meadow grass	paraquat
	Volunteer cereals	dalapon, fluazifop-P-butyl
	Wild oats	fluazifop-P-butyl

Gooseberries

Diseases	Botrytis	benomyl, carbendazim (mbc), dichlofluanid, vinclozolin
	Currant leaf spot	benomyl, Bordeaux mixture, carbendazim (mbc), chlorothalonil, copper hydroxide, copper sulphate + cufraneb, dodine, mancozeb, quinomethionate
	Powdery mildew	benomyl, bupirimate, carbendazim (mbc), dinocap, fenarimol, quinomethionate, sulphur, triadimefon, triforine
Pests	Aphids	chlorpyrifos, demeton-S-methyl, dimethoate, heptenophos, oxydemeton-methyl, pirimicarb, tar oils
	Blackcurrant leaf midge	azinphos-methyl + demeton-S-methyl sulphone, cyfluthrin
	Capsids	chlorpyrifos, cyfluthrin, cypermethrin, dimethoate, fenitrothion
	Caterpillars	carbaryl, chlorpyrifos
	Currant clearwing	carbaryl
	Red spider mites	chlorpyrifos, demeton-S-methyl, dimethoate, malathion, oxydemeton-methyl, quinomethionate, tetradifon

	Sawflies	azinphos-methyl + demeton-S-methyl sulphone, carbaryl, cyfluthrin, cypermethrin, fenitrothion, malathion, nicotine, rotenone
	Scale insects	tar oils
	Winter moth	tar oils
Weeds	Annual dicotyledons	chlorpropham, chlorthal-dimethyl, diuron, isoxaben, lenacil, napropamide, oxadiazon, paraquat, pendimethalin, pentanochlor, propachlor, propyzamide, simazine, sodium monochloroacetate
	Annual grasses	chlorpropham, dalapon, diuron, fluazifop-P-butyl, napropamide, oxadiazon, paraquat, pendimethalin, propachlor, propyzamide, simazine
	Annual meadow grass	lenacil, pentanochlor
	Annual weeds	dichlobenil
	Bindweeds	oxadiazon
	Chickweed	chlorpropham
	Cleavers	napropamide, oxadiazon
	Couch	dalapon
	Creeping bent	paraquat
	Docks	asulam (off-label)
	Groundsel	napropamide
	Perennial dicotyledons	dichlobenil
	Perennial grasses	dalapon, dichlobenil, fluazifop-P-butyl, propyzamide
	Perennial ryegrass	paraquat
	Polygonums	chlorpropham, oxadiazon
	Rough meadow grass	paraquat
	Volunteer cereals	dalapon, fluazifop-P-butyl
	Wild oats	fluazifop-P-butyl

Grapevines

Diseases	Botrytis	chlorothalonil, dichlofluanid, vinclozolin
	Downy mildew	Bordeaux mixture, chlorothalonil, copper hydroxide, copper oxychloride, copper oxychloride + metalaxyl (off-label), copper sulphate + cufraneb, mancozeb (off-label)
	Powdery mildew	copper sulphate + sulphur, fenarimol, sulphur, triadimefon
Pests	Aphids	cypermethrin (off-label), dimethoate (off-label)
	Capsids	cypermethrin (off-label), dimethoate (off-label)
	General insect control	oxydemeton-methyl (off-label)
	Mealy bugs	petroleum oil
	Red spider mites	dimethoate (off-label), petroleum oil, tetradifon
	Scale insects	petroleum oil, tar oils

	Thrips	cypermethrin (off-label), dimethoate (off-label)
Weeds	Annual dicotyledons	diquat (off-label), oxadiazon, paraquat (off-label), propyzamide (off-label), simazine (off-label)
	Annual grasses	oxadiazon, paraquat (off-label), propyzamide (off-label), simazine (off-label)
	Annual weeds	glyphosate (off-label)
	Bindweeds	oxadiazon
	Cleavers	oxadiazon
	Perennial grasses	propyzamide (off-label)
	Perennial weeds	glyphosate (off-label)
	Polygonums	oxadiazon

Hops

Diseases	Alternaria	carbendazim (mbc) (off-label)
	Canker	benomyl (off-label), carbendazim (mbc) (off-label)
	Downy mildew	Bordeaux mixture, chlorothalonil, copper hydroxide, copper oxychloride, copper oxychloride + metalaxyl, copper sulphate + sulphur, fosetyl-aluminium, zineb
	Powdery mildew	bupirimate, copper sulphate + sulphur, dinocap, penconazole, pyrazophos, sulphur, triadimefon, triforine
Pests	Aphids	aldicarb, amitraz, cypermethrin, deltamethrin, demeton-S-methyl, dimethoate, endosulfan, fenpropathrin, fenvalerate, lambda-cyhalothrin, mephosfolan, omethoate
	Caterpillars	fenpropathrin
	Codling moth	amitraz
	Nematodes	1,3-dichloropropene
	Red spider mites	aldicarb, amitraz, demeton-S-methyl, dicofol, dicofol + tetradifon, dimethoate, fenpropathrin, lambda-cyhalothrin, tetradifon
	Soil pests	gamma-HCH
	Virus vectors	1,3-dichloropropene
Weeds	Annual dicotyledons	diquat + paraquat, oxadiazon, paraquat, pendimethalin, propyzamide (off-label), simazine, sodium monochloroacetate
	Annual grasses	diquat + paraquat, oxadiazon, paraquat, pendimethalin, propyzamide (off-label), simazine
	Bindweeds	oxadiazon
	Cleavers	oxadiazon
	Creeping bent	paraquat
	Docks	asulam
	Perennial grasses	diquat + paraquat, propyzamide (off-label)
	Perennial ryegrass	paraquat
	Polygonums	oxadiazon

	Rough meadow grass	paraquat
	Volunteer cereals	diquat + paraquat
Crop control	Chemical stripping	anthracene oil, diquat, diquat + paraquat, paraquat, tar oils

Plums/cherries/damsons

Diseases	Bacterial canker	copper hydroxide, copper oxychloride, copper sulphate + cufraneb
	Blossom wilt	tar oils, vinclozolin
	Brown rot	carbendazim (mbc)
	Canker	Bordeaux mixture
	Pruning wounds	octhilinone
	Rust	benodanil
	Silver leaf	octhilinone
Pests	Aphids	chlorpyrifos, cyfluthrin, deltamethrin, demeton-S-methyl, dimethoate, fenitrothion, malathion, omethoate, oxydemeton-methyl, pirimicarb, tar oils
	Caterpillars	cyfluthrin, cypermethrin, deltamethrin, fenitrothion
	Cherry fruit moth	dimethoate, tar oils
	Plum fruit moth	azinphos-methyl + demeton-S-methyl sulphone, cyfluthrin, deltamethrin, diflubenzuron, pheromones
	Red spider mites	chlorpyrifos, clofentezine, demeton-S-methyl, dimethoate, malathion, oxydemeton-methyl, phosalone, tetradifon
	Rust mite	pirimiphos-methyl
	Sawflies	azinphos-methyl + demeton-S-methyl sulphone, cyfluthrin, deltamethrin, demeton-S-methyl, dimethoate, fenitrothion, oxydemeton-methyl
	Scale insects	tar oils
	Tortrix moths	azinphos-methyl + demeton-S-methyl sulphone, chlorpyrifos, diflubenzuron
	Winter moth	chlorpyrifos, diflubenzuron, fenitrothion, tar oils
Weeds	Annual dicotyledons	isoxaben, napropamide, pendimethalin, pentanochlor, propyzamide, sodium monochloroacetate
	Annual grasses	napropamide, pendimethalin, propyzamide
	Annual meadow grass	pentanochlor
	Annual weeds	amitrole (off-label), glyphosate
	Cleavers	napropamide
	Docks	asulam, asulam (off-label)
	Groundsel	napropamide
	Perennial grasses	propyzamide
	Perennial weeds	amitrole (off-label), glyphosate
Crop control	Sucker control	glyphosate

Plant growth regulation	Sucker inhibition	1-naphthylacetic acid

Quinces

Weeds	Annual weeds	amitrole (off-label)
	Perennial weeds	amitrole (off-label)

Strawberries

Diseases	Botrytis	benomyl, captan, carbendazim (mbc), chlorothalonil, dichlofluanid, iprodione, thiram, vinclozolin
	Powdery mildew	benomyl, bupirimate, carbendazim (mbc), dinocap, fenarimol, sulphur, triadimefon
	Red core	copper oxychloride + metalaxyl, etridiazole (Scotland only), fosetyl-aluminium, fosetyl-aluminium (spring treatment, off-label), propamocarb hydrochloride
	Verticillium wilt	benomyl
Pests	Aphids	aldicarb, chlorpyrifos, demeton-S-methyl, dimethoate, disulfoton, heptenophos, malathion, nicotine, oxydemeton-methyl, phorate, pirimicarb
	Capsids	carbaryl, cyfluthrin, cypermethrin
	Caterpillars	cypermethrin
	Chafer grubs	carbaryl, gamma-HCH
	Cutworms	carbaryl
	Froghoppers	cyfluthrin
	Leafhoppers	demeton-S-methyl, oxydemeton-methyl
	Leatherjackets	carbaryl, gamma-HCH
	Mites	fenbutatin oxide
	Nematodes	1,3-dichloropropene, aldicarb
	Red spider mites	aldicarb, chlorpyrifos, demeton-S-methyl, dicofol, dicofol + tetradifon, dimethoate, fenbutatin oxide, fenpropathrin, oxydemeton-methyl, quinomethionate, triazophos
	Slugs and snails	methiocarb
	Stem nematodes	1,3-dichloropropene, aldicarb
	Strawberry seed beetle	methiocarb
	Tarsonemid mites	aldicarb, dicofol, dicofol + tetradifon, endosulfan
	Tortrix moths	azinphos-methyl + demeton-S-methyl sulphone, Bacillus thuringiensis, carbaryl, chlorpyrifos, cyfluthrin, fenitrothion, triazophos, trichlorfon
	Vine weevil	carbofuran, chlorpyrifos
	Virus vectors	1,3-dichloropropene
	Weevils	carbaryl
	Wireworms	gamma-HCH

Weeds	Annual dicotyledons	chloroxuron, chlorpropham, chlorpropham + fenuron, chlorthal-dimethyl, clopyralid, diphenamid, diquat + paraquat, lenacil, napropamide, pendimethalin, phenmedipham, propachlor, propyzamide, simazine, terbacil, trifluralin
	Annual grasses	alloxydim-sodium, chlorpropham, chlorpropham + fenuron, diphenamid, diquat + paraquat, ethofumesate, fluazifop-P-butyl, napropamide, pendimethalin, propachlor, propyzamide, sethoxydim, simazine, terbacil, trifluralin
	Annual meadow grass	chloroxuron, lenacil
	Blackgrass	sethoxydim
	Chickweed	chlorpropham, chlorpropham + fenuron, ethofumesate
	Cleavers	ethofumesate, napropamide
	Corn marigold	clopyralid
	Couch	sethoxydim, terbacil
	Creeping thistle	clopyralid
	Docks	asulam (off-label)
	Groundsel	napropamide
	Mayweeds	clopyralid
	Perennial dicotyledons	clopyralid
	Perennial grasses	diquat + paraquat, fluazifop-P-butyl, propyzamide, sethoxydim, terbacil
	Polygonums	chlorpropham
	Volunteer cereals	alloxydim-sodium, diquat + paraquat, fluazifop-P-butyl, sethoxydim
	Wild oats	alloxydim-sodium, fluazifop-P-butyl, sethoxydim
Crop control	Runner desiccation	paraquat

Tree fruit

Diseases	Canker	octhilinone
	Pruning wounds	octhilinone
	Silver leaf	octhilinone, Trichoderma viride
Pests	Aphids	fatty acids, rotenone
	Birds	aluminium ammonium sulphate, ziram
	Damaging mammals	aluminium ammonium sulphate, ziram
	Mealy bugs	fatty acids
	Overwintering pests	cresylic acid
	Scale insects	fatty acids
	Thrips	fatty acids
	Tortrix moths	pheromones
	Whitefly	fatty acids

	Woolly aphid	fatty acids
Weeds	Annual dicotyledons	dicamba + MCPA + mecoprop, mecoprop, paraquat
	Annual grasses	paraquat
	Chickweed	mecoprop
	Cleavers	mecoprop
	Creeping bent	paraquat
	Perennial dicotyledons	dicamba + MCPA + mecoprop, glyphosate (wiper application), mecoprop
	Perennial ryegrass	paraquat
	Rough meadow grass	paraquat

Tree nuts

Diseases	Botrytis	iprodione (off-label)
	Gloeosporium	benomyl (off-label)
Pests	Red spider mites	oxydemeton-methyl
Weeds	Annual weeds	glyphosate (off-label)
	Perennial weeds	glyphosate (off-label)

Flowers and ornamentals

General

Diseases	Alternaria	iprodione (seed treatment)
	Black spot	bupirimate + triforine
	Botrytis	captan, iprodione (off-label), tecnazene, vinclozolin
	Damping off	cupric ammonium carbonate, thiabendazole + thiram, thiram (seed treatment), tolclofos-methyl
	Downy mildew	propamocarb hydrochloride
	Fungus diseases	carbendazim (mbc) (off-label)
	Leaf spots	bupirimate + triforine
	Phytophthora	propamocarb hydrochloride
	Powdery mildew	bupirimate, bupirimate + triforine, fenarimol, imazalil, sulphur, triforine
	Rust	triforine
	Seed-borne diseases	thiabendazole + thiram
	Soil-borne diseases	8-hydroxyquinoline sulphate, metam-sodium
Pests	Adelgids	gamma-HCH

Aphids	aldicarb, cypermethrin, deltamethrin, demeton-S-methyl, dimethoate, gamma-HCH, malathion, nicotine, oxydemeton-methyl, permethrin, pirimicarb, pyrethrins + resmethrin, rotenone
Birds	aluminium ammonium sulphate
Browntail moth	Bacillus thuringiensis
Capsids	cypermethrin, deltamethrin, diazinon, gamma-HCH, nicotine
Caterpillars	carbaryl, cypermethrin, deltamethrin, diflubenzuron, permethrin
Chrysanthemum midge	gamma-HCH
Cutworms	cypermethrin
Damaging mammals	aluminium ammonium sulphate
Earwigs	carbaryl
General insect control	oxydemeton-methyl (off-label)
Leaf miners	cypermethrin (off-label), diazinon, dimethoate, gamma-HCH, heptenophos (off-label), nicotine, oxamyl (off-label), permethrin, pyrazophos (off-label), triazophos (off-label)
Leafhoppers	carbaryl, gamma-HCH, nicotine
Leatherjackets	carbaryl
Mealy bugs	diazinon
Midges	carbaryl
Mites	amitraz (off-label), fenbutatin oxide
Nematodes	metam-sodium
Red spider mites	demeton-S-methyl, diazinon, dicofol, dienochlor, dimethoate, fenbutatin oxide, oxydemeton-methyl, tetradifon
Rhododendron bug	gamma-HCH
Sawflies	nicotine
Scale insects	deltamethrin, diazinon
Sciarid flies	fonofos
Slugs and snails	aluminium sulphate + copper sulphate + sodium tetraborate, metaldehyde
Soil pests	metam-sodium
Springtails	carbaryl, diazinon, gamma-HCH
Symphylids	diazinon, gamma-HCH
Thrips	aldicarb, carbaryl, cypermethrin, deltamethrin, diazinon, malathion, nicotine
Vine weevil	aldicarb, fonofos
Whitefly	aldicarb, carbaryl, cypermethrin, diazinon, permethrin, pyrethrins + resmethrin
Woolly aphid	nicotine

Weeds	Annual dicotyledons	chlorthal-dimethyl, diquat (off-label), diquat + paraquat, metazachlor, paraquat (pre-emergence), trifluralin, trifluralin (off-label)
	Annual grasses	diquat + paraquat, paraquat (pre-emergence), trifluralin, trifluralin (off-label)
	Annual meadow grass	metazachlor
	Annual weeds	ammonium sulphamate (pre-planting), linuron (off-label)
	Blackgrass	metazachlor
	General weed control	metam-sodium
	Groundsel, triazine resistant	metazachlor
	Perennial grasses	diquat + paraquat
	Perennial weeds	ammonium sulphamate (pre-planting)
	Volunteer cereals	paraquat (pre-emergence)
Plant growth regulation	Increasing branching	dikegulac
	Rooting of cuttings	1-naphthylacetic acid, 4-indol-3-ylbutyric acid, dichlorophen + 4-indol-3-ylbutyricacid + 1-naphthylacetic acid, indol-3-ylacetic acid

Annuals/biennials

Weeds	Annual dicotyledons	chlorpropham, chlorpropham + fenuron, pentanochlor
	Annual grasses	chlorpropham, chlorpropham + fenuron
	Annual meadow grass	pentanochlor
	Chickweed	chlorpropham, chlorpropham + fenuron
	Polygonums	chlorpropham

Bedding plants

Diseases	Botrytis	benomyl, carbendazim (mbc), iprodione (off-label), quintozene
	Clubroot	cresylic acid
	Damping off	furalaxyl, quintozene
	Powdery mildew	benomyl, carbendazim (mbc)
	Sclerotinia	quintozene
Pests	General insect control	endosulfan (off-label)
	Mushroom flies	pirimiphos-ethyl
	Sciarid flies	diazinon, pirimiphos-ethyl
Weeds	Annual dicotyledons	chloramben, chloroxuron, trifluralin (off-label)
	Annual grasses	chloramben, trifluralin (off-label)
	Annual meadow grass	chloroxuron
	Annual weeds	linuron (off-label)

Plant growth regulation	Increasing flowering	paclobutrazol
	Stem shortening	chlormequat, daminozide, paclobutrazol

Bulbs/corms

Diseases	Botrytis	benomyl, carbendazim (mbc), chlorothalonil, iprodione (off-label), vinclozolin (pre-storage dip), zineb
	Fire	dichlofluanid, ferbam + maneb + zineb, mancozeb, maneb, maneb + zinc, thiram, zineb
	Fungus diseases	carbendazim (mbc) (off-label), thiabendazole (off-label)
	Fusarium diseases	benomyl, carbendazim (mbc), iprodione (off-label), thiabendazole
	Ink disease	chlorothalonil
	Penicillium rot	benomyl, carbendazim (mbc), vinclozolin (pre-storage dip)
	Phytophthora	etridiazole, propamocarb hydrochloride
	Pythium	etridiazole, propamocarb hydrochloride
	Sclerotinia	benomyl, carbendazim (mbc), quintozene
	Stagonospora	benomyl, carbendazim (mbc)
Pests	Aphids	nicotine
	Birds	aluminium ammonium sulphate
	Bulb scale mite	endosulfan
	Capsids	nicotine
	Damaging mammals	aluminium ammonium sulphate
	Large bulb fly	carbofuran (off-label), disulfoton (off-label)
	Leaf miners	nicotine
	Leafhoppers	nicotine
	Sawflies	nicotine
	Soil pests	aldicarb (off-label)
	Stem and bulb nematodes	1,3-dichloropropene
	Thrips	nicotine
	Woolly aphid	nicotine
Weeds	Annual dicotyledons	bentazone, chlorbufam + chloridazon, chlorpropham, chlorpropham + cresylic acid + fenuron, chlorpropham + diuron, chlorpropham + fenuron, chlorpropham + linuron, chlorpropham + linuron (off-label), chlorpropham + pentanochlor, cyanazine, diphenamid, diquat + paraquat, lenacil, paraquat, pentanochlor
	Annual grasses	chlorpropham, chlorpropham + cresylic acid + fenuron, chlorpropham + diuron, chlorpropham + fenuron, chlorpropham + linuron, chlorpropham + linuron (off-label), cyanazine, diphenamid, diquat + paraquat, paraquat
	Annual meadow grass	chlorpropham + pentanochlor, lenacil, pentanochlor

	Chickweed	chlorpropham, chlorpropham + cresylic acid + fenuron, chlorpropham + fenuron
	Creeping bent	paraquat
	Perennial grasses	diquat + paraquat
	Polygonums	chlorpropham
Plant growth regulation	Increasing flowering	paclobutrazol
	Stem shortening	2-chloroethylphosphonic acid, paclobutrazol

Chrysanthemums

Diseases	Blotch	zineb
	Botrytis	benomyl, captan, carbendazim (mbc), gamma-HCH + tecnazene, iprodione, quintozene, tecnazene, thiram
	Damping off	quintozene
	Petal blight	zineb
	Phoma root rot	nabam
	Powdery mildew	benomyl, bupirimate, carbendazim (mbc), cupric ammonium carbonate, dinocap, triforine
	Ray blight	chlorothalonil, mancozeb, zineb
	Rust	benodanil, oxycarboxin, propiconazole (off-label), thiram, triforine, zineb
	Sclerotinia	quintozene
	Soil-borne diseases	metam-sodium
Pests	Ants	gamma-HCH + tecnazene, propoxur
	Aphids	demeton-S-methyl, gamma-HCH + tecnazene, nicotine, permethrin, propoxur, resmethrin
	Capsids	aldicarb, diazinon, gamma-HCH + tecnazene
	Caterpillars	Bacillus thuringiensis, gamma-HCH + tecnazene, permethrin
	Earwigs	gamma-HCH + tecnazene, propoxur
	Leaf miners	aldicarb, diazinon, permethrin
	Leafhoppers	gamma-HCH + tecnazene
	Mealy bugs	diazinon, gamma-HCH + tecnazene
	Nematodes	aldicarb, metam-sodium
	Red spider mites	aldicarb, demeton-S-methyl, diazinon
	Sawflies	gamma-HCH + tecnazene
	Scale insects	diazinon
	Sciarid flies	resmethrin
	Soil pests	metam-sodium
	Springtails	diazinon, gamma-HCH + tecnazene
	Symphylids	aldicarb, diazinon

	Thrips	aldicarb, diazinon, gamma-HCH + tecnazene
	Whitefly	diazinon, gamma-HCH + tecnazene, permethrin, propoxur, resmethrin
	Woodlice	gamma-HCH + tecnazene, propoxur
Weeds	Annual dicotyledons	chloroxuron, chlorpropham, chlorpropham + pentanochlor, pentanochlor
	Annual grasses	chlorpropham
	Annual meadow grass	chloroxuron, chlorpropham + pentanochlor, pentanochlor
	Chickweed	chlorpropham
	General weed control	metam-sodium
	Polygonums	chlorpropham
Plant growth regulation	Increasing flowering	paclobutrazol
	Stem shortening	daminozide, paclobutrazol

Container-grown stock

Diseases	Phytophthora	etridiazole, propamocarb hydrochloride
Pests	Sciarid flies	diazinon
Weeds	Annual dicotyledons	carbetamide + oxadiazon, chloramben, chloroxuron, napropamide, oxadiazon
	Annual grasses	carbetamide + oxadiazon, chloramben, napropamide, oxadiazon
	Annual meadow grass	chloroxuron
	Cleavers	napropamide
	Groundsel	napropamide
	Liverworts	chlorothalonil
	Mosses	chlorothalonil

Dahlias

Diseases	Botrytis	quintozene
	Sclerotinia	quintozene
Pests	Capsids	aldicarb
	Leaf miners	aldicarb
	Nematodes	aldicarb
	Red spider mites	aldicarb
	Symphylids	aldicarb
	Thrips	aldicarb
Weeds	Annual dicotyledons	lenacil
	Annual meadow grass	lenacil

Perennials

Diseases	Botrytis	dichlofluanid
Weeds	Annual dicotyledons	chloramben, chloroxuron, lenacil, propachlor
	Annual grasses	chloramben, propachlor
	Annual meadow grass	chloroxuron, lenacil

Roses

Diseases	Black spot	captan, carbendazim (mbc), chlorothalonil, dichlofluanid, mancozeb, maneb, maneb + zinc, myclobutanil, triforine
	Powdery mildew	benomyl, bupirimate, carbendazim (mbc), dichlofluanid, dinocap, dodemorph, fenarimol, imazalil, myclobutanil, pyrazophos, triforine
	Rust	benodanil, bupirimate + triforine, mancozeb, maneb + zinc, myclobutanil, oxycarboxin, penconazole, triforine
Pests	Aphids	dimethoate, nicotine, oxydemeton-methyl
	Capsids	diazinon
	Caterpillars	trichlorfon
	Leaf miners	diazinon, dimethoate
	Mealy bugs	diazinon
	Red spider mites	aldicarb, diazinon, fenpropathrin, oxydemeton-methyl
	Scale insects	diazinon, malathion
	Slug sawflies	rotenone
	Springtails	diazinon
	Symphylids	diazinon
	Thrips	diazinon
	Whitefly	diazinon
Weeds	Annual dicotyledons	atrazine, chlorthal-dimethyl, diuron + simazine, lenacil, pentanochlor, propyzamide, simazine
	Annual grasses	atrazine, diuron + simazine, propyzamide, simazine
	Annual meadow grass	lenacil, pentanochlor
	Annual weeds	dalapon + dichlobenil, dichlobenil
	Perennial dicotyledons	dichlobenil
	Perennial grasses	dalapon + dichlobenil, dichlobenil, propyzamide
Plant growth regulation	Increasing flowering	paclobutrazol
	Stem shortening	paclobutrazol
	Sucker inhibition	1-naphthylacetic acid

Trees/shrubs

Diseases	Canker	cresylic acid, octhilinone

131

	Crown gall	cresylic acid
	Dutch elm disease	metam-sodium (off-label), thiabendazole (injection)
	Honey fungus	cresylic acid
	Powdery mildew	penconazole
	Pruning wounds	octhilinone
	Scab	penconazole
	Silver leaf	octhilinone, Trichoderma viride
Pests	Adelgids	gamma-HCH
	Aphids	dimethoate, fatty acids, gamma-HCH
	Birds	ziram
	Browntail moth	permethrin
	Capsids	aldicarb, deltamethrin
	Caterpillars	deltamethrin, diflubenzuron
	Damaging mammals	ziram
	Leaf miners	dimethoate
	Leafhoppers	gamma-HCH
	Mealy bugs	fatty acids
	Red spider mites	aldicarb, dimethoate
	Rhododendron bug	gamma-HCH
	Scale insects	deltamethrin, fatty acids
	Thrips	aldicarb, deltamethrin, fatty acids
	Whitefly	aldicarb, fatty acids
	Woolly aphid	fatty acids
Weeds	Annual dicotyledons	atrazine + cyanazine, chloramben, clopyralid, clopyralid + cyanazine (off-label), diquat + paraquat + simazine, diuron, diuron + paraquat, diuron + simazine, lenacil, napropamide, oxadiazon, paraquat, propachlor, propyzamide, simazine
	Annual grasses	alloxydim-sodium, atrazine + cyanazine, chloramben, diquat + paraquat + simazine, diuron, diuron + paraquat, diuron + simazine, napropamide, oxadiazon, paraquat, propachlor, propyzamide, simazine
	Annual meadow grass	lenacil
	Annual weeds	dalapon + dichlobenil, dichlobenil, glyphosate, glyphosate (pre-planting), glyphosate + simazine
	Bindweeds	oxadiazon
	Cleavers	napropamide, oxadiazon
	Corn marigold	clopyralid
	Couch	glyphosate + simazine
	Creeping bent	paraquat
	Creeping thistle	clopyralid
	Groundsel	napropamide

	Mayweeds	clopyralid
	Perennial dicotyledons	clopyralid, clopyralid + cyanazine (off-label), dichlobenil, diquat + paraquat + simazine, diuron + paraquat
	Perennial grasses	atrazine + cyanazine, dalapon + dichlobenil, dichlobenil, diquat + paraquat + simazine, diuron + paraquat, glyphosate + simazine, propyzamide
	Perennial weeds	glyphosate, glyphosate (pre-planting)
	Polygonums	oxadiazon
	Volunteer cereals	alloxydim-sodium
	Wild oats	alloxydim-sodium
Plant growth regulation	Growth retardation	dikegulac
	Sucker inhibition	1-naphthylacetic acid, maleic hydrazide

Hedges

Pests	Browntail moth	trichlorfon
	Small ermine moth	trichlorfon
Weeds	Annual dicotyledons	diquat + paraquat + simazine
	Annual grasses	dalapon, diquat + paraquat + simazine
	Annual weeds	dalapon + dichlobenil
	Bracken	dalapon + dichlobenil
	Couch	dalapon
	Perennial dicotyledons	diquat + paraquat + simazine
	Perennial grasses	dalapon, dalapon + dichlobenil, diquat + paraquat + simazine
	Rushes	dalapon + dichlobenil
Plant growth regulation	Growth retardation	dikegulac, maleic hydrazide

Hardy ornamental nursery stock

Diseases	Botrytis	chlorothalonil, iprodione (off-label)
	Damping off	captan + fosetyl-aluminium + thiabendazole (off-label)
	Downy mildew	captan + fosetyl-aluminium + thiabendazole (off-label)
	Fungus diseases	carbendazim (mbc) (off-label), prochloraz
	Phytophthora	etridiazole, fosetyl-aluminium, furalaxyl, propamocarb hydrochloride
	Powdery mildew	dodemorph (off-label), drazoxolon (off-label)
	Pythium	furalaxyl
Pests	Capsids	deltamethrin
	Codling moth	amitraz (off-label)

	General insect control	azinphos-methyl + demeton-S-methyl sulphone (off-label), deltamethrin (off-label)
	Mites	amitraz (off-label)
	Nematodes	aldicarb
	Scale insects	deltamethrin
	Sciarid flies	fonofos (off-label)
	Thrips	deltamethrin
	Vine weevil	carbofuran, fonofos (off-label)
Weeds	Annual dicotyledons	atrazine + dalapon, chlorpropham + cresylic acid + fenuron, clopyralid (off-label), diphenamid, diuron, diuron + paraquat, linuron + trietazine (off-label), metazachlor, napropamide + trifluralin (off-label), pentanochlor, propyzamide (off-label)
	Annual grasses	atrazine + dalapon, chlorpropham + cresylic acid + fenuron, diphenamid, diuron, diuron + paraquat, linuron + trietazine (off-label), napropamide + trifluralin (off-label), propyzamide (off-label)
	Annual meadow grass	metazachlor, pentanochlor
	Annual weeds	glyphosate + simazine
	Blackgrass	metazachlor
	Chickweed	chlorpropham + cresylic acid + fenuron
	Couch	glyphosate + simazine
	Docks	asulam (off-label)
	Groundsel, triazine resistant	metazachlor
	Perennial dicotyledons	clopyralid (off-label), diuron + paraquat
	Perennial grasses	atrazine + dalapon, diuron + paraquat, glyphosate + simazine, propyzamide (off-label)
Plant growth regulation	Increasing flowering	paclobutrazol
	Stem shortening	paclobutrazol

Standing ground

Weeds	Annual dicotyledons	chlorpropham + cresylic acid + fenuron
	Annual grasses	chlorpropham + cresylic acid + fenuron
	Chickweed	chlorpropham + cresylic acid + fenuron

Water lily

Diseases	Crown rot	carbendazim + metalaxyl (off-label)

Forestry

Forestry plantations

Pests	Adelgids	gamma-HCH
	Birds	ziram
	Browntail moth	diflubenzuron
	Caterpillars	diflubenzuron
	Damaging mammals	aluminium phosphide, sulphonated cod liver oil, ziram
	Grey squirrels	warfarin
	Large pine weevil	gamma-HCH
	Pine looper	diflubenzuron, resmethrin
	Winter moth	diflubenzuron
Weeds	Annual dicotyledons	2,4-D, 2,4-D + dicamba + triclopyr, atrazine, atrazine + cyanazine, atrazine + dalapon, atrazine + terbuthylazine, clopyralid (off-label), diquat + paraquat, paraquat, propyzamide
	Annual grasses	atrazine, atrazine + cyanazine, atrazine + dalapon, atrazine + terbuthylazine, dalapon, dalapon + di-1-p-menthene, diquat + paraquat, paraquat, propyzamide
	Annual weeds	ammonium sulphamate, dalapon + dichlobenil, glyphosate, glyphosate + simazine, hexazinone
	Bracken	asulam, asulam (off-label), dalapon + dichlobenil, dicamba, glyphosate
	Brambles	2,4-D + dicamba + triclopyr, glyphosate, triclopyr
	Broom	triclopyr
	Couch	dalapon, glyphosate
	Creeping bent	paraquat
	Docks	2,4-D + dicamba + triclopyr, triclopyr
	Firebreak desiccation	paraquat
	Gorse	2,4-D + dicamba + triclopyr, triclopyr
	Heather	2,4-D, glyphosate
	Japanese knotweed	2,4-D + dicamba + triclopyr
	Perennial dicotyledons	2,4-D, 2,4-D + dicamba + triclopyr, ammonium sulphamate, atrazine + terbuthylazine, clopyralid (off-label), clopyralid + triclopyr (off-label), glyphosate, glyphosate (wiper application), hexazinone, triclopyr
	Perennial grasses	ammonium sulphamate, atrazine, atrazine + cyanazine, atrazine + dalapon, atrazine + terbuthylazine, dalapon, dalapon + di-1-p-menthene, dalapon + dichlobenil, diquat + paraquat, glyphosate, glyphosate + simazine, hexazinone, propyzamide
	Rhododendrons	2,4-D + dicamba + triclopyr, ammonium sulphamate, glyphosate, triclopyr

Rushes	atrazine + terbuthylazine, dalapon + dichlobenil, glyphosate
Stinging nettle	2,4-D + dicamba + triclopyr, glyphosate, triclopyr
Volunteer cereals	dalapon + di-1-p-menthene, diquat + paraquat
Woody weeds	2,4-D, 2,4-D + dicamba + triclopyr, ammonium sulphamate, clopyralid + triclopyr (off-label), fosamine-ammonium, glyphosate, hexazinone, triclopyr
Crop control	Chemical thinning	glyphosate

Forest nursery beds

Pests	Aphids	pirimicarb
	Clay-coloured weevil	gamma-HCH
	Cutworms	gamma-HCH
	Nematodes	aldicarb (off-label)
	Poplar leaf beetles	gamma-HCH
	Soil pests	gamma-HCH
	Strawberry root weevil	gamma-HCH
Weeds	Annual dicotyledons	benazolin + clopyralid (off-label), diphenamid, paraquat (stale seedbed), simazine
	Annual grasses	diphenamid, paraquat (stale seedbed), simazine

Transplant lines

Pests	Black pine beetle	chlorpyrifos
	Pine weevil	chlorpyrifos
	Vine weevil	chlorpyrifos
Weeds	Annual dicotyledons	paraquat, simazine
	Annual grasses	paraquat, simazine
	Creeping bent	paraquat

Farm woodland

Weeds	Annual dicotyledons	propyzamide
	Annual grasses	propyzamide
	Perennial grasses	propyzamide

Cut logs/timber

Pests	Ambrosia beetle	chlorpyrifos, gamma-HCH
	Elm bark beetle	chlorpyrifos, gamma-HCH
	Great spruce bark beetle	gamma-HCH
	Larch shoot beetle	chlorpyrifos

Pine shoot beetle   chlorpyrifos

Protected crops

General

Diseases

Botrytis	chlorothalonil, iprodione (off-label)
Fungus diseases	formaldehyde
Phytophthora	fosetyl-aluminium
Powdery mildew	imazalil
Soil-borne diseases	cresylic acid, dazomet, formaldehyde, metam-sodium, methyl bromide with amyl acetate, methyl bromide with chloropicrin

Pests

Ants	cresylic acid, gamma-HCH, pirimiphos-methyl
Aphids	cypermethrin, dimethoate, gamma-HCH, nicotine, oxydemeton-methyl, pirimiphos-methyl, rotenone, Verticillium lecanii
Capsids	aldicarb, cypermethrin, gamma-HCH, nicotine, pirimiphos-methyl
Caterpillars	cypermethrin, diflubenzuron, pirimiphos-methyl
Cutworms	cypermethrin
Earwigs	gamma-HCH, pirimiphos-methyl
General insect control	demeton-S-methyl (off-label), oxydemeton-methyl (off-label)
Leaf miners	cypermethrin, deltamethrin (off-label), gamma-HCH, nicotine, pirimiphos-methyl
Leafhoppers	nicotine, pirimiphos-methyl
Mealy bugs	aldicarb, nicotine, pirimiphos-methyl
Millipedes	nicotine
Mites	dicofol + tetradifon, fenbutatin oxide
Nematodes	aldicarb, dazomet, metam-sodium, methyl bromide with amyl acetate, methyl bromide with chloropicrin
Red spider mites	dicofol, dicofol + tetradifon, fenbutatin oxide, pirimiphos-methyl
Rodents	alphachloralose
Sawflies	nicotine, pirimiphos-methyl
Scale insects	aldicarb
Sciarid flies	gamma-HCH
Slugs and snails	aluminium sulphate + copper sulphate + sodium tetraborate, cresylic acid, metaldehyde, methiocarb (off-label)
Soil pests	dazomet, metam-sodium, methyl bromide with amyl acetate, methyl bromide with chloropicrin
Springtails	nicotine

	Tarsonemid mites	aldicarb, dicofol + tetradifon, pirimiphos-methyl
	Thrips	cypermethrin, gamma-HCH, nicotine, pirimiphos-methyl
	Western flower thrips	deltamethrin (off-label), dichlorvos (off-label), endosulfan (off-label)
	Whitefly	aldicarb, cypermethrin, gamma-HCH, pirimiphos-methyl, Verticillium lecanii
	Woodlice	cresylic acid, gamma-HCH, pirimiphos-methyl
	Woolly aphid	nicotine
Weeds	Algae	(alkylaryl) trimethylammonium chloride, benzalkonium chloride, dichlorophen, dodecylbenzyl trimethylammonium chloride
	General weed control	dazomet, metam-sodium, methyl bromide with amyl acetate, methyl bromide with chloropicrin
	Liverworts	(alkylaryl) trimethylammonium chloride
	Mosses	(alkylaryl) trimethylammonium chloride, benzalkonium chloride
	Slime bacteria	dichlorophen

Pot plants

Diseases	Black root rot	carbendazim (mbc)
	Botrytis	benomyl, captan, carbendazim (mbc), gamma-HCH + tecnazene, iprodione, iprodione (off-label), quintozene, thiram
	Damping off	quintozene
	Fusarium diseases	benomyl, carbendazim (mbc)
	Phytophthora	furalaxyl
	Powdery mildew	benomyl, carbendazim (mbc), pyrazophos, triforine
	Pythium	furalaxyl
	Rhizoctonia	benomyl, carbendazim (mbc), quintozene
	Rust	benodanil, mancozeb, oxycarboxin, triforine
	Sclerotinia	quintozene
	Thielaviopsis	benomyl
Pests	Ants	gamma-HCH + tecnazene, propoxur
	Aphids	fatty acids, gamma-HCH + tecnazene, heptenophos, permethrin, pirimicarb, propoxur, resmethrin
	Capsids	aldicarb, diazinon, gamma-HCH + tecnazene
	Caterpillars	gamma-HCH + tecnazene, permethrin
	Earwigs	gamma-HCH + tecnazene, propoxur
	General insect control	endosulfan (off-label)
	Leaf miners	diazinon, permethrin
	Leafhoppers	gamma-HCH + tecnazene, heptenophos

	Mealy bugs	aldicarb, diazinon, fatty acids, gamma-HCH + tecnazene, petroleum oil
	Mushroom flies	pirimiphos-ethyl
	Red spider mites	aldicarb, diazinon, petroleum oil
	Sawflies	gamma-HCH + tecnazene
	Scale insects	aldicarb, diazinon, fatty acids, petroleum oil
	Sciarid flies	pirimiphos-ethyl, resmethrin
	Springtails	diazinon, gamma-HCH + tecnazene
	Symphylids	diazinon
	Tarsonemid mites	aldicarb
	Thrips	aldicarb, diazinon, fatty acids, gamma-HCH + tecnazene
	Vine weevil	aldicarb, carbofuran
	Whitefly	aldicarb, diazinon, fatty acids, gamma-HCH + tecnazene, permethrin, resmethrin
	Woodlice	gamma-HCH + tecnazene
	Woolly aphid	fatty acids
Weeds	Algae	(alkylaryl) trimethylammonium chloride
	Annual dicotyledons	trifluralin (off-label)
	Annual grasses	trifluralin (off-label)
	Annual weeds	linuron (off-label)
	Liverworts	(alkylaryl) trimethylammonium chloride, dichlorophen
	Mosses	(alkylaryl) trimethylammonium chloride, dichlorophen
Plant growth regulation	Flower induction	2-chloroethylphosphonic acid
	Flower life prolongation	sodium silver thiosulphate
	Improving colour	paclobutrazol
	Increasing branching	2-chloroethylphosphonic acid, dikegulac
	Increasing flowering	paclobutrazol
	Prevention of rooting through	(alkylaryl) trimethylammonium chloride, dodecylbenzyl trimethylammonium chloride
	Stem shortening	chlormequat, chlormequat + choline chloride, daminozide, paclobutrazol

Glasshouse cut flowers

Diseases	Alternaria	iprodione
	Botrytis	dicloran, iprodione, quintozene, thiram
	Fusarium diseases	benomyl, carbendazim (mbc)
	Fusarium wilt	benomyl, carbendazim (mbc)
	Powdery mildew	dinocap, imazalil, triforine

	Rhizoctonia	dicloran, quintozene
	Rust	benodanil, mancozeb, oxycarboxin, thiram, triforine, zineb
	Soil-borne diseases	metam-sodium
	Verticillium wilt	benomyl, carbendazim (mbc)
Pests	Ants	propoxur
	Aphids	aldicarb, demeton-S-methyl, dimethoate, malathion, pirimicarb, propoxur, resmethrin, Verticillium lecanii
	Earwigs	propoxur
	General insect control	endosulfan (off-label)
	Leaf miners	dimethoate
	Leafhoppers	malathion
	Mealy bugs	malathion
	Nematodes	metam-sodium
	Red spider mites	aldicarb, demeton-S-methyl, oxydemeton-methyl
	Scale insects	malathion
	Sciarid flies	resmethrin
	Soil pests	metam-sodium
	Thrips	malathion
	Whitefly	malathion, propoxur, resmethrin, Verticillium lecanii
	Woodlice	propoxur
Weeds	Annual dicotyledons	chloroxuron, pentanochlor
	Annual meadow grass	chloroxuron, pentanochlor
	General weed control	metam-sodium
Plant growth regulation	Basal bud stimulation	2-chloroethylphosphonic acid
	Flower life prolongation	sodium silver thiosulphate
	Stem shortening	daminozide (off-label)

Aubergines

Diseases	Botrytis	iprodione (off-label), vinclozolin
Pests	Aphids	fatty acids, resmethrin, Verticillium lecanii
	Leaf miners	cypermethrin (off-label), heptenophos (off-label), nicotine (off-label), oxamyl (off-label)
	Mealy bugs	fatty acids
	Scale insects	fatty acids
	Sciarid flies	resmethrin
	Thrips	fatty acids
	Whitefly	fatty acids, permethrin, resmethrin, Verticillium lecanii

Woolly aphid	fatty acids

Beans

Pests

Aphids	Verticillium lecanii
Whitefly	Verticillium lecanii

Cucumbers

Diseases

Anthracnose	chlorothalonil
Black root rot	benomyl, carbendazim (mbc)
Botrytis	benomyl, carbendazim (mbc), chlorothalonil, dicloran, iprodione, vinclozolin
Damping off	etridiazole, propamocarb hydrochloride, propamocarb hydrochloride (off-label)
Didymella stem rot	chlorothalonil
Downy mildew	copper oxychloride + metalaxyl (off-label)
Gummosis	chlorothalonil
Powdery mildew	benomyl, bupirimate, carbendazim (mbc), chlorothalonil, cupric ammonium carbonate, dinocap, fenarimol, imazalil, nitrothal-isopropyl + zineb-ethylenethiuram disulphide complex (off-label), pyrazophos (off-label), thiophanate-methyl, triforine (off-label)
Rhizoctonia	dicloran, quintozene
Root rot	propamocarb hydrochloride (off-label)
Stem rot	benomyl

Pests

Ants	propoxur
Aphids	demeton-S-methyl, fatty acids, heptenophos, permethrin, pirimicarb, propoxur, pyrethrins + resmethrin, resmethrin, Verticillium lecanii
Caterpillars	Bacillus thuringiensis, Bacillus thuringiensis (off-label), permethrin
Cutworms	cypermethrin
Earwigs	propoxur
French fly	pirimiphos-methyl
Leaf miners	cypermethrin, cypermethrin (off-label), permethrin, pyrazophos (off-label)
Leafhoppers	heptenophos
Mealy bugs	fatty acids, petroleum oil
Mites	fenbutatin oxide
Red spider mites	demeton-S-methyl, diazinon, dicofol, fenbutatin oxide, oxydemeton-methyl, petroleum oil, tetradifon
Scale insects	fatty acids, petroleum oil
Sciarid flies	resmethrin

Springtails	diazinon, gamma-HCH
Symphylids	diazinon, gamma-HCH
Thrips	deltamethrin, diazinon, fatty acids, gamma-HCH, heptenophos
Whitefly	cypermethrin, diazinon, fatty acids, permethrin, propoxur, pyrethrins + resmethrin, resmethrin, Verticillium lecanii
Woodlice	propoxur
Woolly aphid	fatty acids

Glasshouse cucurbits

Diseases	Botrytis	iprodione (off-label)
	Damping off	propamocarb hydrochloride
Pests	Aphids	disulfoton (off-label), fatty acids, pirimicarb (off-label)
	Bean seed flies	bendiocarb (off-label)
	Leaf miners	cypermethrin (off-label), nicotine (off-label)
	Mealy bugs	fatty acids
	Red spider mites	tetradifon
	Scale insects	fatty acids
	Thrips	fatty acids
	Whitefly	fatty acids
	Woolly aphid	fatty acids

Herbs

Pests	Aphids	disulfoton (off-label)

Lettuce

Diseases	Botrytis	dicloran, gamma-HCH + tecnazene, iprodione, tecnazene, thiram, vinclozolin
	Damping off	tolclofos-methyl
	Downy mildew	fosetyl-aluminium, mancozeb, maneb, metalaxyl + thiram, zineb
	Rhizoctonia	dicloran
Pests	Ants	gamma-HCH + tecnazene
	Aphids	dimethoate, gamma-HCH + tecnazene, malathion, oxydemeton-methyl, pirimicarb, Verticillium lecanii
	Capsids	gamma-HCH + tecnazene
	Caterpillars	gamma-HCH + tecnazene
	Cutworms	cypermethrin
	Earwigs	gamma-HCH + tecnazene

142

	Leaf miners	cypermethrin
	Leafhoppers	gamma-HCH + tecnazene, malathion
	Mealy bugs	gamma-HCH + tecnazene, malathion
	Red spider mites	dicofol
	Sawflies	gamma-HCH + tecnazene
	Scale insects	malathion
	Springtails	gamma-HCH + tecnazene
	Thrips	gamma-HCH + tecnazene, malathion
	Whitefly	cypermethrin, gamma-HCH + tecnazene, malathion, permethrin, Verticillium lecanii
	Woodlice	gamma-HCH + tecnazene
Weeds	Annual dicotyledons	cetrimide + chlorpropham, propyzamide (off-label)
	Annual grasses	cetrimide + chlorpropham, propyzamide (off-label)
	Chickweed	cetrimide + chlorpropham
	Polygonums	cetrimide + chlorpropham

Mushrooms

Diseases	Bubble	zineb
	Cobweb	benomyl, carbendazim (mbc), dichlorophen, prochloraz, zineb
	Dry bubble	benomyl, carbendazim (mbc), chlorothalonil, dichlorophen, prochloraz, thiabendazole
	Fungus diseases	dichlorophen, sodium pentachlorophenoxide
	Fusarium diseases	dichlorophen
	Soil-borne diseases	methyl bromide with amyl acetate
	Trichoderma	benomyl, carbendazim (mbc)
	Wet bubble	benomyl, carbendazim (mbc), chlorothalonil, dichlorophen, prochloraz, thiabendazole
Pests	Mushroom flies	pirimiphos-ethyl
	Nematodes	methyl bromide with amyl acetate
	Sciarid flies	chlorfenvinphos, deltamethrin, diazinon, dichlorvos, diflubenzuron, malathion, nicotine, permethrin, pirimiphos-ethyl, pirimiphos-methyl, pyrethrins + resmethrin, pyrethrins + resmethrin (off-label), resmethrin
	Soil pests	methyl bromide with amyl acetate
Weeds	Mosses	sodium pentachlorophenoxide

Onions, leeks and garlic

Diseases	Fusarium diseases	benomyl (off-label)
Pests	Thrips	deltamethrin (off-label)

Peppers

Diseases	Botrytis	carbendazim (mbc), chlorothalonil, iprodione (off-label), vinclozolin
	Damping off	etridiazole, propamocarb hydrochloride
	Powdery mildew	carbendazim (mbc)
Pests	Aphids	fatty acids, heptenophos, permethrin, pirimicarb, resmethrin, Verticillium lecanii
	Caterpillars	Bacillus thuringiensis, cypermethrin, diflubenzuron, permethrin
	Leaf miners	cypermethrin, heptenophos (off-label), nicotine (off-label), oxamyl (off-label), permethrin
	Mealy bugs	fatty acids
	Red spider mites	tetradifon
	Scale insects	fatty acids
	Sciarid flies	deltamethrin, resmethrin
	Thrips	fatty acids
	Whitefly	cypermethrin, fatty acids, permethrin, resmethrin, Verticillium lecanii
	Woolly aphid	fatty acids

Tomatoes

Diseases	Blight	chlorothalonil, copper hydroxide, copper sulphate + cufraneb, copper sulphate + sulphur, cupric ammonium carbonate, maneb, maneb + zinc, zineb
	Botrytis	benomyl, carbendazim (mbc), chlorothalonil, dichlofluanid, dicloran, gamma-HCH + tecnazene, iprodione, iprodione (off-label), quintozene, tecnazene, thiram, vinclozolin
	Damping off	copper oxychloride, etridiazole, propamocarb hydrochloride, propamocarb hydrochloride (off-label), zineb
	Didymella stem rot	benomyl, captan, carbendazim (mbc), maneb (off-label), vinclozolin
	Foot rot	copper oxychloride
	Fusarium wilt	benomyl, carbendazim (mbc)
	Leaf mould	benomyl, carbendazim (mbc), chlorothalonil, cupric ammonium carbonate, dichlofluanid, maneb, maneb + zinc, tecnazene, zineb
	Phytophthora	copper oxychloride, zineb
	Powdery mildew	bupirimate (off-label), fenarimol (off-label), sulphur (off-label)
	Rhizoctonia	dicloran, quintozene
	Root diseases	etridiazole (off-label)
	Root rot	nabam, propamocarb hydrochloride (off-label), zineb

	Soil-borne diseases	metam-sodium, nonylphenoxypoly(ethyleneoxy)ethanol-iodine complex + thiabendazole
	Verticillium wilt	benomyl, carbendazim (mbc)
Pests	Ants	gamma-HCH + tecnazene, propoxur
	Aphids	aldicarb, demeton-S-methyl, diazinon, dimethoate, fatty acids, gamma-HCH, gamma-HCH + tecnazene, heptenophos, malathion, nicotine, oxamyl, permethrin, pirimicarb, propoxur, pyrethrins + resmethrin, resmethrin
	Capsids	gamma-HCH, gamma-HCH + tecnazene
	Caterpillars	carbaryl, cypermethrin, deltamethrin, diflubenzuron, gamma-HCH + tecnazene, permethrin, trichlorfon (off-label)
	Earwigs	gamma-HCH + tecnazene, propoxur
	Leaf miners	aldicarb, cypermethrin, deltamethrin, diazinon, gamma-HCH, nicotine (off-label), oxamyl, permethrin, trichlorfon (off-label)
	Leafhoppers	gamma-HCH, gamma-HCH + tecnazene, heptenophos, malathion
	Mealy bugs	fatty acids, gamma-HCH + tecnazene, malathion, petroleum oil
	Mites	fenbutatin oxide
	Nematodes	aldicarb, metam-sodium, oxamyl
	Potato cyst nematode	oxamyl
	Red spider mites	aldicarb, demeton-S-methyl, diazinon, dicofol, dimethoate, fenbutatin oxide, oxamyl, oxydemeton-methyl, petroleum oil, tetradifon
	Sawflies	gamma-HCH + tecnazene
	Scale insects	fatty acids, malathion, petroleum oil
	Sciarid flies	resmethrin
	Soil pests	metam-sodium
	Springtails	gamma-HCH, gamma-HCH + tecnazene
	Symphylids	gamma-HCH
	Thrips	carbaryl, diazinon, fatty acids, gamma-HCH, gamma-HCH + tecnazene, malathion
	Tomato moth	Bacillus thuringiensis
	Whitefly	cypermethrin, diazinon, fatty acids, gamma-HCH + tecnazene, malathion, oxamyl, permethrin, propoxur, pyrethrins + resmethrin, resmethrin
	Woodlice	carbaryl, gamma-HCH + tecnazene, propoxur
	Woolly aphid	fatty acids
Weeds	Annual dicotyledons	diphenamid, pentanochlor
	Annual grasses	diphenamid
	Annual meadow grass	pentanochlor
	General weed control	metam-sodium

Plant growth regulation	Fruit ripening	2-chloroethylphosphonic acid
	Increasing fruit set	(2-naphthyloxy)acetic acid

Total vegetation control

Non-crop areas

Pests	Rodents	bromadiolone
Weeds	Algae	(alkylaryl) trimethylammonium chloride, benzalkonium chloride, cresylic acid, dichlorophen
	Annual dicotyledons	2,4-D + dicamba + mecoprop, 2,4-D + dicamba + triclopyr, chlorpropham + cresylic acid + fenuron, clopyralid + mecoprop (off-label), dicamba + mecoprop + triclopyr, diquat + paraquat, MCPA + mecoprop, paraquat, picloram
	Annual grasses	chlorpropham + cresylic acid + fenuron, dalapon, diquat + paraquat, paraquat
	Annual weeds	amitrole, glyphosate
	Bracken	asulam, dicamba, imazapyr, picloram
	Brambles	2,4-D + dicamba + mecoprop, 2,4-D + dicamba + triclopyr, dicamba + mecoprop + triclopyr, triclopyr
	Broom	dicamba + mecoprop + triclopyr, triclopyr
	Chickweed	chlorpropham + cresylic acid + fenuron
	Couch	dalapon
	Creeping bent	paraquat
	Docks	2,4-D + dicamba + triclopyr, dicamba + mecoprop + triclopyr, triclopyr
	Gorse	2,4-D + dicamba + triclopyr, dicamba + mecoprop + triclopyr, triclopyr
	Japanese knotweed	2,4-D + dicamba + triclopyr, picloram
	Lichens	cresylic acid
	Liverworts	(alkylaryl) trimethylammonium chloride, cresylic acid
	Mosses	(alkylaryl) trimethylammonium chloride, benzalkonium chloride, chloroxuron, copper sulphate + sodium tetraborate, cresylic acid, dichlorophen
	Perennial dicotyledons	2,4-D + dicamba + mecoprop, 2,4-D + dicamba + triclopyr, clopyralid + mecoprop (off-label), dicamba + mecoprop + triclopyr, MCPA + mecoprop, picloram, triclopyr
	Perennial grasses	dalapon, diquat + paraquat
	Perennial ryegrass	paraquat
	Perennial weeds	amitrole, glyphosate
	Rhododendrons	2,4-D + dicamba + triclopyr
	Rough meadow grass	paraquat

146

Stinging nettle	2,4-D + dicamba + mecoprop, 2,4-D + dicamba + triclopyr, dicamba + mecoprop + triclopyr, triclopyr	
Total vegetation control	2,4-D + dalapon + diuron, amitrole + 2,4-D + diuron, amitrole + 2,4-D + diuron + simazine, amitrole + atrazine, amitrole + atrazine + 2,4-D, amitrole + atrazine + dicamba, amitrole + atrazine + diuron, amitrole + atrazine + MCPA, amitrole + bromacil + diuron, amitrole + diquat + paraquat + simazine, amitrole + simazine, atrazine, atrazine + 2,4-D + sodium chlorate, atrazine + bromacil + diuron, atrazine + imazapyr, atrazine + sodium chlorate, bromacil, bromacil + diuron, bromacil + pentachlorophenol, bromacil + picloram, dalapon + dichlobenil, dichlobenil, diquat + paraquat + simazine, diuron, diuron + paraquat, diuron + simazine, glyphosate, glyphosate + simazine, imazapyr, simazine, sodium chlorate, tebuthiuron	
Volunteer cereals	dalapon, diquat + paraquat	
Volunteer potatoes	dichlobenil	
Woody weeds	2,4-D + dicamba + mecoprop, 2,4-D + dicamba + triclopyr, dicamba + mecoprop + triclopyr, fosamine-ammonium, picloram, tebuthiuron, triclopyr	

Weeds in or near water

Weeds	Aquatic weeds	2,4-D, dalapon, dalapon (off-label), dalapon (off-label, aerial application), dichlobenil, diquat, glyphosate, terbutryn
	Bracken	asulam
	Couch	dalapon
	Docks	asulam
	Perennial dicotyledons	2,4-D
	Perennial grasses	dalapon, glyphosate
	Rushes	glyphosate
	Sedges	glyphosate
	Waterlilies	glyphosate
	Woody weeds	fosamine-ammonium
Plant growth regulation	Growth suppression	maleic hydrazide

Non-crop pest control

Farm buildings/yards

Diseases	Poultry diseases	formaldehyde + gamma-HCH
Pests	Ants	iodofenphos
	Beetles	carbaryl, carbaryl + pyrethrins, dichlorvos
	Cockroaches	iodofenphos

Crickets	iodofenphos	
Ectoparasites	iodofenphos	
Fleas	carbaryl, carbaryl + pyrethrins, iodofenphos	
Flies	azamethiphos, bioallethrin + permethrin, carbaryl + pyrethrins, dichlorvos, gamma-HCH + resmethrin/tetramethrin, iodofenphos, methomyl, phenothrin, phenothrin + tetramethrin, pyrethrins, tetramethrin	
General insect control	dichlorvos + propoxur, propoxur	
Grain storage pests	iodofenphos	
Hide beetle	iodofenphos	
Lice	carbaryl, carbaryl + pyrethrins	
Mealworms	iodofenphos	
Mosquitoes	dichlorvos, phenothrin + tetramethrin	
Poultry ectoparasites	dichlorvos	
Poultry house pests	fenitrothion, formaldehyde + gamma-HCH	
Red mite	carbaryl, carbaryl + pyrethrins	
Rodents	alphachloralose, bromadiolone, calciferol, calciferol + difenacoum, chlorophacinone, coumatetralyl, difenacoum, warfarin	
Silverfish	iodofenphos	
Spiders	carbaryl, carbaryl + pyrethrins	
Wasps	phenothrin + tetramethrin	

Farmland

Pests	Damaging mammals	aluminium phosphide, sodium cyanide
	Rodents	sodium cyanide

Manure heaps

Pests	Flies	trichlorfon

Refuse tips

Pests	Crickets	iodofenphos
	Flies	iodofenphos

Miscellaneous situations

Pests	Wasps	resmethrin + tetramethrin

PESTICIDE PROFILES

1   alachlor

A soil-acting, residual pre-emergence anilide herbicide

Products	Lasso	Monsanto	480 g/l	EC	01185

Uses   Annual grasses, annual dicotyledons in WINTER OILSEED RAPE, FODDER MAIZE.

Notes   **Efficacy**
- Apply before emergence of weeds or crop, in oilseed rape preferably within 2 d after drilling, in maize just prior to weed emergence
- Rainfall needed after application for optimum results
- Do not use on soils with more than 10% organic matter
- Effective residual life about 8 wk
- May be tank-mixed with approved formulations of atrazine (in forage maize), TCA (in winter oilseed rape) or, in sequence with propyzamide in winter oilseed rape to increase duration of control and/or weed spectrum

Crop safety/Restrictions
- Do not use on direct-drilled crops or apply under cold, adverse growing conditions
- Do not apply to cloddy seedbeds

Special precautions/Environmental safety
- Harmful if swallowed
- Irritating to eyes and skin
- Flammable
- Harmful to fish. Do not contaminate ponds, waterways or ditches with chemical or used container. Do not allow spray to fall within 6m of surface waters or ditches.

Protective clothing/Label precautions
- A, C, H
- 5c, 6a, 6b, 12c, 14, 17, 18, 22, 28, 29, 36, 37, 53, 63, 66, 70

2   aldicarb

A soil-applied, systemic carbamate insecticide and nematicide

Products	Temik 10G	Embetec	10% w/w	GR	04339

Uses   Aphids in BEET CROPS, BRASSICAS, ROOT BRASSICAS, POTATOES, PEAS, CARROTS, PARSNIPS, TOMATOES, STRAWBERRIES, HOPS, ORNAMENTALS, GLASSHOUSE FLOWERS. Docking disorder vectors in SUGAR BEET. Cabbage root fly, flea beetles in BRASSICAS, ROOT BRASSICAS. Capsids in CHRYSANTHEMUMS, DAHLIAS, AMENITY TREES AND SHRUBS, GLASSHOUSE CROPS, POT PLANTS. Free-living nematodes in FIELD BEANS, BROAD BEANS, BEET CROPS, ROOT BRASSICAS, WINTER OILSEED RAPE, MAIZE, PEAS, POTATOES, CARROTS, PARSNIPS, STRAWBERRIES. Frit fly in MAIZE. Leaf miners in BEET CROPS, BRASSICAS, PEAS, TOMATOES, CHRYSANTHEMUMS, DAHLIAS. Mealy bugs, scale insects, tarsonemid mites in GLASSHOUSE CROPS, POT PLANTS. Millipedes, pygmy beetle in BEET CROPS.. Nematodes in TOMATOES, CHRYSANTHEMUMS, DAHLIAS, WOODY

NURSERY STOCK, GLASSHOUSE CROPS. Pea and bean weevils in PEAS, FIELD BEANS, BROAD BEANS. Potato cyst nematode in POTATOES. Red spider mites in HOPS, TOMATOES, STRAWBERRIES, CHRYSANTHEMUMS, DAHLIAS, AMENITY TREES AND SHRUBS, ROSES, CARNATIONS, POT PLANTS. Strawberry mite in STRAWBERRIES. Stem nematodes in STRAWBERRIES, BULB ONIONS, GARLIC. Symphylids in CHRYSANTHEMUMS, DAHLIAS. Thrips in PEAS, CHRYSANTHEMUMS, DAHLIAS, AMENITY TREES AND SHRUBS, ORNAMENTALS, POT PLANTS. Vine weevil in POT PLANTS, ORNAMENTALS. Whitefly in AMENITY TREES AND SHRUBS, ORNAMENTALS, POT PLANTS, GLASSHOUSE CROPS. Frit fly in SWEETCORN *(off-label)*. Soil pests in BULBS, BROCCOLI *(off-label)*. Stem nematodes in LEEKS *(off-label)*. Root nematodes in CONIFER SEEDLINGS *(off-label)*. Red spider mites in CURRANT AND GOOSEBERRY ROOTSTOCKS *(off-label)*.

Notes	**Efficacy**

Efficacy
* Must be incorporated into soil by physical means or irrigation. See label for details of application rates, timing, suitable applicators and techniques of incorporation
* Do not use within 14 d of liming
* Persistence and activity may be reduced in soils with pH above 8.0, in seasons of abnormally high rainfall or where soils are saturated for long periods
* Treatment may not give economic yield response where potato cyst nematode level is very high
* Use in potatoes reduces incidence of spraing disease
* Maiden strawberries should be treated within 7 d of planting or as new growth starts. Stem nematodes will not be controlled in plants already infested

Crop safety/Restrictions
* Application should be made to established strawberries in the autumn after completion of picking. Fruiting crops must not be treated in spring
* Treat tomatoes when established but not later than 2 wk after planting. Do not use on tomatoes grown in peat, growing bags or troughs, or on those grown by NFT methods
* Do not mix with potting compost or apply before planting or potting ornamentals
* On ornamentals of unknown susceptibility treat a small number of plants before committing whole batches
* No edible crops other than those listed (see label) should be planted into treated soil for at least 8 wk after application

Special precautions/Environmental safety
* A chemical subject to the Poisons Rules 1982 and the Poisons Act 1972
* Toxic in contact with skin, by inhalation and if swallowed
* Irritating to eyes, skin and respiratory system
* This product contains an anticholinesterase compound. Do not use if under medical advice not to work with such compounds
* Dangerous to game, wild birds and animals. Cover granules completely and immediately after application. Bury spillages. Failure to bury granules immediately and completely is hazardous to wildlife
* Dangerous to fish. Do not contaminate ponds, waterways or ditches with chemical or used container
* Keep unprotected persons out of treated glasshouses for at least 1 d
* Empty container completely and dispose of safely

FOR FULL CONDITIONS OF USE ALWAYS READ THE PRODUCT LABEL

- Do not market pot plants within 4 wk of treatment

Protective clothing/Label precautions
- A, B, C, or D+E, H, K, M
- 2, 4a, 4b, 4c, 6a, 6b, 6c, 14, 16, 18, 22, 25, 28, 29, 36, 37, 38, 40, 45, 52, 64, 68, 79

Withholding period
- Dangerous to livestock. Keep all livestock out of treated areas for at least 13 wk. Bury or remove spillages

Harvest interval
- Tomatoes 6 wk; potatoes 8 wk; brassicas 10 wk; carrots, parsnips 12 wk; onions do not harvest until mature bulb stage

Approval
- Off-label approval to Feb 1992 for use on sweetcorn (OLA 0153/89), bulbs (OLA 0154/89), leeks (OLA 155/89), broccoli (OLA 156/89), conifer seedlings (OLA 0157/89), currant and gooseberry rootstocks (OLA 0158/89)

3   aldicarb + gamma-HCH
A soil and systemic insecticide and nematicide for use in sugar beet

Products	Sentry	Embetec	8.3:1.7% w/w	GR	04419

Uses

Aphids, docking disorder vectors, leaf miners, millipedes, pygmy beetle, springtails, symphylids, wireworms in SUGAR BEET.

Notes

Efficacy
- Apply to soil using recommended types of granule applicator
- Do not use within 14 d of liming or chalking or on sands
- Activity and persistence may be reduced in soils with pH above 8.0 in seasons of abnormally high rainfall or if soil remains saturated for long periods
- For best results apply as a narrow band exactly within the furrow before covering. Granules and seed must be covered completely

Crop safety/Restrictions
- Do not plant edible crops other than sugar beet within 8 wk of treatment
- Do not use if potatoes or carrots are to be planted within 18 mth

Special precautions/Environmental safety
- A chemical subject to the Poisons Rules 1982 and the Poisons Act 1972
- Toxic in contact with skin, by inhalation and if swallowed
- This product contains an anticholinesterase compound. Do not use if under medical advice not to work with such compounds
- Dangerous to game, wild birds and animals. Cover granules completely and immediately after application. Bury spillages
- Dangerous to fish. Do not contaminate ponds, waterways or ditches with chemical or used container

Protective clothing/Label precautions
- A, B, C or D+E, H, K, M
- 2, 4a, 4b, 4c, 14, 16, 18, 22, 25, 28, 29, 36, 37, 40, 45, 52, 64, 68, 70, 79

• Dangerous to livestock. Keep all livestock out of treated areas for at least 13 wk

4   aldrin

A persistent organochlorine insecticide approvals for sale, supply, storage and use of which were revoked in May 1989

5   (alkylaryl) trimethylammonium chloride

A quaternary ammonium horticultural algicide and moss-killer

Products					
1 Gloquat C		Fargro	500 g/l	SL	04093
2 Gloquat C		Flowering Plants	500 g/l	SL	02527

Uses   Algae, mosses, liverworts in POT PLANTS, CAPILLARY BENCHES, SAND BEDS [1, 2]. Algae, mosses, liverworts in PATHS [2]. Prevention of rooting through in POT PLANTS [1, 2].

Notes   **Efficacy**
• For use on capillary beds or benches apply in sufficient water to distribute chemical throughout substrate
• One treatment controls algae from the start to midway through season
• Treatment of paths, stonework etc will control algae and mosses for about 10 mth

Crop safety/Restrictions
• Do not apply to capillary substrates by spraying
• Some unrooted or freshly potted cuttings may be damaged if placed on freshly treated beds. If in doubt allow 2 d between treatment and placing pots on beds
• If re-treating beds or benches remove plants before applying

Special precautions/Environmental safety
• Harmful if swallowed and in contact with skin. Irritating to skin and eyes
• Harmful to fish. Do not contaminate ponds, waterways or ditches with chemical or used container

Protective clothing/Label precautions
• A, C, H
• 5a, 5c, 6a, 6b, 14, 18, 27, 29, 36, 37, 52, 63, 65, 78

6   alloxydim-sodium

A translocated, post-emergence herbicide for control of grass weeds

Products					
Clout		Embetec	75% w/w	SP	03746

FOR FULL CONDITIONS OF USE ALWAYS READ THE PRODUCT LABEL

Annual grasses, volunteer cereals, wild oats in FIELD BEANS, BEET CROPS, SWEDES, TURNIPS, PEAS, POTATOES, CARROTS, PARSNIPS, BULB ONIONS, SALAD ONIONS, DRILLED LEEKS, STRAWBERRIES, WOODY ORNAMENTALS.

Notes

Efficacy
* Apply when weeds are in active growth and have adequate leaf area (annuals from 3-leaf stage) and soil moisture adequate
* Couch can be suppressed by spraying when longest leaves are 30 cm long
* At least 2 h dry weather needed after spraying and do not cultivate for at least 7 d
* Do not use after other post-emergence herbicides
* Annual meadow-grass, ryegrasses and red fescue are not controlled

Crop safety/Restrictions
* Annual crops may be treated from 2-leaf stage onwards
* Carrots, turnips, field beans and peas may be drilled 1 d after application, grass and cereals should not be sown for at least 4 wk
* Peas may be damaged where leaf wax insufficient
* See label for permitted tank mixes and sequential treatments

Special precautions/Environmental safety
* Irritating to eyes and skin

Protective clothing/Label precautions
* A, C
* 6a, 6b, 18, 22, 28, 29, 36, 37, 54, 63, 67

Harvest interval
* beet crops 8 wk; other crops 4 wk

7 alphachloralose

A narcotic rodenticide used to kill mice

Products   Alphachloralose Concentrate   Rentokil   16.6% w/w   CB   01721

Uses   Mice in FARM BUILDINGS, GLASSHOUSES.

Notes

Efficacy
* Apply ready-mixed bait or concentrate mixed with suitable bait material in shallow trays where mouse droppings observed
* Lay bait at several points not more than 1.5 m apart
* Leave in position for several days until mouse activity ceases
* After treatment, clear up poison and bury residue

Special precautions/Environmental safety
* A chemical subject to the Poisons Rules 1982 and Poisons Act 1972
* Toxic if swallowed
* Do not use outside
* Prevent access by children and animals, particularly cats and dogs
* Should a domestic animal be affected keep the animal warm and quiet

Protective clothing/Label precautions
• 4c, 25, 29, 37, 64, 81, 82, 83, 84

8 alphacypermethrin

A contact and ingested pyrethroid insecticide for arable crops

Products Fastac Shell 100 g/l EC 02659

Uses Aphids, yellow cereal fly in WINTER CEREALS. Cabbage stem flea beetle, pollen beetles, seed weevil, bladder pod midge in OILSEED RAPE. Cutworms in POTATOES, SUGAR BEET.

Notes **Efficacy**
• Also provides rapid and lasting control of many other beetles, weevils and caterpillars
• For cabbage stem flea beetle control spray oilseed rape when damage first seen and about 1 mth later. If no adult feeding damage, spray when larval damage seen on leaf petioles and 1 mth later
• For flowering pests on oilseed rape apply at any time during flowering, on pollen beetle best results achieved at green to yellow bud stage (GS 3,3-3,7), on seed weevil between 20 pods set stage and 80% petal fall(GS 4,7-5,8)
• Spray cereals for yellow cereal fly at egg hatch (normally Feb/Mar). Spray in autumn for control of cereal aphids. (See label for details)
• For cutworm control spray at egg hatch and repeat if necessary
• Product has minimum effect on bees and other beneficial insects

Crop safety/Restrictions
• Apply up to 3 sprays on oilseed rape, up to 2 from green bud to end of flowering (GS 3,3-4,9), 1 after yellow bud (GS 3,7)
• Treatment presents minimal hazard to bees but on flowering crops spray in evening, early morning or in dull weather as a precaution
• Apply a maximum of 2 sprays on cereals in autumn or 1 in spring
• Apply a maximum of 2 sprays on potatoes or sugar beet

Special precautions/Environmental safety
• Harmful in contact with skin or if swallowed
• Irritating to skin. Risk of serious damage to eyes
• Flammable
• Dangerous to bees. Do not apply at flowering stage except as directed on oilseed rape
• Extremely dangerous to fish. Do not contaminate ponds, waterways or ditches with chemical or used container
• Do not apply from air within 250 m of any watercourse

Protective clothing/Label precautions
• A, C, H
• 5a, 5c, 6b, 9, 12c, 14, 16, 18, 22, 28, 29, 36, 37, 48, 51, 60, 63, 66, 70, 78

FOR FULL CONDITIONS OF USE ALWAYS READ THE PRODUCT LABEL

Harvest interval
• zero

Approval
• Provisional Approval for aerial application to limited area of pre-flowering or flowering oilseed rape. Contact supplier for allocation before applying

---

9   aluminium ammonium sulphate
An inorganic bird and animal repellent

Products

1 Curb Crop Spray Powder	Sphere	88% w/w	WP	02480
2 Curb Seed Dressing Powder	Sphere	88% w/w	DP	02480
3 Guardsman L Crop Spray	Sphere		SC	03605
4 Guardsman STP Seed Dressing Powder	Sphere	88% w/w	DP	03606
5 Liquid Curb Crop Spray	Sphere		SC	03164
6 Mandops Narsty	Mandops		SC	02860
7 Stay Off	Synchemicals	88% w/w	WP	02019

Uses

Birds, damaging mammals in FIELD CROPS, VEGETABLES, FRUIT CROPS, ORNAMENTALS [1, 3, 5]. Birds, damaging mammals in SEEDS, BULBS, CORMS [2, 4, 7]. Birds, rabbits in CEREALS, OILSEED RAPE, BRASSICAS, PEAS, FRUIT TREES, BUSH FRUIT [6, 7].

Notes

Efficacy
• Apply as overall spray to growing crops before damage starts or mix powder with seed depending on type of protection required
• Spray deposit protects growth present at spraying but only gives limited protection to new growth
• Product must be sprayed onto dry foliage to be effective and must dry completely before dew or frost forms. In winter this may require some wind
• Use of additional sticker (coating agent) recommended to prolong activity [7]

Protective clothing/Label precautions
• 29, 54, 63, 67, [1-7]; 73, 74 [2, 4, 7]

Harvest interval
• fruit crops 6 wk; other crops 4 wk [6]

---

10   aluminium phosphide
A poisonous gassing compound used to kill moles, rabbits etc.

Products

1 Phostoxin	Rentokil	57% w/w	GE	01775
2 Power Phosphine Pellets	Power	57% w/w	GE	04482
3 Talunex	Power	57% w/w	GE	04739

Uses

Rabbits, moles in FARMLAND, FORESTRY.

Efficacy
- Product releases poisonous hydrogen phosphide gas in contact with moisture
- Place pellets in burrows or runs and seal hole by heeling in or covering with turf. Do not cover pellets with soil
- Inspect daily and treat any new or re-opened holes
- Apply only via the 'Rodentex' [2],or Topex [3], applicator

Special precautions/Environmental safety
- A Part 1 Schedule 1 Poison under the Poisons Rules 1982. Only to be used by farmers, growers, foresters or other qualified users in the course of their business
- Very toxic by inhalation, in contact with skin and if swallowed
- Product liberates toxic, highly flammable gas
- Spontaneous combustion can arise due to sudden release of phosphine gas if container, having been opened once, is then re-opened
- Wear suitable protective gloves (synthetic rubber/plastics) when handling product
- Do not open container except for immediate usage. Open container outdoors. Keep away from liquids or water as this causes immediate release of gas. Do not use in wet weather
- Do not use within 10 m of human or animal habitation
- Pellets must not be placed or allowed to remain on ground surface
- Do not use adjacent to watercourses
- Dangerous to fish. Do not contaminate ponds, waterways or ditches with chemical or used container
- Dust remaining after decomposition is harmless and of no environmental hazard
- Highly flammable

Protective clothing/Label precautions
- A
- 3a, 3b, 3c, 12b, 19, 22, 23, 25, 26, 27, 28, 29, 37, 61, 64, 65. See label for full precautions

11 aluminium sulphate + copper sulphate + sodium tetraborate

An inorganic salt mixture for slug and snail control

Products	Nobble	Fieldspray	SL	03498

Uses   Slugs, snails in FIELD CROPS, VEGETABLES, FRUIT CROPS, ORNAMENTALS, GLASSHOUSE CROPS.

Notes   **Efficacy**
- Apply as spray to soil surface. Product has contact effect on slugs and snails and their eggs
- Best results achieved during mild, damp weather when slugs and snails active

FOR FULL CONDITIONS OF USE ALWAYS READ THE PRODUCT LABEL

Special precautions/Environmental safety
* Product of low toxicity to domestic animals and wildlife

Protective clothing/Label precautions
* No information

12   2-aminobutane
A fumigant fungicide for stored potatoes

Products	CSC 2-Aminobutane	CSC	720 g/l	SL	03224

Uses   Skin spot, gangrene, silver scurf in STORED SEED POTATOES.

Notes

Efficacy
* Treatment must only be carried out by trained operators in suitable fumigation chambers under licence from the British Technology Group
* Fumigate within 21 d of lifting

Crop safety/Restrictions
* Do not treat immature tubers and allow a period of healing before treating damaged tubers

Special precautions/Environmental safety
* Harmful by inhalation. Irritating to skin, eyes and respiratory system
* Highly flammable. Keep away from sources of ignition. No smoking
* Do not empty into drains
* Do not contaminate ponds, waterways or ditches with chemical or used container
* Treated potatoes must not be supplied for consumption by humans or lactating dairy cows
* Use must be in accordance with approved Code of Practice for the Control of Substances Hazardous to Health: Fumigation Operations
* The quantity of potatoes to be fumigated in a single stack must not exceed 2000 tonnes

Protective clothing/Label precautions
* A, C,
* 5a, 6a, 6b, 6c, 12b, 14, 16, 18, 24, 25, 28, 29, 36, 37, 54, 64, 65, 78

13   amitraz
A formamidine acaricide and insecticide for top fruit and hops

Products	Mitac 20	Schering	200 g/l	EC	03870

Uses   Aphids, codling moth, red spider mites in APPLES, PEARS, HOPS. Pear sucker in PEARS. Mites in ORNAMENTALS *(off-label)*. Mites, codling moth in HARDY ORNAMENTAL NURSERY STOCK, NURSERY FRUIT TREES AND BUSHES *(off-label)*.

Efficacy
- For red spider mites on apples and pears spray at 60-80% egg hatch and repeat 3 wk later, on hops when numbers build up and repeat every 3 wk as necessary
- For pear sucker control spray when observed and 3 wk later if necessary
- Best results achieved in dry conditions, do not spray if rain imminent

Crop safety/Restrictions
- Maximum total dose per year on apples and pears must not exceed 7 l/ha

Special precautions/Environmental safety
- Harmful in contact with skin and if swallowed
- Flammable
- Harmful to fish. Do not contaminate ponds, waterways or ditches with chemical or used container

Protective clothing/Label precautions
- A, C, H, J, L, M
- 5a, 5c, 12c, 14, 16, 18, 22, 25, 28, 29, 35, 36, 37, 53, 64, 66, 70, 78

Harvest interval
- apples, pears 2 wk; hops 7 wk

Approval
- off-label approval to March 1992 for use on ornamentals and nursery stock (OLA 0292, 0293, 0294/88)

14 amitrole

A translocated, foliar-acting, non-selective triazole herbicide

Products

1 MSS Aminotriazole Technical	Mirfield	98% w/w	TC	04374
2 MSS Aminotriazole 80 WP	Mirfield	80% w/w	WP	04645
3 Weedazol-TL	Bayer	225 g/l	SL	02979

Uses

Annual weeds, perennial weeds, coltsfoot, hogweed, horsetail in NON-CROP AREAS [1, 2]. Couch, docks, annual weeds, perennial weeds in FIELD CROPS *(pre-sowing, autumn stubble)* [3]. Couch, docks, annual weeds, perennial weeds in FALLOWS, HEADLANDS [3]. Couch in WINTER WHEAT *(direct-drilled)* [3]. Volunteer potatoes in BARLEY *(stubble)* [3]. Couch, docks, creeping thistle, annual weeds, perennial weeds in APPLES, PEARS [3]. Annual weeds, perennial weeds in CHERRIES, PLUMS, PEACHES, QUINCES, APRICOTS *(off-label)* [3].

Notes

Efficacy
- In non-crop land may be applied at any time from Apr to Oct. Best results achieved in spring or early summer when weeds growing actively. For coltsfoot, hogweed and horsetail summer and autumn applications are preferred [1, 2]
- In cropland apply when couch in active growth and foliage at least 7.5 cm high [3]

FOR FULL CONDITIONS OF USE ALWAYS READ THE PRODUCT LABEL

- In fallows and stubble plough 3-6 wk after application to depth of 20 cm, taking care to seal the furrow [3]

Crop safety/Restrictions
- Keep off suckers or foliage of desirable trees or shrubs [1, 2]
- Do not spray areas in which the roots of adjacent trees or shrubs extend [1, 2]
- Do not spray on sloping ground when rain imminent and run-off may occur [1, 2]
- Allow specified interval between treatment and sowing crops (see label) [3]
- Apply in autumn on land to be used for spring barley [3]
- Do not sow direct-drilled winter wheat less than 2 wk after application [3]
- Apply round established fruit trees and before 30 June or after harvest. Keep off trees [3]

Special precautions/Environmental safety
- Harmful to fish. Do not contaminate ponds, waterways or ditches with chemical or used container [1, 2]

Protective clothing/Label precautions
- A, C
- 22, 28, 29, 53, 63, 66, [1, 2]; 22, 28, 36, 54, 63, 66 [3]

Approval
- Off-label approval to Feb. 1991 for use in cherry, plum, peach, quince and apricot orchards (OLA 1395/88) [3]

---

15   amitrole + atrazine

A total herbicide mixture of translocated and residual chemicals

Products					
1 Atlazin CDA	Chipman	218:250 g/l	SC	03456
2 Atlazin Flowable	Chipman	218:250 g/l	SC	00159
3 Boroflow A/ATA	ABM	160:270 g/l	SC	04297
4 Chipman Path Weedkiller	Chipman	137:166 g/l	SC	03956
5 New Atraflow Plus	RP Environ.	160:270 g/l	SC	03624

Uses

Total vegetation control in NON-CROP AREAS [1-5].

Notes

Efficacy
- Apply with conventional or CDA sprayers depending on formulation at any time during growing season
- Best results achieved in May-Jun [1-3, 5], Mar-May [4], when weeds growing actively and have sufficient foliage to absorb chemical
- Effectiveness may be reduced if applied in drought conditions
- Do not spray when rain falling or imminent

Crop safety/Restrictions
- Keep off foliage of wanted plants and avoid spraying ground under which roots of valuable trees or shrubs extend
- Heavy rain on slopes may wash chemical onto wanted vegetation
- Do not sow or plant crops or ornamentals for 2-3 yr after treatment

Protective clothing/Label precautions
- A, C, K [1]; A, C, [2-5]

• 21, 28, 29, 54, 65, [1-5]; 14, 66 [1]

Approval
• May be applied through CDA equipment. See introductory notes [1]

16   amitrole + atrazine + 2,4-D
A total herbicide mixture of translocated and residual chemicals

Products

1 CDA Python	CDA Chemicals	105:207:100 g/l	EC	04080
2 Herbazin Special	Fisons	10:30:25% w/w	WP	00884
3 Primatol AD 85WP	Ciba-Geigy	20:40:25% w/w	WP	01636
4 Snapper CDA	ICI Professional	95:190:99 g/l	EC	02943
5 Terramix CDA	Denoon	99:190:95 g/l	EC	04315
6 Terramix CDA	Powaspray	99:190:95 g/l	EC	04315
7 Terranox	Agri-Technics	105:207:100 g/l-	EC	02105
8 Torapron	BP	95:190:99 g/l	EC	02151

Uses   Total vegetation control in NON-CROP AREAS [1-8].

Notes   **Efficacy**
• Apply from Apr onwards when weeds are growing rapidly and have sufficient leaf area to absorb the chemical
• Apply CDA products undiluted with suitable CDA sprayer

Crop safety/Restrictions
• Do not use on ground under which roots of valuable trees or shrubs are growing
• Do not plant on land treated within last 2 yr or longer depending on dose
• Heavy rain on slopes may wash chemical onto wanted vegetation
• Special care needed with CDA formulations to clean sprayer thoroughly after use with detergent or cleaning fluid

Special precautions/Environmental safety
• Irritating to eyes, respiratory system and skin [1, 4-8], to eyes and skin [3], to eyes [2]
• Keep livestock out of treated areas until poisonous weeds such as ragwort have died and become unpalatable
• Harmful to fish. Do not contaminate ponds, waterways or ditches with chemical or used container

Protective clothing/Label precautions
• A, E, M [1, 4-8]; A [2, 3]
• 6a, 6b, 6c, 14, 18, 21, 28, 29, 36, 37, 43, 53, 63, 65 [1, 4-8]; 6a, 18, 21, 28, 29, 43, 53, 63, 67 [2]; 6a, 6b, 18, 22, 28, 29, 34, 36, 37, 43, 53, 63, 67, 70 [3]

Approval
• May be applied through CDA equipment. See introductory notes [1, 4, 5, 6, 8]

FOR FULL CONDITIONS OF USE ALWAYS READ THE PRODUCT LABEL

17   amitrole + atrazine + dicamba
A total herbicide mixture for CDA treatment of established weeds

Products   Sarapron   BP   98.8:197.5:20 g/l   EC   03300

Uses   Total vegetation control in NON-CROP AREAS.

Notes   **Efficacy**
- Apply undiluted with suitable CDA sprayer from Apr onwards when weeds are growing rapidly and have sufficient leaf area to absorb the chemical

Crop safety/Restrictions
- Do not spray under valuable trees or shrubs or close to sensitive plants
- Do not plant on land treated within last 2-3 yr
- Heavy rain may cause run-off on impermeable or compacted surfaces
- After use clean sprayer thoroughly with detergent or cleaning fluid and dispose of washings on non-crop land

Special precautions/Environmental safety
- Irritating to eyes, respiratory system and skin
- Harmful to fish. Do not contaminate ponds, waterways or ditches with chemical or used container

Protective clothing/Label precautions
- A, E, M
- 6a, 6b, 6c, 18, 21, 28, 29, 36, 37, 53, 63, 65

Approval
- May be applied through CDA equipment. See introductory notes

18   amitrole + atrazine + diuron
A total herbicide mixture for early season application

Products

1 Bullseye CDA	ICI Professional	61:139:177 g/l	UL	02930
2 CDA Viper	CDA Chemicals	68:152:189 g/l	UL	04059
3 Duranox	Agri-Technics	68:152:189 g/l	UL	00773
4 Kagolin 5.8FG	Ciba-Geigy	2:0.8:3% w/w	GR	04578
5 Rassamix CDA	Denoon	61:139:177 g/l	UL	04317
6 Rassamix CDA	Powaspray	61:139:177 g/l	UL	04317
7 Rassapron	BP	61:139:177 g/l	UL	01692

Uses   Total vegetation control in NON-CROP AREAS.

Notes   **Efficacy**
- Apply in Feb, Mar or Apr before weeds have become well established
- Apply undiluted with CDA sprayer [1-3, 5-7]

Crop safety/Restrictions
- Avoid drift on to cultivated plants

• Do not plant on land treated within last 2-3 yr
• Heavy rain may cause run-off on impermeable or compacted surfaces

Special precautions/Environmental safety
• Irritating to eyes, respiratory system and skin [1-3, 5-7]
• Keep livestock out of treated areas until poisonous weeds such as ragwort have died and become unpalatable
• Harmful to fish. Do not contaminate ponds, waterways or ditches with chemical or used container

Protective clothing/Label precautions
• A, E, M [1-3, 5-7]; A, C [4]
• 6a, 6b, 6c, 14, 18, 21, 28, 36, 37, 43, 53, 57, 63, 65

Approval
• May be applied through CDA equipment. See introductory notes [1-3, 5-7]

19 amitrole + atrazine + MCPA
A total herbicide mixture of translocated and residual chemicals

Products					
1 Meganox Plus	Agri-Technics	100:250:100 g/l	EC	04305
2 Serramix CDA	Denoon	100:200:100 g/l	EC	04305
3 Serramix CDA	Powaspray	100:200:100 g/l	EC	04305

Uses   Total vegetation control in NON-CROP AREAS.

Notes   **Efficacy**
• Apply through conventional or CDA sprayer at any time during growing season from early spring to autumn

Crop safety/Restrictions
• For use on non-crop land only

Special precautions/Environmental safety
• Irritating to eyes, respiratory system and skin
• Keep livestock out of treated areas until poisonous weeds such as ragwort have died and become unpalatable
• Harmful to fish. Do not contaminate ponds, waterways or ditches with chemical or used container

Protective clothing/Label precautions
• A, C, H, M
• 6a, 6b, 6c, 14, 18, 22, 28, 29, 36, 37, 43, 53, 63, 65

Approval
• May be applied through CDA equipment. See introductory notes [1-3]

FOR FULL CONDITIONS OF USE ALWAYS READ THE PRODUCT LABEL

20 amitrole + bromacil + diuron

A total herbicide mixture of translocated and residual chemicals

Products	BR Destral	ABM	52.8:17.8:17.8% w/w	WP	04286

Uses

Total vegetation control in NON-CROP AREAS, RAILWAY TRACKS.

Notes

Efficacy
* Apply as foliage spray at medium to high volume at any time during growing season
* Ensure continuous mechanical or hydraulic agitation during spraying to prevent settling

Crop safety/Restrictions
* Do not apply or drain or flush equipment on or near young trees, shrubs or other desirable plants or over areas where their roots may extend
* Do not use where chemical may be washed or move into contact with roots of desirable plants

Special precautions/Environmental safety
* Irritating to skin, eyes and respiratory system
* Keep livestock out of treated areas until poisonous weeds such as ragwort have died and become unpalatable
* Harmful to fish. Do not contaminate ponds, waterways or ditches with chemical or used container

Protective clothing/Label precautions
* A, E, F, H
* 6a, 6b, 6c, 16, 18, 23, 24, 28, 29, 36, 37, 43, 53, 63, 67

21 amitrole + 2,4-D + diuron

A total herbicide mixture of translocated and residual chemicals

Products	Trik	Smyth-Morris	WP	02182

Uses

Total vegetation control in NON-CROP AREAS.

Notes

Efficacy
* Apply in spring or late summer/early autumn when weeds are growing actively and have sufficient leaf area to absorb chemical
* Apply maintenance treatment if necessary at lower rate when weeds 7-10 cm high
* Increase rate on areas of peat or high carbon content

Crop safety/Restrictions
* Do not use on ground under which roots of valuable trees or shrubs are growing

Special precautions/Environmental safety
* Keep livestock out of treated areas until poisonous plants such as ragwort have died and become unpalatable

- Harmful to fish. Do not contaminate ponds, waterways or ditches with chemical or used container

Protective clothing/Label precautions
- A
- 28, 29, 43, 53, 63, 65

22 amitrole + 2,4-D + diuron + simazine
A total herbicide mixture of translocated and residual chemicals

Products

1 Hytrol	Agrichem	14:7:23:25% w/w	WP	01104
2 Hytrol	FCC	14:7:23:25% w/w	WP	01104
3 Weedkill	Dermaglen		TW	02356

Uses Total vegetation control in NON-CROP AREAS.

Notes **Efficacy**
- Apply from Apr to Sep when weeds are growing rapidly and have sufficient leaf area to absorb the chemical. Provides control for up to 2 yr
- Do not use on highly organic soils

Crop safety/Restrictions
- Do not use on ground under which roots of valuable trees and shrubs are growing
- Heavy rain may cause run-off on impermeable or compacted surfaces

Special precautions/Environmental safety
- Irritating to eyes, respiratory system and skin
- Not to be used on food crops
- Keep livestock out of treated areas until poisonous plants such as ragwort have died and become unpalatable
- Harmful to fish. Do not contaminate ponds, waterways or ditches with chemical or used container

Protective clothing/Label precautions
- A, C
- 6a, 6b, 6c, 18, 21, 28, 29, 34, 36, 37, 43, 53, 63, 67, 70

23 amitrole + diquat + paraquat + simazine
A translocated, contact and residual total herbicide mixture

Products Groundhog   ICI Professional   3:2.5:2.5:5% w/w   WG   03600

Uses Total vegetation control in NON-CROP AREAS.

FOR FULL CONDITIONS OF USE ALWAYS READ THE PRODUCT LABEL

Efficacy
• Spray to obtain complete cover of weeds. May be applied at any time of year
• Action slower in cool, dull than warm, bright weather but end result the same
• Light rain soon after treatment does not reduce effectiveness

Crop safety/Restrictions
• Keep off all desirable vegetation
• Heavy rain may cause run-off on sloping surfaces

Special precautions/Environmental safety
• Harmful to livestock. Keep all livestock out of treated areas for at least 24 h
• Harmful to fish. Do not contaminate ponds, waterways or ditches with chemical or used container

Protective clothing/Label precautions
• A, C
• 14, 22, 28, 29, 41, 53, 64, 65

---

24 amitrole + diuron

A total herbicide mixture for use in established orchards

Products	Orchard Herbicide	Promark	40:48% w/w	WP	03379

Uses Annual weeds, perennial weeds in APPLES, PEARS.

Notes

Efficacy
• Apply round base of trees or in strips along rows in spring or early summer during active weed growth. May also be used after harvest in autumn
• Moisture required for optimum residual effect
• Do not mix with contact herbicides

Crop safety/Restrictions
• Use only in orchards established for at least 4 yr

Special precautions/Environmental safety
• Harmful to fish. Do not contaminate ponds, waterways or ditches with chemical or used container

Protective clothing/Label precautions
• A
• 6a, 6b, 6c, 16, 18, 21, 28, 29, 36, 37, 53, 63, 67

---

25 amitrole + simazine

A translocated and residual herbicide for non-crop areas and orchards

Products					
1 Boroflow S/ATA	ABM	160:270 g/l	SC	04296
2 CDA Clearway	RP Environ.	100:300 g/l	SC	04584
3 CDA Simflow Plus	RP Environ.	100:300 g/l	SC	04607
4 Clearway	RP Environ.	100:300 g/l	SC	00540

5 Herbazin Plus SC	Fisons	180:300 g/l	SC	03659
6 Mascot Highway Liquid	Rigby Taylor	155:275 g/l	SC	03005
7 Mascot Highway WP	Rigby Taylor	38:48% w/w	WP	03611
8 MSS Simazine/ Aminotriazole 43 FL	Mirfield	155:275 g/l	SC	04361
9 Primatol SE 500FW	Ciba-Geigy	180:300 g/l	SC	01638
10 Simflow Plus	RP Environ.	100:300 g/l	SC	01955
11 Synchemicals Total Weed Killer	Synchemicals	53:110 g/l	SC	02160
12 Syntox Total Weedkiller	Syntex	53:100 g/l	SC	04937
13 Transformer Primatol SE 500FW	Ciba-Geigy	180:300 g/l	SC	04091

Uses   Total vegetation control in NON-CROP AREAS [1-13].

Notes   **Efficacy**
- On non-crop land apply from Apr to Sep when weeds are growing rapidly and have sufficient leaf area to absorb the chemical
- Product specially packed for use with 'Transformer Spraying System' or CDA machines [13]
- Do not spray if foliage wet or rain imminent
- Higher doses needed on soils with high organic matter content or on heavy clays

Crop safety/Restrictions
- Do not use on ground under which roots of valuable trees and shrubs are growing
- Heavy rain may cause run-off on impermeable or compacted surfaces
- Do not plant on land treated within last 2-3 yr

Special precautions/Environmental safety
- Irritating to eyes [3, 9, 10, 13], to eyes and skin [1, 4, 6, 7]
- Not to be used on food crops [5, 9]

Protective clothing/Label precautions
- A, C [1-13]; M, N [2, 3]; A, E, M [6]
- 6a [3, 9, 10, 13]; 6a, 6b [1, 4, 6, 7]; 18, 22, 28, 29, 36, 37, 54, 63, 65 or 66 or 67, 70 [1-13]; 14 [6, 7]; 34, [5, 9, 13]; 41 [13]

Approval
- May be applied through CDA equipment. See introductory notes [2, 6, 9, 13]

26   ammonium sulphamate
A non-selective, inorganic, general purpose herbicide and tree-killer

| **Products** | 1 Amcide | BH&B | 99.5% w/w | CR | 00089 |
| | 2 Root-Out | Dax | 99.5% w/w | CR | 03510 |

FOR FULL CONDITIONS OF USE ALWAYS READ THE PRODUCT LABEL

Uses	Annual weeds, perennial weeds in VEGETABLES, ORNAMENTALS *(pre-planting)*. Annual weeds, perennial grasses, perennial dicotyledons, woody weeds, rhododendrons in FORESTRY.

Notes	**Efficacy**

- Apply as spray to low scrub and herbaceous weeds from Apr to Sep in dry weather when rain unlikely and cultivate after 3-8 wk
- Apply as crystals in frills or notches in trunks of standing trees at any time of year
- Apply as concentrated solution or crystals to stump surfaces within 48 h of cutting. Rhododendrons must be cut level with ground and sprayed to cover cut surface, bark and immediate root area
- Stainless steel or plastic sprayers are recommended. Solutions are corrosive to mild steel, galvanised iron, brass and copper

Crop safety/Restrictions
- Allow 8-12 wk after treatment before replanting
- Keep spray at least 30 cm from growing plants. Low doses may be used under mature trees with undamaged bark

Special precautions/Environmental safety
- Harmful to fish. Do not contaminate ponds, waterways or ditches with chemical or used container

Protective clothing/Label precautions
- 24, 26, 27, 28, 29, 37, 53, 63

27 anthracene oil

A crop desiccant

Products	Sterilite Hop Defoliant	Coventry Chemicals 68% w/w	EC	05060

Uses	Chemical stripping in HOPS.

Notes	**Efficacy**

- Spray when hop bines 1.2 m high and direct spray downwards at 45° onto area to be defoliated. Repeat as necessary until cones are formed

Crop safety/Restrictions
- Do not spray if temperature is above 21 °C or after cones have formed
- Do not drench rootstocks
- Do not spray on windy, wet or frosty days

Special precautions/Environmental safety
- Harmful if swallowed. Irritating to eyes, skin and respiratory system
- Dangerous to fish. Do not contaminate ponds, waterways or ditches with chemical or used container

Protective clothing/Label precautions
- A, C, H
- 5c, 6a, 6b, 6c, 18, 21, 28, 29, 35, 36, 51, 59, 60, 62, 64, 69, 77

28   asulam

A translocated carbamate herbicide for control of docks and bracken

Products	1 Asulox	Embetec	400 g/l	SL	04413
	2 Asulox	RP Environ.	400 g/l	SL	00122

Uses

Bracken in PERMANENT PASTURE, ROUGH GRAZING, FORESTRY LAND, NON-CROP AREAS. Docks in GRASSLAND, APPLES, PEARS, PLUMS, CHERRIES, BLACKCURRANTS, HOPS. Docks, bracken in WATERSIDE AREAS. Bracken in FORESTRY *(off-label)* [2]. Docks in WHITE CLOVER SEED CROPS, STRAWBERRIES, HARDY ORNAMENTAL NURSERY STOCK, NURSERY FRUIT TREES AND BUSHES, DAMSONS, PEACHES, NECTARINES, APRICOTS, GOOSEBERRIES, CURRANTS, CANE FRUIT, BLUEBERRIES, CRANBERRIES *(off-label)* [2]. Docks in MINT, PARSLEY, TARRAGON *(off-label)* [1].

Notes

Efficacy
* Spray bracken when fronds fully expanded but not senescent, usually Jul-Aug, docks in full leaf before flower stem emergence, not less than 3 wk after cutting
* Do not apply in drought or hot, dry conditions
* May be applied in water or with oil (Actipron or Adder) for ULV application on bracken (additives not recommended on forestry land)
* Do not cut or admit stock for 14 d after spraying bracken or 7 d after spraying docks

Crop safety/Restrictions
* In forestry areas some young trees may be checked if sprayed directly (see label)
* Allow at least 6 wk between spraying and planting any crop
* Do not use in pasture before mowing for hay
* In orchards apply as a directed spray
* Do not treat blackcurrant cuttings, hop sets or weak hills
* Cocksfoot and Yorkshire fog may be severely checked, bents, fescues, meadow-grasses and timothy may be checked temporarily

Special precautions/Environmental safety
* Keep livestock out of treated areas until foliage of poisonous weeds such as ragwort has died and become unpalatable

Protective clothing/Label precautions
* A, C, H, M
* 28, 29, 43, 54, 63, 68, 70

Approval
* May be applied through CDA equipment. See introductory notes
* Approved for aerial application on bracken. See introductory notes
* Cleared for aquatic weed control. See introductory notes

FOR FULL CONDITIONS OF USE ALWAYS READ THE PRODUCT LABEL

• Off-label Approval to Sep 1992 for application by 'Ulvaforest' sprayer in forestry (OLA 0750/89) [1] and for use in white clover seed crops (OLA 0751/89), strawberries (OLA 0752/89), hardy ornamental nursery stock, nursery fruit trees and bushes (OLA 0753/89), quinces, stone fruit, cane fruit, currants, gooseberries, blueberries and cranberries (OLA 0754/89) [1]. Off-label approval to Oct 1992 for use in mint, parsley and tarragon (OLA 0915/89) [1]

29   atrazine

A triazine herbicide with residual and foliar activity

See also amitrole + atrazine
               amitrole + atrazine + 2,4-D
               amitrole + atrazine + dicamba
               amitrole + atrazine + diuron
               amitrole + atrazine + MCPA

Products

1	Ashlade 4% At Gran	Ashlade	4% w/w	GR	02890
2	Ashlade Atrazine 50 FL	Ashlade	500 g/l	SC	02891
3	Atlas Atrazine	Atlas	500 g/l	SC	03097
4	Atraflow	RP Environ.	500 g/l	SC	00160
5	BH Atrazine Granules	RP Environ.	4% w/w	GR	00246
6	Borocil A	ABM	4% w/w	GR	04203
7	Boroflow A	ABM	500 g/l	SC	04295
8	FS Atrazine 4% Granules	Ford Smith	4% w/w	GR	04731
9	Gesaprim 500FW	Ciba-Geigy	500 g/l	SC	00982
10	Granular Herbazin Total	Fisons	4% w/w	GR	03661
11	Mascot 4% Atrazine Granular	Rigby Taylor	4% w/w	GR	02437
12	Mascot Gauntlet Liquid	Rigby Taylor	450 g/l	SC	03007
13	Mascot Gauntlet WP	Rigby Taylor	80% w/w	WP	03006
14	MSS Atrazine 50 FL	Mirfield	500 g/l	SC	01398
15	MSS Atrazine 4G	Mirfield	4% w/w	GR	04249
16	MSS Atrazine 80 WP	Mirfield	80% w/w	WP	04360
17	New Chlorea	Chipman	4% w/w	GR	01479
18	Portman Atrazine 50 FL	Portman	500 g/l	SC	01521
19	Unicrop Flowable Atrazine	Unicrop	500 g/l	SC	02268

Uses

Annual dicotyledons, annual grasses in MAIZE, SWEETCORN, RASPBERRIES, ROSES [1, 2, 9, 19]. Annual dicotyledons, annual grasses in MAIZE, RASPBERRIES, ROSES [3]. Annual dicotyledons, annual grasses, couch in MAIZE, SWEETCORN [9, 19]. Annual dicotyledons, annual grasses, couch in MAIZE [3]. Annual dicotyledons, annual grasses, perennial grasses in CONIFER PLANTATIONS [1-4, 9, 14-16, 19]. Total vegetation control in NON-CROP AREAS [1-8, 10-19]. Total vegetation control in FIELD BOUNDARIES [9].

Notes   **Efficacy**

• May be used pre- or early post-weed emergence in maize and sweetcorn
• Root activity enhanced by rainfall soon after application and reduced on high organic soils. Foliar activity effective on weeds up to 3 cm high
• In conifers apply as overall spray in Feb-Apr. May be used in first spring after planting
• Apply to raspberries in spring before new cane emergence, not in season of planting

171

- For total vegetation control apply at any time of year for maintaining weed-free conditions
- Application rates vary with crop, soil type and weed problem. See label for details
- Resistant weed strains may develop with repeated use of atrazine or other triazines

Crop safety/Restrictions
- High rates should not be used on Norway spruce, Western hemlock, European or hybrid larch. See label for details. Do not use on Xmas trees [9]
- Do not apply on ground under which the roots of valuable trees or shrubs extend
- Do not mix with paraquat except as directed on non-crop land
- On slopes heavy rainfall soon after application may cause surface run-off
- No crop other than maize or sweetcorn should be sown for at least 7 mth after application (longer with higher rates - see label). Do not sow oats in autumn following a spring application

Protective clothing/Label precautions
- A, C, M [12, 19]
- 29, 54, 63, 65, 66, 70 [3, 4, 6, 9]; 21, 29, 54, 63, 64, 70 [12, 19]; 29, 54, 63, 67 [5, 8, 10, 11, 13-18]

Approval
- May be applied through CDA equipment. See introductory notes [9, 12]

30 atrazine + bromacil + diuron
A persistent, residual, total herbicide mixture

Products	Borocil Extra	ABM	1.2:0.45:0.45% w/w	GR	04204

Uses Total vegetation control in NON-CROP AREAS.

Notes

Efficacy
- May be applied at any time of year if sufficient rain falls to carry chemical into root zone of weeds
- Best time for initial treatment late winter or early spring, higher rates needed in summer

Crop safety/Restrictions
- Do not use on ground intended for cropping. Growth of vegetation is prevented for 1 yr or more
- Avoid contact with desirable plants and do not treat areas under which roots of desirable plants are growing

Special precautions/Environmental safety
- Harmful to fish. Do not contaminate ponds, waterways or ditches with chemical or used container

FOR FULL CONDITIONS OF USE ALWAYS READ THE PRODUCT LABEL

31   atrazine + cyanazine

A triazine herbicide mixture with residual and foliar activity

Products   Holtox   Shell   250:250 g/l   SC   01064

Uses   Annual grasses, annual dicotyledons in MAIZE, RASPBERRIES, FORESTRY, AMENITY TREES AND SHRUBS. Perennial grasses in FORESTRY, AMENITY TREES AND SHRUBS.

Notes   **Efficacy**
• Apply as soon as possible after planting maize but before crop emerges. Do not cultivate after spraying
• Gives best results if applied to moist soil and rain follows application. If soil moisture low incorporate to 2.5-4 cm before drilling maize
• Apply to raspberries in Feb-Mar before sucker emergence
• In forestry and amenity plantings apply in Jan-Apr (Feb-May in Scotland). May be applied 1 mth after planting provided soil firmed round roots
• Application rates vary with crop, soil type and weed problem - see label for details

Crop safety/Restrictions
• Not recommended for use on raspberry spawn-beds
• Trees should be treated before bud-burst to avoid danger of scorching
• May be used on all forest soils except sands
• Autumn cereals may be planted after use on maize crop provided a 4 mth period since application and soil cultivated thoroughly before drilling

Special precautions/Environmental safety
• Harmful if swallowed or in contact with skin
• Harmful to fish. Do not contaminate ponds, waterways or ditches with chemical or used container

Protective clothing/Label precautions
• A
• 5a, 5c, 18, 21, 28, 29, 36, 53, 63, 66, 70, 78

32   atrazine + 2,4-D + sodium chlorate

A total herbicide mixture with foliar and root action

Products   Atlavar   Chipman   2.54:1.27:55% w/wWP   00157

Uses   Total vegetation control in NON-CROP AREAS.

Efficacy
* Apply as spray at any time during growing season. Best results from application in Apr-May. Persists at least 1 yr
* Do not apply before heavy rain
* Agitation required in tank during spraying to prevent settling out

Crop safety/Restrictions
* Do not apply near cultivated plants or trees or on slopes where run-off may cause damage
* Formulation contains a fire depressant but do not use on sites with high fire risk

Special precautions/Environmental safety
* Harmful if swallowed
* Oxidizing - contact with combustible material may cause fire. If clothes become contaminated wash thoroughly and do not stand near open fire
* Keep livestock out of treated areas unitl poisonous plants such as ragwort have died and become unpalatable

Protective clothing/Label precautions
* A, H, M
* 14, 18, 29, 36, 37, 43, 54, 63, 67, 70, 78

33   atrazine + dalapon

A granular, soil acting herbicide for use in forestry

Products   Atlas Lignum Granules   Atlas   10:10% w/w   GR   03088

Uses   Annual dicotyledons, annual grasses, perennial grasses in FORESTRY PLANTATIONS, HARDWOOD NURSERY STOCK.

Notes   **Efficacy**
* Apply in spring as soon as weed growth starts
* Activity is dependent on the presence of adequate soil moisture
* Application rate varies with soil type and species. See label for details

Crop safety/Restrictions
* Ensure that newly planted trees have roots well covered with soil as accumulation of granules in the rooting zone may cause damage
* Apply when tree foliage is dry and only use on healthy trees
* Heavy rain soon after treatment may cause run-off on sloping ground

Special precautions/Environmental safety
* Irritating to eyes and skin

Protective clothing/Label precautions
* 6a, 6b, 18, 22, 28, 29, 36, 37, 63, 66

FOR FULL CONDITIONS OF USE ALWAYS READ THE PRODUCT LABEL

34   atrazine + imazapyr
A translocated, contact and residual total herbicide

Products					
1 Arsenal XL	Chipman	300:12.5 g/l	SC	04068
2 Arsenal XL	Cyanamid	300:12.5 g/l	SC	04067
3 Arsenal XL CDA	Chipman	300:12.5 g/l	SC	04157
4 Arsenal XL CDA	Cyanamid	300:12.5 g/l	SC	04069
5 Moderator	Chipman	300:12.5 g/l	SC	04846
6 Moderator	Cyanamid	300:12.5 g/l	SC	04845

Uses   Total vegetation control in NON-CROP AREAS.

Notes   **Efficacy**
- For use on railways, industrial sites, footpaths, farm roads, fence lines and around farm buildings
- May be applied at any time of year when weeds actively growing. Best results when applied in Mar-Apr
- Translocates to kill underground storage organs and provides long-term residual control
- Product formulated and packed for use with NOMIX hand lances [5, 6]

Crop safety/Restrictions
- Do not apply near shrubs or trees with trunks less than 10 cm diameter. Do not apply near ash or rowan
- Do not apply higher rates near desirable plants or where their roots may extend
- Do not apply on ground to be planted or on slopes where rain may cause run-off onto planted ground

Special precautions/Environmental safety
- Harmful if swallowed

Protective clothing/Label precautions
- C
- 5c, 14, 18, 29, 36, 37, 54, 63, 66, 78

Approval
- May be applied through CDA equipment. See introductory notes [3, 4, 5, 6]

35   atrazine + sodium chlorate
A total herbicide mixture with foliar and root action

Products				
Atlacide Extra	Chipman	0.32:30% w/w	DP	00124

Uses   Total vegetation control in NON-CROP AREAS.

Notes   **Efficacy**
- Apply by sprinkling dust over foliage
- May be applied at any time during growing season. Best results from application in spring and early summer
- Do not apply before heavy rain

Crop safety/Restrictions
- Formulation contains a fire depressant but do not use on sites with high fire risk

Special precautions/Environmental safety
- Harmful if swallowed
- Oxidizing - contact with combustible material may cause fire. If clothes become contaminated wash thoroughly and do not stand near open fire
- Keep livestock out of treated area until poisonous weeds such as ragwort have died and become unpalatable

Protective clothing/Label precautions
- 5c, 12c, 18, 29, 36, 37, 43, 54, 64, 67, 70, 78

36   atrazine + terbuthylazine

A selective triazine herbicide mixture for use in forestry

Products					
1 Gardoprim A 500FW	Ciba-Geigy	100:400 g/l	SC	04172
2 Transformer Gardoprim A 500FW	Ciba-Geigy	100:400 g/l	SC	04369

Uses

Annual dicotyledons, annual grasses, perennial non-rhizomatous grasses, rushes, perennial dicotyledons in FORESTRY.

Notes

Efficacy
- Apply pre- or post-planting of conifers or broad-leaved trees. Best results achieved by application at planting or before weed growth becomes vigorous
- Specially packed for use with 'Transformer Spraying System' or CDA machines [2]
- Complete control may take up to 6 wk. Exceptionally dry weather after application may reduce effectiveness

Crop safety/Restrictions
- Douglas fir and Sitka spruce may be sprayed at any time of year, other species in dormant season. See label for list of tolerant species
- Do not spray Norway spruce, European larch, Lodge pole pine or Japanese larch after start of bud burst. Do not spray broad-leaved trees in leaf
- Trials on Christmas trees not yet complete and not recommended on this crop
- Heavy rain soon after application on a slope may cause surface run-off

Special precautions/Environmental safety
- Harmful if swallowed and by inhalation
- Harmful to fish. Do not contaminate ponds, waterways and ditches with chemical or used container

Protective clothing/Label precautions
- A, C, H, M
- 5b, 5c, 18, 29, 36, 37, 53, 63, 66, 70, 78

FOR FULL CONDITIONS OF USE ALWAYS READ THE PRODUCT LABEL

37   azamethiphos
A residual organophosphorus insecticide for fly control

Products	Alfacron 10WP	Ciba-Geigy	10% w/w	WP	02832

Uses   Flies in LIVESTOCK HOUSES.

Notes

Efficacy
• Apply as paint to 2.5% of total wall and ceiling surface area at a minimum of 5 points in building or as spray to 30% of surface area

Crop safety/Restrictions
• Only apply to areas out of reach of children and animals

Special precautions/Environmental safety
• Irritating to eyes and skin. May cause sensitization by skin contact
• This product contains an organophosphorus compound. Do not use if under medical advice not to work with such compounds
• Do not apply directly to livestock and poultry
• Do not apply to surfaces on which food or feed is stored, prepared or eaten. Cover feedstuffs and remove exposed milk and eggs before application
• Dangerous to fish. Do not contaminate ponds, waterways or ditches with chemical or used container

Protective clothing/Label precautions
• A, C
• 1, 6a, 6b, 10a, 14, 18, 23, 25, 26, 27, 28, 29, 36, 37, 52, 63, 67

38   azinphos-methyl
An organophosphate insecticide available only in mixtures

39   azinphos-methyl + demeton-S-methyl sulphone
A broad-spectrum contact organophosphate insecticide and acaricide

Products	Gusathion MS	Bayer	25:7.5% w/w	WP	01031

Uses   Aphids in APPLES, PEARS, PEAS. Bryobia mites, codling moth, winter moth, apple blossom weevil, apple sawfly, apple sucker, red spider mites in APPLES, PEARS. Blackcurrant leaf midge, sawflies in BLACKCURRANTS, GOOSEBERRIES. Bramble shoot moth in BLACKBERRIES, LOGANBERRIES. Caterpillars, leaf miners in LEAF BRASSICAS. Pea moth, pea midge, pea and bean weevils, thrips in PEAS. Pear sucker in PEARS. Plum fruit moth, plum sawfly in PLUMS. Pollen beetles, cabbage seed weevil in MUSTARD, SPRING OILSEED RAPE, SEED BRASSICAS. Raspberry beetle in

RASPBERRIES. Tortrix moths in APPLES, PEARS, CHERRIES, PEAS, STRAWBERRIES. Colorado beetle in POTATOES *(off-label)*. General insect control in NURSERY FRUIT TREES AND BUSHES, HARDY ORNAMENTAL NURSERY STOCK *(off-label)*. Caterpillars, thrips in BROCCOLI, CAULIFLOWERS *(off-label)*.

Notes

Efficacy
- Recommended dose, volume and timing vary with pest and crop. See label for details
- High volume application is preferred for apples, pears, blackcurrants and gooseberries
- Unlikely to give satisfactory results if red spider or bryobia mites resistant to organophosphorus compounds are present
- Addition of non-ionic wetter recommended for application on leaf brassicas

Crop safety/Restrictions
- Not more than 5 applications may be made to apples and pears, 4 to strawberries and peas, 3 to plums and leaf brassicas, 2 to blackcurrants, gooseberries, potatoes, mustard and brassica seed crops and 1 to blackberries, raspberries, loganberries and cherries

Special precautions/Environmental safety
- A chemical subject to the Poisons Rules 1982 and the Poisons Act 1972
- Harmful in contact with skin and if swallowed
- This product contains an organophosphorus compound. Do not use if under medical advice not to work with such compounds
- Do not touch cachet with wet hands or gloves
- Dangerous to bees. Do not apply at flowering stage. Keep down flowering weeds
- Harmful to game, wild birds and animals
- Dangerous to fish. Do not contaminate ponds, waterways or ditches with chemical or used container

Protective clothing/Label precautions
- A, C or D+E, H, J
- 1, 5a, 5c, 14, 16, 18, 23, 24, 25, 28, 29, 32, 36, 37, 41, 46, 48, 52, 64, 67, 70, 78

Withholding period
- Harmful to livestock. Keep all livestock out of treated areas for at least 2 wk

Harvest interval
- 3 wk

Approval
- Approved for aerial application on brassicas, brassica seed crops, peas. See introductory notes
- Off-label approval to Nov 1991 for use on potatoes (OLA 1387/88), for aerial application on potatoes (OLA 1391/88); to Feb 1992 for use on nursery fruit trees and bushes and hardy ornamental nursery stock (OLA 0204, 0205/89), on broccoli and cauliflowers (OLA 0218/89)

FOR FULL CONDITIONS OF USE ALWAYS READ THE PRODUCT LABEL

40   aziprotryne
A selective triazine herbicide with foliar and residual activity

Products	Brasoran 50 WP	Ciba-Geigy	50% w/w	WP	00316

Uses   Annual dicotyledons, annual grasses in BRUSSELS SPROUTS, CABBAGES, PEAS, ONIONS, LEEKS.

Notes

Efficacy
- For best results soil should be moist with a fine tilth
- Control lasts 6-8 wk under normal conditions but may be reduced by dry conditions after application
- Emerged weeds are not controlled after 2-leaf stage
- Use as a post-emergence treatment on brassicas, onions and leeks, pre- or post-emergence on peas

Crop safety/Restrictions
- Do not use on very stony soils or apply during rain or frost. Only apply to healthy crops which are growing well
- Do not use insecticide treatment on brassicas for 7 d before or 4 d after application
- Do not use on brassicas under glass
- Crops must have reached correct growth stage before treatment. See label for details
- Spring sown peas may be treated on all soil types as long as covered with at least 25 mm soil
- If necessary to re-drill, any crop may be sown 6 wk after application provided correct dose used and soil thoroughly cultivated

Special precautions/Environmental safety
- Dangerous to fish. Do not contaminate ponds waterways or ditches with chemical or used container

Protective clothing/Label precautions
- 22, 29, 35, 52, 63, 70

Harvest interval
- Brussels sprouts, cabbages 3 wk; peas, onions, leeks 6 wk

41   Bacillus thuringiensis
A bacterial insecticide for control of caterpillars

Products				
1 Atlas Thuricide HP	Atlas	WP	00155
2 Bactospeine	Fargro	WP	00181
3 Bactospeine WP	Koppert	WP	02913
4 Dipel	English Woodland	WP	03214

Uses   Cabbage white butterfly, diamond-back moth, cabbage moth in BRUSSELS SPROUTS, CABBAGES, CAULIFLOWERS, BROCCOLI [1-4]. Strawberry tortrix in STRAWBERRIES [1, 2, 3]. Browntail moth in ORNAMENTALS [2, 3, 4]. Caterpillars in RASPBERRIES [1, 2]. Caterpillars in CHRYSANTHEMUMS [4]. Tomato moth in TOMATOES [1-4]. Caterpillars in CUCUMBERS, PEPPERS [2].

Caterpillars in FRENCH BEANS *(off-label)* [1-4]. Caterpillars in CUCUMBERS *(off-label)* [1, 3, 4].

Notes	**Efficacy**
	• Product affects gut of larvae and must be eaten to be effective. Caterpillars cease feeding and die in 2-3 d (3-5 d for large caterpillars)
	• Apply as soon as larvae appear on crop and repeat every 3-14 d for outdoor crops, every 3 wk under glass
	• Addition of a wetter recommended for use on brassicas
	• Good coverage is essential, especially of undersides of leaves. Use a drop-leg sprayer in field crops

Protective clothing/Label precautions
• 29, 54, 63, 65

Harvest interval
• zero

Approval
• Off-label approval to Feb 1991 for use on french beans, tomatoes and cucumbers (OLA 0070/88) [1]; (OLA 0064/88) [2]; (OLA 0065/88) [3]; (OLA 0068/88) [4]

42 benalaxyl

A phenylamide (acylalanine) fungicide available only in mixtures

43 benalaxyl + mancozeb

A systemic and protectant fungicide mixture

Products	Galben M	DowElanco	8:65% w/w	WP	03809

Uses	Blight in POTATOES.

Notes	**Efficacy**
	• Apply at blight warning prior to crop becoming infected and repeat at 10-21 d intervals depending on risk of infection
	• Spray irrigated crops after irrigation and at 14 d intervals, crops in polythene tunnels at 10 d intervals
	• When active growth ceases in Aug use a different protective fungicide to end of season, starting not more than 10 d after last Galben M application
	• Do not treat crops showing active blight infection

Crop safety/Restrictions
• Do not apply more than 5 sprays during season. Do not use after the end of Aug
• Other acylalanine fungicides should not be used following a Galben M programme

FOR FULL CONDITIONS OF USE ALWAYS READ THE PRODUCT LABEL

Special precautions/Environmental safety
- Irritating to skin, eyes and respiratory system
- Dangerous to fish. Do not contaminate ponds, waterways or ditches with chemical or used container

Protective clothing/Label precautions
- A
- 6a, 6b, 6c, 18, 22, 29, 35, 36, 37, 52, 61, 63, 67

Harvest interval
- 7 d

Approval
- Provisional approval for aerial application on a limited area of potatoes. Contact firm for allocation

---

44 benazolin

A translocated herbicide available only in mixtures

---

45 benazolin + bromoxynil + ioxynil

A post-emergence, HBN herbicide mixture for cereal crops and grass

Products Asset   Schering   50:125:62.5 g/l   EC   03824

Uses Annual dicotyledons in WHEAT, DURUM WHEAT, BARLEY, OATS, RYE, TRITICALE, NEWLY SOWN GRASS.

Notes **Efficacy**
- Commonly used in mixture with mecoprop and other cereal herbicides to extend the range of weeds controlled. Sequential treatments also recommended
- Best results achieved when weeds small and actively growing and crop competitive
- Weeds should be dry when sprayed

Crop safety/Restrictions
- Use in cereals from 2-leaf stage to first node detectable (GS 12-31), in grass from 2-leaf stage when crop growing vigorously (other times may be advised for mixtures)
- Use on oats only in spring, on other cereals in autumn or spring
- Do not apply mixture with mecoprop or MCPA on rye or triticale
- Do not apply to crops undersown with legumes
- Do not use on crops affected by pests, disease, waterlogging or prolonged frost
- Severe frost within 3-4 wk of spraying may scorch crop
- Do not roll or harrow crops for 3 d before or after spraying
- Do not spray grass seed crops less than 5 wk before heading

Special precautions/Environmental safety
- Harmful if swallowed. Irritating to skin and eyes
- Do not apply with hand-held equipment or at less than recommended volume
- Keep livestock out of treated areas until poisonous plants such as ragwort have died and become unpalatable

- Flammable
- Dangerous to fish. Do not contaminate ponds, waterways or ditches with chemical or used container
- Do not allow spray from ground sprayers to fall within 6 m of surface waters or ditches

Protective clothing/Label precautions
- A, C
- 5c, 6a, 6b, 12c, 18, 21, 25, 28, 29, 36, 37, 43, 52, 63, 66, 70, 78

Withholding period
- 6 wk. Period also applies to harvest of grass

46 benazolin + bromoxynil + ioxynil + mecoprop
A post-emergence, HBN herbicide mixture for use in cereal crops

Products	Jaguar	Schering	22.2:55.6:27.8:413 EC g/l	03801

Uses   Annual dicotyledons in WHEAT, BARLEY, OATS.

Notes   **Efficacy**
- Susceptible weeds controlled up to 20 cm in winter, to 10 cm in spring crops. See label for details of species susceptibility
- Best results achieved when weeds growing actively. Do not spray when heavy rain or frost imminent. Weeds should be dry when sprayed

Crop safety/Restrictions
- Apply in spring from 2-leaf stage to first node detectable (GS 12-31)
- Spray cereals undersown with grass when grass has 2 leaves (provided cereal at correct stage)
- Do not spray crops undersown with clover or legume mixtures
- Do not use on crops stressed by waterlogging, mineral deficiency, pests or diseases
- Do not roll or harrow crops for 3 d before or after spraying

Special precautions/Environmental safety
- Harmful in contact with skin and if swallowed
- Irritating to eyes, skin and respiratory system
- Do not apply with hand-held equipment or at less than recommended volume
- Keep livestock out of treated areas until poisonous plants such as ragwort have died and become unpalatable
- Dangerous to fish. Do not contaminate ponds, waterways or ditches with chemical or used container
- Do not allow spray from ground sprayers to fall within 6 m of surface waters or ditches

Protective clothing/Label precautions
- A, C

FOR FULL CONDITIONS OF USE ALWAYS READ THE PRODUCT LABEL

47   benazolin + clopyralid

A post-emergence herbicide mixture for use in oilseed rape

Products	Benazalox	Schering	30:5% w/w	WP	03858

Uses

Mayweeds, chickweed, cleavers, annual dicotyledons in OILSEED RAPE. Mayweeds, chickweed, cleavers, annual dicotyledons in WHITE MUSTARD, EVENING PRIMROSE, SWEDE SEED CROPS *(off-label)*. Annual dicotyledons in CONIFEROUS TREES *(off-label)*.

Notes

Efficacy
● Weeds are controlled from cotyledon stage to 15 cm high
● For best results apply during mild, moist weather when weeds are growing actively and still visible in young crop. Do not spray if rain expected within 4 h
● Frost should not be present on foliage at spraying but after application will not reduce effectiveness
● For grass weed control various tank mixtures are recommended or sequential treatments may be used, allowing specified interval after application of grass-killer. See label for details

Crop safety/Restrictions
● May be used when crop between stages of 3 fully developed leaves to flower buds hidden beneath leaves (GS 1,3-3,1)
● If used on spring sown crops lower rate should be applied
● Oilseed rape may be replanted in the event of crop failure. Wheat, barley or oats may be sown after 1 mth, other crops in the following autumn after ploughing. Winter beans should not be planted in the same year
● Chop and incorporate or burn straw and plant remains in early autumn to release any clopyralid residues. Ensure plant remains completely decayed before planting susceptible crops

Special precautions/Environmental safety
● Irritating to eyes and skin
● Keep livestock out of treated areas until foliage of any poisonous weeds such as ragwort has died and become unpalatable
● Harmful to fish. Do not contaminate ponds, waterways or ditches with chemical or used container

Protective clothing/Label precautions
● A, C
● 6a, 6b, 18, 22, 25, 28, 29, 36, 37, 43, 53, 63, 66

Approval
● Off-label approval to May 1992 for use on white mustard, evening primrose and swedes grown for seed (OLA 0466, 0467, 0468/89); to Aug 1993 for use on coniferous trees (OLA 0434/90)

48 benazolin + 2,4-DB + MCPA

A post-emergence herbicide for undersown cereals, grass and clover

Products

1 Legumex Extra	Schering	27:237:42.3 g/l	SL	03869
2 Setter 33	DowElanco	50:237:43 g/l	SL	04377

Uses

Chickweed, cleavers, knotgrass, annual dicotyledons, perennial dicotyledons in UNDERSOWN CEREALS, CLOVERS, SEEDLING LEYS, GRASSLAND [1]. Chickweed, annual dicotyledons in UNDERSOWN CEREALS, CLOVERS, SEEDLING LEYS [2].

Notes

Efficacy
* Spray when weeds are young and in active growth
* Spray perennial weeds when well developed but before flowering. Higher rates may be used against perennials in established grassland
* Do not spray when heavy rain is imminent or during drought

Crop safety/Restrictions
* Spray clovers after the 1-trifoliate leaf stage and before red clover has more than 3 trifoliate leaves. Do not spray clover seed crops in the year seed is to be taken
* Spray new swards when majority of grasses have at least 2 leaves and clovers as above
* Clovers may be damaged if frost occurs soon after spraying
* Do not spray legumes other than clover
* Spray undersown cereals before 1st node detectable [GS 31]
* Do not roll, harrow, cut or graze crops for 3 d before or after spraying

Special precautions/Environmental safety
* Keep livestock out of treated areas until foliage of poisonous weeds such as ragwort has died and become unpalatable
* Harmful to fish. Do not contaminate ponds, waterways or ditches with chemical or used container

Protective clothing/Label precautions
* 21, 29, 43, 53, 60, 63, 66

Withholding period
* 3 d

49 bendiocarb

A contact, ingested and systemic carbamate insecticide

Products

1 Garvox 3G	Schering	3% w/w	GR	03866
2 Seedox SC	Schering	500 g/l	FS	03856

Uses

Pygmy beetle, springtails, millipedes, symphylids, wireworms in SUGAR BEET [1]. Frit fly, wireworms in MAIZE, SWEETCORN [1,2]. Bean seed flies in PROTECTED

FOR FULL CONDITIONS OF USE ALWAYS READ THE PRODUCT LABEL

COURGETTES, PROTECTED GHERKINS, PROTECTED MARROWS, MELONS, PUMPKINS, SQUASHES *(off-label)* [2].

Efficacy
* Apply in the furrow with seed using a suitable applicator [1] or with suitable seed treatment machinery [2]
* Depending on climatic conditions soil pests are controlled for 6-8 wk [1]
* Calibration of applicator should be checked in the field before drilling
* Use of suitable adjuvant advised to ensure effectiveness. See label for details [2]

Special precautions/Environmental safety
* This product contains an anticholinesterase compound. Do not use if under medical advice not to work with such compounds [1, 2]
* Harmful in contact with skin or if swallowed [2]
* Harmful [1], dangerous [2] to fish. Do not contaminate ponds, waterways or ditches until chemical or used container
* Dangerous to game, wild birds and animals
* Treated seed not to be used as food or feed. Do not re-use sack [2]

Protective clothing/Label precautions
* A, C [1]
* 2, 22, 25, 29, 53, 60, 63, 67 [1]; 2, 5a, 5c, 18, 21, 25, 28, 29, 36, 37, 45, 52, 63, 65, 70, 72, 73, 74, 75, 76, 77, 79

Approval
* Off-label approval to Feb 1991 for use on seeds of courgettes, gherkins, marrows, melons, pumpkins, squashes grown indoors (OLA 0071/88) [2]

---

50   benfuracarb

A soil-applied carbamate insecticide and nematicide for beet crops

Products	Oncol 10G	Farm Protection	10% w/w	GR	04265

Uses

Flea beetles, leaf miners, millipedes, pygmy beetle, springtails, symphylids, wireworms, docking disorder vectors in SUGAR BEET, FODDER BEET, MANGOLDS.

Notes

Efficacy
* Apply at sowing with suitable granule applicator so that granules are mixed with the moving soil closing the seed furrow. See label for recommended applicators
* Where climate and soil conditions favour docking disorder a higher dose rate may be advantageous
* May be used on all soil types

Special precautions/Environmental safety
* Irritating to eyes and skin
* This product contains an anticholinesterase compound. Do not use if under medical advice not to work with such compounds
* Dangerous to livestock. Keep all livestock out of treated areas. Bury or remove spillages
* Dangerous to game, wild birds and animals

- Extremely dangerous to fish. Do not contaminate ponds, waterways or ditches with chemical or used container

Protective clothing/Label precautions
- A, B, C, C or D+E, H, K, M
- 2, 6a, 6b, 14, 16, 18, 22, 24, 25, 28, 29, 36, 37, 40, 45, 51, 63, 67, 78

51 benodanil
A systemic and protectant anilide fungicide

Products					
	1 Calirus	BASF	50% w/w	WP	00368
	2 Mascot Clearing	Rigby Taylor	50% w/w	WP	03422

Uses

Brown rust, yellow rust in WHEAT, BARLEY [1]. Snow rot in WINTER BARLEY [1]. Rust in PLUMS, RASPBERRIES, CARNATIONS, PELARGONIUMS, ROSES [1]. White rust in CHRYSANTHEMUMS [1]. Fairy rings in SPORTS TURF, AMENITY GRASS [2].

Notes

Efficacy
- For rust control in cereals apply at first signs of disease or as protectant treatment where disease potential high. Apply a second spray if necessary. Addition of Citowett wetter advised [1]
- If yellow rust established mix with Calixin and do not add wetter [1]
- For snow rot control apply as protective treatment between mid-Nov and mid-Dec [1]
- Turf should be well spiked before treatment then well soaked with spray. Treat to 1 m either side of fairy rings [2]

Crop safety/Restrictions
- For cereal rust control may be applied from tillering to complete ear emergence (GS 21-59), usually after flag leaf just visible (GS 37) [1]
- Do not spray in hot and/or dry conditions, especially under glass [1]
- Do not apply more than 2 treatments per crop
- Apply in raspberries (summer fruiting cultivars) as a post-harvest spray [1]
- Chrysanthemum plants must be dry before spraying. Deposit on plant must be dry before nightfall [1]

Special precautions/Environmental safety
- Harmful to fish. Do not contaminate ponds, waterways or ditches with chemical or used container

Protective clothing/Label precautions
- 29, 35, 53, 63, 67

Harvest interval
- plums 3 wk

FOR FULL CONDITIONS OF USE ALWAYS READ THE PRODUCT LABEL

52 benomyl

A systemic, MBC fungicide with protectant and eradicant activity

Products

Benlate	Du Pont	50%	WP	00229

Uses

Eyespot, rhynchosporium in CEREALS. Botrytis, light leaf spot in OILSEED RAPE. Chocolate spot in FIELD BEANS. Leaf and pod spot in PEAS, FIELD BEANS *(seed treatment)*. Dollar spot, fusarium patch, red thread in TURF. Botrytis in DWARF BEANS, PEAS, CELERY, LETTUCE, SALAD ONIONS. Light leaf spot, ring spot in BRASSICAS. Powdery mildew in SWEDES, ROSES. Black rot, botrytis, liquorice rot, sclerotinia in CARROTS *(dip)*. Botrytis, liquorice rot in CELERY *(dip)*. Neck rot in ONIONS *(seed treatment)*. Powdery mildew, scab, blossom wilt, canker, storage rots in APPLES, PEARS. American gooseberry mildew, botrytis, currant leaf spot in CURRANTS, GOOSEBERRIES. Botrytis, cane blight, cane spot, purple blotch, powdery mildew, spur blight in CANE FRUIT. Botrytis, powdery mildew, verticillium wilt in STRAWBERRIES. Botrytis, fusarium, penicillium rot, sclerotinia, stagonospora in BULBS, CORMS. Botrytis, powdery mildew in BEDDING PLANTS, CHRYSANTHEMUMS, POT PLANTS. Stem rot, botrytis, powdery mildew, black root rot in CUCUMBERS. Botrytis, leaf mould, didymella stem rot, fusarium wilt, verticillium wilt in TOMATOES. Fusarium wilt, verticillium wilt in CARNATIONS. Fusarium in FREESIAS. Fusarium, rhizoctonia, thielaviopsis in POINSETTIAS. Cobweb, dry bubble, trichoderma, wet bubble in MUSHROOMS. Canker in HOPS *(off-label)*. Gloeosporium in HAZEL NUTS *(off-label)*. Seed-borne diseases in GRASS SEED *(off-label)*. Rhizoctonia in WATERCRESS *(off-label)*. Fusarium in PROTECTED LEEKS *(off-label)*.

Notes

Efficacy
• Apply as overall spray for many diseases. One spray effective for some diseases, repeat sprays at 1-4 wk intervals needed for others. Recommended spray programmes vary with disease and crop. See label for details
• Apply as dip treatment for storage diseases in carrots, celery, apples, pears, bulbs and pre-planting of bulbs
• Apply as drench to control strawberry wilt, root-diseases in cucumbers, tomatoes, bedding plants, carnations, freesias, poinsettias and as drench or incorporated treatment against mushroom diseases
• Apply as dry seed dressing to control onion neck rot
• Addition of non-ionic wetter recommended for many uses. See label for details
• To delay appearance of resistant strains in diseases needing more than 2 applications per season use in programme with fungicide of different mode of action

Crop safety/Restrictions
• Treatment may delay fruit colouring of some early apples such as Worcester Pearmain and Merton Worcester
• Do not use on strawberry runner beds or on mushrooms during picking
• If used against wilt on strawberries do not use as botrytis treatment in same season
• Do not apply high volume sprays on cucumbers where red spider mite predator Phytoseiulus being used

Protective clothing/Label precautions
- 29, 37, 54, 63, 65

Harvest interval
- Hops 7 d; mange-tout peas 12 d; hazel nuts 28 d; peas apply before pod filling stage

Approval
- Approved for aerial application on wheat, barley, dwarf beans, field beans. See introductory notes. Consult firm for details
- Off-label approval to Feb 1991 for use on hops (OLA 0072/88); to Mar 1991 for use on hazel nuts (OLA 0245/88); to Feb 1992 for use on grass seed (OLA 0126/89); to Jun 1990 for use on watercress (OLA 0522/89); to June 1993 for use on protected leeks (OLA 0367/90)

53 benomyl + iodofenphos + metalaxyl

A composite fungicide, insecticide seed coating for onions

Products

1 Polycote Pedigree ingredient 1	Seedcote	50% w/w	WP	03887
2 Polycote Pedigree ingredient 2	Seedcote	50% w/w	WP	03888
3 Polycote Pedigree ingredient 3	Seedcote	350 g/l	SC	03889

Uses

Damping off, neck rot in BULB ONIONS, SALAD ONIONS. Onion fly in BULB ONIONS, SALAD ONIONS.

Notes

Efficacy
- Mix active ingredients with polymer supplied according to label instructions and apply to seed in Polycote Seed Coater machine

Crop safety/Restrictions
- Seed treatment may alter flow characteristics of seed. Recalibrate drill before drilling
- Do not use on seed of low germination vigour and viability

Special precautions/Environmental safety
- Contains an organophosphorus compound. Do not use if under medical advice not to work with such compounds [2]
- Harmful if swallowed [1]. Irritating to eyes and skin [2]
- Dangerous to fish. Do not contaminate ponds, waterways or ditches with chemical or used container [2]

Protective clothing/Label precautions
- A, C [1, 2]
- 5c, 14, 18, 29, 36, 37, 54, 60, 63, 65, 70, 78 [1]; 1, 6a, 6b, 16, 18, 22, 28, 29, 36, 37, 52, 64, 67, 70, 78 [2]; 29, 54, 63, 65 [3]

FOR FULL CONDITIONS OF USE ALWAYS READ THE PRODUCT LABEL

54   bentazone

A post-emergence contact herbicide

Uses   Annual dicotyledons in WINTER FIELD BEANS, SPRING FIELD BEANS, LINSEED, PEAS, POTATOES, DWARF BEANS, NAVY BEANS, RUNNER BEANS, NARCISSI. Annual weeds, mayweeds in BULB ONIONS *(off-label)*.

Notes   **Efficacy**
- Most effective control obtained when weeds are small and growing actively
- Do not apply if rain or frost expected, if foliage wet, in drought or in unseasonably cold weather
- A minimum of 6 h free from rain is required after application
- Various recommendations are made for use in spray programmes with other herbicides or in tank mixes. See label for details
- Addition of Actipron, Adder or Cropspray 11E adjuvant oils recommended for use in dwarf beans and potatoes to improve fat hen control. Do not use under hot or humid conditions or on field beans

Crop safety/Restrictions
- Crops must be treated at correct stage of growth to avoid danger of scorch. See label for details and for varietal tolerances
- Slight scorch may occur on field beans but yield is generally unaffected
- Do not use on crops which have been affected by drought, waterlogging, frost or other stress conditions
- Do not spray at temp above 21°C. Delay spraying until evening if necessary
- Do not tank mix with other than permitted chemicals
- A satisfactory wax test must be carried out before use on peas
- Do not treat potato seed crops, first earlies or varieties being grown for processing

Special precautions/Environmental safety
- Irritating to eyes

Protective clothing/Label precautions
- A, C
- 6a, 18, 21, 28, 29, 36, 37, 54, 63, 66

Harvest interval
- Bulb onions 3 wk

Approval
- Off-label approval to May 1991 for use on bulb onions (OLA 0504/88)

55   bentazone + cyanazine + 2,4-DB

A post-emergence herbicide for undersown cereals and grass

Uses   Chickweed, annual dicotyledons in UNDERSOWN SPRING BARLEY,
UNDERSOWN SPRING WHEAT, UNDERSOWN SPRING OATS, NEWLY
SOWN GRASS.

Notes   **Efficacy**
- Apply at any time in spring, summer or autumn when growth is occurring provided crop at correct stage
- Best results achieved when weeds small and actively growing
- Do not use during drought, waterlogging, extremes of temperature or when frosts are expected
- Do not spray when rainfall imminent or if crop wet

Crop safety/Restrictions
- Apply to cereals from 2-fully expanded leaf stage (wheat from 5-leaf) to first node detectable (GS 12 or 15-31) provided clover has reached 1-trifoliate leaf stage
- Do not treat clover after 3-trifoliate leaf stage
- Apply to newly sown leys after grass has reached 2-leaf stage provided clovers have at least 1 trifoliate leaf and red clover has not passed 3-trifoliate leaf stage
- Do not use on seed crops or cereals undersown with lucerne
- Do not use on crops affected by pest, disease or herbicide damage
- Do not roll or harrow for 7 d before or after spraying
- Growth of clovers may be reduced but effects are normally outgrown

Special precautions/Environmental safety
- Harmful if swallowed. Irritating to eyes and skin
- Keep livestock out of treated areas for at least 4 wk and until poisonous weeds such as ragwort have died and become unpalatable
- Harmful to fish. Do not contaminate ponds, waterways or ditches with chemical or used container

Protective clothing/Label precautions
- A, C
- 5c, 6a, 6b, 18, 21, 28, 36, 37, 43, 53, 63, 66, 70, 78

Withholding period
- 4 wk. Period also applies to harvest of grass

56   bentazone + MCPA + MCPB
A post-emergence herbicide for undersown cereals and grass

Products   Acumen   BASF   200:80:200 g/l   SL   00028

Uses   Annual dicotyledons in UNDERSOWN SPRING BARLEY, UNDERSOWN SPRING
WHEAT, UNDERSOWN SPRING OATS, NEWLY SOWN GRASS.

Notes   **Efficacy**
- Best results when weeds small and actively growing provided crop at correct stage

FOR FULL CONDITIONS OF USE ALWAYS READ THE PRODUCT LABEL

- Good spray cover is essential
- A minimum of 6 h free from rain is required after treatment
- Do not apply if frost expected, if crop wet or in high humidity at above 21 °C
- On first year grass leys use alone on seedling stage weeds (especially chickweed) before end of Sep, use mixture with cyanazine on older chickweed in Oct-Nov

Crop safety/Restrictions
- Apply to cereals from 2-fully expanded leaf stage but before first node detectable (GS 12-30) provided clover has reached 1-trifoliate leaf stage
- Do not treat red clover after 3-trifoliate leaf stage
- Apply to newly sown leys after grass has reached 2-leaf stage provided clovers have at least 1 trifoliate leaf and red clover has not passed 3-trifoliate leaf stage. Grasses must have at least 3 leaves before treating with cyanazine mixture
- Do not use on crops suffering from herbicide damage or physical stress
- Do not use on seed crops or on cereals undersown with lucerne
- Do not roll or harrow for 7 d before or after spraying
- Clovers may be scorched and undersown crop checked but effects likely to be outgrown

Special precautions/Environmental safety
- Harmful if swallowed. Irritating to eyes and skin
- Keep livestock out of treated areas until foliage of poisonous weeds such as ragwort has died and become unpalatable
- Harmful to fish. Do not contaminate ponds, waterways or ditches with chemical or used container

Protective clothing/Label precautions
- A, C
- 5c, 6a, 6b, 14, 18, 21, 28, 29, 36, 37, 43, 53, 63, 66, 70, 78

57   bentazone + MCPB

A post-emergence herbicide mixture for tank mixing with cyanazine

Products	Pulsar	BASF	200:200 g/l	SL	04002

Uses   Annual dicotyledons in PEAS.

Notes

Efficacy
- To be used in tank mix with cyanazine (Fortrol)
- Best results achieved when weeds small and actively growing provided crop at correct stage. Good spray cover is essential
- Do not apply if rain or frost expected or if foliage wet. A minimum of 6 h free from rain required after treatment
- Do not apply during drought or unseasonably cold weather

Crop safety/Restrictions
- Apply only to listed cultivars (see label). Do not treat forage pea cultivars or mange-tout peas
- Apply from 3 fully expanded leaf to before flower buds can be found enclosed in terminal shoot (GS 103 to before GS 210)
- Apply after a satisfactory wax test. Early drilled crops or crops affected by frost or abrasion may not have sufficiently waxy cuticle

- Do not use as tank mix with any other product than cyanazine, nor after use of TCA
- Do not apply to any crop that may have been subjected to stress conditions, where foliage damaged or under hot, sunny conditions when temperature exceeds 21 °C
- Do not apply insecticides or grass herbicides within 7 d of treatment

Special precautions/Environmental safety
- Harmful if swallowed or in contact with skin. Irritating to eyes and skin
- Harmful to fish. Do not contaminate ponds, waterways or ditches with chemical or used container

Protective clothing/Label precautions
- A, C
- 5a, 5c, 6a, 6b, 18, 21, 28, 29, 36, 37, 53, 63, 66, 78

58   benzalkonium chloride
A quaternary ammonium algicide and moss killer for paths, pots etc

| Products | Paramos | Chemsearch | 4.5% w/w | SL | H2060 |

Uses   Algae, mosses in GLASSHOUSE BENCHES, FLOWER POTS, PATHS, WALLS, ROOFS.

Notes   **Efficacy**
- Spray, sprinkle or brush on paths, walls and stonework under dry conditions. Do not use if it has rained within 3-4 d or if rain anticipated within 24 h
- Apply to plant beds or benches before plants are put in place. If applied to pots plants should be removed during spraying or spray directed at base of pots
- Regrowth of moss normally prevented for whole growing season

Crop safety/Restrictions
- Avoid contact with leaves of growing plants

Special precautions/Environmental safety
- Irritating to eyes and skin
- Dangerous to fish. Do not contaminate ponds, waterways or ditches with chemical or used container

Protective clothing/Label precautions
- A, C, H
- 6a, 6b, 14, 16, 18, 21, 29, 36, 37, 52, 63, 67

59   bifenox
A nitrophenyl ether herbicide available only in mixtures

FOR FULL CONDITIONS OF USE ALWAYS READ THE PRODUCT LABEL

bifenox + chlorotoluron

A contact and residual herbicide for blackgrass and other weeds in winter wheat

Products   Dicurane Duo 495FW   Ciba-Geigy   106:389 g/l   SC   03860

Uses   Blackgrass, wild oats, annual grasses, annual dicotyledons in WINTER WHEAT.

Notes

Efficacy
- Weeds are controlled before emergence or as young seedlings. Blackgrass susceptible to 5-leaf, wild oats to 2-leaf stage. See label for other species
- Do not use on soils with more than 10% organic matter
- Autumn application controls most grass weeds germinating in early spring
- Bury or disperse any trash or burnt straw before or during seedbed preparation
- Control may be reduced by prolonged dry conditions after application

Crop safety/Restrictions
- Apply pre-emergence in autumn or from 2-leaf stage to end of tillering (GS 12-29) in autumn or spring
- Use only on listed cultivars sown before 30 Nov. Do not use on undersown crops
- Do not apply to crops severely checked by waterlogging, pest attack, disease, frost, or when frost is imminent
- Risk of crop damage on stony or gravelly soils, especially with heavy rain
- Risk of damage to early sown crops sprayed during rapid growth in autumn
- Do not roll or harrow for 7 d before or after use on emerged crops

Special precautions/Environmental safety
- Irritating to eyes and skin
- Dangerous to fish. Do not contaminate ponds, waterways or ditches with chemical or used container

Protective clothing/Label precautions
- 6a, 6b, 18, 21, 28, 29, 36, 37, 52, 63, 66, 70

---

bifenox + isoproturon

A contact and residual herbicide for use in winter cereals

Products

1 Invicta	Farm Protection	160:400 g/l	SC	04825
2 Invicta	FCC	160:400 g/l	SC	03233

Uses   Annual grasses, annual dicotyledons, blackgrass, field pansy in WINTER WHEAT, WINTER BARLEY.

Notes

Efficacy
- Weeds are controlled from pre-emergence to early tillering of blackgrass, to 2-3 leaf stage of wild oats, dicotyledons to 4 leaf stage
- Best results achieved on a fine, firm seedbed free of clods, trash or straw ash
- Effectiveness may be reduced under dry soil conditions and where waterlogging occurs

- Do not use on soils with more than 10% organic matter
- Do not roll or harrow after application

Crop safety/Restrictions
- Apply pre-emergence to winter wheat only as soon as possible after drilling to at least cm. If emerged weeds present tank mix with Farmon PDQ or Gramoxone 100
- Apply post-emergence to winter wheat from 2-leaf stage to first node detectable (GS 12-31), to winter barley from 3-leaf stage to first node detectable (GS 13-31)
- Do not use pre-emergence on wheat drilled after 30 Nov
- Do not use on durum wheat, undersown crops or those due to be undersown
- Do not spray crops under stress from waterlogging, drought, disease or pest attack
- Do not use post-emergence if frost imminent or after prolonged frosty weather
- Do not roll or harrow during the 7 d before post-emergence application

Special precautions/Environmental safety
- Irritating to eyes and skin

Protective clothing/Label precautions
- 6a, 6b, 18, 21, 25, 28, 29, 36, 37, 54, 60, 63, 66, 70

62 bifenox + isoproturon + mecoprop
A post-emergence herbicide for use in winter cereals

Products Foxstar RP 107:286:143 g/l SC 03274

Uses Blackgrass, wild oats, annual grasses, annual dicotyledons in WINTER WHEAT, WINTER BARLEY.

Notes **Efficacy**
- Grass weeds susceptible to early tillering stage, dicotyledons 5-15 cm (cleavers 10 cm)
- Do not use on soils with more than 10% organic matter
- Do not roll or harrow for 2 mth after spraying
- Do not spray when weeds are fully dormant

Crop safety/Restrictions
- Apply from when crop has at least 1 tiller to first node detectable (GS 21-31)
- Only use on winter wheat and barley, not durum wheat
- Do not use on sands, very light, stony or gravelly soils or soils liable to waterlogging
- Do not treat crops which are under stress for any reason
- Do not treat undersown or broadcast crops
- Crops, especially barley, may show scorch after spraying but effect is soon outgrown
- Do not apply for at least 14 d after rolling or harrowing
- If re-drilling necessary for any reason cultivate to at least 16 cm before drilling

Special precautions/Environmental safety
- Harmful if swallowed. Irritating to eyes and skin

FOR FULL CONDITIONS OF USE ALWAYS READ THE PRODUCT LABEL

- Keep livestock out of treated area until poisonous weeds such as ragwort have died and become unpalatable
- Dangerous to fish. Do not contaminate ponds, waterways or ditches with chemical or used container

Protective clothing/Label precautions
- A, C
- 5c, 6a, 6b, 18, 21, 25, 28, 29, 36, 37, 43, 52, 63, 66, 70, 78

---

63   bifenthrin

A contact and residual insecticide for use in winter cereals, winter oilseed rape and peas

Products	Talstar	DowElanco	100 g/l	EC	04358

Uses
Barley yellow dwarf virus vectors, yellow cereal fly in WINTER CEREALS. Cabbage stem flea beetle, rape winter stem weevil in WINTER OILSEED RAPE. Pea moth, pea aphid in PEAS.

Notes

Efficacy
- Timing of application varies with crop and pest. See label for details
- Treatment may be used on all varieties of peas including vining, combining, forage and edible-podded varieties

Crop safety/Restrictions
- No more than 2 applications should be made on any one crop
- In cereals apply last treatment before end of Mar, in oilseed rape before end of Nov

Special precautions/Environmental safety
- Harmful if swallowed and by inhalation
- Irritating to skin and eyes
- Flammable
- Dangerous to bees. Do not apply at flowering stage except as directed on peas. Keep down flowering weeds in all crops
- Extremely dangerous to fish. Do not contaminate ponds, waterways or ditches with chemical or used container

Protective clothing/Label precautions
- A, C, H
- 5b, 5c, 6a, 6b, 12c, 14, 16, 18, 21, 28, 29, 36, 37, 49, 51, 64, 66, 70, 78

Harvest interval
- zero

---

64   bioallethrin

A pyrethroid insecticide available only in mixtures

65   bioallethrin + permethrin
A residual pyrethroid insecticide mixture for surface application

Products   Insektigun   Spraydex   0.05:0.238% w/w   RH   H3002

Uses   Flies in AGRICULTURAL PREMISES.

Notes   **Efficacy**
- Apply spray to doors, windows and other surfaces covering about 20% of area
- Repeat every 3 wk for continuous control

Crop safety/Restrictions
- Not recommended for use in intensive animal houses as resistance can develop
- Do not spray onto food, grain or animals
- Remove exposed milk and collect eggs before spraying
- Protect milking machinery and containers from contamination

Special precautions/Environmental safety
- Harmful to bees
- Extremely dangerous to fish. Do not contaminate ponds, waterways or ditches with chemical or used container

Protective clothing/Label precautions
- 28, 29, 36, 37, 50, 51, 63

66   Bordeaux mixture
A protectant copper sulphate/lime complex fungicide
See also aluminium sulphate + copper sulphate + sodium tetraborate
    copper sulphate + cufraneb
    copper sulphate + sodium tetraborate
    copper sulphate + sulphur

Products

1 Comac Bordeaux Plus	McKechnie	20% w/w (copper)	WP	00298
2 FS Bordeaux Powder	Ford Smith	8.5% w/w (copper)	WP	04642

Uses   Blight in POTATOES [1, 2]. Celery leaf spot in CELERY [2]. Canker in APPLES [1, 2]. Canker in PEARS [1]. Canker in CHERRIES, PLUMS [2]. Leaf curl in PEACHES [1, 2]. Leaf curl in APRICOTS, NECTARINES [1]. Downy mildew in GRAPEVINES, HOPS [1]. Currant leaf spot in BLACKCURRANTS, GOOSEBERRIES [2]. Rust in BLACKCURRANTS [2].

Notes   **Efficacy**
- Spray interval normally 7-14 d but varies with disease and crop, see label for details

---

FOR FULL CONDITIONS OF USE ALWAYS READ THE PRODUCT LABEL

- Commence spraying potatoes before crop meets in row or immediately first blight period occurs
- For canker control spray monthly from Aug to Oct
- For peach leaf curl control spray at leaf fall in autumn and again in Feb
- Spray when crop foliage dry. Do not spray if rain imminent

Crop safety/Restrictions
- Do not use on copper sensitive cultivars, including Doyenne du Comice pears

Special precautions/Environmental safety
- Harmful if swallowed. Irritating to eyes, skin and respiratory system
- Harmful to fish. Do not contaminate ponds, waterways or ditches with chemical or used container

Protective clothing/Label precautions
- 5c, 6a, 6b, 6c, 25, 27, 29, 41, 53, 63, 67

Withholding period
- Harmful to livestock. Keep all livestock out of treated areas for at least 3 wk

Harvest interval
- 7 d

---

67   bromacil

A soil acting uracil herbicide for non-crop areas and cane fruit
See also amitrole + bromacil + diuron
    atrazine + bromacil + diuron

Products

1 Borocil 1.5	ABM	1.5% w/w	GR	04202
2 Hyvar X	Du Pont	80% w/w	WP	01105
3 Hyvar X	Selectokil	80% w/w	WP	01105

Uses

Annual dicotyledons, annual grasses, couch, perennial grasses in RASPBERRIES, BLACKBERRIES, LOGANBERRIES [2, 3]. Total vegetation control in NON-CROP AREAS [1-3].

Notes

Efficacy
- Apply to established cane fruit as soon as possible in spring after cultivation and before bud break. Avoid further soil disturbance for as long as possible [2, 3]
- If interrow areas are cultivated only a 30 cm band either side of row need be treated [2, 3]
- Best results achieved when soil is moist at time of application
- For total vegetation control apply to bare ground or standing vegetation. Adequate rainfall is needed to carry chemical into root zone
- Against existing vegetation best results achieved in late winter to early spring but application also satisfactory at other times

Crop safety/Restrictions
- May be used on cane fruit established for at least 2 yr [2, 3]
- In Scotland only may be used on newly planted raspberries at a reduced rate immediately after planting followed by light ridging [2, 3]

- Do not use in last 2 yr before grubbing crop to avoid injury to subsequent crops. Carrots, lettuce, beet, leeks and brassicas are extremely sensitive [2, 3]
- When used non-selectively take care not to apply where chemical can be washed into root zone of desirable plants
- Treated land should not be cropped within 3 full yr of treatment and then only after obtaining advice from manufacturer

Special precautions/Environmental safety
- Irritating to eyes, skin and respiratory system [2, 3]

Protective clothing/Label precautions
- 6a, 6b, 6c, 18, 22, 28, 29, 36, 37, 54, 63, 65 [2, 3]

---

68 bromacil + diuron
A root-absorbed residual total herbicide mixture

Products

1 Borocil K	ABM	0.88:0.88% w/w	GR	0428
2 Krovar 1	Du Pont	40:40% w/w	WP	0116
3 Krovar 1	Selectokil	40:40% w/w	WP	0116

Uses

Total vegetation control in NON-CROP AREAS.

Notes

Efficacy
- Spray or apply granules in early stage of weed growth at any time of year, provided adequate moisture to activate chemical supplied by rainfall
- Use higher rates on adsorptive soils or established weed growth
- Do not apply when ground frozen

Crop safety/Restrictions
- Do not apply on or near trees, shrubs, crops or other desirable plants
- Do not apply where roots of desirable plants may extend or where chemical may be washed into contact with their roots
- Do not use on ground intended for subsequent cultivation

Special precautions/Environmental safety
- Irritating to eyes, skin and respiratory system [2, 3]
- Harmful to fish. Do not contaminate ponds, waterways or ditches with chemical or used container [2, 3]

Protective clothing/Label precautions
- 29, 54, 63, 67 [1]; 6a, 6b, 6c, 18, 22, 28, 29, 36, 37, 53, 63, 67 [2, 3]

FOR FULL CONDITIONS OF USE ALWAYS READ THE PRODUCT LABEL

bromacil + pentachlorophenol
A contact and residual total herbicide mixture

Fenocil   Chemsearch   2.5:3.3% w/w   SC   00856

Total vegetation control in NON-CROP AREAS.

Efficacy
* Application at early stage of weed growth kills top growth in 2-3 d and prevents further growth for up to full season

Crop safety/Restrictions
* Do not apply on or near trees, shrubs, crops or other desirable plants
* Do not apply where roots of desirable plants may extend or where chemical may be washed into contact with their roots
* Do not use on ground intended for subsequent cultivation

Special precautions/Environmental safety
* Harmful in contact with skin and if swallowed
* Irritating to skin, eyes and respiratory system
* Flammable
* Keep unprotected persons, especially children, off treated turf and grassland for at least 2 wk or until after watering or heavy rain
* Dangerous to fish. Do not contaminate ponds, waterways or ditches with chemical or used container

Protective clothing/Label precautions
* A, C
* 5a, 5c, 6a, 6b, 6c, 12c, 14, 16, 18, 21, 25, 28, 29, 36, 37, 38, 41, 52, 63, 65, 70, 78

Withholding period
* Harmful to livestock. Keep all livestock out of treated areas for at least 2 wk

70 bromacil + picloram
A persistent residual and translocated herbicide mixture

Products Hydon   Chipman   1.37:1.57% w/w   GR   01088

Uses Total vegetation control in NON-CROP AREAS.

Notes
Efficacy
* Apply at any time from spring to autumn. Best results achieved in Mar-Apr when weeds 50-75 mm high
* Do not apply in very dry weather as moisture needed to carry chemical to roots
* May be used on high fire risk sites

Crop safety/Restrictions
* Do not apply near crops, cultivated plants and trees
* Do not apply on slopes where run-off to cultivated plants or water courses may occur

Special precautions/Environmental safety
• Keep livestock out of treated area until foliage of any poisonous weeds such as ragwort has died and become unpalatable
• Harmful to fish. Do not contaminate ponds, waterways or ditches with chemical or used container

Protective clothing/Label precautions
• 28, 29, 43, 53, 63, 67

71 bromadiolone
An anti-coagulant rodenticide

Products					
1 Rentokil Deadline Liquid Concentrate	Rentokil	0.1% w/w	CB	01737
2 Slaymor	Ciba-Geigy	0.005% w/w	RB	01958
3 Slaymor Bait Bags	Ciba-Geigy	0.005% w/w	RB	03183
4 Slaymor Liquid Concentrate	Ciba-Geigy	0.1% w/w	CB	04826

Uses   Rats, mice in FARM BUILDINGS, FARMYARDS.

Notes   **Efficacy**
• Ready-to-use baits are formulated on a mould-resistant, whole-wheat base [2, 3]
• Use in baiting programme. Place baits in protected situations, sufficient for continuous feeding between treatments
• Chemical is effective against warfarin- and coumatetralyl-resistant rats and mice and does not induce bait shyness
• Place bait bags behind ricks, silage clamps and elsewhere where loose baiting inconvenient [3]
• Mix liquid concentrate with suitable base at recommended rate. See label for details [1 4]

Special precautions/Environmental safety
• Cover bait to prevent access by children, animals and birds

Protective clothing/Label precautions
• 25, 29, 63, 65, 79, 81, 82, 84

72 bromoxynil
A contact acting HBN herbicide available only in mixtures
See also benazolin + bromoxynil + ioxynil
benazolin + bromoxynil + ioxynil + mecoprop

FOR FULL CONDITIONS OF USE ALWAYS READ THE PRODUCT LABEL

73   bromoxynil + chlorsulfuron + ioxynil

A twin pack sulphonyl-urea/HBN herbicide mixture for cereals

Products	Glean TP	Du Pont	x:5:250-x g a.i.	KK	02429

Uses   Annual dicotyledons, mayweeds, knotgrass, chickweed, hemp-nettle in BARLEY, WHEAT.

Notes

Efficacy
- Apply in spring when weeds in cotyledon to 4-leaf stage
- Only treat winter crops when weeds small
- Under moist soil conditions residual action controls weeds germinating after application
- May be used on all soil types but weed control may be reduced when soil very dry
- Add non-ionic wetter to improve control of corn marigold
- Mix two ingredients in spray tank and not before

Crop safety/Restrictions
- Apply to crops from 2-fully expanded leaf stage to first node detectable (GS 12-31)
- Do not use on crops suffering stress as a result of drought, waterlogging, low temperature or other factors
- Do not use on crops undersown with grasses, clovers or other legumes
- Do not apply within 7 d of rolling
- Do not use on any crop previously treated with other sulphonyl-urea herbicides
- In the event of crop failure do not sow any broad-leaved crop within 3 mth
- Only cereals, oilseed rape, field beans or grass may be sown in the same calendar year after a treated crop
- Take extreme care to avoid drift or contamination from spray equipment onto broad-leaved crops, land intended for cropping, ponds, waterways and ditches
- To avoid subsequent damage to crops other than cereals, immediately after spraying throughly clean spraying equipment using specified procedure

Special precautions/Environmental safety
- Do not apply with hand-held equipment or at concentrations higher than those recommended
- Harmful to fish. Do not contaminate ponds, waterways or ditches with chemical or used container

Protective clothing/Label precautions
- A, C
- 18, 21, 25, 28, 29, 36, 37, 53, 57, 60, 63, 66, 70, 78

74   bromoxynil + clopyralid

A post-emergence contact and translocated herbicide used in tank mixes for cereals

Products	Vindex	Quadrangle	240:50 g/l	LI	04049

Uses   Annual dicotyledons in WHEAT, BARLEY, OATS, TRITICALE, LINSEED.

Efficacy
- Recommended for tank mixing with approved formulations of MCPA, mecoprop, fluroxypyr, isoproturon and bentazone
- Best results achieved when weeds small and growing actively in warm, moist weather
- Do not spray during drought, waterlogging, frost, extremes of temperature or if rain is imminent
- Application rate and weed spectrum vary according to other components of tank mixtures. See label for details

Crop safety/Restrictions
- Safe times for spraying crops vary according to other components of tank mixtures. See label for details. Do not apply after first node detectable stage (GS 31)
- Do not spray cereals undersown or to be undersown
- Spraying in frosty weather or when hard frost occurs within 3-4 wk may result in leaf scorch. This effect normally outgrown but may reduce yield of barley of low vigour or under stress on light soils
- Straw from treated crops must not be used in compost or manure for growing tomatoes under glass, but may be used for strawing down strawberries
- Do not roll, harrow or graze crops within 7 d before or after spraying

Special precautions/Environmental safety
- Harmful in contact with skin or if swallowed. Irritating to skin and eyes
- Flammable
- Do not apply with hand-held equipment or at concentrations higher than those recommended
- Dangerous to fish. Do not contaminate ponds, waterways or ditches with chemical or used container

Protective clothing/Label precautions
- A, C
- 5a, 5c, 6a, 6b, 12c, 18, 21, 25, 28, 29, 36, 37, 52, 63, 66, 70, 78

75   bromoxynil + dichlorprop + ioxynil

A post-emergence contact and translocated herbicide for spring cereals

Products	Chafer Certrol-E	BritAg	70:210:105 g/l	EC	00468

Uses   Mayweeds, polygonums, annual dicotyledons in SPRING WHEAT, SPRING BARLEY, SPRING OATS.

Efficacy
- Best results when weeds small and growing actively in a strongly competitive crop
- Do not spray during drought, if rain expected or when night temperatures low [1]
- May be tank mixed with difenzoquat (Avenge 2) for wild oats control. See label for details and other recommended combinations

FOR FULL CONDITIONS OF USE ALWAYS READ THE PRODUCT LABEL

Crop safety/Restrictions
- Apply to crops from 3-fully expanded leaf stage to first node detectable (GS 13-31)
- May be used on crops undersown with grass when grass beyond 4-leaf stage but before crop reaches jointing stage (GS 31)

Special precautions/Environmental safety
- Harmful if swallowed. Irritating to eyes and skin
- Do not apply with hand-held equipment or at concentrations higher than those recommended
- Keep livestock out of treated areas until foliage of any poisonous weeds such as ragwort has died and become unpalatable
- Dangerous to fish. Do not contaminate ponds, waterways or ditches with chemical or used container

Protective clothing/Label precautions
- A, C
- 5c, 6a, 6b, 18, 21, 25, 28, 29, 36, 37, 43, 52, 63, 66, 70, 78

76   bromoxynil + dichlorprop + ioxynil + MCPA

A broad-spectrum, post-emergence contact and translocated herbicide for spring cereals

Products					
1 Actril S	RP	24.3:190:38:235 g/l	SL	04382
2 Atlas Minerva	Atlas	28:336:28:108 g/l	SL	03046

Uses

Mayweeds, chickweed, polygonums, annual dicotyledons in SPRING WHEAT, SPRING BARLEY, SPRING OATS.

Notes

Efficacy
- Best results when weeds small and growing actively in a strongly competitive crop
- Do not spray when rain imminent or when growth is hard from cold or drought

Crop safety/Restrictions
- Apply to crops from 3-fully expanded leaf to before first node detectable (GS 13-30)
- Apply to crops undersown with grass after grass has reached 3- leaf stage [1]
- Do not spray crops undersown with clover mixtures or before undersowing grass or clover
- Scorch may occur if spray applied during periods of night or wind frost
- Do not roll or harrow within a few days before or after spraying

Special precautions/Environmental safety
- Harmful if swallowed [1, 2]. Irritating to eyes [1]
- Do not apply with hand-held equipment or at concentrations higher than those recommended
- Keep livestock out of treated areas until foliage of any poisonous weeds such as ragwort has died and become unpalatable
- Dangerous to fish. Do not contaminate ponds, waterways or ditches with chemical or used container

77 bromoxynil + ethofumesate + ioxynil
A post-emergence herbicide for new grass leys

Products	Nortron Leyclene	Schering	50:200:25 g/l	EC	03831

Uses Annual meadow grass, annual dicotyledons in NEWLY SOWN LEYS.

Notes **Efficacy**
• Best results when weeds small and growing actively in a vigorous crop, soil moist and further rain within 10 d. Mid Oct to end Dec normally suitable
• Annual meadow grass controlled during early crop establishment. Spray weed grasses before fully tillered
• Do not spray in cold conditions or when heavy rain or frost imminent
• Do not use on soils with more than 10% organic matter
• Do not cut for 14 d after spraying or graze in Jan-Feb after spraying in Oct-Dec
• Ash or trash should be burned, buried or removed before spraying

Crop safety/Restrictions
• Apply to healthy ryegrasses and tall fescue when crop has at least 2-3 leaves
• Apply to healthy cocksfoot, timothy and meadow fescue at least 60 d after seedling emergence provided crop has 2-3 leaves
• Do not use on crops under stress, during periods of very dry weather, prolonged frost or waterlogging
• Do not use where clovers or other legumes are valued components of ley
• Do not roll for 7 d before or after spraying
• Any crop may be sown 5 mth after application following ploughing to at least 15 cm

Special precautions/Environmental safety
• Harmful if swallowed. Irritating to skin and eyes
• Flammable
• Do not apply with hand-held equipment
• Do not allow spray from ground sprayer to fall within 6 m of surface water or ditches
• Dangerous to fish. Do not contaminate ponds, waterways or ditches with chemical or used container

Protective clothing/Label precautions
• A, C
• 5c, 6a, 6b, 12c, 18, 21, 25, 29, 36, 52, 63, 66, 70, 78

Withholding period
• Keep livestock out of treated areas for at least 6 wk and until foliage of any poisonous weeds such as ragwort has died and become unpalatable

FOR FULL CONDITIONS OF USE ALWAYS READ THE PRODUCT LABEL

78 bromoxynil + fluroxypyr

A post-emergence contact and translocated herbicide for spring cereals

| Products | Sickle | DowElanco | 300:150 g/l | EC | 03781 |

Uses Annual dicotyledons in SPRING BARLEY, SPRING WHEAT.

Notes **Efficacy**
• Best results when weeds small and growing actively in a strongly competitive crop
• Do not spray when night temperatures low
• Spray is rainfast after 2 h

Crop safety/Restrictions
• Apply from 2-fully expanded leaf stage of crop to first node detectable (GS 12-31)
• Crops undersown with grass may be sprayed provided the grasses are tillering. Do not spray cereals undersown with clover or other legume mixtures
• Do not spray when crops are under stress from cold, drought, waterlogging etc, nor when frost imminent
• Do not roll or harrow for 10 d before or 7 d after spraying

Special precautions/Environmental safety
• Harmful if swallowed. Irritating to eyes and skin
• Flammable
• Do not apply with hand-held equipment or at concentrations higher than those recommended
• Dangerous to fish. Do not contaminate ponds, waterways or ditches with chemical or used container

Protective clothing/Label precautions
• A, C
• 5c, 6a, 6b, 12c, 18, 21, 25, 28, 29, 36, 37, 52, 63, 66, 70, 78

79 bromoxynil + fluroxypyr + ioxynil

A post-emergence contact and translocated herbicide for cereals

| Products | Advance | ICI | 100:90:100 g/l | LI | 03002 |

Uses Cleavers, chickweed, hemp-nettle, speedwells, annual dicotyledons in WHEAT, BARLEY, OATS, RYE, DURUM WHEAT, TRITICALE.

Notes **Efficacy**
• May be used in autumn or spring on winter or spring crops
• Best results when weeds small and growing actively in a highly competitive crop
• Do not spray when weed growth is hard from cold or drought

Crop safety/Restrictions
- Apply from 2-leaf stage of crop to first node detectable (GS 12-31)
- On oats apply only in spring when danger of frost is over
- Do not spray undersown crops or crops to be undersown with clover or legume mixtures
- Do not spray crops stressed by frost, drought, mineral deficiency, pest or disease attack
- Do not roll or harrow for 7 d before or after spraying

Special precautions/Environmental safety
- Harmful if swallowed. Irritating to eyes and skin
- Do not apply with hand-held equipment or at rates higher than those recommended
- Dangerous to fish. Do not contaminate ponds, waterways or ditches with chemical or used container

Protective clothing/Label precautions
- A, C
- 5c, 6a, 6b, 18, 21, 24, 25, 28, 29, 36, 37, 52, 60, 63, 66, 70, 78

80   bromoxynil + ioxynil
A contact acting post-emergence HBN herbicide for cereals

Products					
	1 Briotril Plus	PBI	200:200 g/l	EC	02881
	2 Deloxil	Hoechst	190:190 g/l	EC	00664
	3 Hobane	Farm Protection	240:160 g/l	EC	02704
	4 Novacorn	FCC	240:160 g/l	EC	02816
	5 Oxytril CM	RP	200:200 g/l	EC	04605
	6 Stellox 380EC	Ciba-Geigy	190:190 g/l	EC	04169

Uses

Annual dicotyledons in WHEAT, BARLEY, OATS [1-6]. Annual dicotyledons in RYE [2, 4, 5, 6]. Annual dicotyledons in TRITICALE [2, 5, 6]. Annual dicotyledons in UNDERSOWN CEREALS [2, 5].

Notes

Efficacy
- Best results achieved on young weeds growing actively in a highly competitive crop
- Do not apply during periods of drought or when rain imminent (some labels say 'if likely within 4 or 6 h')
- Recommended for tank mixture with hormone herbicides to extend weed spectrum. See label for details

Crop safety/Restrictions
- Apply to winter or spring cereals from 2-fully expanded leaf stage [1, 2, 3], 1-fully expanded leaf [4, 5, 6] to first node detectable (GS 11 or 12-31)
- Spray oats when danger of frost passed. Do not spray winter oats in autumn [1, 2, 5, 6]
- Timing varies with tank mixtures. See label for details
- Apply to undersown cereals pre-sowing or pre-emergence of legume provided cover crop is at correct stage. Only spray trefoil pre-sowing
- On crops undersown with grasses alone apply from 2-leaf stage of grass
- Do not spray crops stressed by drought, waterlogging or other factors

FOR FULL CONDITIONS OF USE ALWAYS READ THE PRODUCT LABEL

- Do not roll, harrow or graze for several days before or after spraying. Number of days specified varies with product, see label for details

Special precautions/Environmental safety
- Harmful in contact with skin or if swallowed [2, 3]; if swallowed [1, 5, 6]
- Irritating to skin and eyes[1, 3, 5]; to eyes [6]
- Do not apply with hand-held equipment or at concentrations higher than those recommended
- Dangerous to fish. Do not contaminate ponds, waterways or ditches with chemical or used container
- A grazing interval of at least 6 wk must be observed for treated grass [5]

Protective clothing/Label precautions
- A, C
- 5a, 5c [2, 3]; 5c [1, 5, 6]; 6a, 6b [1, 3]; 6a [5, 6]; 18, 21, 25, 28, 29, 36, 37, 63, 66, 70, 78 [1-6]; 16 [2]; 26, 27 [6]; 52 [1, 3, 5, 6]; 53 [2, 4]

---

81 bromoxynil + ioxynil + isoproturon
A contact and residual herbicide for use in cereals

Products Astrol   Embetec   5:5:60% w/w   SG   04417

Uses Annual dicotyledons, annual grasses, annual meadow grass in WHEAT, BARLEY.

Notes **Efficacy**
- Spray as early as possible in autumn or spring provided crop at correct stage when weeds small and adequate soil moisture present
- Control may be reduced if very dry or wet conditions persist after application
- Bury any trash or burnt straw during preparation of seedbed
- On highly organic soils weeds only controlled after emergence
- Blackgrass and wild oats can be controlled in winter crops at higher rates
- Recommended for tank mixing with hormone and other herbicides to extend weed spectrum. See label for details

Crop safety/Restrictions
- Ensure crop seed well covered at sowing
- Apply from 2-fully expanded leaf stage to first node detectable (GS 12-31)
- Do not treat triticale, durum wheat, rye, oats or cereals undersown or to be undersown
- Do not treat crops severely damaged by pest or disease attack or suffering stress resulting from drought or prolonged frost
- Winter wheat or barley may suffer damage if hard frost follows application but crop normally recovers
- Do not roll or harrow within 7 d before or after spraying

Special precautions/Environmental safety
- Do not apply with hand-held equipment or at less than recommended volumes
- Dangerous to fish. Do not contaminate ponds, waterways or ditches with chemical or used container

Protective clothing/Label precautions
- A, C

82 bromoxynil + ioxynil + isoproturon + mecoprop
A contact, translocated and residual herbicide for use in cereals

Products	Terset	RP	5.4:5.4:32.5:19% w/w	SG	02944

Uses   Annual dicotyledons, annual meadow grass in WHEAT, BARLEY.

Notes   **Efficacy**
● Spray as early as possible in autumn or spring provided crop at correct stage when weeds in active growth and adequate soil moisture present
● Control may be reduced if very dry or wet conditions persist after application
● On highly organic soils weeds only controlled after emergence
● Do not spray when weeds fully dormant

Crop safety/Restrictions
● Apply to winter or spring crops from 3-fully expanded leaf stage to first node detectable (GS 13-31)
● Do not treat triticale, durum wheat, rye, oats, broadcast crops, undersown cereals or cereals about to be undersown
● Do not spray crops suffering from stress, pests, diseases or deficiency
● If hard frost occurs within 3-4 wk of spraying winter barley of low vigour on light soils or subject to stress yield may not be optimum
● Do not roll or harrow within 7 d before or after spraying

Special precautions/Environmental safety
● Irritating to eyes and skin
● Do not apply with hand-held equipment or at less than recommended volume
● Dangerous to fish. Do not contaminate ponds, waterways or ditches with chemical or used container

Protective clothing/Label precautions
● A, C
● 6a, 6b, 18, 21, 25, 28, 29, 36, 37, 52, 63, 66, 70, 78

83 bromoxynil + ioxynil + mecoprop
A post-emergence contact and translocated herbicide for cereals

Products	Swipe 560 EC	Ciba-Geigy	56:56:448 g/l	EC	02057

Uses   Annual dicotyledons in WHEAT, BARLEY, OATS, TRITICALE, DURUM WHEAT, WINTER RYE, NEWLY SOWN RYEGRASS.

FOR FULL CONDITIONS OF USE ALWAYS READ THE PRODUCT LABEL

Efficacy
- Best results when weeds small and growing actively in a strongly competitive crop
- Do not spray in rain or when rain imminent. Control may be reduced by rain in 6 h
- Recommended for tank-mixing with approved MCPA-amine for hemp nettle control. See label for other recommended mixtures

Crop safety/Restrictions
- Apply to winter crops from 3 leaves unfolded (GS 13) or in spring from 3 leaves unfolded to first node detectable (GS 31)
- Apply to spring crops from 3 leaves unfolded to first node detectable (GS 13-31)
- Do not spray oats or durum wheat in autumn
- Apply to crops undersown with ryegrass from 2-3 leaf stage of grass provided crop at correct stage
- Do not spray crops undersown with clover or other legumes
- Do not spray crops under stress from frost, waterlogging, drought or other causes
- Do not roll or harrow within 5 d after spraying

Special precautions/Environmental safety
- Harmful in contact with skin or if swallowed. Irritating to eyes and skin
- Do not apply with hand-held equipment or at less than recommended volumes
- Keep livestock out of treated areas until foliage of any poisonous weeds such as ragwort has died and become unpalatable
- Dangerous to fish. Do not contaminate ponds, waterways or ditches with chemical or used container

Protective clothing/Label precautions
- A, C
- 5a, 5c, 6a, 6b, 18, 21, 25, 26, 27, 28, 29, 35, 36, 37, 41, 52, 63, 66, 70, 78

Withholding period
- Where used on grassland intended for animal consumption there must be a grazing/harvest interval of at least 6 wk

84 bupirimate

A systemic pyrimidine fungicide active against powdery mildew

Products	Nimrod	ICI	250 g/l	EC	01498

Uses

Powdery mildew in COURGETTES, CUCUMBERS, MARROWS, APPLES, PEARS, RASPBERRIES, BLACKCURRANTS, GOOSEBERRIES, STRAWBERRIES, HOPS, ROSES, CHRYSANTHEMUMS, ORNAMENTALS. Powdery mildew in TOMATOES *(off-label)*.

Notes

Efficacy
- Apply before or at first signs of disease and repeat at 5-14 d intervals. Timing and maximum dose vary with crop. See label for details
- On apples lower rates weekly give better results than higher rates fortnightly during periods conducive to disease
- Addition of Agral wetter recommended on roses when foliage mature and shiny
- Product has negligible effect on Phytoseiulus and Encarsia and may be used in conjunction with biological control of red spider mite

Crop safety/Restrictions
- With apples, hops and ornamentals cultivars may vary in sensitivity to spray. See label for details
- If necessary to spray cucurbits in winter or early spring spray a few plants 10-14 d before spraying whole crop to test for likelihood of leaf spotting problem
- On roses some leaf puckering may occur on young soft growth in early spring or under low light intensity. Avoid use of high rates or wetter
- Never spray flowering begonias (or buds showing colour) as this can scorch petals
- Do not mix with other chemicals for application to begonias, cucumbers or gerberas

Special precautions/Environmental safety
- Irritating to eyes and skin
- Flammable
- Harmful to fish. Do not contaminate ponds, waterways or ditches with chemical or used container

Protective clothing/Label precautions
- A, C
- 6a, 6b, 12c, 18, 29, 35, 36, 37, 51, 63, 66, 70

Harvest interval
- apples, pears, strawberries 1 d; cucurbits, tomatoes 2 d; blackcurrants 7 d; raspberries 8 d; gooseberries, hops 14 d

Approval
- Off-label approval to Feb 1991 for use on tomatoes (OLA 0969/88)

---

85   bupirimate + triforine

A systemic protectant and eradicant fungicide for ornamental crops

Products   Nimrod-T   ICI Professional   62.5:62.5 g/l   EC   01499

Uses   Black spot, leaf spots, powdery mildew in ORNAMENTALS. Rust in ROSES.

Notes   **Efficacy**
- If disease already present on roses give double strength spray at first application
- To prevent infection on roses spray in early May and repeat every 10-14 d. If infected with blackspot in previous season commence at bud burst

Crop safety/Restrictions
- Spraying in high glasshouse temperatures may cause temporary leaf damage
- Test varietal susceptibility of roses by spraying a few plants and allow 14 d for any symptoms to develop

Special precautions/Environmental safety
- Irritating to eyes
- Harmful to fish. Do not contaminate ponds, waterways or ditches with chemical or used container

---

FOR FULL CONDITIONS OF USE ALWAYS READ THE PRODUCT LABEL

86 calciferol

A hypercalcaemic rodenticide

Products Hyperkil Bait Antec 0.1% w/w RB 04310

Uses Rats, mice in FARM BUILDINGS.

Notes **Efficacy**
• Place small baits for mice, larger baits for rats, on trays in protected situations where rodents known to be active
• Check baits within 7 d and if more than half bait eaten replace with twice the quantity
• May be used indoors and outdoors
• Controls warfarin resistant rats and mice. Does not induce bait shyness

Crop safety/Restrictions
• Prevent access to baits by children and domestic animals

Protective clothing/Label precautions
• 25, 29, 63, 67, 81, 82, 83, 84

87 calciferol + difenacoum

A mixture of rodenticides

Products 1 Sorexa CD Concentrate Sorex 2:0.1% w/w CB 03513
2 Sorexa CD Ready to Use Sorex 0.1:0.0025% w/w RB 03514

Uses Mice in FARM BUILDINGS.

Notes **Efficacy**
• Mix two components of concentrate with cereal base thoroughly as directed and leave for 2-3 d before using
• Lay baits in mouse runs in many locations throughout infested area
• It is important to lay many small baits as mice are sporadic feeders
• Cover baits to protect from moisture in outdoor situations
• Inspect baits frequently and replace or top up with fresh material until they remain untouched

Special precautions/Environmental safety
• Harmful in contact with skin or if swallowed [1]
• Protect baits from access by children, domestic or other animals

Protective clothing/Label precautions
• A, C [1]

• 5a, 5c, 14, 18, 21, 29, 36, 37, 63, 67, 70, 81, 83, 84 [1]; 25, 29, 63, 67, 81, 82, 83, 84 [2]

88   captafol

A protectant dicarboximide fungicide no further use of which will be permitted after 31 December 1990

89   captan

A protectant dicarboximide fungicide

Products					
1 Captan 83	DowElanco	83% w/w	WP	04231
2 Fargro Captan 83	Fargro	83% w/w	WP	02919
3 Hortag Captan 83	Avon	83% w/w	WP	02788
4 Hortag Captan Dust	Avon	15% w/w	DP	01069
5 PP Captan 83	ICI	83% w/w	WP	01619

Uses

Scab in APPLES, PEARS [1-3, 5]. Gloeosporium rot in APPLES [1-5]. Phytophthora fruit rot in APPLES [1, 2, 5]. Botrytis in STRAWBERRIES [1-4]. Didymella in TOMATOES [2, 3, 5]. Black spot in ROSES [1-3, 5]. Botrytis in CHRYSANTHEMUMS, CYCLAMENS, RASPBERRIES, ORNAMENTALS [4].

Notes

Efficacy
• Spray every 7-10 d to maintain protection. Rate, timing and interval vary with crop and disease. See label for details
• Sprays may be applied to apples and pears during blossom stage. Late summer and autumn sprays reduce incidence of gloeosporium rot
• Apply as post-harvest dip to reduce phytophthora rot in stored apples
• Thorough wetting is essential for botrytis control in strawberries. One spray must be applied immediately after strawing
• Spray stems thoroughly at ground level for didymella control in tomatoes
• Do not leave diluted material for more than 2 h. Agitate well before spraying
• Do not mix with strongly alkaline materials or sprays containing oils

Crop safety/Restrictions
• Do not use on apple cultivars Bramley, Monarch, Winston, Edward VII, Kidd's Orange or Red Delicious or on pear cultivar D'Anjou
• Do not use on strawberries grown for canning

Special precautions/Environmental safety
• Irritating to eyes, skin and respiratory system
• Harmful to fish. Do not contaminate ponds, waterways or ditches with chemical or used container

Protective clothing/Label precautions
• 6a, 6b, 6c, 18, 28, 29, 35, 36, 37, 53, 63, 65, 70

FOR FULL CONDITIONS OF USE ALWAYS READ THE PRODUCT LABEL

• apples, pears 7 d [1, 4], zero [6]; strawberries zero [3]

90 captan + fosetyl-aluminium + thiabendazole
A fungicide seed dressing for use on peas

Products   Aliette Extra   Embetec   17.2:52.8:12.9%   WS   04410
                                                                    w/w

Uses   Damping off, ascochyta, downy mildew in PEAS. Downy mildew, damping off in HARDY ORNAMENTAL NURSERY STOCK *(off-label)*.

Notes   **Efficacy**
• Apply seed treatment as slurry through continuous flow seed dresser or as drench as appropriate
• Calibrate equipment with product before treating seed and calibrate seed drill for treated seed before sowing
• Keep dressed seed in a dry, draught-free store and sow as soon as possible after dressing. Do not store for more than 6 mth

Crop safety/Restrictions
• Consult processor before treating seed for processing crops

Special precautions/Environmental safety
• Irritating to eyes and skin
• Harmful to fish. Do not contaminate ponds, waterways or ditches with chemical or used container
• Treated seed not to be used as food or feed. Do not re-use sack

Protective clothing/Label precautions
• 6a, 6b, 18, 28, 29, 36, 37, 53, 63, 65, 70, 72, 73, 74, 76

Approval
• Off-label Approval to Sep 1993 for use on hardy ornamental nursery stock (OLA 0509/90)

91 captan + gamma-HCH
A fungicide and insecticide seed dressing for brassicas and linseed

Products   Gammalex   ICI   10:75% w/w   DS   00965

Uses   Flea beetles in BRASSICAS, LINSEED, OILSEED RAPE. Cabbage stem flea beetle in OILSEED RAPE. Seed-borne diseases in BRASSICAS, LINSEED, OILSEED RAPE.

Notes   **Efficacy**
• Apply to seed by shaking in closed container or end-over-end drum treater
• To prolong activity against flea beetles mix seed thoroughly with linseed oil or paraffin before adding Gammalex and drill within 1 wk

- Provides protection from sowing to first true leaf stage

Crop safety/Restrictions
- Do not treat seed with more than 12% moisture content
- Treated seed without oil may be stored for 6 mth provided moisture content does not exceed 12%
- When sowing treated seed, drill may need adjusting to maintain seeding rate

Special precautions/Environmental safety
- Harmful in contact with skin, by inhalation and if swallowed
- Irritating to skin, eyes and respiratory system
- Harmful to fish. Do not contaminate ponds, waterways or ditches with chemical or used container
- Treated seed not to be used as food or feed. Do not re-use sack

Protective clothing/Label precautions
- F
- 5a, 5b, 5c, 6a, 6b, 6c, 16, 18, 29, 36, 37, 53, 63, 65, 70, 72, 73, 74, 78

92 captan + nuarimol
A systemic and protectant fungicide for apples and pears

Products Kapitol DowElanco 72.6:2.4% w/w SP 03063

Uses Powdery mildew in APPLES. Scab in APPLES, PEARS.

Notes **Efficacy**
- Spray at 7-10 d intervals from bud burst until terminal growth ceases
- A full spray programme will reduce primary mildew in the next season
- May be tank mixed with many insecticides, acaricides and nutrients. See label for details

Crop safety/Restrictions
- May be used on all commonly grown dessert and Bramley apples and pears
- Up to 15 treatments per yr may be applied with a minimum interval of 7 d between treatments

Special precautions/Environmental safety
- Harmful to fish. Do not contaminate ponds, waterways or ditches with chemical or used container

Protective clothing/Label precautions
- 22, 25, 28, 29, 35, 36, 37, 53, 61, 63, 65

Harvest interval
- apples, pears 14 d

FOR FULL CONDITIONS OF USE ALWAYS READ THE PRODUCT LABEL

captan + penconazole
A protectant fungicide for use on apple trees

roducts	Topas C 50WP	Ciba-Geigy	47.5:2.5% w/w	WP	03232

ses Powdery mildew, scab in APPLES.

otes **Efficacy**
* Use as a protective spray every 7-14 d from bud burst until extension growth ceases
* High antisporulant activity reduces development of primary mildew and controls spread of secondary mildew
* Does not affect beneficial insects so can be used in integrated control programmes
* Penconazole is recommended alone for mildew control if scab is not a problem

Crop safety/Restrictions
* Do not use more than 10 sprays containing penconazole per season

Special precautions/Environmental safety
* Irritating to eyes, skin and respiratory system
* Dangerous to fish. Do not contaminate ponds, waterways or ditches with chemical or used container

Protective clothing/Label precautions
* A, C
* 6a, 6b, 6c, 14, 18, 22, 29, 35, 36, 37, 52, 63, 70

Harvest interval
* apples 14 d

carbaryl
A contact carbamate insecticide, worm killer and fruit thinner

Products	1 Carbaryl 45 Turf Wormkiller	Synchemicals	450 g/l	SC	03732
	2 Microcarb	Micro-Bio	50% w/w	WP	01339
	3 Murvin 85	DowElanco	85% w/w	WP	04234
	4 Thinsec	ICI	450 g/l	SC	02463
	5 Tornado	ICI Professional	450 g/l	SC	03731

Uses Apple leaf midge, apple sucker, capsids, codling moth, earwigs, leafhoppers, rhynchites, tortrix moths, weevils, winter moth in APPLES [3, 4]. Sawflies, chafer grubs in APPLES [3]. Fruit thinning in APPLES [3, 4]. Codling moth, earwigs, pear leaf midge, pear sawfly, pear sucker, rhynchites, tortrix moths, weevils, winter moth in PEARS [3]. Sawflies, earwigs, leaf midges, tortrix moths, capsids, clay-coloured weevil in BLACKCURRANTS [3]. Sawflies, caterpillars, currant clearwing in GOOSEBERRIES [3]. Raspberry beetle, raspberry moth, rhynchites in RASPBERRIES [3]. Chafer grubs, cutworms, leatherjackets, tortrix moths, capsids, weevils in STRAWBERRIES [3]. Flea beetles in BRASSICAS, OILSEED RAPE [3]. Caterpillars, leatherjackets in BRASSICAS, ORNAMENTALS [3]. Caterpillars, cutworms, leatherjackets in LETTUCE [3]. Pea and bean weevils, pea midge, pea moth, thrips in PEAS [3]. Thrips, caterpillars, woodlice in TOMATOES [3]. Earwigs, leatherjackets, midges, leafhoppers, springtails, thrips, whitefly in ORNAMENTALS [3]. Earthworms in TURF, AMENITY GRASS [1, 3, 4, 5]. Red mite, spiders, beetles, fleas, lice in POULTRY HOUSES [2]. Colorado beetle in POTATOES *(off-label)* [3].

<table>
<tr><td>Notes</td><td>

Efficacy
- Application rate and timing vary with crop and pest. See label for details
- Do not use in conjunction with oil-containing sprays or highly alkaline materials
- Effective thinning of apples depends on thorough wetting of foliage and fruitlets. Dose and timing vary with cultivar
- Fruit from treated trees may need picking 5-7 d earlier than from untreated trees
- For earthworm control apply in autumn or spring when worms active. On dry soil apply a light watering after treatment
- For use in poultry houses spray or dust where insects congregate [2]

Crop safety/Restrictions
- Carbaryl is dangerous to pollinating insects and particular care must be taken when spraying close to blossom period or when orchard adjacent to other flowering crops
- Do not spray Laxton's Fortune between pink bud stage and end of first week in Jun
- Do not use on apples between late pink bud and end of first week in Jun except for purpose of fruit thinning
- Do not use on young tomato plants or Delmarvel chrysanthemums [3]
- On plants of unknown sensitivity test first on a small sample
- Do not spray onto eggs and keep off birds [2]

Special precautions/Environmental safety
- Harmful if swallowed
- This product contains an anticholinesterase compound. Do not use if under medical advice not to work with such compounds
- Harmful to fish. Do not contaminate ponds, waterways or ditches with chemical or used container
- Dangerous to bees. Do not apply at flowering stage. Keep down flowering weeds

Protective clothing/Label precautions
- A, C [1-3, 5]
- 2, 5c, 22, 28, 29, 53, 63, 65, 78 [2]; 2, 5c, 18, 22, 29, 35, 36, 37, 48, 53, 63, 67, 70, 78 [1, 3-5]; 60 [4]

Harvest interval
- apples, pears, brassicas, oilseed rape, sugar beet 7 d; raspberries 2 wk; strawberries, blackcurrants 6 wk

Approval
- Approved for aerial application on brassicas, lettuce, peas, apples, pears, blackcurrants, gooseberries, raspberries, strawberries. See introductory notes. Consult firm for details [3]
- Off-label approval to Nov 1991 for use on potatoes for Colorado beetle control as required by Statutory Notice (OLA 1386/88) [3]

</td></tr>
</table>

FOR FULL CONDITIONS OF USE ALWAYS READ THE PRODUCT LABEL

95   carbaryl + pyrethrins
An insecticide formulation for ULV spraying in poultry houses

Products   Microcarb-T   Micro-Bio   UL   01343

Uses   Red mite, spiders, beetles, lice, fleas, flies in POULTRY HOUSES.

Notes   **Efficacy**
* Apply with Turbair sprayer only
* Direct spray towards concentrations of insects from a distance of 3-4.5 m

Crop safety/Restrictions
* Use well away from naked flame or hot surfaces and do not smoke while spraying
* Do not spray on eggs

Special precautions/Environmental safety
* Harmful if swallowed
* This product contains an anticholinesterase compound. Do not use if under medical advic not to work with such compounds
* Inflammable

Protective clothing/Label precautions
* No information

Approval
* Product formulated for ULV application with Turbair sprayer. See introductory notes

96   carbendazim (mbc)
A systemic fungicide with curative and protectant activity

Products

1 Ashlade Carbendazim Flowable	Ashlade	500 g/l	SC	02662
2 Ashlade Carbendazim WP	Ashlade	50% w/w	WP	02807
3 Battal FL	Farm Protection	500 g/l	SC	00215
4 Bavistin	BASF	50% w/w	WP	00217
5 Bavistin DF	BASF	50% w/w	WG	03848
6 Bavistin FL	BASF	500 g/l	SC	00218
7 Campbell's Carbendazim 50% Flowable	MTM Agrochem.	500 g/l	SC	02681
8 Carbate Flowable	PBI	500 g/l	SC	03341
9 Clifton Carbendazim	Clifton	500 g/l	SC	03032
10 Delsene 50 DF	Du Pont	50% w/w	WG	02692
11 Derosal WDG	Hoechst	80% w/w	WG	03404
12 Hinge	Quadrangle	500 g/l	SC	03060
13 Mascot Systemic Turf Fungicide	Rigby Taylor	500 g/l	SC	02839
14 Maxim	FCC	500 g/l	SC	02712
15 Stempor DG	ICI	50% w/w	WG	02021
16 Tripart Defensor FL	Tripart	500 g/l	SC	02752
17 Turfclear	Fisons	500 g/l	SC	02253

Uses	Eyespot, rhynchosporium in CEREALS [1-12, 14-16]. Light leaf spot in OILSEED RAPE [3-6, 8-12, 14, 15]. Chocolate spot in FIELD BEANS [1-12, 14-16]. Dollar spot, fusarium patch, red thread in TURF [13, 17]. Earthworms in TURF [17]. Botrytis in DWARF BEANS, NAVY BEANS [1-7, 10, 11, 14, 15]. Anthracnose in DWARF BEANS, NAVY BEANS [3-7]. Alternaria, light leaf spot, ring spot in BRUSSELS SPROUTS, CABBAGES [6]. Celery leaf spot in CELERY [2, 4-6, 8]. Big vein in LETTUCE [4]. Botrytis in ONIONS [4-6, 8-10]. Powdery mildew in SWEDES [10]. Canker, powdery mildew, scab, storage rots in APPLES [4-7, 11, 16]. Scab, storage rots in PEARS [4-7, 11]. American gooseberry mildew, botrytis, currant leaf spot in BLACKCURRANTS, GOOSEBERRIES [1, 2, 4-6, 11, 16]. American gooseberry mildew, botrytis, currant leaf spot in BLACKCURRANTS [10]. Brown rot in PLUMS, CHERRIES [4-6]. Botrytis, cane spot, powdery mildew, spur blight in CANE FRUIT [1, 2, 4-6, 11, 16]. Botrytis, powdery mildew in STRAWBERRIES [1, 2, 4-6, 11, 16]. Botrytis, fusarium, penicillium rot, sclerotinia, stagonospora in BULBS, CORMS [2, 4-6, 8, 10, 11]. Fusarium wilt, verticillium wilt in CARNATIONS [4-6, 8]. Fusarium in FREESIAS [4-6, 8]. Black root rot, fusarium, rhizoctonia in POINSETTIAS [4-6, 8]. Botrytis, powdery mildew in POT PLANTS, BEDDING PLANTS [4-6, 8, 11]. Black spot, mildew in ROSES [4-6]. Botrytis, powdery mildew, black root rot in CUCUMBERS [4-6, 8]. Botrytis, powdery mildew in PEPPERS, CHRYSANTHEMUMS [4-6, 8]. Botrytis, leaf mould, didymella, fusarium wilt, verticillium wilt in TOMATOES [2, 4-6, 8, 11]. Dactylium, dry bubble, trichoderma, wet bubble in MUSHROOMS [4-6, 8]. Canker, alternaria in HOPS *(off-label)* [1-4, 6, 7, 10, 15, 16]. Fungus diseases in HARDY ORNAMENTAL NURSERY STOCK, NURSERY FRUIT TREES AND BUSHES, ORNAMENTALS, ORNAMENTAL BULBS *(off-label)* [4, 6].

| Notes | **Efficacy**
• Different products vary in the diseases listed as controlled for a number of the crops. Labels must be consulted for full details and for rates and timings
• Mostly applied as spray or drench. Spray treatments normally applied at first sign of disease and repeated after 10-14 d if required
• Apply as a drench to control soil-borne diseases in cucumbers and tomatoes and as a pre-planting dip treatment for bulbs
• For big-vein disease in lettuce incorporate into peat blocks [4]
• Apply by incorporation into casing to control mushroom diseases
• Addition of non-ionic wetter to spray recommended for many uses
• Do not apply if rain or frost expected or crop wet
• To delay appearance of resistant strains alternate treatment with non-MBC fungicide. Eyespot in cereals is now widely resistant
• Not compatible with alkaline products such as lime sulphur

Crop safety/Restrictions
• For use in Brussels sprouts and cabbage use only in tank mixture with Corbel [6]
• Do not treat crops suffering from drought or other physical or chemical stress
• Do not use on strawberry runner beds
• Apply as drench rather than spray where red spider mite predators are being used
• Consult processors before using on crops for processing
• Establish safety of spray before using on ornamentals of unknown tolerance
• Use drench for tomatoes, cucumbers and peppers on soil-grown crops only |
|---|---|

FOR FULL CONDITIONS OF USE ALWAYS READ THE PRODUCT LABEL

Special precautions/Environmental safety

- Number of sprays should not exceed 1 per crop of french beans, 2 per crop of cereals, 4 per annum on gooseberries, 5 per crop of Brussels sprouts or cabbages, 6 per annum on blackcurrants
- Harmful to fish. Do not contaminate ponds, waterways or ditches with chemical or used container

Protective clothing/Label precautions

- 29, 35, 36, 37, 53, 54, 60, 63, 65, 66, 67, 70 [1-10, 12-17]; 29, 53, 67 [11]

Harvest interval

- zero [10-11]; cereals not after full ear emergences (GS 59) [4-9, 12, 14]; cereals zero, field and dwarf beans, oilseed rape 7 d [15]; Brussels sprouts, cabbage 14 d [6]

Approval

- Approved for aerial application on cereals [4, 6, 8, 10, 11, 15]; barley, winter wheat [3]; field beans [3, 4, 6, 8, 10, 11, 15]; dwarf beans [3, 4, 6, 11, 15]; oilseed rape [4, 6, 8, 10]; onions [4, 6, 8]. See introductory notes. Provisional approval for aerial application on cereals, dwarf beans, field beans, oilseed rape, onions [5] (written permission must first be obtained from BASF)
- Off-label approval to Feb 1991 for use on hops (OLA 0076/88) [1]; (OLA 0077/88) [2]; (OLA 0078, 0079/88) [3]; (OLA 0080/88) [4]; (OLA 0081/88) [6]; (OLA 0082/88) [7]; (OLA 0083/88) [10]; (OLA 0087/88) [15]; (OLA 0075, 0088/88) [16]
- Off-label approval to Apr 1992 for use on hardy ornamental nursery stock, nursery fruit trees and bushes, ornamentals (OLA 0360, 0361/89) [4, 6], to May 1992 for use on ornamental bulbs (OLA 0458, 0459) [4, 6]

---

97   carbendazim + chlorothalonil
A systemic and protectant fungicide mixture

Products   Bravocarb   Fermenta   100:450 g/l   SC   02623

Uses   Eyespot in WINTER BARLEY, WINTER WHEAT. Rhynchosporium in BARLEY. Septoria in WINTER WHEAT. Alternaria, botrytis, light leaf spot in OILSEED RAPE. Ascochyta, botrytis, downy mildew in PEAS. Chocolate spot in FIELD BEANS.

Notes   **Efficacy**
- Apply to winter wheat and barley between leaf sheath lengthening and first node detectable stages (GS 30-31) for eyespot/early septoria control. Further application up to ear emergence (GS 51) may be needed for rhynchosporium control
- To protect wheat against septoria spray immediately infection visible between flag leaf just visible and ear just fully emerged (GS 37-59)
- Treatment gives good control of cereal ear diseases particularly when applied directly to the ear. It also suppresses rusts, mildew and net blotch
- Tank-mix with a specific fungicide if rust/mildew levels become high and with prochloraz if MBC-resistant strains of eyespot are present
- Spray peas and beans at flowering and 2-4 wk later
- Spray oilseed rape in autumn, in spring and at petal-fall. Tank-mix with a specific fungicide for the petal-fall spray if alternaria risk high or disease established in crop

Special precautions/Environmental safety
- Irritating to eyes, skin and respiratory system

- Dangerous to fish. Do not contaminate ponds, waterways or ditches with chemical or used container

Protective clothing/Label precautions
- A, C
- 6a, 6b, 6c, 18, 21, 26, 29, 36, 37, 52, 60, 63, 65

Approval
- Approved for aerial application on wheat, barley. See introductory notes. Consult firm for details

---

98 carbendazim + chlorothalonil + maneb
A broad spectrum protectant and eradicant fungicide for cereals

Products					
1 Mancarb Plus	Ashlade	80:150:200 g/l	SC	04440
2 Tripart Victor	Tripart	80:150:200 g/l	SC	04359

Uses
Powdery mildew, yellow rust, brown rust, septoria diseases, brown foot rot and ear blight, botrytis, eyespot, sooty moulds in WINTER WHEAT, SPRING WHEAT. Rhynchosporium, yellow rust, brown rust, eyespot, net blotch, botrytis in WINTER BARLEY, SPRING BARLEY.

Notes

Efficacy
- Apply at any stage, dependent on disease pressure, up to ears fully emerged (GS 59)
- Highly effective against late diseases on flag leaf and ear of winter cereals

Crop safety/Restrictions
- Do not use more than 2 applications per crop

Special precautions/Environmental safety
- Irritating to eyes, skin and respiratory system
- Dangerous to fish. Do not contaminate ponds, waterways or ditches with chemical or used container

Protective clothing/Label precautions
- A, C
- 6a, 6b, 6c, 18, 21, 28, 29, 35, 36, 37, 52, 63, 67

Harvest interval
- 7 d

---

99 carbendazim + flusilazole
A broad-spectrum systemic and protectant fungicide for cereals

Products					
Punch C	Du Pont	12:25% w/w	EC	04690

FOR FULL CONDITIONS OF USE ALWAYS READ THE PRODUCT LABEL

Mildew, septoria, yellow rust, brown rust, eyespot in WINTER WHEAT, WINTER
BARLEY, SPRING BARLEY. Rhynchosporium, net blotch in WINTER BARLEY,
SPRING BARLEY.

Notes

Efficacy
- Apply at early stage of disease development or in routine programme of preventive
sprays
- Most effective timing of treatment varies with disease - see label for details
- Higher rate active against both MBC sensitive and MBC resistant eyespot
- Rain occurring within 2 h of spraying may reduce effectiveness
- Do not apply during frosty weather
- To prevent build-up of resistant strains of mildew tank mix with approved morpholine
fungicide (e.g. fenpropimorph, tridemorph)

Crop safety/Restrictions
- Lower rate may be applied at any time up to watery-ripe stage of crop (GS 71), higher
rate once only and up to 2nd node stage (GS 32)
- Punch C (or any other formulation containing flusilazole) must not be applied to winter
wheat more than 3 times per season, or to winter or spring barley more than 2 times
per season

Special precautions/Environmental safety
- Irritating to eyes
- Flammable
- Dangerous to fish. Do not contaminate ponds, waterways or ditches with chemical or
used container

Protective clothing/Label precautions
- A, C
- 6a, 12c, 18, 24, 28, 29, 36, 37, 52 63, 68, 70

100   carbendazim + flutriafol

A systemic fungicide with protectant and eradicant activity

Products   Early Impact   ICI   150:94 g/l   SC   02915

Uses   Brown rust, leaf blotch, net blotch, powdery mildew, yellow rust in WINTER BARLEY,
SPRING BARLEY. Eyespot in WINTER BARLEY. Brown rust, eyespot, glume blotch,
leaf spot, powdery mildew, yellow rust in WINTER WHEAT.

Notes

Efficacy
- Apply at early stage of disease development or as routine in high risk situations
- For MBC sensitive eyespot control apply between leaf sheath erect stage and before
second node detectable (GS 30-31). Control normally persists 4-6 wk. See label for
timing details
- When mildew well established up to 10% level mix with tridemorph (Calixin). Do not
delay treatment beyond this level. Seed treatment may be preferable on spring barley

Crop safety/Restrictions
- Do not use more than 2 sprays containing flutriafol on any cereal crop

- Tank-mix with Calixin may cause transient crop scorch, especially when crops are under stress. Avoid spraying during cold weather or periods of frost

Special precautions/Environmental safety
- Irritating to eyes and skin
- Harmful to fish. Do not contaminate ponds, waterways or ditches with chemical or used container

Protective clothing/Label precautions
- A, C
- 6a, 6b, 14, 18, 21, 28, 29, 36, 37, 53, 63, 66, 70

Harvest interval
- Do not apply after grain watery ripe stage (GS 71)

101   carbendazim + mancozeb
A broad-spectrum systemic and protectant fungicide for cereals

Products					
1 Kombat	Hoechst	10:54% w/w	WP	01163
2 Kombat WDG	Hoechst	12.4:63.3% w/w	WG	04344
3 Septal WDG	Schering	12.4:63.3% w/w	WG	04279

Uses

Brown rust, yellow rust, mildew, sooty moulds in WINTER WHEAT, WINTER BARLEY. Eyespot, septoria in WINTER WHEAT. Net blotch, rhynchosporium in WINTER BARLEY. Downy mildew, light leaf spot in WINTER OILSEED RAPE.

Notes

Efficacy
- Best results achieved by application at beginning of disease development
- Do not spray when crops wet, if rain imminent or if temperature exceeds 30°C
- Spray wheat between leaf sheath erect and first spikelets visible (GS 30-51) and barley between flag leaf ligule visible and complete ear emergence (GS 39-59)
- Will protect against mildew/rusts but tank-mix with a specific rust/mildew fungicide if these diseases are already present in the crop
- MBC resistant eyespot not controlled. If present use products containing prochloraz
- May be applied at low volume from flag leaf ligule visible (GS 39) up to 7 d before harvest [1]
- Spray winter oilseed rape pre-flowering up to yellow bud stage (GS 3, 7)

Crop safety/Restrictions
- Do not spray crops suffering stress from drought

Special precautions/Environmental safety
- Irritating to eyes, skin and respiratory system [1], to respiratory system [2,3]
- May cause sensitization by skin contact
- Harmful to fish. Do not contaminate ponds, waterways or ditches with chemical or used container

---

FOR FULL CONDITIONS OF USE ALWAYS READ THE PRODUCT LABEL

- A
- 6a, 6b, 6c [1]; 6c, 10a [2, 3]; 18, 21, 29, 35, 36, 37, 63, 67 [1-3]; 16, 28, 54, 71 [1]; 53 [2, 3]

Harvest interval
- 28 d

Approval
- Approved for aerial application on cereals [1]; winter wheat, winter barley, winter oilseed rape [2, 3]. See introductory notes

102 carbendazim + maneb
A broad-spectrum systemic and protectant fungicide

Products

1 Ashlade M	Ashlade	10:64% w/w	WP	02879
2 Ashlade Mancarb FL	Ashlade	50:320 g/l	SC	02928
3 Campbell's MC Flowable	MTM Agrochem.	62:400 g/l	SC	03467
4 Delsene M Flowable	Du Pont	50:320 g/l	SC	03041
5 Headland Dual	WBC Technology	62:400 g/l	SC	03782
6 MaxiMate	FCC	62:400 g/l	SC	03468
7 Multi-W FL	PBI	50:320 g/l	SC	04131
8 Squadron	Quadrangle	100:275 g/l	SC	03324
9 Tripart Legion	Tripart	50:320 g/l	SC	02997

Uses

Yellow rust, brown rust in WHEAT, BARLEY [3-7]. Eyespot in WINTER WHEAT, WINTER BARLEY [3, 5-7, 8, 9]. Sooty moulds in WINTER WHEAT, WINTER BARLEY [3-7, 8, 9]. Septoria in WINTER WHEAT [1-5, 7, 8, 9]. Powdery mildew in WHEAT [7]. Rhynchosporium in BARLEY [3, 5-7, 8]. Net blotch, powdery mildew, sooty moulds in BARLEY [7]. Botrytis, downy mildew, light leaf spot in OILSEED RAPE [7]. Light leaf spot in OILSEED RAPE [8].

Notes

Efficacy
- Apply in cereals from flag leaf just visible (GS 37) until first ears visible (GS 51), best when flag leaf ligules visible (GS 39) or, if not possible to spray earlier, from ear emergence complete (GS 59) until grain watery ripe (GS 71)
- Yield response greatest in lush, well grown crops when Jun-Jul period wet
- To avoid development of strains of eyespot resistant to MBC-type fungicides only make one application per season
- Treatment gives useful control of late attacks of rust or mildew but for early attacks or established infection tank-mix with specific mildew or rust fungicide
- Apply to oilseed rape as soon as disease appears in autumn or early spring and repeat if necessary between green bud and early flowering (GS 3,3-4,0) [7], when 25% of plants infected and disease spreading [8]
- Do not spray if frost or rain expected

Crop safety/Restrictions
- Do not apply after grain watery ripe stage (GS 71)

Special precautions/Environmental safety
- Harmful if swallowed [3, 5]
- Irritating to eyes and skin [6, 7]

- Irritating to eyes, skin and respiratory system [3, 5]

Protective clothing/Label precautions
- A, C
- 5c, 6a, 6b, 6c, 14, 16, 18, 21, 28, 29, 35, 36, 37, 53, 54, 63, 66, 70, 78 [1-9]

Harvest interval
- All crops 7 d

Approval
- Approved for aerial application on wheat, barley [4, 5]; cereals [2, 8]; winter wheat, winter barley [9]; wheat, barley, oats, rye, triticale [1]. See introductory notes

103 carbendazim + maneb + sulphur
A broad-spectrum systemic and protectant fungicide for cereals

Products Bolda FL   Farm Protection   50:320:100 g/l   SC   03463

Uses Brown rust, yellow rust, mildew, septoria, sooty moulds in WHEAT. Eyespot in WHEAT, WINTER BARLEY. Rhynchosporium in BARLEY. Alternaria, botrytis, light leaf spot in OILSEED RAPE. Fusarium ear blight in WINTER WHEAT.

Notes **Efficacy**
- Apply as late season spray to give protection during grain filling period
- Treat cereals from flag leaf just visible to ear emergence (GS 37-51). Later applications may control ear disease but are less likely to improve yield
- If rust or mildew pressure high or infection already established tank-mix with specific rust or mildew fungicide
- Treat oilseed rape at full flower and again 3 wk later if necessary
- The combination of systemic and multi-site inhibitor components reduces the risk of resistant strains of fungi developing
- Add suitable non-ionic wetting agent if applying from air

Special precautions/Environmental safety
- Irritating to eyes, skin and respiratory system
- Flammable. Keep away from fire or sparks

Protective clothing/Label precautions
- A
- 6a, 6b, 6c, 12c, 18, 21, 28, 29, 35, 36, 37, 54, 63, 67

Harvest interval
- 7 d

Approval
- Approved for aerial application on winter wheat, spring wheat, winter barley, oilseed rape. See introductory notes

FOR FULL CONDITIONS OF USE ALWAYS READ THE PRODUCT LABEL

104 carbendazim + maneb + tridemorph
A protectant and systemic fungicide for use in cereals

Products					
1 Ashlade Cosmic FL	Ashlade	40:320:90 g/l	SC	03607
2 Cosmic FL	BASF	40:320:90 g/l	SC	03473

Uses   Eyespot, powdery mildew, brown rust, yellow rust, sooty moulds in BARLEY, WINTER WHEAT. Septoria in WINTER WHEAT. Net blotch, rhynchosporium in BARLEY.

Notes

Efficacy
- Apply before leaf disease becomes established
- In winter barley autumn application is particularly recommended on disease susceptible cultivars but further treatment may be needed in spring
- Up to 2 applications may be made in spring between leaf sheath lengthening and complete ear emergence (GS 30-59), allowing at least 4 wk before harvest
- Tank mixtures with other fungicides recommended for increased protection against net blotch, rusts, powdery mildew and septoria. See label for details
- To delay appearance of resistant strains alternate treatment with non-MBC fungicide. Eyespot in cereals is now widely resistant
- Do not apply if rain expected or crop wet
- Systemic effect reduced under conditions of severe drought stress

Crop safety/Restrictions
- Do not apply to wheat during periods of temperature above 21 °C or high light intensity. Under such conditions spray in late evening

Special precautions/Environmental safety
- Irritating to eyes, skin and respiratory system
- Harmful to fish. Do not contaminate ponds, waterways or ditches with chemical or used container

Protective clothing/Label precautions
- A, C
- 6a, 6b, 6c, 16, 18, 21, 29, 35, 36, 37, 41, 53, 60, 63, 66, 70

Withholding period
- Keep livestock out of treated areas for at least 14 d

Harvest interval
- 4 wk

Approval
- Approved for aerial application on barley and winter wheat. See introductory notes [1, 2]

105 carbendazim + metalaxyl
A protectant fungicide for use in fruit and cabbage storage

Products					
Ridomil mbc 60WP	Ciba-Geigy	50:10% w/w	WP	01804

| Uses | Storage rots in APPLES, PEARS. Phytophthora, botrytis in STORED CABBAGES. Crown rot in WATER LILY *(off-label)*. |

| Notes | **Efficacy** |

- Apply as post-harvest drench or dip to prevent spread of storage diseases
- Crop must be treated immediately after harvest
- Little curative activity on crop infected at or before harvest
- Best results achieved when crop stored in a controlled environment
- Use with calcium chloride for bitter pit control in susceptible apple cultivars

Crop safety/Restrictions
- Allow crop to drain thoroughly after treatment

Special precautions/Environmental safety
- Irritating to eyes and skin
- Treated fruit must not be processed or sold for at least 4 wk, cabbage for 7 wk
- Harmful to fish. Do not contaminate ponds, waterways or ditches with chemical or used container

Protective clothing/Label precautions
- A
- 6a, 6b, 18, 22, 29, 35, 36, 37, 53, 63, 67, 70

Approval
- Off-label approval to May 1992 for use on water-lily (OLA 0420/89)

106 carbendazim + prochloraz
A broad-spectrum systemic and contact fungicide for cereals

| Products | Sportak Alpha | Schering | 100:266 g/l | SC | 03872 |

| Uses | Powdery mildew, rhynchosporium in WINTER BARLEY, WINTER RYE. Eyespot, net blotch in WINTER BARLEY. Eyespot, leaf spot in WINTER RYE. Dark leaf spot, light leaf spot, canker, white leaf spot, botrytis, sclerotinia stem rot in OILSEED RAPE. |

| Notes | **Efficacy** |

- To protect against eyespot in high risk situations apply to winter cereals from when leaf sheaths begin to become erect to first node detectable (GS 30-31). May also be used to control eyespot already in crop (up to 10% of tillers affected)
- Applied for eyespot control, product will also control rhynchosporium (barley, rye) or mildew (barley) and will protect against new infections of net blotch (barley), mildew (rye) and septoria (rye). Treatment will often suppress sharp eyespot and fusarium if they are developing at time of application
- Tank-mix with tridemorph, fenpropimorph or fenpropidin to control established mildew infections
- May be applied up to full ear emergence (GS 59), but if used earlier in season, prochloraz alone or plus a non-MBC fungicide is preferred treatment

FOR FULL CONDITIONS OF USE ALWAYS READ THE PRODUCT LABEL

- Apply in oilseed rape at first signs of disease and repeat if necessary. Timing varies with disease - see label for details.
- A period of at least 3 h without rain should follow spraying

Crop safety/Restrictions
- Do not apply after ears fully emerged (GS 59)
- Do not apply more than twice in any crop

Special precautions/Environmental safety
- Harmful in contact with skin. Irritating to eyes and skin
- Flammable
- Dangerous to fish. Do not contaminate ponds, waterways or ditches with chemical or used container

Protective clothing/Label precautions
- A, C
- 5a, 6a, 6b, 12c, 18, 21, 29, 35, 36, 37, 52, 60, 63, 66, 70, 78

Harvest interval
- Oilseed rape 6wk

Approval
- Approved for aerial application on barley. See introductory notes.

---

107 carbendazim + propiconazole
A contact and systemic fungicide for winter cereals

Products

Hispor 45WP	Ciba-Geigy	20:25% w/w	SP	01050

Uses

Eyespot, powdery mildew, yellow rust, brown rust in WINTER WHEAT, WINTER BARLEY. Rhynchosporium, net blotch in WINTER BARLEY. Septoria in WINTER WHEAT.

Notes

Efficacy
- Major benefit obtained from spring treatment. Control provided for about 30 d
- Apply in spring from stage when leaf sheath begins to lengthen until first node detectable (GS 30-31). Further sprays may be applied up to fully emerged ear (GS 59)

Crop safety/Restrictions
- Apply a maximum of 4 propiconazole-containing products per year in wheat, up to 3 per year in barley

Special precautions/Environmental safety
- Irritating to eyes and skin
- Dangerous to fish. Do not contaminate ponds, waterways or ditches with chemical or used container
- Harmful to bees. Do not apply at flowering stage. Keep down flowering weeds

Protective clothing/Label precautions
- A, C
- 6a, 6b, 18, 22, 28, 29, 35, 36, 37, 50, 52, 63

Approval
• Approved for aerial application in wheat and barley. See introductory notes

108 carbendazim + tecnazene
A protectant fungicide and sprout suppressant for stored potatoes

Products					
1 Hickstor 6 + MBC	Hickson & Welch	2:6% w/w	DP	04176
2 Hortag Carbotec	Avon	1.8:4% w/w	DP	04135
3 Hortag Tecnacarb Dust	Avon	1.8:6% w/w	DP	02929
4 Tripart Arena Plus	Tripart	2:6% w/w	DP	04258

Uses   Dry rot in STORED POTATOES. Sprout suppression in STORED POTATOES.

Notes   **Efficacy**
• Potatoes should be dormant, have a mature skin and be dry and free from dirt
• Treat tubers as they go into store using a dusting machine and not by hand [3], measured amounts may be applied by hand [1, 4]
• Ensure even cover of tubers. Cover clamps with straw etc to aid vapour-phase transmission of a.i. Pack boxes as tightly together as possible
• Effectiveness of treatment is reduced if ventilation in store is inadequate or excessive
• Treatment can give protection for 3-4 mth [3] or up to 6 mth [1, 4]. It also suppresses skin spot, gangrene and silver scurf but does not cure blemishes already present on tubers. It will not control sprouting if tubers have already broken dormancy

Crop safety/Restrictions
• Air seed potatoes for 6 wk and ensure that chitting has commenced before planting out Treatment may delay emergence and possibly slightly reduce ware yield

Protective clothing/Label precautions
• 28, 29, 35, 54, 63, 64, 67

Withholding period
• Minimum period between treatment and removal from store 6wk

Stop-press
• An advisory MRL for tecnazene of 5mg/kg was announced in PR No 7, Aug 1989

109 carbetamide
A residual pre- and post-emergence carbamate herbicide

Products					
Carbetamex	Embetec	70% w/w	WP	04415

FOR FULL CONDITIONS OF USE ALWAYS READ THE PRODUCT LABEL

Annual grasses, volunteer cereals, some annual dicotyledons in WINTER OILSEED RAPE, SPRING CABBAGE, SPRING GREENS, SEED BRASSICAS, SUGAR BEET SEED CROPS, LUCERNE, SAINFOIN, WINTER FIELD BEANS, RED CLOVER, WHITE CLOVER.

Notes

Efficacy
- Best results achieved pre- or early post-emergence of weeds under cool, moist conditions. Adequate soil moisture is essential
- Dicotyledons controlled include chickweed, cleavers and speedwell
- Weed growth stopped rapidly though full effects may take 6-8 wk to develop
- Do not use on soils with more than 10% organic matter
- Do not apply during prolonged periods of cold weather when weeds fully dormant
- Any ash from straw burning or trash must be dispersed or buried before spraying
- Various tank mixes effective against a wider range of dicotyledons are recommended. See label for details and for mixtures with other pesticides and sequential treatments

Crop safety/Restrictions
- Apply to brassicas from late autumn to late winter provided crop has at least 4 true leaves (spring cabbage, spring greens), 3-4 true leaves (seed crops, oilseed rape)
- Apply to established lucerne and sainfoin from mid-Oct to end Feb, to established red and white clover from Feb to mid-Mar
- After treatment do not sow brassicas or field beans for 2 wk, peas or runner beans for 8 wk, cereals or maize for 16 wk
- Do not apply after using dalapon or for at least 14 d after treating with another grass-killer or pyrethroid insecticide

Protective clothing/Label precautions
- 29, 54, 63, 67, 70

Harvest interval
- 6 wk

410 carbetamide + oxadiazon
A residual herbicide for container-grown nursery stock

Products Ronstar TX   Hortichem   1.5:2% w/w   GR   03524

Uses Annual dicotyledons, annual grasses in CONTAINER-GROWN STOCK.

Notes

Efficacy
- Apply pre-weed emergence, weeds are killed as they emerge from soil
- Adequate soil moisture needed to achieve effective results. Irrigate after application if necessary but do not mix into soil
- Capillary sand beds and gravel or plastic covered standing-out beds may be treated before stocking

Crop safety/Restrictions
- Apply as soon as possible after potting, before crop plants making soft growth
- Safe on a wide range of ornamental shrubs. See label for details and for list of species to be avoided
- For species of unknown sensitivity test first on a small number of plants

- Do not treat plants under glass and only treat plants in plastic tunnels under cool conditions
- Do not use on plants rooted in high sand or non-organic media
- Do not apply to plants with wet foliage or allow granules to collect on leaves

Special precautions/Environmental safety
- Irritating to eyes
- Not to be used on food crops
- Dangerous to fish. Do not contaminate ponds, waterways or ditches with chemical or used container

Protective clothing/Label precautions
- A, C
- 6a, 18, 22, 26, 27, 28, 30, 34, 36, 37, 52, 63, 70

111   carbofuran

A systemic carbamate insecticide and nematicide for soil treatment

Products					
1 Nex	Tripart	5% w/w	GR	03417
2 Rampart	Sipcam	5% w/w	GR	04435
3 Rampart	Unicrop	5% w/w	GR	04532
4 Sipcam UK Carbosip 5G	Sipcam	5% w/w	GR	04535
5 Yaltox	Bayer	5% w/w	GR	02371

Uses

Flea beetles, pygmy beetle, beet leaf miner, millipedes, springtails, wireworms, free-living nematodes in SUGAR BEET, FODDER BEET, MANGOLDS [1-5]. Tortrix moths in SUGAR BEET [5]. Cabbage root fly, cabbage stem weevil, diamond-back moth, flea beetles in BROCCOLI, BRUSSELS SPROUTS, CABBAGES, CALABRESE, CAULIFLOWERS [1-5]. Aphids, cabbage root fly, cabbage stem weevil, flea beetles, turnip root fly in SWEDES, TURNIPS [5]. Cabbage stem flea beetle, rape winter stem weevil, cabbage root fly in WINTER OILSEED RAPE [5]. Aphids, carrot fly, free-living nematodes in CARROTS [1-5]. Aphids, carrot fly, free-living nematodes in PARSNIPS [5]. Frit fly in MAIZE, SWEETCORN [1-5]. Onion fly, stem nematodes in BULB ONIONS [1-5]. Potato cyst nematode in POTATOES [1-5]. Vine weevil in STRAWBERRIES, HARDY ORNAMENTAL NURSERY STOCK, POT PLANTS [5]. Large bulb fly in OUTDOOR NARCISSI *(off-label)* [5]. Cabbage root fly, soil and seedling pests in CHINESE CABBAGE, SAVOY CABBAGE, COLLARDS, KALE, KOHLRABI *(off-label)* [5].

Notes

Efficacy
- May be used on all types of mineral and organic soil
- Apply through a suitably calibrated granule applicator and incorporate into the soil
- Method of application, rate and timing vary with pest and crop. See label for details
- Performance is reduced in dry soil conditions
- Do not apply to potatoes in very wet or water-logged soils
- Controls only first generation carrot fly, use a follow-up application of a suitable specific insecticide for later attacks

FOR FULL CONDITIONS OF USE ALWAYS READ THE PRODUCT LABEL

- May be applied in mixture with granular fertilizers
- Do not apply to any site more than once in 2 yr to reduce the risk of enhanced biodegradation

Crop safety/Restrictions
- When applied at drilling do not allow granules to come into contact with crop seed
- Use on carrots is limited to crops growing in mineral soils
- Use only on onions intended for harvest as mature bulbs
- Apply on ornamentals as surface treatment to moist soil after planting or potting up and follow immediately by thorough watering. Do not treat plants grown under cover, see label for details of species which have been treated safely
- On strawberries apply after last harvest of year

Special precautions/Environmental safety
- This product contains an anticholinesterase carbamate compound. Do not use if under medical advice not to work with such compounds
- Harmful in contact with skin and if swallowed
- Wear suitable protective gloves if handling treated pots/soil within 4 wk of treatment
- Dangerous to livestock. Bury or remove spillages [1-4]
- Dangerous to fish. Do not contaminate ponds, waterways or ditches with chemical or used container
- Dangerous to game, wild birds and animals. Bury or remove spillages

Protective clothing/Label precautions
- A, B, C, C or D+E, H, K, M
- 2, 5a, 5c, 14, 16, 18, 22, 25, 28, 29, 36, 37, 40, 45, 52, 64, 67, 70, 78

Withholding period
- Dangerous to livestock. Keep all livestock out of treated areas for at least 2 wk [5]

Harvest interval
- 6 wk. At least 4 wk must elapse between application and release of treated plants for sale or supply [5]

Approval
- Off-label approval to July 1992 for use on outdoor narcissi (OLA 0540/89) [5]; to Nov 1992 for use on chinese cabbage, savoy, collards, kale and kohl rabi (OLA 1006/89) [5]

112   carbosulfan

A systemic carbamate insecticide for control of soil pests

| **Products** | Marshal 10G | RP | 10% w/w | GR | 04646 |

Uses   Aphids, flea beetles, mangold fly, millipedes, pygmy mangold beetle, springtails, symphylids, wireworms, free-living nematodes in SUGAR BEET. Cabbage root fly, cabbage stem weevil, flea beetles in CABBAGES, CAULIFLOWERS, BRUSSELS SPROUTS, BROCCOLI, CALABRESE, COLLARDS, TURNIPS, SWEDES. Carrot fly, aphids, free-living nematodes in CARROTS, PARSNIPS.

Notes   **Efficacy**
- At recommended rates seed is not damaged by contact with product

- Apply with suitable granule applicator feeding directly into seed furrow or immediately behind drill coulter (behind seed drill boot for brassicas) or use bow-wave technique
- For transplanted brassicas apply sub-surface with 'Leeds' coulter
- See label for details of suitable applicators and settings. Correct calibration is essential
- Where applied annually in intensive brassica growing areas enhanced biodegradation by soil organisms may lead to unsatisfactory control

Crop safety/Restrictions
- Do not mix with compost for blockmaking or use in Hassy trays for brassicas

Special precautions/Environmental safety
- This product contains an anticholinesterase compound. Do not use if under medical advice not to work with such compounds
- Harmful if swallowed
- Dangerous to game, wild birds and animals. Bury spillages
- Dangerous to fish. Do not contaminate ponds, waterways or ditches with chemical or used container

Protective clothing/Label precautions
- A, B, C, C or D+E, H, K, M
- 2, 5c, 14, 16, 18, 22, 25, 28, 29, 36, 37, 45, 52, 64, 67, 70, 78

Harvest interval
- leaf brassicas 60 d; sugar beet, swedes, turnips, carrots, parsnips 100 d

113   carboxin

An oxathiin fungicide available only in mixtures

114   carboxin + gamma-HCH + thiram

A fungicide and insecticide dressing for rape seed

Products   Vitavax RS Flowable   Uniroyal   3.44:50.8:6.92% w/w   FS   02310

Uses   Canker, damping off in FODDER RAPE, OILSEED RAPE. Flea beetles in FODDER RAPE, OILSEED RAPE.

Notes   **Efficacy**
- Prior to and during application product must be thoroughly agitated with drum agitator to ensure uniform mixing
- Keep at temperatures above 10°C prior to and during application

Crop safety/Restrictions
- Do not store treated seed for more than 3 mth

FOR FULL CONDITIONS OF USE ALWAYS READ THE PRODUCT LABEL

115 carboxin + imazalil + thiabendazole
A fungicide seed dressing for barley and oats

Products	Cerevax Extra	ICI	300:20:25 g/l	FS	02501

Uses

Brown foot rot and ear blight, covered smut, leaf stripe, loose smut, net blotch in SPRING BARLEY. Brown foot rot and ear blight, pyrenophora leaf spot, loose smut in OATS.

Notes

Efficacy
- Best applied through specially designed seed treatment machinery
- Drill seed within 8 wk of treatment

Crop safety/Restrictions
- Do not treat seed with moisture content above 16%. Store treated seed in a cool, dry, well ventilated place
- Do not apply if seed coated with fungicide other than Milstem for mildew control

Special precautions/Environmental safety
- Irritating to eyes and skin
- Treated seed not to be used as food or feed. Do not re-use sack
- Harmful to fish. Do not contaminate ponds, waterways or ditches with chemical or used container

Protective clothing/Label precautions
- A, C
- 6a, 6b, 18, 22, 29, 36, 37, 53, 63, 65, 70, 72, 73, 74, 76, 78

116 carboxin + thiabendazole
A fungicide seed dressing for wheat and rye

Products	Cerevax	ICI	360:20 g/l	FS	02500

Uses

Brown foot rot and ear blight, bunt, loose smut in WHEAT, RYE. Loose smut in WHEAT. Stripe smut in RYE.

Efficacy
• Best applied through specially designed seed treatment machines
• Drill within 8 wk of treatment

Crop safety/Restrictions
• Do not treat seed with moisture content above 16%. Store treated seed in a cool, dry and well ventilated place
• Do not apply if seed already treated with a fungicide
• Treatment may lower germination capacity if seed not of good quality

Special precautions/Environmental safety
• Irritating to eyes, skin and nose
• Treated seed not to be used as food or feed. Do not re-use sack
• Harmful to fish. Do not contaminate ponds, waterways or ditches with chemical or used container

Protective clothing/Label precautions
• A
• 6a, 6b, 6c, 18, 23; 24, 29, 36, 37, 53, 63, 65, 70, 72, 73, 74, 76

117 cetrimide + chlorpropham
A soil-acting herbicide for lettuce under cold glass

Products

1	Atlas Herbon Pabrac	Atlas	80:80 g/l	SC	03997
2	Croptex Pewter	Hortichem	80:80 g/l	SC	02507

Uses Annual grasses, annual dicotyledons, chickweed, polygonums in PROTECTED LETTUCE.

Notes **Efficacy**
• Apply to drilled lettuce under cold glass pre- or within 24 h post-drilling, to transplanted crops pre-planting
• Adequate irrigation must occur before or after treatment
• Best results achieved on firm soil of fine tilth, free from clods and weeds
• Light cultivation after treatment not detrimental to control but untreated soil should not be brought to surface

Crop safety/Restrictions
• Do not use where seed has chitted or on crop foliage
• Excess irrigation may cause temporary check to crop under certain circumstances
• Do not apply where tomatoes or brassicas are growing in same house
• In the event of crop failure only lettuce should be grown within 2 mth

Special precautions/Environmental safety
• Irritating to eyes, skin and respiratory system
• Highly flammable
• Harmful to fish. Do not contaminate ponds or ditches with chemical or used container

FOR FULL CONDITIONS OF USE ALWAYS READ THE PRODUCT LABEL

• A, C

• 6a, 6b, 6c, 12b, 18, 21, 28, 29, 36, 37, 53, 60, 63, 65

118   chloramben

A residual benzoic acid herbicide for use in ornamentals

Products	Granular Naptol	Synchemicals	4% w/w	GR	00999

Uses   Annual dicotyledons, annual grasses in NEWLY PLANTED SHRUBS, PERENNIALS, BEDDING PLANTS, CONTAINER-GROWN STOCK.

Notes   **Efficacy**
• Apply to newly planted shrub borders, perennial borders and container-grown stock in spring and summer
• Apply to damp soil surface free of weeds. If no rain falls within 48 h irrigate lightly
• Weeds controlled for 8-10 wk. Treatment may be repeated if necessary

Crop safety/Restrictions
• Apply to bedding plants 5-10 d after planting out, avoiding excess contact with plants
• Do not apply to pansies or begonias. Leaf symptoms have been noted on some portulaca, petunia and alyssum
• For species of unknown sensitivity test first on a small number of plants
• Do not use on light, sandy soils or on seeded crops prior to emergence

Special precautions/Environmental safety
• Not to be used on food crops

Protective clothing/Label precautions
• 29, 34, 54, 63, 67

119   chlorbufam

A carbamate herbicide available only in mixtures

120   chlorbufam + chloridazon

A residual and contact herbicide for onions, leeks and bulbs

Products	Alicep	BASF	20:25% w/w	WP	00077

Uses   Annual dicotyledons in ONIONS, LEEKS, NARCISSI, GLADIOLI, TULIPS. Annual dicotyledons in CHIVES *(off-label)*.

Efficacy
- Best results from pre-emergence use when soil moist, firm and clod free and rainfall sufficient to carry chemical to germinating weed zone. Residual activity reduced when soils too dry, too wet or with more than 5% organic matter
- For pre-emergence use adjust rate according to soil type. See label for details
- Contact action effective up to 2-leaf stage of weeds. For repeat low dose programme on direct drilled onions spray at cotyledon stage
- On highly organic soils only contact action is effective

Crop safety/Restrictions
- Apply to drilled onions or leeks after drilling to shortly before emergence or after crook stage (lower rates at any crop stage), to transplanted crops when established
- Apply to tulips, narcissi and gladioli pre- or shortly post-emergence
- On sands or very light soils do not apply pre-emergence to drilled onions or leeks and delay post-emergence treatment as long as possible
- Excessive rain after application may cause crop damage
- Scorch may result from post-emergence application during very warm weather
- In case of crop failure sprayed fields may only be replanted with onions, leeks, tulips, narcissi or gladioli

Special precautions/Environmental safety
- Irritating to eyes, skin and respiratory system
- May cause sensitization by skin contact
- Harmful to fish. Do not contaminate ponds, waterways or ditches with chemical or used container

Protective clothing/Label precautions
- A, C
- 6a, 6b, 6c, 10a, 18, 22, 28, 29, 36, 37, 53, 63, 67

Approval
- Off-label Approval to Sep 1993 for use on chives (OLA 0522/90)

121   chlordane
A persistent organochlorine lumbricide for use on turf

Products					
1 Sydane 25		Synchemicals	250 g/l	EC	02059
2 Sydane Granular		Synchemicals	20% w/w	GR	02060

Uses   Earthworms in TURF.

Notes   **Efficacy**
- Spray or apply granules to turf at any time of year but preferably in spring or autumn when worms active. Effects of treatment persist for up to 12 mth
- Mow closely before treatment and do not cut for 3 d afterwards
- Compacted turf should be spiked and irrigated before application and a light watering is recommended after treatment of dry turf

FOR FULL CONDITIONS OF USE ALWAYS READ THE PRODUCT LABEL

Crop safety/Restrictions
- Do not treat newly sown grass until 6 wk after germination

Special precautions/Environmental safety
- Harmful in contact with skin and if swallowed
- Harmful to fish. Do not contaminate ponds, waterways or ditches with chemical or used container

Protective clothing/Label precautions
- A
- 5a, 5c, 14, 18, 22, 28, 29, 36, 37, 41, 53, 63, 66, 70, 78

Withholding period
- Harmful to livestock. Keep all livestock out of treated areas for at least 14 d

Approval
- Approvals for sale and supply ceased on 31 Dec 1990; storage and use permitted until 31 Dec 1992

---

122 chlorfenvinphos
A contact and ingested soil-applied organophosphorus insecticide

Products

1	Birlane 24	Shell	240 g/l	EC	00275
2	Birlane Granules	Shell	10% w/w	GR	00276
3	Birlane Liquid Seed Treatment	Embetec	322 g/l	LS	00278
4	Sapecron 240EC	Ciba-Geigy	240 g/l	EC	01861
5	Sapecron 10FG	Ciba-Geigy	10% w/w	GR	01860
6	Sedanox	Bayer	10% w/w	GR	05033

Uses
Wheat bulb fly in WINTER WHEAT [1, 3, 4, 5]. Yellow cereal fly in WINTER WHEAT [4]. Cabbage root fly in LEAF BRASSICAS, SWEDES, TURNIPS [1, 2, 4, 6]. Cabbage root fly in BRUSSELS SPROUTS, BROCCOLI, CABBAGES, CAULIFLOWERS, SWEDES, TURNIPS [5]. Cabbage root fly in RADISHES [1, 4]. Frit fly in MAIZE [2, 4, 5]. Frit fly in SWEETCORN [2, 5]. Carrot fly in CARROTS [1, 2, 4, 5, 6]. Carrot fly in PARSNIPS, PEAT BLOCK CELERY [4]. Mushroom flies in MUSHROOMS [1, 2, 4, 5]. Colorado beetle in POTATOES *(off-label)* [1, 4]. Insect control in MOOLI, RADISHES, KOHLRABI *(off-label)* [2, 5]. Cabbage root fly in KOHLRABI *(off-label)* [1]. Insect control in CELERIAC *(off-label)* [2].

Notes

Efficacy
- Soil incorporation recommended for most treatments. Application method, timing, rate and frequency vary with pest, crop and soil type. See labels for details
- Overall pre-planting sprays are less effective than band treatment and should not be used in areas of heavy cabbage root fly infestation [4]
- Cabbage and cauliflower grown in peat blocks can be protected from cabbage root fly by incorporation into the peat used for blocking [2, 5, 6]
- Efficacy reduced on highly organic or very dry soils unless crop irrigated
- Apply as seed treatment for wheat bulb fly control on highly organic soils [3]

Crop safety/Restrictions
- Band sprays on sown brassicas may cause damage if applied post-emergence [1]

- Danger of scorch if applied to maize during very hot weather or if crop is under stress
- Do not apply in conjunction with seed treatments containing gamma-HCH
- Treated seed can be stored for 1 mth provided moisture content is 16% or less. Do not exceed drilling depth of 4 cm [3]

Special precautions/Environmental safety
- A chemical subject to the Poisons Rules 1982 and the Poisons Act 1972 [1, 3, 4]
- This product contains an organophosphorus compound. Do not use if under medical advice not to work with such compounds
- Toxic in contact with skin and if swallowed [1, 3, 4]. Irritating to eyes and skin [1, 4]
- Harmful if swallowed [2, 5, 6]
- Flammable [1, 3, 4]
- Harmful to game, wild birds and animals. Bury spillages [3]
- Dangerous to fish. Do not contaminate ponds, waterways or ditches with chemical or used container
- Treated seed not to be used as food or feed [3]

Protective clothing/Label precautions
- A, C, H, J, M [1]; B, C, H [2, 5, 6]; A, C, H [3, 4]
- 1, 4a, 4c, 6a, 6b, 12c, 14, 16, 18, 21, 25, 26, 27, 28, 29, 35, 36, 37, 52, 64, 66, 70, 79 [1, 4]; 1, 5c, 14, 18, 22, 25, 29, 35, 36, 37, 52, 64, 67, 70, 78 [2, 5, 6]; 1, 4a, 4c, 14, 16, 18, 21, 25, 28, 29, 36, 37, 52, 64, 67, 70, 72, 73, 74, 75, 76, 77, 79, 80 [3]

Harvest interval
- Brassicas, carrots, maize 21 d; radish 28 d; celery, parsnips 12 wk

Approval
- Approved for aerial application on wheat (Jan and Feb only) [1]; winter cereals (Jan and Feb only) [4]. See introductory notes
- Off-label approval to Nov 1991 for use in potatoes only as required by Statutory Notice for Colorado beetle control (OLA 1384, 1385/88) [4, 1], to May 1992 for use in mooli, radish, kohlrabi (OLA 0405, 0408/89) [2, 5], to Jun 1993 for use in celeriac (OLA 0377/90) [2], kohlrabi (OLA 0381/90)[1]

123 chloridazon

A residual pyridazinone herbicide for beet crops
See also chlorbufam + chloridazon

Products					
1 Ashlade Chloridazon FL	Ashlade	430 g/l	SC	02875
2 Atlas Silver	Atlas	430 g/l	SC	03547
3 Better Flowable	Sipcam	430 g/l	SC	04924
4 Chiltern Pyrazol	Chiltern	430 g/l	SC	04651
5 Portman Weedmaster	Portman	430 g/l	SC	02550
6 Power Chloridazon	Power	430 g/l	SC	02717
7 Pyramin DF	BASF	65% w/w	SG	03438
8 Starter Flowable	Truchem	430 g/l	SC	03421
9 Tripart Gladiator	Tripart	430 g/l	SC	00986
10 Trojan SC	Schering	430 g/l	SC	03926

FOR FULL CONDITIONS OF USE ALWAYS READ THE PRODUCT LABEL

Uses	Annual dicotyledons, annual meadow grass in SUGAR BEET, FODDER BEET, MANGOLDS.

Notes	**Efficacy**

- Absorbed by roots of germinating weeds and best results achieved pre-emergence of weeds or crop when soil moist and adequate rain falls after application
- Apply pre-emergence as soon as possible after drilling in mid-Mar to mid-Apr on fine, firm, clod-free seedbed
- Where crop drilled after mid-Apr or soil dry apply pre-drilling and incorporate to 2.5 cm immediately afterwards
- Application rate depends on soil type. See label for details
- Various tank mixes recommended on sugar beet for pre- and post-emergence use and as repeated low dose treatments. See label for details

Crop safety/Restrictions
- Up to 4 treatments per crop can be made on sugar beet, one treatment per crop on fodder beet and mangolds
- Do not use on coarse sands, sands or fine sands or where organic matter exceeds 5%
- Crop vigour may be reduced by treatment of crops growing under unfavourable conditions including poor tilth, drilling at incorrect depth, soil capping, physical damage, pest or disease damage, excess seed dressing, trace-element deficiency or a sudden rise in temperature after a cold spell
- Seed should be drilled at 20-25 mm
- In the event of crop failure only sugar or fodder beet, mangolds or maize may be re-drilled on treated land after cultivation
- Winter cereals may be sown in autumn after ploughing. Any spring crop may follow treated beet crops harvested normally

Special precautions/Environmental safety
- Harmful if swallowed. May cause sensitization by skin contact [4]
- Irritant. May cause sensitization by skin contact [7]
- Irritating to eyes, skin and respiratory system [1-5]
- Harmful to fish. Do not contaminate ponds, waterways or ditches with chemical or used container

Protective clothing/Label precautions
- A [1-6, 8-10]; A, C [7]
- 18, 21, 29, 36, 37, 53, 63, 66, 67, 70, 78 [1-5, 7-9]; 5c, 10a [4, 5, 8, 9]; 6a, 6b, 6c [1, 2, 5]; 6, 10a [7]; 21, 29, 53, 63, 66, 70 [6, 10]

Approval
- May be applied through CDA equipment. See introductory notes [10]

124 chloridazon + chlorpropham + fenuron + propham
A residual pre-emergence herbicide for beet crops

Products	Atlas Electrum	Atlas	200:30:20:120 g/l SC	03548

Uses	Annual dicotyledons, annual grasses in SUGAR BEET, FODDER BEET, MANGOLDS.

Efficacy

- Best results achieved from application to firm, moist, clod-free seedbed when adequate rain falls afterwards
- Apply immediately after drilling pre-emergence of crop or weeds
- Do not disturb soil surface after application
- Dose rate depends on soil type - see label. Do not use on highly organic soils

Crop safety/Restrictions

- Crops affected by poor growing conditions may be checked by treatment
- Excessive rainfall after application may check crop
- Do not incorporate chemical into soil
- In the event of crop failure only sugar or fodder beet or mangolds should be re-drilled
- After lifting treated crops the land should be ploughed or cultivated repeatedly to 15 cm to dissipate residues

Special precautions/Environmental safety

- Irritating to skin, eyes and respiratory system
- Harmful to fish. Do not contaminate ponds, waterways or ditches with chemical or used container

Protective clothing/Label precautions

- A
- 6a, 6b, 6c, 18, 21, 28, 29, 36, 37, 53, 63, 66

125   chloridazon + ethofumesate

A residual pre-emergence herbicide for beet crops

Products

1 Magnum	BASF	275:170 g/l	SC	01237
2 Spectron	Schering	211:200 g/l	SC	03828

Uses

Annual dicotyledons, annual meadow grass, blackgrass in SUGAR BEET, FODDER BEET, MANGOLDS [2]. Annual dicotyledons, annual meadow grass in SUGAR BEET, FODDER BEET [1].

Notes   **Efficacy**

- Apply at or immediately after drilling pre-emergence of crop or weeds [2], pre- or post-emergence up to cotyledon stage of weeds [1], on a fine, firm, clod-free seedbed
- Best results achieved when soil moist and adequate rain falls after spraying
- May be used on soil classes 'loamy sand' - 'silty clay loam' [1], all soils except sands and those with more than 5% organic matter [2]
- May be applied by conventional or repeat low dose methods. See label for details
- Recommended for tank mixing with triallate on sugar beet incorporated pre-drilling and with various other beet herbicides early post-emergence. See label for details

FOR FULL CONDITIONS OF USE ALWAYS READ THE PRODUCT LABEL

* Crop vigour may be reduced by treatment of crops growing under unfavourable conditions including poor tilth, drilling at incorrect depth, soil capping, physical damage, excess nitrogen, excess seed dressing, trace element deficiency or a sudden rise in temperature after a cold spell
* In the event of crop failure only sugar beet, fodder beet or mangolds may be re-drilled
* Any crop may be sown 3 mth after spraying following ploughing to 15 cm
* Frost after pre-emergence treatment may check crop growth

Special precautions/Environmental safety
* May cause sensitization by skin contact [1]
* Harmful to fish. Do not contaminate ponds, waterways or ditches with chemical or used container

Protective clothing/Label precautions
* A, C [1]
* 10a, 18, 21, 28, 29, 36, 37, 53, 63, 66, 78 [1]; 21, 28, 29, 53, 63, 66 [2]

Approval
* May be applied through CDA equipment. See introductory notes [2]

126 chloridazon + fenuron + propham
A residual pre-emergence herbicide for beet crops

Products	Barrier	Truchem	300:30:170 g/l	SC	03124

Uses Annual dicotyledons, annual grasses in SUGAR BEET, FODDER BEET, MANGOLDS.

Notes

Efficacy
* Apply before crop chits and before weed emergence, up to 10 d after drilling in Mar, 7 d in early Apr and 3 d in mid-Apr, or band spray at drilling
* Best results achieved on a fine, firm, moist, clod-free seedbed
* Some rainfall is essential after spraying to activate the chemical
* For season long control follow up with a post-emergence herbicide. See label for recommendations

Crop safety/Restrictions
* Do not use on sands or very heavy or peaty soils
* Drill at depth of 20 mm. Crop may be damaged if very heavy rain follows treatment
* Do not apply at temperatures below 5°C or use water near freezing for mixing
* Crops affected by poor growing conditions may be checked by treatment
* Do not tank mix with Avadex BW and incorporate as crop check may occur. Application on very light and light soils after pre-planting incorporation of Avadex BW may cause check to crop which is normally outgrown

Special precautions/Environmental safety
* Irritating to eyes, skin and respiratory system
* Harmful to fish. Do not contaminate ponds, waterways or ditches with chemical or used container

Protective clothing/Label precautions
* A, C

127   chloridazon + lenacil

A residual pre-emergence herbicide for beet crops

Products	Advizor	Farm Protection	200:133 g/l	SC	00032

Uses   Annual dicotyledons, annual meadow grass in SUGAR BEET, FODDER BEET, MANGOLDS.

Notes

Efficacy
- Apply at or immediately after drilling before emergence of crop or weeds
- Best results achieved from application to firm, moist, weed-free seedbed when adequate rain falls afterwards
- Application rate depends on soil type. See label for details
- May be used on very light, light or medium soils. Not on sands, stony or gravelly soils, heavy soils or those with more than 10% organic matter
- Recommended for tank-mixing with Avadex BW incorporated pre-drilling on sugar beet and with various other sugar beet herbicides pre- or early post-emergence. See label for details

Crop safety/Restrictions
- Crops should be drilled to at least 13 mm and the seed well covered
- Heavy rainfall after spraying may reduce crop stand especially when rain is followed by very hot weather
- Only use in post-emergence mixtures on crops growing vigorously and not stressed by drought, pest attack, deficiency or other factors
- In the event of crop failure only beet crops may be re-drilled on treated land but any crop may be sown 4 mth after treatment following ploughing to 15 cm

Special precautions/Environmental safety
- Irritating to eyes, skin and respiratory system
- Harmful to fish. Do not contaminate ponds, waterways or ditches with chemical or used container

Protective clothing/Label precautions
- 6a, 6b, 6c, 18, 21, 28, 29, 36, 37, 53, 63, 65, 70

128   chloridazon + propachlor

A residual pre-emergence herbicide for use in onions and leeks

Products	Ashlade CP	Ashlade	86:400 g/l	SC	02852

Uses   Annual grasses, annual dicotyledons in BULB ONIONS, SALAD ONIONS, LEEKS.

FOR FULL CONDITIONS OF USE ALWAYS READ THE PRODUCT LABEL

Efficacy
- Best results achieved from application to firm, moist, weed-free seedbed when adequate rain falls afterwards
- Apply pre-emergence of sown crops, preferably soon after drilling, before weeds emerge. Loose or fluffy seedbeds must be consolidated before application
- Apply to transplanted crops when soil has settled after planting
- Do not use on soils with more than 10% organic matter

Crop safety/Restrictions
- Ensure crops are drilled to 20 mm depth
- Crops stressed by nutrient deficiency, pests or diseases, poor growing conditions or pesticide damage may be checked by treatment, especially on sandy or gravelly soils
- In the event of crop failure only onions, leeks or maize should be planted
- Any crop can follow a treated onion or leek crop harvested normally as long as the ground is cultivated thoroughly before drilling

Special precautions/Environmental safety
- Harmful in contact with skin and if swallowed. Irritating to eyes and skin
- Harmful to fish. Do not contaminate ponds, waterways or ditches with chemical or used container

Protective clothing/Label precautions
- A, C
- 5a, 5c, 6a, 6b, 18, 21, 28, 29, 36, 37, 53, 63, 66, 70, 78

129   chlormequat

A plant-growth regulator for reducing stem growth and lodging

Products

1	ABM Chlormequat 40	ABM	400 g/l	SL	00003
2	ABM Chlormequat 72.5	ABM	725 g/l	SL	00004
3	Ashlade 4-60 CCC	Ashlade	460 g/l	SL	02894
4	Ashlade 700 CCC	Ashlade	700 g/l	SL	02912
5	Atlas Chlormequat 46	Atlas	460 g/l	SL	03092
6	Atlas Chlormequat 700	Atlas	700 g/l	SL	03402
7	CCC 460	FCC	460 g/l	SL	00454
8	CCC 700	FCC	700 g/l	SL	03366
9	Cleanacres PDR 675	Cleanacres	675 g/l	SL	04037
10	Clifton Chlormequat 46	Clifton	460 g/l	SL	03019
11	Fargro Chlormequat	Fargro	460 g/l	SL	02600
12	Farmacel	Farm Protection	460 g/l	SL	00821
13	Farmacel 645	Farm Protection	645 g/l	SL	03796
14	Headland Swift	WBC Technology	750 g/l	SL	04537
15	Hyquat	Agrichem	460 g/l	SL	01094
16	Hyquat 70	Agrichem	700 g/l	SL	03364
17	Hyquat 75	Agrichem	750 g/l	SL	03365
18	Mandops Barleyquat B	Mandops	620 g/l	SL	01256
19	Mandops Bettaquat B	Mandops	620 g/l	SL	01257
20	Mandops Chlormequat 700	Mandops	700 g/l	SL	01264
21	Mandops Spring Podquat	Mandops	590 g/l	SL	01274
22	MSS Chlormequat 40	Mirfield	400 g/l	SL	01401
23	MSS Chlormequat 46	Mirfield	460 g/l	SL	03935
24	MSS Chlormequat 60	Mirfield	600 g/l	SL	03936
25	MSS Chlormequat 70	Mirfield	700 g/l	SL	03937

26 Portman Chlormequat 400	Portman	400 g/l	SL	01523
27 Portman Chlormequat 460	Portman	460 g/l	SL	02549
28 Portman Chlormequat 600	Portman	600 g/l	SL	01525
29 Portman Chlormequat 700	Portman	700 g/l	SL	03465
30 Power Chlormequat 64	Power	640 g/l	SL	03562
31 Power Chlormequat 640	Power	640 g/l	SL	04282
32 Power Chlormequat 700	Power	700 g/l	SL	04765
33 Quadrangle Chlormequat 700	Quadrangle	700 g/l	SL	03401
34 Star Chlormequat	Star	700 g/l	SL	03133
35 Titan	Schering	667 g/l	SL	03925
36 Tripart Brevis	Tripart	700 g/l	SL	03754
37 Tripart Chlormequat 460	Tripart	460 g/l	SL	03685

Uses Lodging control in WINTER WHEAT, SPRING WHEAT [1-3, 5-10, 12-17, 19-37]. Lodging control in WINTER OATS, SPRING OATS [1-10, 12, 13, 15, 17, 19-37 21-37]. Lodging control in SPRING OATS [4, 14, 16]. Lodging control, increasing yield in WINTER BARLEY, SPRING BARLEY [3, 4, 8, 10, 12, 14-18, 20, 36, 37]. Lodging control, increasing yield in WINTER BARLEY [3, 6-8, 9, 13, 23, 24, 25, 34]. Lodging control in WINTER RYE, SPRING RYE [18]. Lodging control in WINTER RYE [3, 4, 12, 36, 37]. Increasing yield in OILSEED RAPE, PEAS, BEANS [21]. Stem shortening in POINSETTIAS, PELARGONIUMS, BEDDING PLANTS, POT PLANTS [11].

Notes **Efficacy**
- Most effective results on cereals normally achieved from spring application, on wheat and rye from leaf sheath erect to first node detectable (GS 30-31), on oats at second node detectable (GS 32), on barley from mid-tillering to leaf sheath erect (GS 25-30). However, recommendations vary with product. See label for details
- In oilseed rape and beans apply to winter crops after 3-leaf stage and before start of active growth (usually by early Apr), to spring crops between 3-leaf stage and 35 d after emergence [21]
- In peas apply when crop 12-18 cm high [21]
- In tank mixes with pesticides optimum timing for herbicide action may differ from that for growth reduction. See label for details of tank mix recommendations
- Addition of approved non-ionic wetter recommended on oats
- At least 6, preferably 24 h, required before rain for maximum effectiveness. Do not apply to wet crops
- May be used on cereals undersown with grass or clovers

Crop safety/Restrictions
- Do not use on very late sown spring wheat or oats or on crops under stress
- Mixtures with liquid nitrogen fertilizers may cause scorch
- Maximum number of treatments per crop and latest time of treatment permitted vary with product - See label for details

Special precautions/Environmental safety
- Harmful in contact with skin and if swallowed

Protective clothing/Label precautions
- A

FOR FULL CONDITIONS OF USE ALWAYS READ THE PRODUCT LABEL

• 5a, 5c, 18, 21, 25, 28, 29, 34, 44, 37, 54, 60, 63, 65, 70, 78

Approval
• Approved for aerial application on wheat, oats [6, 8, 10, 14, 15, 16, 19, 20, 26, 27, 29, 33, 37]; wheat, spring oats [14]; barley [6, 8, 18, 20]; wheat [31]; cereals [35]. See introductory notes

130   chlormequat + 2-chloroethylphosphonic acid
A plant growth regulator for use in winter wheat

Products					
1 Pacer	Farm Protection	360:180 g/l	SL	04454
2 Upgrade	RP	360:180 g/l	SL	04624

Uses   Lodging control in WINTER WHEAT.

Notes   **Efficacy**
• Apply before lodging has started
• Best results obtained when crops growing vigorously
• Only crops growing under conditions of high fertility should be treated
• Recommended dose varies with growth stage. See label for details and recommendations for use of sequential treatments
• Do not spray when crop wet or rain imminent

Crop safety/Restrictions
• Do not spray after leaf sheaths have split and ear is visible (GS 47)
• Do not spray during cold weather or periods of night frost, when soil is very dry, when crop diseased or suffering pest damage, nutrient deficiency or herbicide stress
• If used on seed crop grown for certification inform seed merchant before using

Special precautions/Environmental safety
• Harmful if swallowed and in contact with skin. Irritating to eyes
• Harmful to fish. Do not contaminate ponds, waterways or ditches with chemical or used container

Protective clothing/Label precautions
• A, C
• 5a, 5c, 6a, 18, 21, 25, 28, 29, 36, 37, 53, 60, 63, 66, 70, 78

131   chlormequat + choline chloride
A plant growth regulator for use in cereals and certain ornamentals

Products				
1 ABM 5C Chlormequat Plus	ABM		SL	03583
2 Arotex Extra	ICI	644:32.2 g/l	SL	00117
3 Ashlade 5C	Ashlade	460:- g/l	SL	04727
4 Atlas 5C Chlormequat	Atlas	460:320 g/l	SL	03084
5 Chafer 5C Chlormequat 640	Britag	640:64 g/l	SL	02587
6 New 5C Cycocel	BASF	645:32 g/l	SL	01482
7 New 5C Cycocel	Cyanamid	645:32 g/l	SL	01483
8 Portman 5C Chlormequat	Portman	460:- g/l	SL	02921

| 9 Portman Supaquat | Portman | 640:- g/l | SL | 03466 |
| 10 Tripart 5C | Tripart | 460:- g/l | SL | 04726 |

Uses Lodging control in WINTER WHEAT, SPRING WHEAT, WINTER OATS [1-10]. Increasing yield in SPRING BARLEY, WINTER BARLEY [1-10]. Lodging control in SPRING OATS [2, 3, 6-10]. Lodging control in WINTER RYE, TRITICALE [2, 6, 7] Increasing yield in WINTER OILSEED RAPE [6, 7]. Stem shortening in POINSETTIAS, PELARGONIUMS [6].

Notes **Efficacy**
- Influence on growth varies with crop and growth stage. Risk of lodging reduced by application at early stem extension. Root development and yield can be improved by earlier treatment
- Most effective results normally achieved from spring application. On winter barley an autumn treatment may also be useful. Timing of spray is critical and recommendations vary with product. See label for details
- Often used in tank-mixes with pesticides. Recommendations for mixtures and sequential treatments vary with product. See label for details
- Add authorised non-ionic wetter when spraying oats or oilseed rape
- At least 6 h required before rain for maximum effectiveness. Do not apply to wet crops
- May be used on cereals undersown with grass or clovers

Crop safety/Restrictions
- Do not spray very late sown spring crops, crops on soils of low fertility, crops under stress from any cause or if frost expected
- Mixtures with liquid nitrogen fertilizers may cause scorch

Special precautions/Environmental safety
- Harmful if swallowed or in contact with skin [2], harmful if swallowed [1, 3-10]

Protective clothing/Label precautions
- A [1, 3-10]; A, C [2]
- 5c, 18, 21, 28, 29, 36, 37, 54, 63, 66, 70, 78 [1-10]; 5a [2]

Approval
- Approved for aerial application on wheat, oats, barley [2, 4, 6, 7]; wheat, oats [3, 10]; rye [2, 6]; triticale [2, 6, 7]; oilseed rape [6, 7]. See introductory notes

132 chlormequat + di-1-p-menthene
A plant-growth regulator for reducing stem growth and lodging

Products
1 Mandops Halloween	Mandops	440 g/l	SL	01268
2 Mandops Hele Stone	Mandops	470 g/l	SL	01269
3 Mandops Podquat	Mandops	470 g/l	SL	03003

Uses Lodging control in WINTER WHEAT, SPRING WHEAT, WINTER OATS, SPRING OATS [1]. Lodging control in WINTER BARLEY, SPRING BARLEY, WINTER

FOR FULL CONDITIONS OF USE ALWAYS READ THE PRODUCT LABEL

RYE, SPRING RYE [2]. Lodging control, increasing yield in OILSEED RAPE, PEAS, FIELD BEANS, BROAD BEANS [3].

Efficacy
• May be applied in autumn, spring or as a combined autumn and spring treatment
• Apply to cereals in autumn from 2-3 leaf stage (GS 12-13) to before growth ceases, in spring from start of growth to second node detectable (GS 32) for winter crops, from 2-3 to 4-5 leaf stage (GS 12-13 to 14-15) for spring crops [1, 2]
• On winter oilseed rape and beans either apply as soon as possible after 3-leaf stage (GS 1, 3) until growth ceases followed by spring treatment in mid-Mar to early Apr or use a single spring spray. See label for details [3]
• See label for tank mix recommendations
• May be used on undersown crops
• May be used at temperatures down to 1°C provided spray dries on leaves before rain, frost or snow occurs

Crop safety/Restrictions
• Do not apply to plants covered by frost
• Mixtures with liquid nitrogen fertilizers may cause scorch

Special precautions/Environmental safety
• Harmful if swallowed

Protective clothing/Label precautions
• A
• 5c, 18, 21, 28, 29, 36, 37, 54, 63, 70, 78

133   2-chloroethylphosphonic acid

A plant growth regulator for cereals and various horticultural crops
See also chlormequat + 2-chloroethylphosphonic acid

1 Cerone	Embetec	480 g/l	SL	04416
2 Cerone	ICI	480 g/l	SL	00463
3 Ethrel C	ICI	480 g/l	SL	00810
4 Power Ethephon 48	Power	480 g/l	SL	04944

Lodging control in WINTER WHEAT, WINTER BARLEY, SPRING BARLEY, WINTER RYE, TRITICALE [1, 2, 4]. Lodging control in SPRING WHEAT [1, 2]. Lodging control in DURUM WHEAT [2]. Fruit ripening in APPLES, TOMATOES [3]. Stem shortening in DAFFODILS [3]. Basal bud stimulation in GLASSHOUSE ROSES [3]. Increasing branching in GERANIUMS [3]. Inducing flowering in BROMELIADS [3].

Efficacy
• Best results achieved on crops growing vigorously under conditions of high fertility
• May be used on barley from second node detectable to first awns visible (GS 32-49), on other cereals from flag leaf just visible to boots swollen (GS 37-45) preferably following chlormequat treatment. See label for details
• Do not spray crops when wet or if rain imminent
• Best results on horticultural crops when temperature does not fall below 10°C
• Use with naphthylacetic acid to accelerate apple ripening [3]

- Use on tomatoes 17 d before planned pulling date [3]
- Apply as drench to daffodils when stems average 15 cm [3]
- Apply to glasshouse roses when new growth started after pruning [3]
- See label for tank mixes. Do not mix Ethrel C with any spray other than Phyomone

Crop safety/Restrictions
- Do not spray crops suffering from stress caused by any factor, during cold weather or periods of night frost nor when soil very dry
- Do not apply to cereals within 10 d of herbicide or liquid fertilizer application
- Do not spray Triumph barley
- Only one application may be made on any one crop

Special precautions/Environmental safety
- Irritating to eyes and skin
- Harmful [1, 2, 4], dangerous [3], to fish. Do not contaminate ponds, waterways or ditches with chemical or used container

Protective clothing/Label precautions
- A, C
- 6a, 6b, 18, 22, 25, 29, 36, 37, 53, 63, 66, 70, 78 [1-4]; 14, 52 [3]

Harvest interval
- apples 5 d; sprayed tomatoes 3 d, fogged tomatoes 5 d

Approval
- Approved for aerial application on winter barley [1, 2]. See introductory notes

134   2-chloroethylphosphonic acid + mepiquat chloride
A plant growth regulator for reducing lodging in cereals

Products					
1 Power Platoon	Power	155:305 g/l	SL	04894
2 Terpal	BASF	155:305 g/l	SL	02103
3 Terpal	Clifton	155:305 g/l	SL	05026

Uses

Lodging control in WINTER WHEAT, WINTER BARLEY, SPRING BARLEY, WINTER RYE, TRITICALE.

Notes

Efficacy
- Best results achieved on crops growing vigorously under conditions of high fertility
- Recommended dose and timing vary with crop, cultivar, growing conditions, previous treatment and desired degree of lodging control. See label for details
- Use one treatment per crop, on barley up to first awns visible (GS 49), on winter wheat and triticale up to boots swollen (GS 45), on winter rye up to flag leaf just visible (GS 37)
- Add an authorised non-ionic wetter to spray solution
- May be applied to crops undersown with grass or clovers
- Do not apply to crops if wet or rain expected

FOR FULL CONDITIONS OF USE ALWAYS READ THE PRODUCT LABEL

Crop safety/Restrictions
* Do not spray crops damaged by herbicides or stressed by drought, waterlogging etc
* Do not treat crops on soils of low fertility unless adequately fertilized
* Late tillering may be increased with crops subject to moisture stress and may reduce quality of malting barley
* Do not apply to winter cultivars sown in spring or treat barley, triticale or winter rye on soils with more than 10% organic matter (winter wheat may be treated)
* Do not apply at temperatures above 21°C
* Do not use straw from treated cereals as a mulch or growing medium

Protective clothing/Label precautions
* 29, 54, 63, 66

---

135 chlorophacinone
An anticoagulant rodenticide

Products					
1 Drat	RP Environ.	2.5 g/l	CB	04418
2 Karate	Lever	0.006% w/w	RB	02479
3 Ridento Ready-to-Use Rat Bait	Ace	0.006% w/w	RB	03804
4 Sakarat Special	Killgerm	0.006% w/w	RB	02844
5 Skaterpax	Lever	0.006% w/w	RB	02479

Uses

Rats, mice, voles in FARM BUILDINGS.

Notes

Efficacy
* Chemical formulated with oil, thus improving weather resistance of bait
* Mix concentrate with any convenient bait, such as grain, apple, carrot or potato [2]
* Use in baiting programme. Lay small baits for mice, larger baits for rats
* Replenish baits every few d and remove unused bait when take ceases or after 7-10 d

Special precautions/Environmental safety
* Harmful in contact with skin and if swallowed [1]
* Prevent access to baits by children, domestic animals and birds; see label precautions for other precautions required

Protective clothing/Label precautions
* A [1]
* 5a, 5c, 14, 16, 18, 22, 25, 26, 27, 29, 36, 37, 46, 54, 63, 65, 78, 81, 82, 83, 84, 85 [1]; 25, 29, 63, 67, 81, 82, 83, 84 [2-5]

---

136 chloropicrin
A highly toxic and lachrymatory insecticide available only in mixtures

---

137   chlorothalonil

A protectant dicarbonitrile fungicide for use in many crops

See also   carbendazim + chlorothalonil

Products

1 BB Chlorothalonil	Brown Butlin	500 g/l	SC	03320
2 Bombardier	Farm Protection	500 g/l	SC	02675
3 Bombardier	Unicrop	500 g/l	SC	02675
4 Bravo 500	BASF	500 g/l	SC	04939
5 Bravo 500	BASF	500 g/l	SC	04945
6 Chiltern Olé	Chiltern	500 g/l	SC	04004
7 Contact 75	Fermenta	75% w/w	WP	04772
8 Daconil Turf	ICI Professional	500 g/l	SC	03658
9 Jupital	Fermenta	500 g/l	SC	04946
10 Power Chlorothalonil 50	Power	500 g/l	SC	03419
11 Repulse	ICI	500 g/l	SC	02724
12 Sipcam UK Rover 500	Sipcam	500 g/l	SC	04165
13 Tripart Faber	Tripart	500 g/l	SC	04338
14 Tripart Faber	Tripart	500 g/l	SC	04549

Uses

Glume blotch, leaf spot in WINTER WHEAT [1-6, 9, 10, 12-14]. Leaf blotch in BARLEY [1-3, 6, 10, 12-14]. Blight in POTATOES [1-7, 9, 11-14]. Chocolate spot in FIELD BEANS [1-7, 9, 11-14]. Alternaria, botrytis, downy mildew, damping off and wirestem, ring spot in LEAF BRASSICAS [1-3, 6, 11-14]. Ring spot, downy mildew in LEAF BRASSICAS [4, 5, 9]. Botrytis, downy mildew in OILSEED RAPE [4, 5, 9, 13]. Botrytis, downy mildew, ascochyta in PEAS [4, 5, 9, 11, 13]. Anthracnose, fusarium patch, melting out, take-all patch, brown patch, grey snow mould in TURF [8]. Celery leaf spot in CELERY [1-7, 9, 11-14]. Onion leaf blotch, leaf rot, neck rot in ONIONS [1-3, 6, 11, 12, 14]. Leaf rot, neck rot in ONIONS [4, 5, 9, 13]. Leaf rot, neck rot in LEEKS [13]. Botrytis in CANE FRUIT [11]. Botrytis in RASPBERRIES [4, 5, 9]. Botrytis, cane spot, powdery mildew in CANE FRUIT [13]. Currant leaf spot in BLACKCURRANTS [1-5, 9, 11-14]. Currant leaf spot in REDCURRANTS [13]. Currant leaf spot in GOOSEBERRIES [4, 5, 9, 13]. Botrytis in STRAWBERRIES [4, 5, 9, 11, 13]. Botrytis, downy mildew in GRAPEVINES [4, 5, 9, 11, 13]. Downy mildew in HOPS [4, 5, 9, 11, 13]. Botrytis in HARDY ORNAMENTAL NURSERY STOCK, BULBS [11]. Black spot in ROSES [11]. Ray blight in CHRYSANTHEMUMS [11]. Ink disease in IRISES [4, 5, 9, 11, 13]. Blight, botrytis, leaf mould in TOMATOES [4, 5, 9, 11, 13]. Botrytis in PEPPERS [4, 5, 9, 11, 13]. Anthracnose, botrytis, gummosis, didymella in CUCUMBERS [11]. Botrytis, powdery mildew in CUCUMBERS [4, 5, 9]. Botrytis in CUCUMBERS [13]. Botrytis in GLASSHOUSE ORNAMENTALS [4, 5, 9, 11]. Dry bubble, wet bubble in MUSHROOMS [4, 5, 9, 11, 13]. Mosses, liverworts in CONTAINER-GROWN STOCK [11]. Ascochyta in COMBINING PEAS, VINING PEAS *(off-label)* [1, 3, 5, 6, 11, 13]. Celery leaf spot in CELERIAC *(off-label)* [1, 3, 5, 6, 11, 13].

Notes

Efficacy

• For some crops products differ in diseases listed as controlled. See label for details and for application rates, timing and number of sprays

• Apply as protective spray or as soon as disease appears and repeat at 7-21 d intervals

---

FOR FULL CONDITIONS OF USE ALWAYS READ THE PRODUCT LABEL

- Activity against septoria may be reduced where serious mildew or rust present. In such conditions mix with suitable mildew or rust fungicide
- Do not mow or water turf for 24 h after treatment. Do not add surfactant or mix with liquid fertilizer
- For botrytis control in strawberries important to start spraying early in flowering period and repeat at least 3 times at 10 d intervals
- In addition to disease control provides control of mosses and liverworts in container-grown nursery stock

Crop safety/Restrictions
- Do not apply more than 2 treatments per crop on winter wheat, brassicas, peas, onions, leeks or mushrooms, more than 3 treatments per year on blackcurrants, 4 treatments per year on strawberries
- Do not use on dessert grapes as russetting may occur. Do not use on wine grapes within 1 mth of harvest as fermentation may be affected
- On strawberries some scorching of calyx may occur with protected crops

Special precautions/Environmental safety
- Irritating to eyes, skin and respiratory system [1-5, 7-14], to skin [6]
- May cause temporary allergy to sensitive persons, characterized by redness of eyes, bronchial irritation and skin rash. Persons so affected should contact a physician immediately [11]
- Dangerous to fish. Do not contaminate ponds, waterways or ditches with chemical or used container

Protective clothing/Label precautions
- A, C
- 6a, 6b, 6c, 14, 18, 21, 25, 26, 27, 28, 29, 36, 37, 52, 60, 63, 66, 70, 78

Harvest interval
- potatoes zero; tomatoes, cucumbers, peppers 12 h; mushrooms 24 h; currants, gooseberries, cane fruit, grapevines, strawberries 3 d; brassicas, celery 7 d; hops 10 d; field beans, peas, onions, leeks 14 d; wheat may be sprayed up to ears just completely emerged (GS 59)

Approval
- Approval for aerial spraying on wheat, barley, potatoes, field beans [1-3, 6, 12, 14]; on winter wheat, potatoes [4, 5, 9]; on peas, beans [11, 12]. See introductory notes
- Off-label approval to Feb 1991 for use on combining and vining peas (where pod not intended for consumption) (OLA 0100/88) [1]; (OLA 0101/88) [3]; (OLA 0102/88) [5]; (OLA 0098/88) [6]; (OLA 0092/88) [11]; (OLA 0091/88) [12]
- Off-label approval to Feb 1991 for use on celeriac (OLA 0112/88) [1]; (OLA 0113/88) [3]; (OLA 0114/88) [5]; (OLA 0115/88) [6]; (OLA 0106/88) [13]

138   chlorothalonil + fenpropimorph

A systemic and protectant fungicide for use in winter wheat

Products					
1 BAS 438	BASF	250:187 g/l	SC	03451
2 Corbel CL	BASF	250:187 g/l	SC	04196
3 Mistral CT	RP	250:187 g/l	SC	04543

Uses   Powdery mildew, yellow rust, brown rust, septoria diseases in WINTER WHEAT.

Efficacy
- Apply before disease established from flag leaf visible to ear just completely emerged (GS 39-59)
- Addition of Bavistin, Bavistin DF or Bavistin FL recommended for additional sooty mould control and for later applications during ear emergence. See label for details and other recommended tank mixes

Crop safety/Restrictions
- Up to 2 applications per crop may be made
- Some crop scorch may occur if applied during high temperatures

Special precautions/Environmental safety
- Harmful by inhalation. Irritating to eyes, skin and respiratory system
- Dangerous to fish. Do not contaminate ponds, waterways or ditches with chemical or used container

Protective clothing/Label precautions
- A, C
- 5a, 6a, 6b, 6c, 18, 21, 26, 27, 28, 29, 36, 52, 60, 63, 66, 70, 78

Harvest interval
- 5 wk

139 chlorothalonil + flutriafol
A systemic, eradicant and protectant fungicide for winter wheat

Products	Impact Excel	ICI	300:47 g/l	SC	03758

Uses

Brown rust, glume blotch, leaf spot, powdery mildew, sooty moulds, yellow rust in WINTER WHEAT.

Notes

Efficacy
- Best results achieved by applying protective spray at flag leaf emergence (GS 37) and full ear emergence (GS 59). If disease already present spray before it reaches the top 2 leaves and ears
- Prior use of fenpropidin (Patrol) advised where mildew serious. See label for recommended tank mixes

Crop safety/Restrictions
- May be used at all stages before beginning of flowering (GS 61)
- Do not apply more than 2 sprays of products containing flutriafol in any crop
- On certain cultivars with erect leaves high transpiration can result in flag leaf tip scorch. This may be increased by treatment but does not affect yield

Special precautions/Environmental safety
- Irritating to eyes and skin. May cause sensitization by skin contact
- Harmful to fish. Do not contaminate ponds, waterways an ditches with chemical or used container

FOR FULL CONDITIONS OF USE ALWAYS READ THE PRODUCT LABEL

140 chlorothalonil + metalaxyl
A systemic and protectant fungicide for various field crops

| Products | Folio 575FW | Ciba-Geigy | 500:75 g/l | SC | 04122 |

Uses   Downy mildew in FIELD BEANS, BROAD BEANS, OILSEED RAPE, BULB ONIONS, SALAD ONIONS. White blister in BRUSSELS SPROUTS. White tip in LEEKS.

Notes   **Efficacy**
• Apply at first signs of disease (oilseed rape, beans), at first signs of disease or when weather conditions favourable to disease (Brussels sprouts, onions, leeks)
• Repeat treatment at 14-21 d (14 d for oilseed rape and beans) intervals if necessary
• Up to 2 applications may be made on oilseed rape, beans, onions and leeks, up to 3 on Brussels sprouts
• Oilseed rape crops most likely to benefit from treatment are infected crops between cotyledon and 3-leaf stage (GS 1,0-1,3)

Special precautions/Environmental safety
• Irritating to skin and eyes
• Dangerous to fish. Do not contaminate ponds, waterways or ditches with chemical or used container

Protective clothing/Label precautions
• A, C
• 6a, 6b, 18, 22, 29, 35, 36, 37, 52, 63, 65

Harvest interval
• Field beans, broad beans, Brussels sprouts 14 d; onions, leeks 21 d

141 chlorothalonil + propiconazole
A systemic and protectant fungicide for winter wheat

| Products | Sambarin TP | Ciba-Geigy | 500:250 g/l | KL | 04094 |

Uses   Septoria diseases, yellow rust, brown rust, powdery mildew in WINTER WHEAT.

Notes   **Efficacy**
• Apply from start of flag leaf emergence up to and including when ears just fully emerged (GS 37-59)

Crop safety/Restrictions
• Do not apply more than once to any crop
• Do not apply after the ear just fully emerged stage (GS 59)

Special precautions/Environmental safety
- Irritating to eyes, skin and respiratory system (Sambarin A), to eyes and skin (Sambarin B)
- Dangerous to fish. Do not contaminate ponds, waterways or ditches with chemical or used container (Sambarin A & B)

Protective clothing/Label precautions
- A, C
- 6a, 6b, 16, 18, 21, 28, 29, 36, 37, 52, 63, 66, 70 [A and B]; 6c, 15 [A]; 14, 35 [B]

Harvest interval
- 35 d

---

142 chlorotoluron
A contact and residual urea herbicide for cereals
See also bifenox + chlorotoluron

Products

1 Chiltern Chlortoluron	Chiltern	500 g/l	SC	04490
2 Chlortoluron 500	PBI	500 g/l	SC	03686
3 Dicurane 500 FW	Ciba-Geigy	500 g/l	SC	00698
4 Portman Chlortoluron	Portman	500 g/l	SC	03068
5 Power Chlortoluron 500 FC	Power	500 g/l	SC	03499
6 Talisman	FCC	500 g/l	SC	03109
7 Toro	Sipcam	500 g/l	SC	04734
8 Tripart Ludorum	Tripart	500 g/l	SC	03059
9 Tripart Ludorum 700	Tripart	700 g/l	SC	03999

Uses

Blackgrass, wild oats, rough meadow grass, annual grasses, annual dicotyledons in WINTER BARLEY, WINTER WHEAT, DURUM WHEAT, TRITICALE [1, 3, 6-9]. Blackgrass, wild oats, rough meadow grass, annual grasses, annual dicotyledons in WINTER BARLEY, WINTER WHEAT, TRITICALE [2]. Blackgrass, wild oats,rough meadow grass, annual grasses, annual dicotyledons in WINTER BARLEY, WINTER WHEAT [4, 5].

Notes

Efficacy
- Best results achieved by application soon after drilling. Application in autumn controls most weeds germinating in early spring
- For wild oat control apply within 1 wk of drilling, not after 2-leaf stage. Blackgrass and meadow grasses controlled to 5 leaf, ryegrasses to 3 leaf stage
- Any trash or burnt straw should be buried and dispersed during seedbed preparation
- Do not use on soils with more than 10% organic matter
- Control may be reduced if prolonged dry conditions follow application
- Harrowing after treatment may reduce weed control

Crop safety/Restrictions
- Use only on listed crop varieties. See label. Ensure seed well covered at drilling
- Apply in autumn to wheat, barley or triticale sown before 30 Nov as pre-emergence spray or post-emergence from 1-unfolded leaf stage (GS 11) to end Dec

---

FOR FULL CONDITIONS OF USE ALWAYS READ THE PRODUCT LABEL

- Only apply pre-emergence to durum wheat
- Apply in spring at any stage up to the end of tillering (GS 29)
- Do not apply pre-emergence to crops sown after 30 Nov
- Do not apply to crops severely checked by waterlogging, pests or other factors or during prolonged frost
- Do not use on undersown crops or those due to be undersown
- Do not apply post-emergence in mixture with liquid fertilizers
- Do not roll for 7 d before or after application to an emerged crop
- Crops on stony or gravelly soils may be damaged, especially after heavy rain

Special precautions/Environmental safety
- Irritating to skin and eyes

Protective clothing/Label precautions
- A, C [2]; A [5] 6a, 6b, 18, 21, 28, 29, 36, 37, 54, 60, 63, 65, 66, 70

Approval
- May be applied through CDA equipment. See label for details. See introductory notes [2, 3, 7, 8]
- Approved for aerial application on cereals [1, 3]. See introductory notes

---

143 chloroxuron
A residual urea herbicide for strawberries and ornamentals

Products   Tenoran 50WP   Ciba-Geigy   50%   WP   02100

Uses   Annual dicotyledons, annual meadow grass in STRAWBERRIES, HERBACEOUS ORNAMENTALS, VIOLAS, PANSIES, CHRYSANTHEMUMS, CONTAINER-GROWN STOCK, FREESIAS. Mosses in HARD SURFACES.

Notes   **Efficacy**
- Apply to clean, weed-free soil to control annual weeds for 6-8 wk
- On maiden strawberries apply 2 wk after planting, on fruiting beds 6-8 wk before strawing (not later than mid-Apr and not between flowering and harvest) and in autumn immediately after cleaning up beds
- Rainfall or up to 6 mm irrigation must follow spraying before cloching strawberries
- Apply to chrysanthemums at least 3 d after planting
- Do not use on highly organic soils. Activity is reduced in very dry weather
- On container-grown stock irrigate before spraying and again immediately after to wash chemical off foliage
- For moss control apply under warm, dry conditions from Mar to Oct

Crop safety/Restrictions
- Do not use on sands or very light soils
- Test treat subjects of unknown tolerance on small batches in first instance
- Do not use on American spray chrysanthemums
- Plants under glass or polythene must be hardened off before spraying

Protective clothing/Label precautions
- 29, 54, 63, 67, 70

144   chloroxuron + ferrous sulphate
A lawn sand herbicide for moss control in turf

Products   Ashlade D-Moss   Ashlade   0.45:7.1% w/w   SA   02831

Uses   Mosses in TURF.

Notes   **Efficacy**
• Apply to established turf in spring to early autumn when moss growing actively. Late Mar-early Apr often gives best results
• Apply when grass dry and water if no rain for 2 d after applying
• Mow 3 d before applying and allow at least 4 d before mowing after treatment
• Remove dead moss 4-6 wk after application

Crop safety/Restrictions
• Do not apply within 12 mth of sowing
• Do not apply in long dry spells, when frost imminent or when rain likely within 24 h
• Do not walk on treated areas until after rain or irrigation
• Do not use first mowing after treatment for compost or mulch

Protective clothing/Label precautions
• 26, 29, 37, 54, 63, 67

145   chlorpropham
A residual carbamate herbicide and potato sprout suppressant
See also cetrimide + chlorpropham
          chloridazon + chlorpropham + fenuron + propham

Products
1	Atlas CIPC 40	Atlas	400 g/l	EC	03049
2	Campbell's CIPC 40%	MTM Agrochem.	400 g/l	EC	00389
3	Mirvale 500HN	Ciba-Geigy	500 g/l	HN	01360
4	MSS CIPC 40 EC	Mirfield	400 g/l	EC	01403
5	MSS CIPC 5 G	Mirfield	5% w/w	GR	01402
6	MSS CIPC 50 LF	Mirfield	500 g/l	EC	03285
7	MSS CIPC 50 M	Mirfield	500 g/l	EC	01404
8	Warefog 25	Dean	600 g/l	HN	02323

Uses   Annual grasses, annual dicotyledons, chickweed, polygonums in ONIONS, BULBS [1, 2, 4]. Annual grasses, annual dicotyledons, chickweed, polygonums in LEEKS [1, 4]. Annual grasses, annual dicotyledons, chickweed, polygonums in CARROTS [1, 2]. Annual grasses, annual dicotyledons, chickweed, polygonums in LETTUCE [2, 4]. Annual grasses, annual dicotyledons, chickweed, polygonums in LUCERNE, CELERY, PARSLEY, BLACKCURRANTS, GOOSEBERRIES, STRAWBERRIES, CHRYSANTHEMUMS, ANNUAL FLOWERS [2]. Sprout suppression in STORED POTATOES [3, 5, 6, 7, 8]. Volunteer ryegrass in GRASS SEED CROPS *(off-label)* [1, 2, 4]. Growth retardation in GRASS SEED CROPS *(off-label)* [1, 2, 4].

FOR FULL CONDITIONS OF USE ALWAYS READ THE PRODUCT LABEL

Efficacy
- Apply to freshly cultivated soil. Adequate rainfall must occur after spraying. Activity is greater in cold, wet than warm, dry conditions
- For sprout suppression apply with suitable fogging or rotary atomizer equipment or sprinkle granules over tubers before sprouting commences. Repeat applications may be needed. See label for details
- Cure potatoes according to label instructions before treatment and allow 3 wk between completion of loading into store and first treatment [3]

Crop safety/Restrictions
- Apply to seeded crops pre-emergence of crop or weeds, to onions as soon as first crop seedlings visible, to planted crops a few days before planting, to bulbs immediately after planting, to fruit crops in late autumn-early winter. See label for further details
- Excess rainfall after application may result in crop damage
- Do not use on sands, very light soils or soils low in organic matter
- Poor conditions at drilling or planting, soil compaction, surface capping, waterlogging or attack by pests may result in crop damage
- On crops under glass high temperatures and poor ventilation may cause crop damage
- Only clean, mature, disease-free potatoes should be treated for sprout suppression
- Do not use on potatoes for seed. Do not handle, dry or store seed potatoes or any other seeds or bulbs in boxes or buildings in which potatoes are being or have been treated [3]

Special precautions/Environmental safety
- Harmful if swallowed and in contact with skin [1, 2, 6, 7]
- Harmful in contact with skin or by inhalation [3]
- Irritating to eyes, skin and respiratory system [1, 6]
- Flammable [6, 7]
- When working in confined spaces attention should be paid to protective clothing requiements specified on the label [3, 8]
- Harmful to fish. Do not contaminate ponds, waterways or ditches with chemical or used container [4-8]

Protective clothing/Label precautions
- A
- 5a, 5c [1, 2, 6, 7]; 5a, 5b [3]; 6a, 6b, 6c [1, 6]; 12c [6, 7]; 18, 22, 28, 29, 36, 37, 54, 60, 63, 65, 70, 78 [1, 2, 3]; 18, 22, 28, 29, 36, 37, 53, 63, 65, 70 [4-8]

Harvest interval
- Potatoes must not be removed for sale or processing for at least 21 d after treatment

Approval
- Some products are formulated for application by thermal fogging. See label for details. See introductory notes [3, 8]
- Off-label approval to Aug 1991 for use in grass seed crops (OLA 0892/88) [1]; (OLA 0888/88) [2]; OLA (0889/88) [4]

146   chlorpropham + cresylic acid + fenuron
A residual herbicide for vegetables and ornamentals

Products	Atlas Red	Atlas	200:-:50 g/l	SC	03091

Annual grasses, annual dicotyledons, chickweed in FIELD BEANS, BROAD BEANS, PEAS, ONIONS, LEEKS, RHUBARB, BULBS, HARDY ORNAMENTAL NURSERY STOCK, STANDING GROUND, PATHS AND DRIVES.

Notes

Efficacy
- Apply to soil freshly cultivated and free of established weeds. Adequate rainfall must occur after spraying. Activity is greater in cold, wet than warm, dry conditions

Crop safety/Restrictions
- Apply to seeded crops pre-emergence of crop and weed, to transplanted onions and leeks 10-14 d post-planting, to transplanted flowers and strawberries 5 d pre-transplanting, to rhubarb and nursery stock in dormant season, avoiding foliage of conifers or evergreens
- Do not use on sands, very light soils or soils low in organic matter
- Poor conditions at drilling or planting, soil compaction, surface capping, waterlogging or pest attack may result in crop damage
- In the event of crop failure only recommended crops should be planted in treated soil
- Plough or cultivate to 15 cm after harvest to dissipate any residues

Special precautions/Environmental safety
- Harmful if swallowed and in contact with skin
- Irritating to eyes, skin and respiratory system
- Flammable
- Dangerous to fish. Do not contaminate ponds, waterways or ditches with chemical or used container

Protective clothing/Label precautions
- A
- 5a, 5c, 6a, 6b, 6c, 12c, 18, 21, 25, 28, 29, 36, 37, 52, 60, 63, 67, 70

147 chlorpropham + diuron
A residual herbicide for bulbs, peas and beans

Products Residuren Extra   Farm Protection   200:80 g/l   EC   01793

Uses Annual grasses, annual dicotyledons in BULBS, SPRING PEAS, SPRING BROAD BEANS, TICK BEANS.

Notes

Efficacy
- Apply to clean land as, with exception of chickweed, emerged weeds not controlled
- Best results achieved by application to a fine firm seedbed when soil moist
- Apply after drilling sown crops before weeds germinate
- Apply to autumn or spring planted bulbs after planting to shortly before emergence, to established beds after late summer cleaning to shortly before emergence
- Do not cultivate after treatment unless weed growth makes this necessary

FOR FULL CONDITIONS OF USE ALWAYS READ THE PRODUCT LABEL

Crop safety/Restrictions
• On sown crops apply at least 1 wk before crop emergence
• Do not treat crops sown by broadcasting followed by harrowing
• After harvesting treated legumes the soil should be ploughed before a sensitive crop such as kale is sown

Special precautions/Environmental safety
• Irritating to skin, eyes and respiratory system
• Harmful to fish. Do not contaminate ponds, waterways or ditches with chemical or used container

Protective clothing/Label precautions
• 6a, 6b, 6c, 18, 21, 28, 29, 36, 37, 53, 63, 66

148 chlorpropham + diuron + propham
A residual herbicide for lettuce

Products	Atlas Pink C	Atlas	25:6:100 g/l	SC	03095

Uses Annual grasses, annual dicotyledons in OUTDOOR LETTUCE.

Notes

Efficacy
• Best results achieved from application to firm, moist, weed-free seedbed when adequate rain falls afterwards. Activity is greater in cold wet than warm, dry conditions
• Apply to soil after drilling or prior to transplanting
• Incorporation to 2.5-5 cm prior to sowing or planting can improve results under dry soil conditions or on organic soils

Crop safety/Restrictions
• Poor conditions at drilling or planting, soil compaction, surface capping, waterlogging and attack by pests may result in crop damage
• Treatment of lettuce under frames or polythene tunnels may cause injury during sunny periods
• In the event of crop failure only lettuce should be replanted in treated soil
• Plough or cultivate to 10 cm after harvest to dissipate any residues

Special precautions/Environmental safety
• Harmful if swallowed and in contact with skin
• Irritating to eyes, skin and respiratory system

Protective clothing/Label precautions
• 5a, 5c, 6a, 6b, 6c, 12c, 18, 21, 28, 29, 36, 37, 54, 63, 65, 78

149 chlorpropham + fenuron
A residual herbicide for vegetables and ornamentals

Products	Croptex Chrome	Hortichem	80:15 g/l	SC	02415

Annual grasses, annual dicotyledons, chickweed in FIELD BEANS, BROAD BEANS, PEAS, ONIONS, LEEKS, SPINACH, STRAWBERRIES, BULBS, DRILLED FLOWERS [1-3].

Notes

Efficacy
• Apply to soil freshly cultivated and free of established weeds. Adequate rainfall must occur after spraying. Activity is greater in cold, wet than warm, dry conditions

Crop safety/Restrictions
• Apply to seeded crops pre-emergence of crop or weeds, to transplanted onions and leeks 10-14 d post-planting, to transplanted flowers and strawberries 5 d pre-transplanting, to rhubarb and nursery stock in dormant season avoiding foliage of conifers or evergreens
• Do not use on sands, very light soils or soils low in organic matter
• Poor conditions at drilling or planting, soil compaction, surface capping, waterlogging or attack by pests may result in crop damage
• In the event of crop failure only recommended crops should be replanted in treated soil
• Plough or cultivate to 15 cm after harvest to dissipate any residues

Special precautions/Environmental safety
• Harmful if swallowed and in contact with skin
• Irritating to eyes, skin and respiratory system
• Flammable
• Dangerous to fish. Do not contaminate ponds, waterways or ditches with chemical or used container

Protective clothing/Label precautions
• A
• 5a, 5c, 6a, 6b, 6c, 12c, 18, 21, 25, 28, 29, 36, 37, 52, 60, 63, 67, 70, 78

150   chlorpropham + fenuron + propham
A residual herbicide for beet crops

Products

1 Atlas Gold	Atlas	37.5:25:150 g/l	SC	03086
2 Campbell's Sugar Beet Herbicide	MTM Agrochem.	27.5:28.5:147 g/l	EC	00423
3 MSS Sugar Beet Herbicide	Mirfield	37.5:25:150 g/l	SC	02447
4 Truchem Quintex	Truchem	30:30:130 g/l	EC	03552

Uses

Annual grasses, annual dicotyledons in SUGAR BEET, FODDER BEET, RED BEET, MANGOLDS.

Notes

Efficacy
• Apply to firm, moist, weed-free seedbed at drilling or up to 10 d after drilling in Mar, up to 7 d in early Apr, up to 3 d in mid-Apr pre-emergence of crop and weeds
• Activity is greater in cold, wet than warm, dry conditions
• Application rate varies with soil type and weather conditions. See label for details
• Avoid disturbance of soil surface after application

FOR FULL CONDITIONS OF USE ALWAYS READ THE PRODUCT LABEL

- Treatment recommended as particularly useful on organic/fen soils [1]

Crop safety/Restrictions
- Do not use on very light or light soils
- When crop vigour is reduced by poor soil conditions, waterlogging, pest or disease attack, drilling too deep, rapid temperature change etc some crop damage may occur
- In the event of crop failure only beet crops should be re-drilled within 12 wk of treatment

Special precautions/Environmental safety
- Harmful if swallowed and in contact with skin [1-3]
- Irritating to skin, eyes and respiratory system [1-4]
- Dangerous to fish [2, 3], harmful to fish [4]. Do not contaminate ponds, waterways or ditches with chemical or used container

Protective clothing/Label precautions
- A, C [1, 3, 4]; A [2]
- 5a, 5c [1-3] 6a, 6b, 6c, 12c, 18, 21, 25, 28, 29, 36, 37, 54, 60, 63, 65, 70, 78 [1-4]; 52 [2, 3]; 53 [4]

151   chlorpropham + linuron

A residual and contact herbicide for use in bulb crops

Products	Profalon	Hoechst	200:100 g/l	EC	01640

Uses   Annual dicotyledons, annual grasses in DAFFODILS, NARCISSI, TULIPS. Annual dicotyledons, annual grasses in ORNAMENTAL BULBS *(off-label)*.

Notes   **Efficacy**
- Weeds controlled by combined residual action, requiring adequate soil moisture, and contact effect on young seedlings, requiring dry leaf surfaces for good control
- Do not spray during or immediately prior to rainfall
- Do not cultivate after spraying unless necessary

Crop safety/Restrictions
- Spray daffodils and narcissi pre-emergence or post-emergence before flower buds show. Spray tulips pre-emergence only
- Do not apply to very light soils or silts low in humus or clay content

Special precautions/Environmental safety
- Irritating to skin, eyes and respiratory system
- Flammable

Protective clothing/Label precautions
- A, C
- 6a, 6b, 6c, 12c, 16, 18, 21, 28, 29, 36, 37, 54, 60, 63, 66, 70

Approval
- Approved for aerial application on bulbs and corms. See introductory notes
- Off-label approval to Jan 1993 for use on ornamental bulbs (OLA 0051/90)

152   chlorpropham + pentanochlor
A contact and residual herbicide for carrots, bulbs and chrysanthemums

Products   Atlas Brown   Atlas   150:300 g/l   EC   03835

Uses   Annual dicotyledons, annual meadow grass in CARROTS, CELERIAC, CELERY, FENNEL, PARSLEY, PARSNIPS, NARCISSI, TULIPS, CHRYSANTHEMUMS.

Notes   **Efficacy**
- Apply as pre- or post-weed emergence spray
- Best results by application to weeds up to 2-leaf stage on fine, firm, moist seedbed
- Greatest contact action achieved under warm, moist conditions, the short residual action greatest in earlier part of year

Crop safety/Restrictions
- Apply to carrots and related crops pre- or post-emergence after fully expanded cotyledon stage, to narcissi and tulips at any time before emergence
- Apply to chrysanthemums either pre-planting or after planting as carefully directed spray, avoiding foliage
- Only treat healthy crops growing well
- Any crop may be sown or planted after 4 wk following ploughing and cultivation

Special precautions/Environmental safety
- Irritating to eyes

Protective clothing/Label precautions
- 6a, 18, 21, 28, 29, 36, 37, 54, 63, 66

153   chlorpropham + propham
A plant growth regulator suppressing sprouting of potatoes

Products
1 Atlas Indigo	Atlas	220:30 g/l	HN	03087
2 Pommetrol M	Fletcher	452:48 g/l	HN	01615
3 Power Gro-Stop	Power	260:40 g/l	HN	02719

Uses   Sprout suppression in STORED POTATOES.

Notes   **Efficacy**
- Apply to clean tubers after curing but before any growth of shoots has commenced, normally 3-6 wk after entering storage
- Treatment effective for 70-100 d and up to 3 applications can be made to a stored crop
- Apply with thermal fog generator according to manufacturers recommendations
- Apply into vent ducts in well ventilated store where forced draft ventilation can be used

FOR FULL CONDITIONS OF USE ALWAYS READ THE PRODUCT LABEL

- Treatment is most effective at 7-9°C, above 14°C sprout suppression will not occur
- Do not use on outdoor clamps unless built with correct system of ducts [1]

Crop safety/Restrictions
- Do not treat potatoes intended for seed
- Do not treat potatoes stored in wooden boxes
- Do not use on potatoes with a high level of skin spot infection
- Do not handle, store or dry seed potatoes, seed grain, bulbs or other seed in buildings where potatoes are being treated or have previously been treated

Special precautions/Environmental safety
- Harmful if swallowed and in contact with skin
- Irritating to eyes, skin and respiratory system
- Allow at least 3 wk between application and marketing or processing of treated potatoes

Protective clothing/Label precautions
- A
- 5a, 5c, 6a, 6b, 6c, 18, 21, 28, 29, 36, 37, 54, 60, 63, 65, 78

Approval
- Products formulated for ULV application by thermal fogging. See introductory notes

---

154   chlorpyrifos
A contact and ingested organophosphate insecticide and acaricide

Products

1 Crossfire	RP Environ.	228 g/l	EC	03598
2 Dursban 4	DowElanco	480 g/l	EC	00775
3 Dursban 5G	DowElanco	5% w/w	GR	00776
4 Spannit	PBI	480 g/l	EC	01992
5 Spannit Granules	PBI	6% w/w	GR	04048
6 Talon	FCC	480 g/l	EC	03375

Uses

Aphids, frit fly, leatherjackets, thrips, wheat-blossom midges, wheat bulb fly in CEREALS [2, 3, 4]. Leatherjackets in CEREALS [3]. Frit fly in MAIZE, SWEETCORN [2, 4, 6]. Frit fly, leatherjackets in AMENITY GRASS, GOLF COURSES [1]. Frit fly, leatherjackets in GRASSLAND [2, 4, 6]. Aphids, caterpillars in BRASSICAS [2, 4, 6]. Leatherjackets, cabbage root fly in BRASSICAS [2-5]. Cutworms in BRASSICAS [2-6]. Cutworms in POTATOES, VEGETABLES [2, 4, 6]. Leatherjackets, pygmy mangold beetle in SUGAR BEET [2, 4]. Leatherjackets in PEAS [4]. Rosy rustic moth in RHUBARB [2]. Aphids, apple blossom weevil, capsids, caterpillars, codling moth, red spider mites, sawflies, suckers, tortrix moths, winter moth, woolly aphid in APPLES, PEARS [2, 4]. Aphids, red spider mites, tortrix moths, winter moth in PLUMS [2, 4]. Aphids, red spider mites, tortrix moths, vine weevil in STRAWBERRIES [2, 4]. Aphids, raspberry beetle, raspberry cane midge, red spider mites in RASPBERRIES [2, 4]. Aphids, capsids, caterpillars, red spider mites in CURRANTS, GOOSEBERRIES [2, 4]. Black pine beetle, pine weevil, vine weevil in FORESTRY TRANSPLANT LINES [2, 4]. Ambrosia beetle, larch shoot beetle, pine shoot beetle, elm bark beetle in CUT LOGS [2, 4]. Colorado beetle in POTATOES *(off-label)* [2].

Efficacy
- Apply as a foliar or soil treatment for most uses, as granules or a drench for soil pests of brassicas, as a drench for vine weevil control, as a dip for forestry transplants. Number and timing of applications vary with pest and crop. See label for details
- Brassicas raised in plant-raising beds must be re-treated at transplanting
- Activity may be reduced when soil temperature below 5°C or on organic soils
- In dry conditions the effect of granules applied as a surface band may be reduced [3, 5

Crop safety/Restrictions
- Do not apply to young lettuce plants [2] or treat potatoes under severe drought stress. The variety Desiree is particularly susceptible [2, 4, 6]
- Do not apply to sugar beet under stress or within 4 d of applying a herbicide [2, 4]
- Do not mix with highly alkaline materials. Not compatible with zineb [2]
- In apples use pre-blossom up to pink/white bud and post-blossom after petal fall [2]
- Cuttings from treated grass should not be used as a mulch for at least 12 mth [1]

Special precautions/Environmental safety
- This product contains an organophosphorus compound. Do not use if under medical advice not to work with such compounds
- Harmful in contact with skin and if swallowed [1, 2, 4, 6]
- Irritating to eyes and skin [1, 2, 4, 6]
- Flammable [1, 2, 4, 6]
- Dangerous to bees. Do not apply at flowering stage. Keep down flowering weeds [1, 2, 4, 6]
- Dangerous to fish. Do not contaminate ponds, waterways or ditches with chemical or used container [1-6]

Protective clothing/Label precautions
- A, C, H [1, 2, 4, 6]; A [3, 5]
- 1, 2, 5a, 5c, 6a, 6b, 12c, 14, 18, 21, 23, 28, 29, 35, 36, 37, 48, 52, 63, 66, 70, 78 [1, 2, 4, 6]; 1, 2, 14, 29, 35, 37, 52, 63, 67 [3, 5]

Withholding period
- Lactating cows should not be grazed on treated pasture within 14 d [2, 4, 6]

Harvest interval
- cane fruit, strawberries 7 d; apples, pears, plums, currants, gooseberries, carrots, cereals 14 d;other crops 21 d [2]; brassicas, cereals 6 wk [3]; apples, pears, carrots, cereals 14 d; other crops 21 d [4]; brassicas 6 wk [5]; carrots, cereals 14 d; brassicas, maize, sweetcorn, potatoes 21 d [6]

Approval
- Approved for aerial application on wheat, barley, oats for control of wheat bulb fly and leatherjackets from early Jan to end of Feb. See introductory notes [2]
- Off-label approval to Nov 1991 for use on potatoes for Colorado beetle control (OLA 1383/88) [2]

FOR FULL CONDITIONS OF USE ALWAYS READ THE PRODUCT LABEL

155   chlorpyrifos + dimethoate

A contact, systemic and fumigant insecticide for brassica crops

Products   Atlas Sheriff   Atlas   3.6:3.6% w/w   GR   04114

Uses   Cabbage root fly, aphids, wireworms, cutworms, leatherjackets in OILSEED RAPE, SWEDES, TURNIPS, BROCCOLI, BRUSSELS SPROUTS, CABBAGES, CAULIFLOWERS, KALE.

Notes   **Efficacy**
- Apply with suitable granule applicator (see label for details) as surface or sub-surface band treatment. Check calibration before applying
- May be used on drilled or transplanted crops and on nursery beds
- On nursery beds apply as overall treatment
- Apply by mid-Apr or at time of drilling or planting and repeat if necessary

Special precautions/Environmental safety
- Irritating to eyes
- Product contains organophosphorus compounds. Do not use if under medical advice not to work with such compounds
- Dangerous to fish. Do not contaminate ponds, waterways or ditches with chemical or used container
- Harmful to game, wild birds and animals

Protective clothing/Label precautions
- A, C
- 1, 6a, 14, 18, 29, 35, 36, 37, 46, 52, 63, 67, 78

Harvest interval
- 4 wk

156   chlorpyrifos + disulfoton

A systemic and contact organophosphate insecticide for brassicas

Products   Twinspan   PBI   4:6% w/w   GR   02255

Uses   Aphids, cabbage root fly in BROCCOLI, BRUSSELS SPROUTS, CABBAGES, CAULIFLOWERS.

Notes   **Efficacy**
- Apply with suitable band applicator as bow-wave treatment at drilling followed if necessary by surface band treatment 2 d after singling, or as sub-surface band treatment at transplanting. See label for details
- Can be used on all mineral and organic soils
- Crops treated in plant-raising beds must be treated again at transplanting
- Effect of surface band application may be reduced in dry weather

Crop safety/Restrictions
- Do not treat mini-cauliflowers

Special precautions/Environmental safety

- This product contains an organophosphorus compound. Do not use if under medical advice not to work with such compounds
- Harmful if swallowed and by inhalation
- Dangerous to game, wild birds and animals
- Dangerous to fish. Do not contaminate ponds, waterways or ditches with chemical or used container

Protective clothing/Label precautions

- A, B, C or D+E, H, K, M
- 1, 5b, 5c, 14, 16, 18, 22, 25, 26, 27, 28, 29, 35, 36, 37, 45, 52, 64, 67, 70, 78

Harvest interval

- 6 wk

157   chlorpyrifos-methyl

An organophosphate insecticide and acaricide for grain store use

Products					
1 Cooper Graincote	Wellcome	450 g/l	EC	02409
2 Reldan 50	DowElanco	500 g/l	EC	02556

Uses

Grain storage pests in STORED GRAIN [1, 2]. Pre-harvest hygiene in GRAIN STORES [1, 2]. Storage pests in STORED OILSEED RAPE [1, 2].

Notes

Efficacy

- May be applied pre-harvest to surfaces of empty store and grain handling machinery and as admixture with grain
- Apply to grain after drying to moisture content below 14%, cooling and cleaning
- Insecticide may become depleted at grain surface if grain or rapeseed is being cooled by continuous extraction of air from the base leading to reduced mite control
- Resistance to organophosphorus compounds sometimes occurs in insect and mite pests of stored products

Special precautions/Environmental safety

- This product contains an organophosphorus compound. Do not use if under medical advice not to work with such compounds
- Irritating to eyes and skin
- Flammable [1]
- Dangerous to fish. Do not contaminate ponds, waterways or ditches with chemical or used container

Protective clothing/Label precautions

- A, C, H, M
- 1, 6a, 6b, 12c, 18, 21, 25, 28, 29, 36, 37, 52, 63, 66, 70

Harvest interval

- barley for malting 8 wk

FOR FULL CONDITIONS OF USE ALWAYS READ THE PRODUCT LABEL

158 chlorsulfuron

A sulphonyl-urea herbicide available only in mixtures
See also bromoxynil + chlorsulfuron + ioxynil

159 chlorthal-dimethyl

A residual herbicide for vegetables, fruit, ornamentals and turf

Products	Dacthal	Fermenta	75% w/w	WP	02629

Uses Annual dicotyledons in BRASSICAS, FRENCH BEANS, RUNNER BEANS, ONIONS, LEEKS, SAGE, BLACKCURRANTS, GOOSEBERRIES, RASPBERRIES, STRAWBERRIES, ROSES, ORNAMENTALS. Slender speedwell in TURF.

Notes **Efficacy**
- Best results on fine firm weed-free soil when adequate rain or irrigation follows
- Recommended alone on roses, french and runner beans, various ornamentals, strawberries and turf, in tank-mix with propachlor on brassicas, onions, leeks, sage, ornamentals, established strawberries and newly planted soft fruit
- Apply after drilling or planting prior to weed emergence. Rates and timing vary with crop and soil type. See label for details
- Do not use on organic soils
- For control of slender speedwell in turf apply when weed growing actively

Crop safety/Restrictions
- Apply to brassicas pre-emergence or after 3-4 true leaf stage
- Do not apply mixture with propachlor to newly planted strawberries after rolling or application of other herbicides
- Many types of ornamental have been treated successfully. See label for details. For species of unknown susceptibility treat a small number of plants first
- Do not use where bent grasses form a major constituent of turf
- Do not plant lettuce within 6 mth of application, seeded turf within 2 mth, other crops within 3 mth. In the event of crop failure deep plough before re-drilling or planting

Protective clothing/Label precautions
- 29, 54, 63, 67

160 chlorthal-dimethyl + propachlor

A contact and residual herbicide for onions and brassicas

Products	Decimate	Fermenta	225:216 g/l	SC	04858

Uses Annual dicotyledons, annual grasses in ONIONS, BROCCOLI, BRUSSELS SPROUTS, CABBAGES, CAULIFLOWERS, MUSTARD.

Efficacy
- Best results achieved by application to fine, firm weed-free soil when adequate rain or irrigation follows
- Apply after drilling or planting prior to weed emergence
- Do not use on soils with more than 10% organic matter

Crop safety/Restrictions
- Apply to onions pre-emergence or from post-crook to young plant stage
- Apply to brassicas after drilling but pre-emergence, after 3-4 leaf stage or at any time after transplanting
- Pre-emergence treatment may check brassica growth but effect normally outgrown
- Young brassicas raised under glass should be hardened off before treatment. Do not use on protected crops
- Do not plant lettuce within 6 mth of application, seeded turf within 2 mth, other crops within 3 mth. In the event of crop failure deep plough before re-drilling or planting

Special precautions/Environmental safety
- Irritating to eyes and skin

Protective clothing/Label precautions
- A, C
- 6a, 6b, 14, 16, 18, 21, 25, 28, 29, 36, 37, 54, 63, 66

161   clofentezine

A selective ovicidal acaricide for use in top fruit

Products | Apollo 50 SC | Schering | 500 g/l | SC | 03996

Uses   Red spider mites in APPLES, PEARS, PLUMS, CHERRIES.

Notes   **Efficacy**
- Acts on eggs and early motile stages of mites. For effective control total cover of plants is essential, particular care being needed to cover undersides of leaves
- For red spider mite control spray apples and pears between bud burst and pink bud, plums and cherries between white bud and first flower. Rust mite is also suppressed
- Treatment effective for 10-12 wk. On established infestations apply in conjunction with an adult acaricide
- Product safe on predatory mites, bees and other predatory insects

Protective clothing/Label precautions
- 21, 29, 36, 37, 54, 60, 63, 67, 71

FOR FULL CONDITIONS OF USE ALWAYS READ THE PRODUCT LABEL

162   clopyralid

A foliar, translocated picolinic herbicide for beets, brassicas etc
See also benazolin + clopyralid
        bromoxynil + clopyralid

Products	Dow Shield	DowElanco	200 g/l	SL	03880

Uses

Annual dicotyledons, mayweeds, corn marigold, perennial dicotyledons, creeping thistle in SUGAR BEET, FODDER BEET, RED BEET, MANGOLDS, OILSEED RAPE, FODDER RAPE, CEREALS, SWEDES, TURNIPS, MAIZE, SWEETCORN, LINSEED, GRASSLAND, CABBAGES, KALE, CALABRESE, BROCCOLI, CAULIFLOWERS, BRUSSELS SPROUTS, ONIONS, STRAWBERRIES, WOODY ORNAMENTALS. Creeping thistle in ESTABLISHED GRASSLAND *(off-label)*. Annual dicotyledons, perennial dicotyledons in CLOVER SEED CROPS, GRASS SEED CROPS *(off-label)*. Annual dicotyledons, perennial dicotyledons in FORESTRY, SAGE, HARDY ORNAMENTAL NURSERY STOCK, NURSERY FRUIT TREES AND BUSHES *(off-label)*.

Notes

Efficacy
- Best results achieved by application to young weed seedlings. Treat creeping thistle at rosette stage and repeat 3-4 wk later
- High activity on weeds of Compositae. For most crops recommended for use in tank mixes. See label for details
- Do not apply when crop damp or when rain expected within 6 h

Crop safety/Restrictions
- Timing of application varies with weed problem, crop and other ingredient of tank mixes. See label for details
- Do not apply to cereals later than the second node detectable stage (GS 32)
- Do not use straw from treated cereals in compost or any other form for glasshouse crops. Straw may be used for strawing down strawberries
- Do not use on onions at temperatures above 20°C
- Do not treat maiden strawberries or runner beds or apply to early leaf growth during blossom period or within 4 wk of picking. Aug or early Sep sprays may reduce yield
- Apply as directed spray in woody ornamentals, avoiding leaves, buds and green stems. Do not apply in root zone of families Compositae or Papilionaceae
- Do not plant susceptible autumn-sown crops in same year as treatment. Do not apply later than Jul where susceptible crops are to be planted in spring. See label for details

Special precautions/Environmental safety
- Livestock must be kept out of treated areas for at least 7 d after application and until foliage of any poisonous weeds such as ragwort has died down and become unpalatable

Protective clothing/Label precautions
- A, C
- 6a, 6b, 18, 21, 28, 29, 35, 36, 37, 54, 60, 63, 66

Harvest interval
- grassland 7d; strawberries 4 wk; maize, sweetcorn, onions, Brussels sprouts, broccoli, cabbage, cauliflowers, calabrese, kale, fodder rape, oilseed rape, swedes, turnips, sugar beet, red beet, fodder beet, mangolds 6 wk

• Off-label approval to Apr 1992 for use in established grassland (wiper application), clover seed crops, grass seed crops (OLA 0391, 0392, 0393/89); to May 1992 for use in forestry, sage, nursery stock of hardy ornamentals, fruit trees and bushes (OLA 0449, 0450, 0454/89)

163 clopyralid + cyanazine
A contact and translocated herbicide mixture for cereals

Products	Coupler SC	Shell	60:350 g/l	SC	03393

lunseed 1-0/ha 'Tsn of beft fbl

Uses Annual dicotyledons in WHEAT, BARLEY, OATS. Annual dicotyledons, perennial dicotyledons in CONIFERS AND BROADLEAVED TREES *(off-label)*.

Notes **Efficacy**
• Best results achieved by application in good growing conditions, when weeds small and growing actively in a strongly competitive crop
• Do not spray when rain imminent
• Recommended for use in tank mixes with various other cereal herbicides to extend range of weeds controlled. See label for details

Crop safety/Restrictions
• Treat winter crops from 3 leaves unfolded to before third node detectable (GS 13-32), spring crops to before second node detectable (GS 13-31)
• May be used on crops undersown with ryegrass after ryegrass has reached 4-leaf stage
• Do not apply to crops suffering stress from adverse weather, soil or other factors
• Hard frost or application in hot weather may cause transient effects on crop
• Do not roll or harrow within 7 d before or after spraying
• Do not use straw from treated cereals in compost or any other form for glasshouse crops. Straw may be used for strawing down strawberries

Special precautions/Environmental safety
• Harmful if swallowed. Irritating to eyes and skin
• Harmful to fish. Do not contaminate ponds, waterways or ditches with chemical or used container

Protective clothing/Label precautions
• A, C
• 5c, 6a, 6b, 18, 21, 28, 29, 36, 37, 53, 63, 65, 70, 78

Approval
• Off-label approval to Sep 1993 for use on conifers and broadleaved trees (OLA 0508/90)

FOR FULL CONDITIONS OF USE ALWAYS READ THE PRODUCT LABEL

164   clopyralid + dichlorprop + MCPA
A translocated herbicide mixture for cereals

Products   Lontrel Plus   ICI   15:420:175 g/l   SL   01226

Uses   Annual dicotyledons, mayweeds, chickweed, hemp-nettle, redshank in SPRING WHEAT, SPRING BARLEY, SPRING OATS. Annual dicotyledons in GRASS SEED CROPS *(off-label)*.

Notes   **Efficacy**
- Best results achieved by application in good growing conditions, when weeds growing actively in strongly competitive crop
- Do not spray when rain is imminent or weed growth is hard from cold or drought

Crop safety/Restrictions
- Apply to cereal crops from 4-leaf stage to first node detectable (GS 14-31)
- Do not spray when second node can be detected (GS 32)
- Do not spray rye, crops undersown with clover mixtures or crops to be undersown
- Do not roll or harrow within a few days before or after spraying
- Crops may be scorched if application made during periods of night or wind frost
- Do not use straw from treated crops in compost or any other form for glasshouse crops other than cucumbers. Avoid drift onto susceptible crops. Tomatoes are particularly sensitive and may be affected at a considerable distance

Special precautions/Environmental safety
- Harmful if swallowed. Irritating to eyes and skin
- Keep livestock out of treated areas until foliage of any poisonous weeds such as ragwort has died and become unpalatable

Protective clothing/Label precautions
- A, C
- 5c, 6a, 6b, 18, 21, 28, 29, 36, 37, 43, 54, 60, 63, 70, 78

Approval
- Off-label approval to May 1992 for use on grass seed crops (OLA 0451/89)

165   clopyralid + fluroxypyr + ioxynil
A contact and translocated herbicide mixture for use in cereals

Products   Hotspur   Farm Protection   45:150:200 g/l   LI   04099

Uses   Annual dicotyledons, cleavers, chickweed, mayweeds, hemp-nettle, speedwells in WHEAT, BARLEY, OATS, TRITICALE.

Notes   **Efficacy**
- Best results achieved by application in good growing conditions, when weeds small and growing actively in a strongly competitive crop
- Do not spray if rain falling or imminent

Crop safety/Restrictions
- Apply to winter or spring cereals from 2-leaf stage to first node detectable (GS 12-31)
- Late application on winter crops may reduce yield, especially on barley during periods of frost or stress
- Crops undersown with grass may be treated from tillering stage of grass. Do not spray crops undersown or about to be undersown with clover or other legumes
- Do not spray during drought, waterlogging, frost or extremes of temperature
- Do not use straw from treated crops in compost or any other form for glasshouse crops
- Do not apply with hand-held equipment or at concentrations higher than those recommended

Special precautions/Environmental safety
- Harmful if swallowed. Irritating to eyes and skin
- Flammable
- Keep livestock out of treated areas until foliage of any poisonous weeds such as ragwort has died and become unpalatable
- Harmful to fish. Do not contaminate ponds, waterways or ditches with chemical or used container

Protective clothing/Label precautions
- A, C
- 5c, 6a, 6b, 12c, 18, 21, 25, 28, 29, 36, 37, 43, 53, 60, 63, 66, 70, 78

166   clopyralid + mecoprop
A translocated herbicide mixture for cereals and grassland

Products	Seloxone	ICI	15:510 g/l	SL	01915

Uses

Annual dicotyledons, mayweeds, corn marigold, docks in WHEAT, BARLEY, OATS, DURUM WHEAT, RYE, TRITICALE, GRASSLAND. Annual dicotyledons, perennial dicotyledons in GRASS SEED CROPS, NON-CROP AREAS *(off-label)*.

Notes

Efficacy
- May be used in autumn or spring. Best results achieved by application when weeds growing actively in a strongly competitive crop
- Do not spray when rain is imminent or weed growth is hard from cold or drought
- For suppression of docks and thistles cut in mid-summer and spray regrowth in Aug-Sep. Clover in sward will be damaged

Crop safety/Restrictions
- Apply to cereals from 1-leaf stage to first node detectable (GS 11-31). Do not spray when second node can be detected (GS 32)
- Do not spray cereals undersown with clover or legumes. Undersown grasses must have at least 3 fully expanded leaves and 2 tillers
- Do not spray winter barley if frost imminent or after prolonged frosty weather
- Do not roll or harrow within a few days before or after spraying

FOR FULL CONDITIONS OF USE ALWAYS READ THE PRODUCT LABEL

- Do not use straw from treated crops in compost or any other form for glasshouse crops. The straw may be used for strawing down strawberries or for mushroom compost
- Avoid drift onto susceptible crops. Tomatoes may be affected at a considerable distance

Special precautions/Environmental safety
- Irritating to eyes and skin
- Keep livestock out of treated areas until foliage of any poisonous weeds such as ragwort has died and become unpalatable

Protective clothing/Label precautions
- A, C
- 6a, 6b, 18, 21, 28, 29, 36, 37, 43, 54, 63, 66, 70

Approval
- Off-label approval to May 1992 for use in grass seed crops (OLA 0453/89), to July 1992 for use as an industrial herbicide (OLA 0614/89)

---

167   clopyralid + propyzamide
A post-emergence herbicide for winter oilseed rape

Products					
1 Matrikerb	PBI	4.3:43% w/w	WP	01308
2 Matrikerb	Rohm & Haas	4.3:43% w/w	WP	02443

Uses

Annual dicotyledons, mayweeds, annual grasses, barren brome in WINTER OILSEED RAPE. Annual dicotyledons, mayweeds, annual grasses in EVENING PRIMROSE *(off-label)* [1, 2].

Notes

Efficacy
- Apply from Oct to end Jan. Mayweeds and groundsel may be controlled after emergence but before crop large enough to shield seedlings
- Best results achieved on fine, firm, moist soils when weeds germinating or small
- Do not use on soils with more than 10% organic matter
- Effectiveness reduced by surface organic debris, burnt straw or ash
- Do not apply if rainfall imminent or frost present on foliage

Crop safety/Restrictions
- Apply to crop as soon as possible after 3-true leaf stage (GS 1,3)
- Minimum period between spraying and drilling following crop varies from 10 to 40 wk with different crops. See label for details

Protective clothing/Label precautions
- A, C
- 22, 28, 29, 35, 37, 54, 63, 67, 70

Harvest interval
- oilseed rape 6 wk; evening primrose 14 wk

Approval
- Off-label approval to June 1992 for use on evening primrose (OLA 0513, 0514/89)

---

168   clopyralid + triclopyr
A perennial and woody weed herbicide for grassland

Products   Grazon 90   DowElanco   60:240 g/l   EC   02900

Uses   Perennial dicotyledons, docks, brambles, broom, gorse in ESTABLISHED GRASSLAND, GRASSY AREAS AROUND FARM. Perennial dicotyledons, woody weeds in CONIFER PLANTATIONS *(off-label)*.

Notes   **Efficacy**
* For good results must be applied to actively growing weeds
* Spray stinging nettle before flowering, docks in rosette stage in spring, creeping thistle before flower stems 15 cm high, brambles, broom and gorse in Jun-Aug
* Allow 2-3 wk regrowth after grazing or mowing before spraying perennial weeds
* Do not cut grass for 7 d before or 28 d after spraying

Crop safety/Restrictions
* Only use on permanent pasture or leys established for at least 1 yr
* Do not apply overall where clover is an important constituent of sward
* Do not roll or harrow within 7 d before or after spraying. Do not direct drill kale, swedes, turnips, grass or grass mixtures within 6 wk of spraying. Do not plant susceptible autumn-sown crops (eg winter beans) in same year as treatment. Do not spray after end July where susceptible crops to be planted next spring
* Do not allow drift onto other crops, amenity plantings or gardens

Special precautions/Environmental safety
* Irritating to eyes and skin
* Not to be used on food crops
* Dangerous to fish. Do not contaminate ponds, waterways or ditches with chemical or used container
* Keep livestock out of treated areas for at least 7 d and until foliage of any poisonous weeds such as ragwort or buttercup has died down and become unpalatable

Protective clothing/Label precautions
* A, C, H, M
* 6a, 6b, 14, 18, 21, 28, 29, 34, 36, 37, 43, 57, 63, 70

Approval
* Off-label approval to May 1992 for use in coniferous forestry crops (OLA 0460/89)

169   copper hydroxide
A protectant copper fungicide and bactericide

Products   1 Chiltern Kocide 101   Chiltern   77% w/w   WP   03316
                                                   (hydroxide)
           2 Comac Parasol   McKechnie   50% w/w (copper) WP   01543

FOR FULL CONDITIONS OF USE ALWAYS READ THE PRODUCT LABEL

| 3 Wetcol 3 | Ford Smith | 30 g/l (copper) | SC | 02360 |

Uses Leaf spot, glume blotch in BARLEY, WHEAT [1]. Cercospora leaf spot in SUGAR BEET [1]. Halo blight in FRENCH BEANS [1]. Downy mildew in BRASSICAS [1]. Blight in POTATOES [1]. Alternaria blight in CARROTS [1]. Blight in POTATOES, TOMATOES [2, 3]. Bacterial blight in CELERY [1]. Celery leaf spot in CELERY [3]. Downy mildew in ONIONS [1]. Canker in APPLES, PEARS [1, 2, 3]. Collar rot in APPLES *(drench)* [1]. Scab in APPLES, PEARS [3]. Leaf curl in PEACHES, APRICOTS, NECTARINES [2, 3]. Currant leaf spot in CURRANTS, GOOSEBERRIES [1, 2]. Currant leaf spot in BLACKCURRANTS [3]. Cane spot in LOGANBERRIES [3]. Cane spot, spur blight in RASPBERRIES [2, 3]. Downy mildew in HOPS [1, 2, 3]. Downy mildew in GRAPEVINES [1, 2]. Bacterial canker in CHERRIES [2, 3].

Notes **Efficacy**
* Number and timing of sprays vary with disease and crop, see label for details
* On celery and sugar beet add Cropspray 11E oil [1]

Crop safety/Restrictions
* Do not use on copper sensitive cultivars, including Doyenne du Comice pears
* Do not use for collar rot on apples if soil pH below 5.6 [1]
* If spraying cherries in Aug or Sep add cotton seed oil to prevent leaf scorch

Special precautions/Environmental safety
* Harmful if swallowed. Irritating to skin, eyes and respiratory system [2]
* Harmful to fish. Do not contaminate ponds, waterways or ditches with chemical or used container

Protective clothing/Label precautions
* 5c, 6a, 6b, 6c [2]; 29, 41, 53, 63, 65 [1-3]

Withholding period
* Harmful to livestock. Keep all livestock out of treated areas for at least 3 wk

Harvest interval
* 7 d [2]

Approval
* Approved for aerial application on barley, wheat, beans, potatoes. See introductory notes. Consult firm for details [1]

170 copper oxychloride
A protectant copper fungicide and bactericide

Products
1 Cuprokylt	Unicrop	50% w/w	WP	00604
2 Cuprokylt L	Unicrop	270 g/l	SC	02769
3 Cuprosana H	Unicrop	6% w/w	DP	00605
4 FS Dricol 50	Ford Smith	87.7% w/w	WP	04644
5 Headland Inorganic Liquid Copper	WBC Technology	435 g/l	SC	04919

Uses	Blight in POTATOES, OUTDOOR TOMATOES [1, 4, 5]. Buck-eye rot, damping off, foot rot in TOMATOES [1, 4, 5]. Celery leaf spot in CELERY [1, 4, 5]. Canker in APPLES, PEARS [1, 2, 4, 5]. Bacterial canker in CHERRIES [1, 4, 5]. Bacterial canker in PLUMS [1]. Leaf curl in PEACHES [1, 4]. Leaf curl in APRICOTS, NECTARINES [4]. Rust in BLACKCURRANTS [1, 4, 5]. Cane spot in CANE FRUIT [1, 4, 5]. Downy mildew in HOPS [1-5]. Downy mildew in GRAPEVINES [1, 5]. Collar rot in APPLES *(off-label)* [2]. Spear rot in CALABRESE *(off-label)* [1].

Notes	**Efficacy**
	• Spray crops when foliage dry. Do not spray if rain expected soon
	• Spray interval commonly 10-14 d but varies with crop, see label for details
	• If buck-eye rot occurs, spray soil surface and lower parts of tomato plants to protect unaffected fruit [1]
	• To break cycle of bacterial canker infection apply drenching spray mid-Aug, mid-Sep and mid-Oct [1]
	• Increase dose and volume with growth of hops. Complete cover is essential
	Crop safety/Restrictions
	• Some peach cultivars are sensitive to copper. Seek advice before spraying [1]
	• Slight damage may occur to leaves of cherries and plums [1]
	Special precautions/Environmental safety
	• Harmful to livestock. Keep all livestock out of treated areas for at least 3 wk [1, 2, 5]
	• Harmful to fish.Do not contaminate ponds, waterways or ditches with chemical or used container
	Protective clothing/Label precautions
	• 29, 41, 53, 63, 65 [1, 2, 5]; 28, 29, 53, 63, 65 [3]
	Harvest interval
	• calabrese 3 d [1]; hops 7 d [2]
	Approval
	• Approved for aerial application on potatoes. See introductory notes [1]
	• Off-label approval to Feb 1991 for use on apples to control collar rot (OLA 0117/88) [2], to Nov 1991 for use on calabrese (OLA 1380/88) [1]

171   copper oxychloride + maneb + sulphur

A protectant fungicide and yield stimulant for wheat and barley

Products	1 Ashlade SMC	Ashlade	1:16:64% w/w	WP	03640
	2 Ashlade SMC Flowable	Ashlade	10:160:640 g/l	SC	04560
	3 Rearguard	Unicrop	10:160:640 g/l	SC	04938
	4 Tripart Senator Flowable	Tripart	10:160:640 g/l	SC	04561

Uses	Mildew, leaf spot, glume blotch in WHEAT. Mildew, leaf blotch in BARLEY.

FOR FULL CONDITIONS OF USE ALWAYS READ THE PRODUCT LABEL

Efficacy
- Use in a protectant programme of up to 3 sprays per season covering period from first node detectable to beginning of ear emergence (GS 31-51), see label for details
- Treatment has little effect on established disease
- In addition to fungicidal effects treatment can also give nutritional benefits
- Do not apply when foliage is wet or rain imminent

Crop safety/Restrictions
- Do not apply to crops suffering stress from any cause
- Do not roll or harrow within 7 d of treatment
- Do not apply with any trace elements, liquid fertilizers or wetting agent
- Latest time of application before ear fully emerged (GS 59)
- A maximum of 3 applications may be made in any one crop

Special precautions/Environmental safety
- Harmful if swallowed and by inhalation. Irritating to skin and eyes
- Harmful to fish. Do not contaminate ponds, waterways or ditches with chemical or used container

Protective clothing/Label precautions
- A [1]; A, C [2-4]; 5b, 5c, 6a, 6b, 18, 21, 28, 29, 36, 37, 41, 53, 63, 67, 70, 78

Withholding period
- Harmful to livestock. Keep livestock out of treated areas for at least 3 wk [1]

172 copper oxychloride + metalaxyl
A systemic and protectant fungicide mixture

Products   Ridomil Plus 50WP   Ciba-Geigy   35:15% w/w   WP   01803

Uses   Collar rot in APPLES. Root rot in RASPBERRIES, LOGANBERRIES, TAYBERRIES. Red core in STRAWBERRIES. Downy mildew in HOPS. Downy mildew in BRUSSELS SPROUTS, CABBAGES, BROCCOLI, CAULIFLOWERS, GRAPEVINES, CUCUMBERS *(off-label)*.

Notes

Efficacy
- Drench soil at base of apples in Sep to Dec or Mar in first 2 yr after planting only
- Apply drench as band along sides of raspberry rows between mid-Sep and mid-Oct and in Mar. Control reduced if advanced attack at time of spraying
- Apply drench to maiden strawberries immediately after planting and to established plants immediately new growth begins in autumn
- Spray hops to run-off. Increase dose and volume with hop growth

Crop safety/Restrictions
- Do not use on fruiting apple trees
- A maximum of 2 sprays per year may be used on cane fruit, only one spray per year on strawberries
- Use only on established hops. Apply up to 8 sprays at 10-14 d intervals

Special precautions/Environmental safety
- Irritating to eyes and skin

• Harmful to fish. Do not contaminate ponds, waterways or ditches with chemical or used container

Protective clothing/Label precautions
• A
• 6a, 6b, 18, 22, 28, 29, 35, 36, 37, 41, 53, 63, 67, 70

Withholding period
• Harmful to livestock. Keep all livestock out of treated areas for at least 3 wk. Bury or remove spillages

Harvest interval
• 14 d

Approval
• Off-label Approval to Feb 1991 for use in grapevines (OLA 0164/88); to Mar 1991 for use in Brussels sprouts, cabbages, broccoli, cauliflowers (OLA 0263/88); to Aug 1991 for use in protected cucumbers (OLA 0861, 0862/88)

173 copper sulphate
See Bordeaux Mixture
See also aluminium sulphate + copper sulphate + sodium tetraborate

174 copper sulphate + cufraneb
A protectant copper fungicide

Products   Comac Macuprax   McKechnie   16:7% w/w   WP   01229

Uses   Blight in POTATOES, TOMATOES. Canker in APPLES, PEARS. Bacterial canker in CHERRIES. Leaf curl in PEACHES, NECTARINES, APRICOTS. Currant leaf spot in BLACKCURRANTS, GOOSEBERRIES. Cane spot in RASPBERRIES. Downy mildew in GRAPEVINES.

Notes   **Efficacy**
• Spray interval normally 10-14 d but varies with disease and crop, see label for details
• Spray when crop foliage dry. Do not spray if rain imminent

Crop safety/Restrictions
• Do not use on copper sensitive cultivars including Doyenne du Comice pears

Special precautions/Environmental safety
• Harmful if swallowed. Irritating to eyes, skin and respiratory system
• Harmful to fish. Do not contaminate ponds, waterways or ditches with chemical or used container

Protective clothing/Label precautions
• 5c, 6a, 6b, 6c, 29, 41, 53, 63, 65

FOR FULL CONDITIONS OF USE ALWAYS READ THE PRODUCT LABEL

Withholding period
- Harmful to livestock. Keep all livestock out of treated areas for at least 3 wk

Harvest interval
- 7 d

175 copper sulphate + sodium tetraborate
A contact and residual herbicide for moss control on hard surfaces

| Products | Corry's Moss Remover | Synchemicals | 16.6:16.6% w/w | DP | 02411 |

Uses Mosses in GRAVEL PATHS, HARD TENNIS COURTS.

Notes **Efficacy**
- Best results achieved when sprinkled evenly over slightly damp surface and lightly watered the next day
- Sweep or rake off moss after it is dead
- Also kills weeds and removes slime and fungus

Special precautions/Environmental safety
- Harmful to fish. Do not contaminate ponds, waterways or ditches with chemical or used container

Protective clothing/Label precautions
- 28, 29, 36, 53, 63, 67

176 copper sulphate + sulphur
A contact fungicide for powdery and downy mildew control

| Products | Top-Cop | Stoller | 6:670 g/l | SC | 04553 |

Uses Blight in POTATOES, TOMATOES. Powdery mildew in SUGAR BEET. Mildew in SWEDES, TURNIPS, LEAF BRASSICAS. Downy mildew, powdery mildew in HOPS. Mildew in GRAPEVINES.

Notes **Efficacy**
- Apply at first blight warning or when signs of disease appear and repeat every 7-10 d as necessary
- Timing varies with crop and disease, see label for details

Crop safety/Restrictions
- Do not apply to hops at or after the burr stage

Special precautions/Environmental safety
- Irritating to skin, eyes and respiratory system
- Dangerous to fish. Do not contaminate ponds, waterways or ditches with chemical or used container

Withholding period
• Harmful to livestock. Keep livestock out of treated areas for at least 3 wk

177 coumatetralyl
An anticoagulant rodenticide

Products					
	1 Racumin Bait	Bayer	0.0375% w/w	RB	01679
	2 Racumin Tracking Powder	Bayer	0.75% w/w	TP	01681

Uses Rats in FARM BUILDINGS.

Notes **Efficacy**
• Chemical is available as a bait and tracking powder and it is recommended that both formulations are used together
• Place bait in runs in sheltered positions near rat holes. Use at least 250 g per baiting point and examine at least every 2 d. Replenish as long as bait being eaten
• When no signs of activity seen for about 10 d remove unused bait
• Apply tracking powder in a 2.5-5 cm wide layer inside and across entrance to holes or blow into holes with dusting machine. Lay a 3 mm thick layer along runs in 30 cm long patches. Only use on exposed runs if well away from buildings

Special precautions/Environmental safety
• Cover bait or powder to prevent access by children, animals or birds

Protective clothing/Label precautions
• 28, 29, 63, 66, 67, 81, 82, 83, 84

178 cresylic acid
A contact fungicide, soil sterilant and insecticide
See also chlorpropham + cresylic acid + fenuron

Products					
	1 Armillatox	Armillatox	30% (phenol)	EC	00115
	2 Bray's Emulsion	ICI Professional	48% v/v	EC	00323

Uses Honey fungus in TREES AND SHRUBS [1]. Honey fungus, canker, crown gall in TREES AND SHRUBS [2]. Soil-borne diseases in GLASSHOUSE CROPS [2]. Clubroot in BRASSICAS, WALLFLOWERS [1]. Rose replant disease in ROSES [1]. Overwintering pests in FRUIT TREES, CURRANTS, CANE FRUIT [2]. Ants, slugs, woodlice in GLASSHOUSE STRUCTURES AND SURROUNDS [1, 2]. Mosses, lichens, liverworts, algae in PATHS [2]. Mosses in TURF [1].

FOR FULL CONDITIONS OF USE ALWAYS READ THE PRODUCT LABEL

Efficacy
- To control honey fungus drench around the collar region of woody subjects. Avoid treating waterlogged or frozen soil. Apply a foliar and/or root feed in following season
- Plants should be re-treated annually to prevent reinvasion
- For canker control prune back branch to healthy wood, cut out diseased tissues and paint wounds and surrounding area [2]
- For crown gall control loosen soil round collar and saturate area for 30 cm around plant [2]
- To control soil-borne diseases apply as drench in autumn as soon as crop removed
- Apply as winter wash in Dec or Jan to cover entire plant, with particular attention to cracks and crevices in bark [2]

Crop safety/Restrictions
- Do not apply honey fungus control treatment to new plantings until established at least 12 mth. Minimize contact with feeding roots
- Do not replant sterilized areas around treated trees for 6-8 mth
- Do not allow spray to contact any green tissues
- Soil sterilized against damping off and foot rot may be planted after 7 wk
- Protect plants or grass under trees treated with winter wash to avoid scorch

Special precautions/Environmental safety
- Irritating to eyes, skin and respiratory system
- Dangerous to fish. Do not contaminate ponds, waterways or ditches with chemical or used container

Protective clothing/Label precautions
- A
- 6a, 6b, 6c, 18, 21, 28, 29, 36, 37, 52, 63, 65

179 cufraneb

A copper fungicide available only in mixtures
See also copper sulphate + cufraneb

180 cupric ammonium carbonate

A protectant copper fungicide

Products	Croptex Fungex	Hortichem	8.2% w/w (copper)	SL	02888

Uses

Leaf mould, blight in TOMATOES. Powdery mildew in CHRYSANTHEMUMS, CUCUMBERS. Celery leaf spot in CELERY. Currant leaf spot in BLACKCURRANTS. Cane spot in LOGANBERRIES, RASPBERRIES. Leaf curl in PEACHES. Damping off in SEEDLINGS OF ORNAMENTALS.

Notes

Efficacy
- Apply spray to both sides of foliage
- With protected crops keep foliage dry before and after spraying

Crop safety/Restrictions
- Do not spray plants which are dry at the roots
- Ventilate glasshouse immediately after spraying

Special precautions/Environmental safety
- Harmful if swallowed
- Harmful to fish. Do not contaminate ponds, waterways or ditches with chemical or used container

Protective clothing/Label precautions
- 5c, 18, 29, 36, 37, 41, 53, 63, 65, 70, 78

Withholding period
- Harmful to livestock. Keep all livestock out of treated areas for at least 3 wk. Bury or remove spillages

181   cyanazine
A contact and residual triazine herbicide
See also atrazine + cyanazine
           bentazone + cyanazine + 2,4-DB
           clopyralid + cyanazine

Products					
1 Fortrol	Shell	500 g/l	SC	00924
2 Match	Shell	500 g/l	SC	04769

Uses
Annual dicotyledons, annual grasses in WINTER WHEAT, WINTER BARLEY, WINTER OILSEED RAPE, SPRING FIELD BEANS, POTATOES, MAIZE, SWEETCORN, ONIONS, NARCISSI, TULIPS [1, 2]. Annual dicotyledons, annual grasses in BROAD BEANS *(Scotland only)* [1]. Annual dicotyledons, annual grasses in SPRING BROAD BEANS, LEEKS [2]. Annual dicotyledons, annual grasses in PEAS [1]. Annual dicotyledons, annual grasses in CALABRESE, SPRING GREENS *(off-label)* [1].

Notes

Efficacy
- Weeds controlled before emergence or at young seedling stage. Best results achieved when applied to moist soil with rain falling within 6 h
- Residual activity normally persists for about 2 mth
- Do not use as pre-emergence treatment on soils with more than 10% organic matter

Crop safety/Restrictions
- Apply pre-emergence to peas (see label for susceptible cultivars) and beans. Do not spray crops on soils classed as sands. Recommended for use on potatoes as tank mixture with pendimethalin (Stomp 330), alone or as tank mixture on cereals and oilseed rape. See label for details of timings, split applications, rates etc
- Apply pre- or post-emergence to winter cereals, post-emergence to winter hardened oilseed rape from 5-leaf stage to before flower buds appear above leaves (GS 1,5-3,1), but not on light soils or sands
- Do not spray for 7 d after harrowing cereals

FOR FULL CONDITIONS OF USE ALWAYS READ THE PRODUCT LABEL

- On onions only apply post-emergence to crops on fen soils with more than 10% organic matter after 2-true leaf stage
- On maize, sweetcorn, broad beans and field beans use as pre-emergence spray, on flower bulbs pre- or early post-emergence
- Do not treat crops stressed by frost, waterlogging, drought, chemical damage etc. Heavy rain shortly after treatment may lead to damage, especially on lighter soils

Special precautions/Environmental safety
- Harmful if swallowed and in contact with skin
- Harmful to fish. Do not contaminate ponds, waterways or ditches with chemical or used container

Protective clothing/Label precautions
- A, C
- 5a, 5c, 18, 21, 28, 29, 36, 37, 53, 63, 65, 70, 78

Harvest interval
- Onions, leeks 8 wk

Approval
- Off-label Approval to Feb 1991 for use on calabrese and spring greens (OLA 0119/88)

---

182   cyanazine + fluroxypyr

A post-emergence herbicide for use in cereals

Products	Spitfire	DowElanco	45.4:20.6% w/w	KL 04324

Uses

Annual dicotyledons, cleavers, chickweed, speedwells, dead nettle in WINTER WHEAT, WINTER BARLEY, SPRING BARLEY.

Notes

Efficacy
- Best results achieved by early spraying when weeds small and growing actively but, when weather cold and dry, delay until new growth starts
- Weeds emerging after application are not controlled
- Do not spray if rain imminent

Crop safety/Restrictions
- Winter cereals may be sprayed from 2-leaf stage (GS 12) to end of December. In spring, winter cereals may be sprayed from leaf sheath erect to second node detectable (GS 30-32), spring cereals from 2-leaf stage (GS 12-32)
- Do not spray crops undersown with clover or other legumes. Crops undersown with grasses may be sprayed provided grasses tillering
- Do not spray during prolonged frosty weather or when crop under stress from any cause
- Do not roll or harrow for 7 d before or after spraying

Special precautions/Environmental safety
- Harmful if swallowed and in contact with skin. Irritating to eyes
- Flammable
- Harmful to fish. Do not contaminate ponds, waterways or ditches with chemical or used container

183 cyanazine + isoproturon
A contact and residual herbicide for winter cereals

Products	Quiver	Shell	200:350 g/l	SC	04244

Uses Annual dicotyledons, annual grasses, barren brome in WINTER WHEAT, WINTER BARLEY.

Notes

Efficacy
• Weeds controlled before emergence or at young seedling stage
• Apply from early post-emergence of crop to end of March. Soil moisture is necessary for effective weed control by root uptake
• Very wet or dry conditions after application will reduce effectiveness
• Reduced control can result from presence of surface ash and trash
• Do not use on soils with more than 5% organic matter
• Apply as sequential treatment after pre-emergence treatment with Fortrol for control of barren brome

Crop safety/Restrictions
• Do not treat crops after start of stem erection (GS 30)
• Do not treat durum wheats or broadcast or undersown crops
• Ensure crops are drilled to 2.5-4 cm
• Do not treat crops on sands, very light or stony soils
• Heavy rain shortly after treatment may lead to damage, especially on lighter soils
• Do not apply in frosty weather or treat crops suffering from frost damage or any other stress factor
• Do not roll within 7 d before or after spraying

Special precautions/Environmental safety
• Harmful in contact with skin and if swallowed. Irritating to skin and eyes
• Harmful to fish. Do not contaminate ponds, waterways or ditches with chemical or used container

Protective clothing/Label precautions
• A, C
• 5a, 5c, 6a, 6b, 18, 21, 25, 26, 27, 28, 29, 36, 37, 53, 63, 65, 70, 78

FOR FULL CONDITIONS OF USE ALWAYS READ THE PRODUCT LABEL

cyanazine + mecoprop
A contact and translocated herbicide for cereals

Products	Cleaval	Shell	60:400 g/l	SC	00541

Uses

Annual dicotyledons, perennial dicotyledons in WHEAT, OATS, BARLEY, RYEGRASS LEYS.

Notes

Efficacy
* Best results achieved when weeds small and growing actively under warm, moist conditions. Overwintered weeds should be making new growth, biennials and perennials at the flowerbud stage
* Do not spray when rain imminent
* See label for recommended tank mixes

Crop safety/Restrictions
* Apply to winter crops in autumn from 2 leaves unfolded (GS 12) to end Dec, in spring from leaf sheaths erect to before second node detectable (GS 30-31)
* Treat spring crops from 2 leaves unfolded to before first node detectable (GS 12-30)
* Apply to ryegrass after 4-leaf stage. Do not use on undersown clover
* Do not spray crops suffering stress from adverse weather, soil or other conditions. Do not spray winter barley on light soils and subject to stress in frosty weather
* Do not roll or harrow for 7 d before or after spraying

Special precautions/Environmental safety
* Harmful if swallowe. Irritating to eyes
* Keep livestock out of treated areas until foliage of any poisonous weeds such as ragwort has died and become unpalatable
* Harmful to fish. Do not contaminate ponds, waterways or ditches with chemical or used container

Protective clothing/Label precautions
* A, C
* 5c, 6a, 18, 21, 28, 29, 36, 37, 43, 53, 63, 65, 70, 78

185 cycloxydim
A translocated post-emergence herbicide for control of grass weeds

Products	Laser	BASF	200 g/l	EC	05251

Uses

Blackgrass, wild oats, couch, creeping bent, volunteer cereals in SUGAR BEET, FODDER BEET, MANGOLDS, FIELD BEANS, WINTER OILSEED RAPE, PEAS, POTATOES, BRUSSELS SPROUTS, CABBAGES, CAULIFLOWERS, SWEDES.

Notes

Efficacy
* Best results achieved when weeds small and have not begun to compete with crop. Weeds emerging after application are not controlled

- Perennial grasses should have sufficient foliage to absorb spray and should not be cultivated for at least 14 d after treatement
- On established couch pre-planting cultivation recommended to fragment rhizomes and encourage uniform emergence
- Apply to dry foliage when rain not expected for at least 2 h
- Product must be used with Actipron adjuvant

Crop safety/Restrictions
- Recommended time of application varies with crop. See label for details
- On peas a crystal violet wax test should be done if leaf wax likely to have been affected by weather conditions or other chemical treatment. The wax test is essential if other products are to be sprayed after treatment
- Do not apply to crops damaged or stressed by adverse weather, pest or disease attack or other pesticide treatment
- A second treatment may be applied at a reduced dose on crops other than winter oilseed rape
- Prevent drift onto other crops, especially cereals and grass

Special precautions/Environmental safety
- Irritating to skin and eyes

Protective clothing/Label precautions
- A, C
- 6a, 6b, 18, 21, 29, 36, 37, 54, 63, 66

Harvest interval
- cabbage, cauliflower 4 wk; peas 5 wk; sugar and fodder beet, mangolds, potatoes, field beans, swedes, Brussels sprouts 8 wk; winter oilseed rape 12 wk

186   cyfluthrin
A pyrethroid insecticide with contact and residual activity

Products   Baythroid   Bayer   50 g/l   LI   04273

Uses   Barley yellow dwarf virus vectors in WINTER CEREALS. Cabbage stem flea beetle, rape winter stem weevil in OILSEED RAPE. Caterpillars, aphids, flea beetles in LEAF BRASSICAS, FLOWERHEAD BRASSICAS. Pea and bean weevils in PEAS, BEANS, DWARF BEANS, RUNNER BEANS. Pea moth, pea aphid in PEAS. Winter moth, tortrix moths, codling moth, aphids, capsids, apple sucker, apple sawfly, apple blossom weevil in APPLES. Pear sucker in PEARS. Caterpillars, aphids, plum sawfly, plum fruit moth in PLUMS. Sawflies, capsids, blackcurrant leaf midge in BLACKCURRANTS, GOOSEBERRIES. Capsids, strawberry tortrix, froghoppers in STRAWBERRIES.

Notes   **Efficacy**
- Mostly applied at early signs of infestation, sometimes as a preventative treatment, with a follow-up treatment if necessary
- Timing, rate and volume vary with pest and crop, see label for details

FOR FULL CONDITIONS OF USE ALWAYS READ THE PRODUCT LABEL

Crop safety/Restrictions
- Do not apply after end of March on winter cereals
- Up to 3 applications may be made on vining peas and apples, up to 2 on oilseed rape, pears and plums
- Do not apply to pears during blossom period

Special precautions/Environmental safety
- Harmful if swallowed. Irritating to eyes and skin
- Flammable
- Dangerous to fish. Do not contaminate ponds, waterways or ditches with chemical or used container
- Dangerous to bees. Do not apply at flowering stage. Keep down flowering weeds

Protective clothing/Label precautions
- A, C, H
- 5c, 6a, 6b, 12c, 16, 18, 23, 24, 29, 36, 37, 48, 52, 63, 66, 70, 78

Harvest interval
- zero

187 cyhexatin

An organotin acaricide approvals for use of which were revoked in Nov 1987

188 cymoxanil

An aliphatic nitrogen fungicide available only in mixtures

189 cymoxanil + mancozeb

A protectant and systemic fungicide for potato blight control

Products					
1	Ashlade Blight Fungicide	Ashlade	4.5:68% w/w	WP	03209
2	Curzate M	Du Pont	4.5:68% w/w	WP	04343
3	Fytospore	Farm Protection	5.25:71.6% w/w	WP	00960
4	Systol M	Quadrangle	4.5:68% w/w	WP	03480

Uses

Blight in POTATOES.

Notes

Efficacy
- Apply immediately after blight warning or as soon as local conditions dictate and repeat at 10-14 d intervals until haulm dies down or is burnt off
- Spray interval may need to be less than 10 d when growth rapid and should not be more than 10 d in irrigated crops

Special precautions/Environmental safety
- Irritating to eyes, skin and respiratory system

- Harmful to fish. Do not contaminate ponds, waterways or ditches with chemical or use container

Protective clothing/Label precautions
- A [1, 2, 4]; A, C [3]
- 6a, 6b, 6c, 18, 21, 28, 29, 36, 37, 53, 63, 67 [1-4]; 35 [1, 4]; 54 [3]

Harvest interval
- zero [2]; 7d [1, 4]

Approval
- Approved for aerial application on potatoes. See introductory notes. Consult firm for details [1-4]

---

190 cymoxanil + mancozeb + oxadixyl
A systemic and contact protective fungicide for potatoes

Products   Trustan   Du Pont   3.2:56:8% w/w   WP   0502⸛

Uses   Blight in POTATOES.

Notes   **Efficacy**
- Apply first spray as soon as risk of blight infection or official warning. In absence of warning, spray just before crop meets within row
- Repeat spray every 10-14 d according to blight risk
- Do not treat crops already showing blight infection
- Do not apply within 2-3 h of rainfall or irrigation and only apply to dry foliage
- Use fentin based fungicide for final spray in blight control programme

Crop safety/Restrictions
- Up to 5 sprays may be applied to a crop. No more than 5 sprays of phenylamide based products should be applied in one season
- Do not apply after end of Aug

Special precautions/Environmental safety
- Irritating to eyes, skin and respiratory system
- May cause sensitization by skin contact
- Harmful to fish. Do not contaminate ponds, waterways or ditches with chemical or use container

Protective clothing/Label precautions
- A
- 6a, 6b, 6c, 10a, 14, 18, 23, 24, 28, 29, 36, 37, 53, 63, 67

Harvest interval
- zero

FOR FULL CONDITIONS OF USE ALWAYS READ THE PRODUCT LABEL

191   cypermethrin
A contact and stomach acting pyrethroid insecticide

Uses Aphids, barley yellow dwarf virus vectors, yellow cereal fly in WINTER CEREALS [1-13]. Cabbage stem flea beetle, rape winter stem weevil in OILSEED RAPE [1-13]. Pollen beetles, seed weevil, pod midge in FLOWERING OILSEED RAPE *(restricted area)* [1, 13]. Aphids, cabbage stem flea beetle, caterpillars in BRASSICAS [2, 7-11, 13]. Aphids, cabbage stem flea beetle, caterpillars in CABBAGES, BRUSSELS SPROUTS [1, 3, 4, 6, 12]. Aphids, cabbage stem flea beetle, caterpillars in CABBAGES, BRUSSELS SPROUTS, CAULIFLOWERS, KALE [5]. Aphids, cabbage stem flea beetle, caterpillars in SWEDES [1, 3, 4, 12]. Aphids, pea moth, pea and bean weevils in PEAS [1-13]. Pea and bean weevils in BEANS [1, 3, 4, 6, 7, 11, 12]. Caterpillars, whitefly in FRENCH BEANS [1, 3, 4]. Cutworms in ROOT CROPS [7, 9, 11]. Cutworms in POTATOES, CARROTS, PARSNIPS, SUGAR BEET [1, 4, 6, 12]. Cutworms in POTATOES [3]. Aphids, caterpillars, cutworms in CELERY, LETTUCE [1, 3, 4]. Cutworms in LEEKS [6]. Frit fly, leatherjackets in GRASS RE-SEEDS [1, 3, 4, 11]. Aphids, capsids, caterpillars, codling moth, sawflies, tortrix moths in APPLES [1-8, 11-13]. Aphids, capsids, caterpillars, suckers in PEARS [1-8, 11-13]. Caterpillars in CHERRIES, PLUMS [1, 3, 4, 8, 11-13]. Capsids, sawflies in BLACKCURRANTS, REDCURRANTS [1, 3, 4, 12]. Capsids, sawflies in BLACKCURRANTS [2, 5, 8, 11]. Capsids, sawflies in CURRANTS, GOOSEBERRIES [1, 2, 4-7, 11, 12]. Caterpillars, capsids in STRAWBERRIES [1-8, 11, 12]. Aphids in HOPS [1-5, 8, 11-13]. Aphids, capsids, caterpillars, cutworms, thrips, whitefly in ORNAMENTALS, PROTECTED ORNAMENTALS [1, 3, 4]. Cutworms, leaf miners, whitefly in CUCUMBERS, PROTECTED LETTUCE, PROTECTED CELERY [1, 3, 4]. Whitefly, leaf miners, caterpillars in TOMATOES, PEPPERS [7]. Aphids, capsids, thrips in GRAPEVINES *(off-label)* [1]. American serpentine leaf miner in AUBERGINES, CELERY, ARTICHOKES, FENNEL, ASPARAGUS, RHUBARB, LETTUCE, ENDIVE, CHICORY, SPINACH BEET, SPINACH, CUCUMBERS, COURGETTES, GHERKINS, MARROWS, MELONS, PUMPKINS, SQUASHES, ONIONS, SHALLOTS, GARLIC, LEEKS, ORNAMENTALS *(off-label)* [3]. Colorado beetle in POTATOES *(off-label)* [1]. Asparagus beetle in ASPARAGUS *(off-label)* [1].

Notes **Efficacy**
● Combines rapid action, good persistence, and high activity on Lepidoptera

289

- As effect is mainly via contact good coverage is essential for effective action. Spray volume should be increased on dense crops
- A repeat spray after 10-14 d is needed for some pests of outdoor crops, several sprays at shorter intervals for whitefly and other glasshouse pests
- Rates and timing of sprays vary with crop and pest. See label for details
- Most organophosphate and organochlorine-resistant pests are susceptible
- Add a non-ionic wetter to improve results on leaf brassicas. In Brussels sprouts use of a drop-leg sprayer may be beneficial

Crop safety/Restrictions
- Test spray sample of new or unusual ornamentals before committing whole batches
- Before using on flowering oilseed rape obtain area allocation from supplier

Special precautions/Environmental safety
- Harmful in contact with skin or if swallowed [2, 8, 10, 13]; if swallowed [6, 7]
- Irritating to eyes and skin [1-13]. May cause sensitization by skin contact [6]
- Flammable
- Dangerous to bees. Do not apply at flowering stage except as directed for oilseed rape. Keep down flowering weeds
- Extremely dangerous to fish [1-7, 9-13]. Dangerous to fish [8]. Do not contaminate ponds, waterways or ditches with chemical or used container. Avoid contamination of waterways by spray drift

Protective clothing/Label precautions
- A, C
- 6a, 6b, 12c, 14, 18, 21, 24, 28, 29, 36, 37, 48, 63, 66, 70, 78 [1-13]; 5a, 5c [2, 8, 10, 13]; 5c [6, 7]; 9 [6]; 51 [8]; 52 [1-7, 9-13]

Harvest interval
- zero

Approval
- Off-label Approval to Feb 1991 for use on grapevines (OLA 0120/88), french beans (OLA 0121/88) [1]; to Apr 1992 for use on aubergines, celery, artichoke, fennel, asparagus, rhubarb, lettuce, endive, chicory, spinach beet, spinach, cucumbers, cucurbits, onions, leeks, shallots, garlic, potatoes, beans, chinese cabbage (OLA 0403/89) [7]; to Nov 1991 for use on potatoes as required by Statutory Notice for Colorado beetle control by ground or aerial application (OLA 1388, 1389/88) [1]; to Feb 1992 for use on asparagus (OLA 0144/90) [1]

192 2,4-D

A translocated phenoxy herbicide for use in cereals and grass
See also amitrole + atrazine + 2,4-D
 amitrole + 2,4-D + diuron
 amitrole + 2,4-D + diuron + simazine
 atrazine + 2,4-D + sodium chlorate

Products					
1 Agrichem 2,4-D	Agrichem	500 g/l	SL	04098
2 Agricorn D	FCC	500 g/l	SL	00056

FOR FULL CONDITIONS OF USE ALWAYS READ THE PRODUCT LABEL

3 Atlas 2,4-D	Atlas	480 g/l	SL	03052
4 BH 2,4-D Ester 50	RP Environ.	465 g/l	EC	00240
5 Campbell's Destox	MTM Agrochem.	500 g/l	EC	00397
6 Campbell's Dioweed 50	MTM Agrochem.	500 g/l	SL	00401
7 CDA Dicotox Extra	R P Environ.	75 g/l	UL	04757
8 Dicotox Extra	RP Environ.	400 g/l	EC	00696
9 Dormone	RP Environ.	465 g/l	SL	00751
10 Farmon 2,4-D	Farm Protection	480 g/l	SL	00826
11 For-Ester	Synchemicals	500 g/l	EC	00914
12 MSS 2,4-D Amine	Mirfield	500 g/l	SL	01391
13 MSS 2,4-D Ester	Mirfield	500 g/l	EC	01393
14 Shell D 50	Shell	500 g/l	SL	01931
15 Silvapron D	BP		UL	01935
16 Syford	Synchemicals	500 g/l	SL	02062

Uses

Annual dicotyledons, perennial dicotyledons in WHEAT, BARLEY [1-3, 5, 6, 10, 12-14, 16]. Annual dicotyledons, perennial dicotyledons in WINTER RYE [1-3, 5, 6, 10, 12, 13]. Annual dicotyledons, perennial dicotyledons in SPRING RYE [1, 2, 6, 12, 13]. Annual dicotyledons, perennial dicotyledons in WINTER OATS [3, 10, 12, 13]. Annual dicotyledons, perennial dicotyledons in ESTABLISHED GRASSLAND [1-3, 5, 6, 10-14, 16]. Annual dicotyledons, perennial dicotyledons in GRASS SEED CROPS [5, 13, 14, 16]. Annual dicotyledons, perennial dicotyledons in AMENITY TURF, SPORTS TURF [2, 4-9, 11, 14, 16]. Annual dicotyledons, perennial dicotyledons, heather, willows in CONIFER PLANTATIONS, FORESTRY [4, 15]. Aquatic weeds, perennial dicotyledons in WATER OR WATERSIDE AREAS [3, 9].

Notes

Efficacy
- Best results achieved by spraying weeds in seedling to young plant stage when growing actively in a strongly competing crop
- Most effective stage for spraying perennials varies with species. See label for details
- Spray aquatic weeds when in active growth between May and Sep
- Do not spray if rain falling or imminent
- Do not cut grass or graze for at least 7 d after spraying

Crop safety/Restrictions
- Spray winter cereals in spring when leaf-sheath erect but before first node detectable (GS 30), spring cereals from 5-leaf stage to before first node detectable (GS 15-30)
- Do not use on undersown cereals or on newly sown leys containing clover
- Do not spray grass seed crops after ear emergence
- Do not spray within 6 mth of laying turf or sowing fine grass
- Selective treatment of resistant conifers can be made in Aug when growth ceased and plants hardened off, spray must be directed if applied earlier. See label for details
- Do not plant conifers until at least 1 mth after treatment
- Do not spray crops stressed by cold weather or drought or if frost expected
- Do not use shortly before or after sowing of any crop
- Do not roll or harrow within 7 d before or after spraying
- Do not direct drill brassicas or grass/clover mixtures within 3 wk of application

Special precautions/Environmental safety
- Harmful in contact with skin and if swallowed [1-6, 10-12, 14-16]
- Irritating to eyes
- Keep livestock out of treated areas until foliage of any poisonous weeds such as ragwort has died and become unpalatable

291

- Harmful to fish. Do not contaminate ponds, waterways or ditches with chemical or use container
- May be used to control aquatic weeds in presence of fish if used in strict accordance with directions for waterweed control and precautions needed for aquatic use [3, 9]

Protective clothing/Label precautions
- A, C, H [1]
- 5a, 5c, 18, 21, 29, 36, 37, 43, 53, 63, 65, 66, 70, 78 [1-16]; 55, 56 [3, 9]

Approval
- May be applied through CDA equipment. See label for details. See introductory notes on ULV application [14]
- Approved for aquatic weed control. See introductory notes on use of herbicides in or near water [3, 9]

193 2,4-D + dalapon + diuron
A total herbicide mixture of translocated and residual chemicals

Products	Destral	ABM	9.8:41.7:19.5% w/w	WP	04288

Uses Total vegetation control in NON-CROP AREAS, FENCELINES.

Notes **Efficacy**
- Best results achieved by application in May-Jun when weeds growing rapidly and have sufficient leaf area to absorb chemical
- Rain shortly after application may reduce effectiveness

Crop safety/Restrictions
- Application once a year at recommended rate for footpath joints should not affect established trees or shrubs but contact with bark or foliage must be avoided
- Use at reduced rate in tree and shrub nurseries
- Do not apply or drain or flush equipment on or near young trees or other desirable plants or on areas where chemical may come into contact with roots

Special precautions/Environmental safety
- Irritating to skin, eyes and respiratory system
- Keep livestock out of treated areas until foliage of any poisonous weeds such as ragwort has died and become unpalatable
- Harmful to fish. Do not contaminate ponds, waterways or ditches with chemical or used container

Protective clothing/Label precautions
- A, E, F, H
- 6a, 6b, 6c, 16, 18, 21, 24, 28, 29, 36, 37, 43, 53, 63, 67

FOR FULL CONDITIONS OF USE ALWAYS READ THE PRODUCT LABEL

194   2,4-D + dicamba

A translocated herbicide for use on turf

Products				
1 Green Up Weedfree Spot Weedkiller for Lawns	Synchemicals		AE	03253
2 Keychem SWK 333	Keychem	200:35 g/l	EC	05045
3 Longlife Plus	ICI Professional	0.72:0.1% w/w	GR	03929
4 New Estermone	Synchemicals	200:35 g/l	EC	02939

Uses

Annual dicotyledons, perennial dicotyledons in LAWNS, AMENITY TURF.

Notes

Efficacy
- Best results achieved by application when weeds growing actively from Apr to Sep [1-3], in spring or early summer (later with irrigation and feeding) [4]
- More resistant weeds may need repeat treatment after 3 wk
- Granules contain NPK fertilizer which aids filling of gaps left after weeds killed [3]
- Use aerosol formulation as a spot treatment [1]
- Do not use during drought conditions or mow for 3 d before or after treatment

Crop safety/Restrictions
- Do not treat turf established for less than 6 wk [1], newly sówn or turfed areas [2-4]
- Avoid spray drift onto cultivated crops or ornamentals
- Do not re-seed for 6 wk after application [4]

Special precautions/Environmental safety
- Keep livestock out of treated areas until foliage of any poisonous weeds such as ragwort has died and become unpalatable
- Harmful to fish. Do not contaminate ponds, waterways or ditches with chemical or used container

Protective clothing/Label precautions
- 21, 29, 37, 53 [1]; 29, 36, 37, 43, 53, 63, 65, 67 [3]; 21, 29, 43, 53, 63, 65 [2, 4]

195   2,4-D + dicamba + ioxynil

A translocated and contact herbicide for use on turf

Products				
Super Verdone	ICI Professional	72:12:48 g/l	SL	02051

Uses

Annual dicotyledons, perennial dicotyledons, speedwells in AMENITY TURF.

Notes

Efficacy
- Best results achieved by application when weeds growing actively from Apr to Sep
- Do not use during drought unless irrigation carried out before and after treatment
- Do not mow for 3 d before or after treatment
- Repeat application after 4-6 wk if necessary

Crop safety/Restrictions
- Do not treat new turf until well established (6-9 mth after seeding or turfing)
- Avoid spray running or drifting onto cultivated plants

- Do not use first mowings after treatment as mulch unless composted
- Do not re-seed or turf within 8 wk of last treatment

Special precautions/Environmental safety
- Irritating to skin
- Do not apply with hand-held equipment or at concentrations higher than those recommended
- Keep livestock out of treated areas for at least 2 wk if poisonous weeds such as ragwort are present
- Harmful to fish. Do not contaminate ponds, waterways or ditches with chemical or used container

Protective clothing/Label precautions
- A, C
- 6b, 18, 21, 28, 29, 36, 37, 43, 53, 63, 66

196   2,4-D + dicamba + mecoprop
A translocated herbicide for perennial and woody weed control

Products   Weed and Brushkiller (New Formula)   Synchemicals   144:32:144 g/l   EC   02961

Uses   Annual dicotyledons, perennial dicotyledons, stinging nettle, brambles, woody weeds in ROUGH GRAZING, AMENITY GRASS, NON-CROP AREAS.

Notes   **Efficacy**
- Mix with water for application as foliar spray. Apply Mar-Jul on perennials when in active growth, Jun-Sep on woody plants, May-Jul on gorse
- Do not spray in drought conditions
- Mix with paraffin or other light oil for application as basal bark treatment on established trees in Jul-Sep or as stump treatment in Jan-Mar

Crop safety/Restrictions
- Avoid spray drift onto crops or ornamentals. Do not use in hot or windy conditions
- Do not cultivate or replant land for at least 6 wk after treatment
- Do not cut or graze for at least 7 d after treatment

Special precautions/Environmental safety
- Irritating to eyes
- Keep livestock out of treated areas until foliage of any poisonous weeds such as ragwort has died and become unpalatable
- Harmful to fish. Do not contaminate ponds, waterways or ditches with chemical or used container

Protective clothing/Label precautions
- A, C, H
- 6a, 18, 21, 29, 36, 37, 43, 53, 63, 65

FOR FULL CONDITIONS OF USE ALWAYS READ THE PRODUCT LABEL

2,4-D + dicamba + triclopyr
A translocated herbicide for perennial and woody weed control

Products Broadshot Shell 200:85:65 g/l EC 03056

Uses Annual dicotyledons, perennial dicotyledons, stinging nettle, docks, thistles, japanese knotweed, woody weeds, brambles, gorse, rhododendrons in ESTABLISHED GRASSLAND, FORESTRY, NON-CROP AREAS.

Notes **Efficacy**
- Apply as foliar spray to herbaceous or woody weeds. Timing and growth stage for best results vary with species. See label for details
- Dilute with water for stump treatment, with paraffin for basal bark treatment
- May be applied at 1/3 or 1/4 dilution with ropewick applicators
- Do not roll or harrow grassland for 7 d before or after spraying

Crop safety/Restrictions
- Do not use on pasture established less than 1 yr
- Where clover a valued constituent of sward only use as a spot treatment
- Do not use on grass being grown for seed
- Do not graze for 7 d or mow for 14 d after treatment
- Do not direct drill grass, clover or brassicas for at least 6 wk after treatment
- Sprays may be applied in pines, spruce and fir providing drift is avoided. Optimum time is mid-autumn when tree growth ceased but weeds not yet senescent
- Do not plant trees for 1-3 mth after spraying depending on dose applied. See label
- Avoid spray drift into greenhouses or onto crops or ornamentals. Vapour drift may occur in hot conditions

Special precautions/Environmental safety
- Harmful if swallowed. Irritating to eyes and skin
- Keep livestock out of treated area for at least 7 d or until foliage of any poisonous weeds such as ragwort has died and become unpalatable
- Dangerous to fish. Do not contaminate ponds, waterways or ditches with chemical or used container

Protective clothing/Label precautions
- A, C
- 5c, 6a, 6b, 18, 21, 29, 36, 37, 43, 52, 63, 66, 70, 78

198 2,4-D + dichlorprop + MCPA + mecoprop
A translocated herbicide for use in cereals

Products Campbell's New Camppex MTM Agrochem. 34:133:53:164 g/l SL 00414

Uses Annual dicotyledons, perennial dicotyledons, chickweed, cleavers in APPLES, PEARS.

Efficacy
• Use in established apple and pear orchards (from 1 yr after planting) as directed application
• Spray when weeds in active growth and at recommended growth stage. See label for details
• Effectiveness may be reduced by rain within 12 h
• Do not roll, harrow or cut grass crops on orchard floor within at least 3 d before or after spraying

Crop safety/Restrictions
• Do not spray trees or allow drift onto trees
• Do not spray during blossom period. Do not spray to run-off
• Avoid spray drift onto neighbouring crops

Special precautions/Environmental safety
• Harmful in contact with skin and if swallowed. Irritating to eyes
• Keep livestock out of treated areas for at least 2 wk and until foliage of any poisonous weeds such as ragwort has died and become unpalatable
• Harmful to fish. Do not contaminate ponds, waterways or ditches with chemical or used container

Protective clothing/Label precautions
• A, C
• 5a, 5c, 6a, 18, 21, 29, 36, 37, 43, 53, 63, 66, 70, 78

199 2,4-D + mecoprop
A translocated herbicide for use in turf and grassland

Products

1 BH CMPP/2,4-D	RP Environ.	116:250 g/l	SL	00249
2 CDA Supertox 30	RP Environ.	56:114 g/l	UL	04664
3 Com-Trol	Certified	6:6% w/w	SL	03343
4 Mascot Selective	Rigby Taylor	80:200 g/l	SL	03423
5 Select-Trol	Chemsearch	6:6% w/w	SL	03342
6 Supertox 30	RP Environ.	90:190 g/l	SL	02050
7 Sydex	Synchemicals	125:250 g/l	SL	02061
8 Verdone CDA	ICI Professional	6.7:13.3%	EC	03410
9 Zennapron	BP	200 g/l (total)	EC	02377

Uses

Annual dicotyledons, perennial dicotyledons in AMENITY TURF, SPORTS TURF, LAWNS [1-9]. Annual dicotyledons, perennial dicotyledons in ESTABLISHED GRASSLAND, GRASS SEED CROPS [7].

Notes **Efficacy**
• May be applied from Apr to Sep, best results in May-Jun when weeds in active growth
• Some species may need repeat treatment after 4-8 wk
• Do not spray during drought conditions or when rain imminent
• Apply CDA formulations undiluted with suitable equipment

FOR FULL CONDITIONS OF USE ALWAYS READ THE PRODUCT LABEL

- Do not mow for 2-4 d before or 1 d after treatment (3 d after on some labels)
- For best results apply fertilizer 1-2 wk before treatment

Crop safety/Restrictions
- Do not use on grass established for less than 6 mth (not in year of sowing) [2]
- Do not spray EC formulations at temperatures above 25-26°C
- Do not use first 4 mowings for mulching unless composted for at least 6 mth (do not use first 2 mowings for composting on some labels)
- Do not resow treated areas for at least 6 wk (8 wk on some labels)
- Avoid spray drift onto nearby trees, shrubs, vegetables or flowers

Special precautions/Environmental safety
- Harmful in contact with skin [1, 8]; harmful in contact with skin and if swallowed [4, 7]
- Irritating to eyes [6, 7]; Irritating to eyes and skin [2-5]; irritating to skin, eyes and respiratory system [9]
- Keep livestock out of treated areas until foliage of any poisonous weeds such as ragwort has died and become unpalatable
- Harmful to fish. Do not contaminate ponds, waterways or ditches with chemical or used container [1-3, 5, 7-9]

Protective clothing/Label precautions
- A [8]; A, C [3, 5, 6]; A, M [8]; A, C, H, M [2]
- 5a [1, 8]; 5a, 5c [4, 7]; 6a [6, 7]; 6a, 6b [2-5]; 6a, 6b, 6c [9]; 18, 21, 29, 36, 37, 43, 63, 65, 70 [1-9]; 53 [1-3, 5, 7-9]; 54 [4, 6]; 78 [1, 4, 7, 8]

Approval
- May be applied through CDA equipment. See label for details. See introductory notes on ULV application [2, 8]

200   2,4-D + picloram
A persistent translocated herbicide for non-crop land

Products   Atladox HI   Chipman   240:65 g/l   SL   00126

Uses   Annual dicotyledons, perennial dicotyledons, docks, creeping thistle, Japanese knotweed, ragwort, woody weeds, brambles, scrub in AMENITY GRASS, NON-CROP GRASS, ROAD VERGES.

Notes   **Efficacy**
- Apply as overall foliar spray during period of active growth when foliage well developed

Crop safety/Restrictions
- Do not apply around desirable trees or shrubs where roots may absorb chemical
- Prevent leaching into areas where desirable plants are present
- Avoid drift of spray onto desirable plants
- Do not use cuttings from treated grass for mulching or composting

Special precautions/Environmental safety
- Irritating to eyes
- Flammable

- Keep livestock out of treated areas until foliage of any poisonous weeds such as ragwor has died and become unpalatable
- Harmful to fish. Do not contaminate ponds, waterways or ditches with chemical or usec container

Protective clothing/Label precautions
- 6a, 12c, 21, 29, 43, 53, 63, 66

201 dalapon

A translocated herbicide for grass weed control
See also *atrazine + dalapon*
  2,4-D + dalapon + diuron

Products					
1 Atlas Dalapon	Atlas	85% w/w	SP	03100
2 BH Dalapon	RP Environ.	85% w/w	SP	03047
3 Synchemicals Dalapon	Synchemicals	85% w/w	SP	03047

Uses

Annual grasses, volunteer cereals, perennial grasses, couch in BEET CROPS, OILSEED RAPE, POTATOES, ASPARAGUS, CARROTS, APPLES, PEARS, BLACKCURRANTS, GOOSEBERRIES, NON-CROP AREAS [1, 2]. Annual grasses, perennial grasses, couch in ASPARAGUS, APPLES, PEARS, BLACKCURRANTS, GOOSEBERRIES, RASPBERRIES, RHUBARB, NON-CROP AREAS [3]. Annual grasses, perennial grasses, couch in CEREALS *(stubble treatment)* [2]. Annual grasses, perennial grasses, couch in LUCERNE, RASPBERRIES, BLACKBERRIES, LOGANBERRIES, RHUBARB, FORESTRY [1]. Perennial grasses, couch, reeds, bulrushes in DITCHES, BANKS OF WATERCOURSES [2]. Annual grasses, perennial grasses, couch in KALE, ROAD VERGES, PARKS AND GARDENS, HEDGE BOTTOMS [2]. Common reed in WATER OR WATERSIDE AREAS *(off-label, aerial application)* [2]. Spartina in ESTUARIES *(off-label)* [2].

Notes

Efficacy
- Best results by using when good growing conditions continue for at least 2 wk
- Do not spray in drought conditions or when grasses in flower
- For stubble treatment of couch spray as soon as possible after harvest before early Nov when shoots at least 10 cm high and plough within 2-4 wk [2]

Crop safety/Restrictions
- After autumn stubble treatment do not sow winter cereals; spring cereals may be sown after 4 mth. Do not sow spring cereals after spring treatment
- After spring treatment of couch plough within 2-4 wk. Arable crops listed may be sown after a further 20 d, potatoes immediately after ploughing (red skinned cultivars may suffer some loss of colour)
- Low rates may be used on beet crops, carrots and oilseed rape [1], sugar beet and oilseed rape [2], as post-emergence treatments. See label for details
- Apply in fruit crops (established at least 3-4 yr), perennial crops and forestry plantations as directed sprays during dormant season. Do not use under plums or other stone fruit
- Do not spray under young hawthorn hedges until well established [2]

FOR FULL CONDITIONS OF USE ALWAYS READ THE PRODUCT LABEL

Special precautions/Environmental safety
- Irritating to eyes, skin and respiratory system
- Do not use treated water for irrigation purposes within 5 wk of treatment [2]

Protective clothing/Label precautions
- A, C
- 6a, 6b, 6c, 16, 18, 22, 29, 36, 37, 54, 63, 66, 70 [1-4]; 55, 56 [2]

Approval
- Cleared for aquatic weed control. See introductory notes on use of herbicides in or near water [2]
- Off-label Approval to Sep 1992 for aerial application on common reed (only under supervision of appropriate water authority) (OLA 0745/89) [2], on Spartina anglica in estuaries (in consultation with Nature Conservancy and after obtaining a 'Disposal at Sea' licence from MAFF Fisheries Laboratory) (OLA 0393, 0394/90) [2]

202 dalapon + dichlobenil

A persistent herbicide for use in woody plants and non-crop areas

Products	Fydulan	Chipman	10:6.75% w/w	GR	00958

Uses

Annual weeds, perennial grasses in WOODY ORNAMENTALS, ROSES. Annual weeds, perennial grasses, rushes, bracken in HEDGES, FORESTRY. Total vegetation control in NON-CROP AREAS.

Notes

Efficacy
- Apply evenly with suitable granule applicator. Best results achieved by application in late winter/early spring to slightly moist soil, especially if rain falls soon after
- May be used on all soil types. Do not disturb soil after application
- For forestry use apply as a 1 m wide band
- In non-crop situations apply before end of spring

Crop safety/Restrictions
- Use in ornamentals before onset of spring growth, usually Feb-early Mar. See label for list of tolerant species
- Use in roses established for at least 2 yr before bud growth starts, usually early Feb
- Use in forestry before end of winter. Allow at least 2 mth before planting. Apply after planting as soon as soil settled
- Do not apply to areas underplanted with bulbs or herbaceous stock
- Do not apply near hops or glasshouses, nor if nearby crop foliage wet
- Store well away from bulbs, corms, tubers or seeds

Special precautions/Environmental safety
- Irritating to eyes, skin and respiratory system

Protective clothing/Label precautions
- 6a, 6b, 6c, 22, 29, 36, 37, 63, 67

Approval
- Cleared for aquatic weed control. See introductory notes on use of herbicides in or near water

203   dalapon + di-1-p-menthene

A grass weed herbicide plus a coating agent

Products   Volunteered   Mandops   KL   02836

Uses   Annual grasses, volunteer cereals, perennial grasses in OILSEED RAPE, FODDER
RAPE, KALE, TURNIPS, SUGAR BEET, MANGOLDS, CARROTS, CONIFER
PLANTATIONS. Annual grasses, perennial grasses in CONIFER PLANTATIONS.

Notes   **Efficacy**
- Apply to emerged grass weeds. Best results on volunteer cereals and wild oats up to
  jointing, on blackgrass to 4-5 leaf stage
- After drying on crop 30 min required for spray to become rainfast
- Product recommended for use as a pigeon repellant as well as for weed control

Crop safety/Restrictions
- Recommended crop stages for spraying are winter oilseed rape 2-4 leaf (GS 1,2-1,4)
  (not after end Nov), spring oilseed rape 3-4 leaf (GS 1,3-1,4), carrots after pencil stage,
  kale, turnips and fodder rape 2-4 leaf, sugar beet and mangolds pre-emergence or 3-6
  leaf. Application rates vary, see label for details
- Apply to conifers as a directed spray, do not wet trees
- Oilseed rape may be damaged by treatment in very cold weather, especially Mikado
- Do not use during severe frost
- Do not apply during flowering or within 4 wk of harvest

Special precautions/Environmental safety
- Irritating to skin and eyes

Protective clothing/Label precautions
- 6a, 6b, 16, 18, 21, 29, 36, 37, 54, 63, 65

Harvest interval
- 4 wk

204   daminozide

A plant-growth regulator for use in certain ornamentals

Products   1 B-Nine   Fargro   85% w/w   SP   04468
                2 Dazide   Fine   85% w/w   SP   02691

Uses   Internode reduction in CHRYSANTHEMUMS, AZALEAS, HYDRANGEAS,
BEDDING PLANTS [1, 2]. Internode reduction in POINSETTIAS [1]. Internode
reduction in POT PLANTS [2]. Internode reduction in GLASSHOUSE FLOWERS
(off-label) [2].

FOR FULL CONDITIONS OF USE ALWAYS READ THE PRODUCT LABEL

Efficacy
- Best results obtained by application in late afternoon when glasshouse has cooled down
- Spray when foliage dry

Crop safety/Restrictions
- Apply only to turgid, well watered plants. Do not water for 24 h after spraying
- Do not use on chrysanthemum Fandango
- Do not mix with other spray chemicals

Special precautions/Environmental safety
- Irritating to eyes [2]
- Harmful to fish. Do not contaminate ponds, waterways or ditches with chemical or used container

Protective clothing/Label precautions
- A, C, H
- 6a, 18, 22, 28, 29, 36, 37, 53, 63, 65

Approval
- Sales of daminozide for use on food crops were halted worldwide by the manufacturer in October 1989. Sales for use on flower crops were not affected
- Off-label approval to Jan 1993 for use on flower crops (OLA 0069/90) [2]

205   dazomet
A soil fumigant for field and glasshouse use

Products					
1 Basamid	BASF	98-99% w/w	GR	00192
2 Boszamet	Bos	98-99% w/w	GR	02976
3 Dazomet	DowElanco	98-99% w/w	GR	03733

Uses

Soil-borne diseases in FIELD CROPS, VEGETABLES, GLASSHOUSE CROPS, SOIL LOAM OR TURF FOR COMPOST. Nematodes, soil insects in FIELD CROPS, VEGETABLES, GLASSHOUSE CROPS, SOIL LOAM OR TURF FOR COMPOST. Weed seeds in FIELD CROPS, VEGETABLES, GLASSHOUSE CROPS, SOIL LOAM OR TURF FOR COMPOST.

Notes

Efficacy
- Product acts by releasing methyl isothiocyanate in contact with moist soil
- Soil sterilization is carried out after harvesting one crop and before planting the next
- The soil must be of fine tilth, free of clods and evenly moist to the depth of sterilization
- Soil moisture must not be less than 50% of water-holding capacity or oversaturated. If too dry, water at least 7-14 d before treatment
- Do not treat ground where water table may rise into treated layer
- In order to obtain short treatment times it is recommended to treat soils when soil temperature is above 7°C. Treatment should be used outdoors before winter rains make soil too wet to cultivate - usually early Nov
- For club root control treat only in summer when soil temperature above 10°C
- Where onion white rot a problem unlikely to give effective control where inoculum level high or crop under stress [1]

- Apply granules with suitable applicators, mix into soil immediately to desired depth and seal surface with polythene sheeting, by flooding or heavy rolling. See label for suitable application and incorporation machinery

Crop safety/Restrictions
- 14-28 d after treatment cultivate lightly to allow gas to disperse and after a further 14-28 d (timing depends on soil type and temperature) do safety test on soil samples with cress. See label for details
- Do not treat structures containing live plants or any ground within 1 m of live plants

Special precautions/Environmental safety
- Harmful in contact with skin and if swallowed
- Irritating to eyes, skin and respiratory system

Protective clothing/Label precautions
- A, M
- 5a, 5c, 6a, 6b, 6c, 16, 18, 22, 28, 29, 36, 37, 54, 63, 67, 70, 78 [1-3]; see label for additional precautions

---

206   2,4-DB

A translocated phenoxy herbicide for use in lucerne

See also benazolin + 2,4-DB + MCPA
     bentazone + cyanazine + 2,4-DB

| **Products** | Campbell's DB Straight | MTM Agrochem. | 300 g/l | SL | 00394 |

Uses   Annual dicotyledons, thistles in LUCERNE, CEREALS UNDERSOWN LUCERNE.

Notes   **Efficacy**
- Best results achieved on young seedling weeds under good growing conditions. Treatment less effective in cold weather and dry soil conditions
- Rain within 12 h may reduce effectiveness

Crop safety/Restrictions
- Apply after first trifoliate leaf stage of lucerne. Optimum time 3-4 trifoliate leaves
- Spray lucerne undersown in cereals as soon as possible after first trifoliate leaf stage, while barley or oats from 2-leaf stage (GS 12), wheat from 3-leaf stage (GS 13), to before start of jointing (GS 30)
- Do not allow spray drift onto neighbouring crops

Special precautions/Environmental safety
- Keep livestock out of treated areas until foliage of any poisonous weeds such as ragwort has died and become unpalatable
- Harmful to fish. Do not contaminate ponds, waterways or ditches with chemical or used container

---

FOR FULL CONDITIONS OF USE ALWAYS READ THE PRODUCT LABEL

Protective clothing/Label precautions
• 21, 29, 43, 53, 63, 66, 70

207 2,4-DB + linuron + MCPA

A translocated herbicide for undersown cereals and grass

Products	1 Alistell	Farm Protection	220:30:30 g/l	EC	00080
	2 Clovacorn Extra	FCC	220:30:30 g/l	EC	02596

Uses

Annual dicotyledons in UNDERSOWN CEREALS, CLOVERS, SEEDLING GRASSLAND [1, 2]. Annual dicotyledons in GRASS SEED CROPS *(off-label)* [1].

Notes

Efficacy
• Best results achieved on young seedling weeds growing actively in warm, moist weather
• May be applied at any time of year provided crop at correct stage and weather suitable
• Avoid spraying if rain falling or imminent

Crop safety/Restrictions
• Spray winter cereals when fully tillered but before first node detectable (GS 29-30)
• Spray spring wheat from 5-fully expanded leaf stage (GS 15), barley and oats from 2-fully expanded leaves (GS 12), to before first node detectable (GS 30)
• Do not spray cereals undersown with lucerne, peas or beans
• Apply to clovers after 1-trifoliate leaf, to grasses after 2-fully expanded leaf stage. Do not spray established clover seed crops
• Do not spray in conditions of drought, waterlogging or extremes of temperature
• In frosty weather clover leaf scorch may occur but damage normally outgrown
• Do not use on sands or soils with more than 10% organic matter
• Do not roll or harrow within 7 d before or after spraying
• Avoid drift of spray or vapour onto susceptible crops

Special precautions/Environmental safety
• Harmful in contact with skin and if swallowed. Irritating to eyes and skin
• Keep livestock out of treated areas until foliage of any poisonous weeds such as ragwort has died and become unpalatable
• Harmful to fish. Do not contaminate ponds, waterways or ditches with chemical or used container

Protective clothing/Label precautions
• A, C
• 5a, 5c, 6a, 6b, 18, 21, 25, 28, 29, 36, 37, 43, 53, 63, 66, 78

Approval
• Off-label Approval to May 1992 for use on grass seed crops (OLA 0457/89) [1]

208   2,4-DB + MCPA

A translocated herbicide for cereals and direct or undersown clovers

Uses

Annual dicotyledons, perennial dicotyledons in CEREALS, UNDERSOWN CEREALS, LEYS [1]. Annual dicotyledons, polygonums, perennial dicotyledons in UNDERSOWN CEREALS, NEWLY SOWN LEYS [2, 3]. Annual dicotyledons, polygonums, perennial dicotyledons in CEREALS, UNDERSOWN CEREALS, CLOVERS, NEWLY SOWN LEYS [4].

Notes

Efficacy
* Best results achieved on young seedling weeds under good growing conditions
* Spray thistles and other perennials when 10-20 cm high provided clover at correct stage
* Effectiveness may be reduced by rain within 12 h, by very cold conditions or drought

Crop safety/Restrictions
* Apply in spring to winter cereals from leaf sheath erect stage to before first node detectable (GS 30)
* Apply to spring barley or oats from 2-leaf stage (GS 12), spring wheat 5-leaf stage (GS 15), to before first node detectable (GS 30)
* Spray clovers as soon as possible after first trifoliate leaf, grasses after 2-3 leaf stage
* Red clover may suffer temporary distortion after treatment
* Do not spray established clover crops or lucerne
* Do not roll or harrow within 7 d before or after spraying
* Do not spray immediately before or after sowing any crop
* Avoid drift onto neighbouring sensitive crops

Special precautions/Environmental safety
* Keep livestock out of treated areas until foliage of any poisonous weeds such as ragwort has died and become unpalatable
* Harmful to fish. Do not contaminate ponds, waterways or ditches with chemical or used container

Protective clothing/Label precautions
* 18, 21, 29, 37, 43, 53, 60, 63, 66, 70

209   2,4-DB + mecoprop

A translocated herbicide for undersown cereals and new leys

FOR FULL CONDITIONS OF USE ALWAYS READ THE PRODUCT LABEL

Uses Annual dicotyledons, perennial dicotyledons, chickweed in UNDERSOWN CEREALS, ESTABLISHED GRASSLAND.

Notes **Efficacy**
- Best results achieved on young seedling weeds growing well in a vigorous crop
- Effectiveness may be reduced by rain within 12 h, by very cold conditions or drought

Crop safety/Restrictions
- Spray winter cereals in spring from fully tillered to before first node detectable (GS 30)
- Spray spring barley from 2-leaf stage (GS 12), spring wheat from 5-leaf stage (GS 15), to before first node detectable (GS 30)
- Spray clovers as soon as possible after development of first trifoliate leaf
- Red clover may suffer temporary distortion after treatment
- Do not spray established clover crops, lucerne, peas or beans
- Seedling grasses are unharmed by treatment
- Do not roll or harrow for 7 d before or after treatment
- Avoid drift onto neighbouring sensitive crops

Special precautions/Environmental safety
- Keep livestock out of treated areas until foliage of any poisonous weeds such as ragwort has died and become unpalatable
- Harmful to fish. Do not contaminate ponds, waterways or ditches with chemical or used container

Protective clothing/Label precautions
- 18, 21, 29, 36, 37, 43, 53, 63, 66, 70

210 deltamethrin

A pyrethroid insecticide with contact and residual activity

Products

1 Decis	Hoechst	25 g/l	EC	00657
2 Thripstick	Aquaspersions	0.125 g/l	AL	02134

Uses Aphids in CEREALS, OILSEED RAPE, PEAS, APPLES, PLUMS, HOPS, ORNAMENTALS [1]. Barley yellow dwarf virus vectors, yellow cereal fly in CEREALS [1]. Caterpillars in LEAF BRASSICAS, ROOT BRASSICAS, APPLES, PLUMS, WOODY ORNAMENTALS, FLOWERS [1]. Cabbage stem flea beetle, rape winter stem weevil in WINTER OILSEED RAPE [1]. Pollen beetles, cabbage stem weevil in OILSEED RAPE [1]. Flea beetles in LEAF BRASSICAS, ROOT BRASSICAS, SUGAR BEET [1]. Pea and bean weevils in PEAS, FIELD BEANS, BROAD BEANS [1]. Pea moth in PEAS [1]. Capsids, codling moth, tortrix moths, sawflies in APPLES [1]. Suckers in APPLES, PEARS [1]. Plum fruit moth, sawflies in PLUMS [1]. Raspberry beetle in RASPBERRIES [1]. Caterpillars in OUTDOOR LETTUCE [1]. Capsids, scale insects, thrips in NURSERY STOCK, FLOWERS, WOODY ORNAMENTALS [1]. Leaf miners, caterpillars in TOMATOES [2]. Thrips in CUCUMBERS [2]. Sciarid flies in PEPPERS [2]. Mushroom flies in MUSHROOMS [2]. Western flower thrips, american serpentine leaf miner in NON-EDIBLE GLASSHOUSE ORNAMENTALS, GLASSHOUSE VEGETABLE CROPS *(off-label)* [1]. Insect control in SALAD ONIONS, BULB ONIONS, LEEKS, GARLIC *(off-label)* [1]. Colorado beetle in POTATOES *(off-label)* [1]. Aphids in GRASS SEED CROPS *(off-label)* [1]. Insect control in HARDY ORNAMENTAL NURSERY STOCK, NURSERY FRUIT TREES AND BUSHES *(off-label)* [1]. Alder flea beetle, pollen beetles in EVENING PRIMROSE *(off-label)* [1]. Thrips in PROTECTED ONIONS, PROTECTED LEEKS,PROTECTED GARLIC *(off-label)* [1]. Flea beetles in SORREL *(off-label)* [1]. Leafhoppers in MARJORAM *(off-label)* [1]. Aphids, caterpillars in FENNEL, RADICCHIO *(off-label)* [1]. Insect pests in LUCERNE *(off-label)* [1]. Non-indigenous leaf miners in HERBS *(off-label)* [1]. Aphids, caterpillars in CELERY, LETTUCE, CHINESE CABBAGE, CELERIAC *(off-label)* [1].

Notes

Efficacy
- A contact and stomach poison with 3-4 wk persistence, particularly effective on caterpillars and sucking insects. Good spray coverage is essential for effective control [1]
- Normally applied at first signs of damage with follow-up treatments where necessary at 10-14 d intervals. Rates, timing and recommended combinations with other pesticides vary with crop and pest. See label for details [1]
- Spray is rainfast within 1 h. May be applied in frosty weather provided foliage not covered in ice [1]
- Temperatures above 35°C may reduce effectiveness or persistence
- Spray Thripstick without dilution onto floor covering of cucumbers in soil-less culture. Thrips killed when they fall to ground to pupate. Use within 2 mth of purchase [2]
- For western flower thrips control in glasshouses apply with fogging machine. See OLA for details [1]

Crop safety/Restrictions
- For aphid control in cereals apply one spray per crop in spring/summer up to and including flowering complete stage (GS 69) [1]
- Do not spray cereals in spring/summer within 6 m of the edge of the crop [1]
- Do not spray crops suffering from drought or other physical stress [1]
- Test spray small numbers of ornamentals before committing whole batches [1]
- Consult processer before treating crops for processing

Special precautions/Environmental safety
- Harmful in contact with skin or if swallowed. Irritating to eyes and skin
- Flammable
- Approved respirator must be worn for application by fogging. See OLA for details [1]
- Dangerous to bees. Do not apply at flowering stage except as directed on cereals and peas. Keep down flowering weeds in all crops
- Extremely dangerous to fish. Do not contaminate ponds, ditches or waterways with chemical or used container

Protective clothing/Label precautions
- A, C, H
- 5a, 5c, 6a, 12c, 16, 18, 21, 28, 29, 36, 37, 48, 51, 60, 61, 63, 66, 70, 78

Harvest interval
- zero

FOR FULL CONDITIONS OF USE ALWAYS READ THE PRODUCT LABEL

- Off-label Approval to Aug 1991 for use in onions, leeks and garlic (OLA 0887/88) [1]; to Nov 1991 for use in potatoes in a statutory emergency against Colorado beetle (OLA 1390/88) [1]; to Feb 1992 for use on non-edible glasshouse crops and glasshouse vegetables (OLA 0127, 0219/89) [1]; to Apr 1992 for use on grass seed crops (OLA 0341/89) [1]; to May 1992 for use on hardy ornamental nursery stock, nursery fruit trees and bushes (OLA 0424/89) [1], evening primrose (OLA 0465/89) [1]; to July 1992 for use on protected onions, leeks and garlic (OLA 0616/89) [1], sorrel (OLA 0617/89) [1], fennel and radicchio (OLA 663/89) [1], marjoram (OLA 664/89); to Nov 1992 for use on lucerne (OLA 0938/89); to Feb 1992 for use on glasshouse herbs (only as required by Statutory Notice under Plant Health Order) (OLA 0271/90); indoor and outdoor celery, lettuce and Chinese cabbage (OLA 0435, 0436/90) [1]; to Sep 1993 for use on celeriac (OLA 0489/90) [1]
- Approved for aerial application to restricted areas of cereals, peas, leaf brassicas, winter oilseed rape and ornamental trees [1]. Written permission must first be obtained from Hoechst

211 deltamethrin + heptenophos

A systemic aphicide with knockdown, vapour and anti-feeding activity

Products	Decisquick	Hoechst	25:400 g/l	EC	03117

Uses

Aphids, pea moth, virus vectors in PEAS. Leaf roll virus vectors, potato mosaic virus vectors in POTATOES. Virus yellows vectors in SUGAR BEET. Aphids, caterpillars in FENNEL, RADICCHIO *(off-label)*. Insect pests in BROCCOLI, BRUSSELS SPROUTS, CABBAGES, CALABRESE, CAULIFLOWERS, KOHLRABI, ROSCOFF CAULIFLOWER *(off-label)*.

Notes

Efficacy
- Works by combination of contact, systemic, vapour and anti-feeding activity and persists for 10-14 d, longer in cool and shorter in hot weather
- Spray peas as soon as aphids seen and repeat once if necessary
- Spray potatoes before aphid build-up and repeat if necessary. Spray seed crops at 80% emergence, whether aphids present or not, with up to 3 further sprays at 14 d intervals
- Spray sugar beet immediately warning issued or when aphids first seen and repeat once if necessary
- Product rainfast within 2 h. Do not spray wet crops

Crop safety/Restrictions
- Do not spray crops suffering from severe moisture stress

Special precautions/Environmental safety
- This product contains an organophosphorus compound. Do not use if under medical advice not to work with such compounds
- Toxic if swallowed. Harmful in contact with skin. Irritating to eyes and skin
- Flammable
- Dangerous to bees. Do not apply at flowering stage except as directed on peas. Keep down flowering weeds in all crops
- Extremely dangerous to fish. Do not contaminate ponds, ditches or waterways with chemical or used container

Protective clothing/Label precautions

● A, C, H

● 1, 4c, 5a, 6a, 6b, 12c, 14, 16, 18, 21, 28, 29, 35, 36, 37, 49, 51, 63, 66, 70, 79

Harvest interval

● 24 h

Approval

● Off-label approval to Nov 1992 for use on fennel and radicchio (OLA 0939/89); to Sep 1993 for use on leaf and flowerhead brassicas and Kohlrabi (OLA 0510/90)

212   demeton-S-methyl

A systemic and contact organophosphorus insecticide and acaricide

Products					
1 Ashlade Persyst	Ashlade	580 g/l	EC	02801
2 Campbell's DSM	MTM Agrochem.	580 g/l	EC	00405
3 Chafer Azotox 580	Britag	580 g/l	EC	02482
4 Chiltern Demeton-S-methyl	Chiltern	500 g/l	EC	00488
5 Metasystox 55	Farm Protection	580 g/l	EC	01331
6 Mifatox	FCC	580 g/l	EC	01350
7 Power DSM	Power	580 g/l	EC	03627
8 Quad DSM Systemic Insecticide	Quadrangle	500 g/l	EC	01668
9 Star DSM	Star	580 g/l	EC	03334
10 Tripart Systemic Insecticide	Tripart	500 g/l	EC	02211
11 Vassgro DSM	Vass	500 g/l	EC	03306

Uses

Aphids in CEREALS, BEET CROPS, POTATOES, FIELD BEANS, BROAD BEANS, DWARF BEANS, RUNNER BEANS, PEAS, BRASSICAS, CARROTS, PARSNIPS, CELERY, LETTUCE, APPLES, PEARS, PLUMS, DAMSONS, CHERRIES, CURRANTS, GOOSEBERRIES, RASPBERRIES, STRAWBERRIES [1-11]. Thrips in FIELD BEANS, BROAD BEANS, DWARF BEANS, RUNNER BEANS [5]. Mangold fly in BEET CROPS [1-11]. Leafhoppers in POTATOES [5, 7]. Leafhoppers in STRAWBERRIES [5]. Aphids in GRASS SEED CROPS [2-8, 10, 11]. Pea midge in PEAS [1-5, 7-9]. Red spider mites in APPLES, PEARS, PLUMS, DAMSONS, CHERRIES, CURRANTS, GOOSEBERRIES, RASPBERRIES, STRAWBERRIES [1-11]. Bryobia mites, leafhoppers, pear sucker, sawflies, woolly aphid in APPLES, PEARS [1-6, 8-11]. Sawflies in PLUMS, CHERRIES [1-6, 8-11]. Aphids in APRICOTS, PEACHES [4, 5, 8, 10, 11]. Blackcurrant leaf midge in BLACKCURRANTS [5]. Aphids, red spider mites in HOPS [3, 6]. Aphids, red spider mites in CUCUMBERS, TOMATOES, CARNATIONS, CHRYSANTHEMUMS, ORNAMENTALS [1-11]. Aphids, thrips in MANGE-TOUT PEAS *(off-label)* [2]. Insect control in BEANS, BRASSICAS, CELERIAC, COURGETTES, GARLIC, HERBS, LEEKS, LUCERNE, MANGE-TOUT PEAS, MARROWS, ONIONS, RADICCHIO, SPINACH, SWEETCORN, PROTECTED FRUIT, PROTECTED VEGETABLES *(off-label)* [3, 5].

FOR FULL CONDITIONS OF USE ALWAYS READ THE PRODUCT LABEL

Efficacy
- For several crops different products vary in the range of pests listed as controlled by treatment. See label for details
- Apply as spray when pest appears and repeat as necessary. Dose, number and timing of sprays vary with pest and crop. See label for details
- May also be applied by soil watering on many ornamentals. Do not apply to dry soil
- Do not spray waxy-leaved plants such as brassicas or sugar beet under hot, dry conditions. Delay until late evening or early morning. Do not spray wilting plants
- For bryobia mite control spray at temperatures above 13°C
- Strains of red spider mite, peach-potato aphid in sugar beet and some other pests resistant to organophosphorus compounds have developed in certain areas

Crop safety/Restrictions
- Do not treat brassicas after the end of Oct. Do not spray cauliflower curds.
- Do not spray carnations in bloom as residues may cause unpleasant odours
- Do not use more than 1 spray per season on currants or gooseberries [5]
- Do not mix with fentin compounds or alkaline sprays

Special precautions/Environmental safety
- A chemical subject to the Poisons Rules 1982 and Poisons Act 1972
- Toxic in contact with skin, by inhalation and if swallowed
- Flammable
- This product contains an organophosphorus compound. Do not use if under medical advice not to work with such compounds
- Do not apply with hand-held ULV equipment
- Harmful to game, birds and animals
- Harmful to bees. Do not apply at flowering stage. Keep down flowering weeds
- Harmful to fish. Do not contaminate ponds, waterways or ditches with chemical or used container

Protective clothing/Label precautions
- A, C, H, M
- 1, 4a, 4b, 4c, 12c, 14, 16, 18, 21, 23, 24, 25, 26, 27, 28, 29, 36, 37, 41, 46, 50, 53, 64, 66, 70, 79

Withholding period
- Harmful to livestock. Keep all livestock out of treated areas for at least 2 wk

Harvest interval
- cereals 2 wk; mangolds, fodder beet for clamping 10 d; mangolds, fodder beet for fodder 3 wk; other crops 3 wk

Approval
- Approved for aerial application on cereals, peas, potatoes, sugar beet, carrots, brassicas, field beans [2, 6]; on cereals, peas, potatoes, sugar beet [3]; on cereals, peas, sugar beet, beans [4]; on cereals, peas, potatoes, beet crops, carrots, brassicas, beans, parsnips, celery, top fruit, soft fruit, herbage seed, maize, mangolds [5, 7]; on cereals, peas, potatoes, sugar beet, beans [8, 10, 11]; on grass seed [11]. See introductory notes
- Off-label approval to Mar 1992 for use on mange-tout peas (OLA 0246/89) [2]; to July 1992 for use on range of crops as listed in Uses section (OLA 0628/89) [3, 5]

213 demeton-S-methyl sulphone

An organophosphate insecticide available only in mixtures
See also azinphos-methyl + demeton-S-methyl sulphone

214 desmetryn

A contact triazine herbicide for use on brassica crops

Products   Semeron 25WP   Ciba-Geigy   25% w/w   WP   01916

Uses   Annual dicotyledons, fat-hen in FODDER RAPE, KALE, BRUSSELS SPROUTS, CABBAGES, SPRING GREENS.

Notes   **Efficacy**
- May be applied from Mar to mid-Nov to weeds up to 5-10 cm high (fat-hen to 35 cm)
- A higher dose required in drier eastern than western areas. See label for details
- Do not apply when rain imminent. Rain within 24 h reduces effectiveness
- May be tank-mixed with aziprotryne or clopyralid. See label for details

Crop safety/Restrictions
- Apply to drilled crops after plants have 3 true leaves and are at least 12.5 cm high, to transplants at least 2 wk after planting and after reaching above stage
- Do not spray broccoli, cauliflower, oilseed rape, turnips, swedes, undersown fodder rape, poor, thin crops or seed crops intended for producing Maris Kestrel kale
- Do not spray in frosty weather or if crops affected by frost
- Do not spray crops affected by cabbage root fly
- Do not spray for at least 7 d after applying insecticide
- Do not respray for at least 5 d until full effects of first spray can be seen
- Crop scorch may occur after spraying but crop recovers quickly without effect on yield

Protective clothing/Label precautions
- 22, 29, 35, 54, 63, 67

Harvest interval
- 4 wk

Approval
- Approved for aerial application on brassicas. See introductory notes

215 diazinon

A contact organophosphorus insecticide for soil and foliar treatment

Products   1 Basudin 5FG   Ciba-Geigy   5% w/w   GR   00213
              2 Basudin 40WP   Ciba-Geigy   40% w/w   WP   00214

FOR FULL CONDITIONS OF USE ALWAYS READ THE PRODUCT LABEL

| 3 Darlingtons Diazinon Granules | Darmycel | 5% w/w | GR | 02755 |
| 4 Diazinon Liquid | DowElanco | 160 g/l | EC | 04232 |

Uses

Cabbage root fly in LEAF BRASSICAS, ROOT BRASSICAS [1, 2]. Carrot fly in CARROTS [1, 2]. Carrot fly in CELERY [2]. Lettuce root aphid in LETTUCE [1, 2, 4]. Leaf miners, scale insects in LETTUCE [2]. Onion fly in SALAD ONIONS [2]. Leaf miners, mealy bugs, capsids, scale insects, thrips, red spider mites, springtails, symphylids, whitefly in POT PLANTS, ORNAMENTALS, ROSES, FLOWERS, CHRYSANTHEMUMS [2, 4]. Sciarid flies in BEDDING PLANTS, CONTAINER-GROWN ORNAMENTALS [2]. Aphids, leaf miners, red spider mites, thrips, whitefly in TOMATOES [2, 4]. Springtails, symphylids, red spider mites, thrips, whitefly in CUCUMBERS [2, 4]. Mushroom flies in MUSHROOMS [1-4]. Carrot fly in CELERY *(off-label)* [4].

Notes

Efficacy

- Apply as granules or spray to drilled brassicas after 2-leaf stage, as granules or drench to transplants within 4 d of planting or in mid-Apr
- Apply to carrots as granules before or at drilling against early carrot fly, or as drench against second and third generation attacks on late-lifted crops
- For lettuce root aphid control apply pre-planting or sowing [1, 2, 4] or incorporate into compost prior to blocking [2]
- Use as drench for control of springtails, symphylids and sciarid flies
- Incorporate in mushroom compost for mushroom fly control [1-3] or mix with casing material [4]. May also be used for spraying empty mushroom houses [4]
- Number and timing of applications varies with pest and crop. See label for details

Crop safety/Restrictions

- Do not use on maidenhair fern, some scorch can occur on fuchsia [4]
- Do not treat maidenhair fern, gloxinia, geranium, begonia, pelargonium or chrysanthemum varieties Mermaid or Fred Shoesmith [2]
- On subjects of unknown tolerance treat a small number of plants in first instance [2, 4]

Special precautions/Environmental safety

- This product contains an organophosphorus compound. Do not use if under medical advice not to work with such compounds
- Harmful in contact with skin, by inhalation or if swallowed [2, 4]
- Flammable [4]
- Harmful to game, wild birds and animals
- Dangerous to bees. Do not apply at flowering stage. Keep down flowering weeds [2, 4]
- Harmful to fish. Do not contaminate ponds, waterways or ditches with chemical or used container

Protective clothing/Label precautions

- A [1-3]; A, C [4]
- 1, 22, 29, 35, 36, 37, 46, 53, 63, 67, 70 [1, 3]; 1, 5a, 5b, 5c, 18, 22, 29, 35, 36, 37, 46, 48, 53, 63, 70, 78 [2, 4]; 67 [2]; 12c, 41, 66 [4]

Withholding period

- Harmful to livestock. Keep all livestock out of treated areas for at least 2 wk [4]

Harvest interval

- cucumbers 2 d;other edible crops 2 wk

311

• Off-label approval to Feb 1991 for use on celery (OLA 0123/88) [4]

216 dicamba

A translocated benzoic herbicide for control of bracken etc
See also amitrole + atrazine + dicamba
  2,4-D + dicamba
  2,4-D + dicamba + ioxynil
  2,4-D + dicamba + mecoprop
  2,4-D + dicamba + triclopyr

Products	Tracker	Shell	480 g/l	SL	03847

Uses Bracken in FORESTRY, AMENITY GRASS, NON-CROP AREAS. Perennial dicotyledons in GRASSLAND, AMENITY GRASS *(wiper application)*.

Notes **Efficacy**
• For bracken control apply to soil over litter as a concentrated solution which must be washed down to roots by rainfall
• Apply in Mar, Apr or early May in season of planting or one year after planting as 2-5 cm band midway between tree rows
• Where trees planted in furrows delay application until second year
• Frond production affected up to 1 m from band of application
• Control may be reduced where rhizomes broken up by ploughing
• For dock and perennial weed control in grassland use in diluted form with a ropewick applicator when weeds growing vigorously but before senescence of flower heads

Crop safety/Restrictions
• Avoid contact with tree foliage or application in rooting zone of trees
• Do not use where run-off likely to occur after heavy rain
• Avoid drift onto susceptible crops. Tomatoes may be affected by vapour drift at a considerable distance

Special precautions/Environmental safety
• Keep livestock out of treated areas for at least 2 wk and until foliage of any poisonous weeds such as ragwort and buttercups have died and become unpalatable

Protective clothing/Label precautions
• 21, 29, 43, 54, 63, 65

217 dicamba + dichlorprop + MCPA

A translocated herbicide for use in cereal crops

Products	Chafer Mephetol Extra	BritAg	20.8:333:167 g/l	SL	01323

FOR FULL CONDITIONS OF USE ALWAYS READ THE PRODUCT LABEL

Annual dicotyledons, polygonums, scentless mayweed, perennial dicotyledons in WHEAT, BARLEY, OATS, UNDERSOWN CEREALS, NEWLY SOWN GRASS.

Efficacy
- Best results achieved on seedling weeds growing under warm dry conditions in a vigorous crop
- Spray perennials after growth of adequate leaf area but before flowering
- Do not spray in drought or if rain expected

Crop safety/Restrictions
- Spray winter crops in spring from leaf sheath erect stage (GS 30-31)
- Spray spring crops from 5-leaf stage to first node detectable (GS 15-31)
- Spray grass at 2-3 leaf stage, provided cereal is at correct stage if undersown
- Do not use on cereals to be undersown with mixtures containing clover or legumes
- Do not roll or harrow for 10 d before or 7 d after spraying
- Avoid damage by spray or vapour drift on susceptible crops, tomato especially sensitive

Special precautions/Environmental safety
- Harmful in contact with skin and if swallowed
- Keep livestock out of treated areas until foliage of any poisonous weeds such as ragwort has died and become unpalatable

Protective clothing/Label precautions
- 5a, 5c, 18, 21, 29, 36, 37, 43, 54, 60, 63, 66, 70, 78

218   dicamba + maleic hydrazide

A herbicide/plant growth regulator mixture for amenity grass

Products   BH Dockmaster   RP Environ.   125:125 g/l   SL   04001

Uses   Docks in ROAD VERGES, NON-CROP GRASS.

Notes   **Efficacy**
- Apply during growing season once dock foliage well developed and growth active
- Spray before seed head formation is well advanced. If docks have gone to seed cut and spray regrowth in late summer/early autumn

Crop safety/Restrictions
- Prevent spray drift onto susceptible crops or other valuable plants

Special precautions/Environmental safety
- Keep livestock out of treated areas for at least 2 wk or until foliage of any poisonous weeds such as ragwort has died and become unpalatable
- Harmful to fish. Do not contaminate ponds, waterways or ditches with chemical or used container

Protective clothing/Label precautions
- 21, 29, 43, 53, 60, 63, 67

219   dicamba + maleic hydrazide + MCPA
A herbicide/plant growth regulator mixture for amenity grass

Products   Mazide Selective   Synchemicals   6:200:75 g/l   SL   02070

Uses   Annual dicotyledons, perennial dicotyledons in ROAD VERGES, AMENITY GRASS. Growth retardation in ROAD VERGES, AMENITY GRASS.

Notes   **Efficacy**
- Best results achieved by application in Apr-May when grass and weeds growing actively but before weeds have started to flower
- May be used either as one annual spray or as spring spray repeated after 8-10 wk

Crop safety/Restrictions
- Do not use on fine turf

Special precautions/Environmental safety
- Keep livestock out of treated areas until foliage of any poisonous weeds such as ragwort has died and become unpalatable
- Harmful to fish. Do not contaminate ponds, waterways or ditches with chemical or used container

Protective clothing/Label precautions
- 21, 29, 43, 53, 63, 66

220   dicamba + MCPA + mecoprop
A translocated herbicide for cereals and grassland

Products

1 Banlene Plus	Schering	18:252:84 g/l	SL	03851
2 Campbell's Field Marshal	MTM Agrochem.	18:360:160 g/l	LI	00406
3 Campbell's Grassland Herbicide	MTM Agrochem.	25:200:400 g/l	LI	00407
4 Dock-Ban	Quadrangle	19.5:245:86.5 g/l	SL	01671
5 Docklene	Schering	84:84:336 g/l	SL	03863
6 Headland Relay	WBC Technology	25:200:400 g/l	LI	03778
7 Herrisol	Bayer	18:252:84 g/l	SL	01048
8 Hyprone	Agrichem	18.7:100:194 g/l	SL	01093
9 Hysward	Agrichem	1.6:9.3:17.0% w/w	SL	01096
10 Mascot Super Selective	Rigby Taylor	15:100:200 g/l	SL	03621
11 MSS Mircam Plus	Mirfield	19.5:245:86.5 g/l	SL	01416
12 Paddox	Farm Protection	18:237:80 g/l	SL	03646
13 Pasturol	FCC	25:200:400 g/l	SL	01545
14 Quadban	Quadrangle	19.5:245:86.5 g/l	SL	03114
15 Selectrol	Adams	18:252:84 g/l	SL	04461
16 Springcorn Extra	FCC	18:360:160 g/l	SL	02004
17 Tetralex Plus	Shell	18:252:84 g/l	SL	02115
18 Tribute	Chipman	354 g/l (total)	SL	03470

FOR FULL CONDITIONS OF USE ALWAYS READ THE PRODUCT LABEL

Uses

Annual dicotyledons, chickweed, cleavers, mayweeds, polygonums, perennial dicotyledons, docks in WHEAT, OATS, BARLEY, RYE, LEYS, PERMANENT PASTURE [1, 7, 12, 14]. Annual dicotyledons, chickweed, cleavers, mayweeds, polygonums, perennial dicotyledons in WHEAT, OATS, BARLEY [2, 8, 11, 16]. Annual dicotyledons, chickweed, cleavers, mayweeds, polygonums, perennial dicotyledons in WHEAT, OATS, BARLEY, RYE, LEYS [17]. Annual dicotyledons, chickweed, cleavers, mayweeds, polygonums, perennial dicotyledons, docks in LEYS, PERMANENT PASTURE [3-6, 13]. Annual dicotyledons, perennial dicotyledons, docks in ESTABLISHED GRASSLAND [9]. Annual dicotyledons, chickweed, cleavers, mayweeds, polygonums, perennial dicotyledons, docks in GRASS SEED CROPS [7, 8, 12, 14, 17]. Annual dicotyledons, perennial dicotyledons in ORCHARDS [1]. Annual dicotyledons, perennial dicotyledons, docks in APPLES, PEARS [7, 17]. Annual dicotyledons, perennial dicotyledons in APPLES [5]. Annual dicotyledons, perennial dicotyledons in TURF [10, 15, 18, 19].

Notes

Efficacy
- Best results achieved on young, actively growing weeds up to 15 cm high in a strongly competitive crop, perennials when well developed but before flowering. See label for details of susceptibility
- Spray grassland and turf from early spring to Oct when grasses growing actively
- Do not spray in rain, when rain imminent or in drought

Crop safety/Restrictions
- Spray winter cereals from main leaf sheath erect and at least 5 cm high but before first node detectable (GS 30), spring cereals from 5-leaf stage to before first node detectable (GS 15-30) (some formulations only recommend treatment in spring)
- Do not spray cereals undersown with clovers or legumes, to be undersown with grass or legumes or grassland where clovers or other legumes are important
- Spray newly sown grass or undersown crops when grass seedlings have 2-3 leaves
- Do not spray leys established less than 18 mth or orchards established less than 3 yr
- Do not spray grass seed crops later than 5 wk before emergence of seed heads
- In orchards avoid drift of spray onto tree foliage. Do not spray during blossom period
- Do not roll, harrow, cut, graze for 3-7 d before or after treatment (products vary)
- Do not use first mowing after treatment for mulch unless composted for at least 6 mth. See label for other precautions [10, 15, 18, 19]

Special precautions/Environmental safety
- Irritating to eyes [1-12, 14, 15, 17-19]. Irritating to skin and eyes [13, 16]
- Harmful if swallowed [2]. Harmful in contact with skin and if swallowed [3, 6, 12]
- Keep livestock out of treated areas until foliage of any poisonous weeds such as ragwort has died and become unpalatable
- Harmful to fish. Do not contaminate ponds, waterways or ditches with chemical or used container

Protective clothing/Label precautions
- A, C
- 5a, 5c, 6a, 6b, 18, 21, 29, 36, 37, 43, 53, 63, 66, 70, 78 [1-19]

221   dicamba + mecoprop

A translocated post-emergence herbicide for cereals and grassland

Products

1 Di-Farmon	Farm Protection	21:319 g/l	SL	03882
2 Endox	FCC	112:265 g/l	SL	00798
3 Farmon Condox	Farm Protection	112:265 g/l	SL	03883
4 Hyban	Agrichem	18.7:300 g/l	SL	01084
5 Hygrass	Agrichem	18.7:300 g/l	SL	01090
6 MSS Mircam	Mirfield	18.7:300 g/l	SL	01415

Uses

Annual dicotyledons, mayweeds, chickweed, cleavers, polygonums, perennial dicotyledons in WHEAT, BARLEY, OATS [1, 4, 6]. Annual dicotyledons, perennial dicotyledons, docks, thistles in LEYS, PERMANENT PASTURE [1, 2, 3, 5].

Notes

Efficacy
- Best results by application to young weed seedlings under good growing conditions
- Do not spray in cold, frosty or windy weather, when rain expected or in drought
- Spray perennial weeds in grassland just before flowering, docks in rosette stage in late spring with repeat spray in Aug-Oct if necessary
- Allow 10-14 d regrowth of docks after grazing

Crop safety/Restrictions
- Apply to winter or spring cereals in spring from 5-fully expanded leaf stage to first node detectable (GS 15-31)
- Do not spray cereals undersown or to be undersown
- Do not apply to leys established for less than 2 yr or to permanent pastures where clovers are an essential part of the sward
- Do not roll, harrow, cut or graze within 7 d before or after spraying

Special precautions/Environmental safety
- Harmful if swallowed and in contact with skin [1, 6], in contact with skin [2, 3]
- Irritating to eyes [1, 3, 4, 5], to skin and eyes [2]
- Keep livestock out of treated areas until foliage of any poisonous weeds such as ragwort has died and become unpalatable
- Harmful to fish. Do not contaminate ponds, waterways or ditches with chemical or used container

Protective clothing/Label precautions
- A, C
- 5a, 5c, 6a, 6b, 18, 21, 29, 36, 37, 43, 53, 63, 65, 70, 78 [1-6]

222   dicamba + mecoprop + triclopyr

A translocated herbicide for perennial and woody weeds in grassland

Products   Fettel   Farm Protection   78:130:72 g/l   EC   02516

FOR FULL CONDITIONS OF USE ALWAYS READ THE PRODUCT LABEL

Annual dicotyledons, perennial dicotyledons, docks, stinging nettle, thistles, woody weeds, brambles, broom, gorse in ESTABLISHED LEYS, PERMANENT PASTURE, NON-CROP AREAS.

Efficacy
- Apply to actively growing perennials before flowering shoots appear, usually late spring or summer. Later application can be made to brambles and regrowth after cutting
- With well established weeds retreatment may be needed in following season
- Do not spray in drought, frost, extremes of temperature or if rain falling or imminent

Crop safety/Restrictions
- Do not use on leys established for less than 2 yr
- Do not apply overall where clovers are an essential part of sward
- Do not roll or harrow within 7 d before or after treatment
- Avoid drift onto nearby susceptible crops. Drift of vapour may occur in hot conditions
- Do not apply near desirable plants where chemical may move and contact roots
- Do not sow grass, clovers or brassicas by direct drilling or minimum cultivation for at least 6 wk after spraying

Special precautions/Environmental safety
- Irritating to eyes and skin
- Keep livestock out of treated areas for 7d and until foliage of any poisonous weeds such as ragwort has died and become unpalatable
- Dangerous to fish. Do not contaminate ponds, waterways or ditches with chemical or used container
- Not to be used on food crops

Protective clothing/Label precautions
- A, C
- 6a, 6b, 18, 21, 28, 29, 34, 36, 37, 43, 52, 63, 65

223 dicamba + paclobutrazol
A plant growth regulator/herbicide mixture for amenity grass

Products   Holdfast D   ICI Professional   25:250 g/l   SC   02157

Uses   Annual dicotyledons, perennial dicotyledons in AMENITY GRASS. Growth retardation in AMENITY GRASS.

Notes   ### Efficacy
- Uptake mainly via roots and aided by rain after application
- Apply 2 wk before grass growth starts in spring (mid-Mar to mid-May). Can also be applied up to Jul if grass growing rapidly and in Aug/Sep to reduce autumn flush
- Grass growth retarded for 3 mth or more
- Do not apply during periods of drought or frost
- Where coarse grasses (eg cocksfoot) dominant addition of maleic hydrazide or mefluidide recommended

Crop safety/Restrictions
- Use only on permanent grassland, not on grass within 1 yr after sowing

- Use only on areas of restricted or limited public access
- Do not use more than 2 applications per season
- Do not use where food crops or ornamentals to be grown or grass resown within 2 yr
- Fine-leaved grasses can be retarded more than others leading to an uneven appearance
- Do not use cuttings from treated grass for composting or mulch

Special precautions/Environmental safety
- Not to be used on food crops
- Harmful to fish. Do not contaminate ponds, waterways or ditches with chemical or used container

Protective clothing/Label precautions
- 21, 29, 53, 63, 66

224   dichlobenil

A residual nitrile herbicide for use in woody crops and non-crop areas
See also dalapon + dichlobenil

Products					
	1 BH Prefix D	RP Environ.	6.75% w/w	GR	03688
	2 Casoron G	Chipman	6.75% w/w	GR	00448
	3 Casoron G	ICI	6.75% w/w	GR	00448
	4 Casoron G	ICI Professional	6.75% w/w	GR	00448
	5 Casoron G	Synchemicals	6.75% w/w	GR	00449
	6 Casoron G4	Chipman	4% w/w	GR	02406
	7 Casoron G4	ICI Professional	4% w/w	GR	02406
	8 Casoron G4	Synchemicals	4% w/w	GR	00450
	9 Casoron G-SR	ICI Professional	20% w/w	GR	00451
	10 Prefix D	Promark	6.75% w/w	GR	01631
	11 Prefix D	Shell	6.7% w/w	GR	01631

Uses   Annual weeds, perennial grasses, perennial dicotyledons in ROSES, ESTABLISHED TREES AND SHRUBS [1, 2, 4-8, 10, 11]. Annual weeds, perennial grasses, perennial dicotyledons in APPLES, PEARS, BLACKCURRANTS, REDCURRANTS, GOOSEBERRIES, RASPBERRIES, BLACKBERRIES, LOGANBERRIES, ESTABLISHED TREES AND SHRUBS [3, 10, 11]. Volunteer potatoes in NON-CROP AREAS [3, 10, 11]. Total vegetation control in NON-CROP AREAS [1-8, 10, 11]. Aquatic weeds in AQUATIC SITUATIONS [4, 9].

Notes   **Efficacy**
- Best results achieved by application in winter to moist soil during cool weather, particularly if rain follows soon after. Do not disturb treated soil by hoeing
- Lower rates control annuals, higher rates perennials, see label for details
- Residual activity lasts 3-6 mth with selective, up to 12 mth with non-selective rates
- Apply to potato clamp sites before emergence of potatoes for blight prevention
- For control of emergent, floating and submerged aquatics apply to water surface in early spring. Intended for use in still or sluggish flowing water [4, 9]
- Do not use on fen, peat or moss soils

FOR FULL CONDITIONS OF USE ALWAYS READ THE PRODUCT LABEL

Crop safety/Restrictions
- Apply to crops in dormant period. See label for details of timing and rates
- Do not treat crops established for less than 2 yr. See label for lists of resistant and sensitive species. Do not treat stone fruit
- Do not treat Norway spruce being grown for Christmas trees
- Do not apply in or near glasshouses, near hops or areas underplanted with bulbs, annuals or herbaceous stock
- Do not apply when crop foliage wet or to frozen, snow-covered or waterlogged ground
- Do not apply to sites less than 18 mth before replanting or sowing
- Store well away from corms, bulbs, tubers and seeds

Special precautions/Environmental safety
- Keep poultry out of treated areas for 7 d [1, 10, 11]
- Do not use treated water for irrigation purposes within 2 wk of treatment [3, 9]
- Harmful to fish. Do not contaminate ponds, waterways or ditches with chemical or used container

Protective clothing/Label precautions
- 22, 29, 36, 37, 63, 67 [2, 6]; 29, 53, 58, 63, 67, 70, [1, 3, 5, 7, 8, 10, 11]; 29, 53, 55, 56, 63, 67, 70 [4, 9]; 42 [1, 10, 11]

Approval
- Cleared for aquatic weed control. See introductory notes on use of herbicides in or near water [4, 9]

---

225 dichlofluanid

A protectant phenylsulfamide fungicide for horticultural crops

Products Elvaron Bayer 50% w/w WP 00789

Uses Botrytis in CANE FRUIT, CURRANTS, GOOSEBERRIES, STRAWBERRIES, GRAPEVINES, TOMATOES, PAEONIES. Cane spot, cane blight, mildew in RASPBERRIES. Downy mildew in CAULIFLOWERS. Leaf mould in TOMATOES. Black spot, mildew in ROSES. Fire in TULIPS.

Notes **Efficacy**
- Apply as a protective spray and repeat at 7-14 d intervals. Timing, number and dose of sprays vary with crop and disease. See label for details
- Treatment also gives reduction of mildew and spur blight on raspberries and useful control of mildew on strawberries
- Apply to cauliflower seedlings under cover as soon as first seedlings appear and repeat after 3, 5, 7 and 10 d using a wetting agent. Thereafter apply weekly

Crop safety/Restrictions
- Do not use on strawberries under glass or polythene. Plants which have been under cover recently may still be susceptible to leaf scorch
- On raspberries up to 7 applications may be made, up to 6 on currants
- On grapes for winemaking apply no more than 8 sprays in season

Special precautions/Environmental safety
- Harmful to fish. Do not contaminate ponds, waterways or ditches with chemical or used container

Protective clothing/Label precautions
• 22, 28, 29, 53, 63, 67

Harvest interval
• Raspberries 1 wk; strawberries 2 wk; blackcurrants, redcurrants, gooseberries, loganberries, blackberries, cauliflowers, grapes (for fresh consumption) 3 wk; grapes (for winemaking) 5 wk

226 dichlorophen
A moss-killer, fungicide, bactericide and algicide

Products

1 Algofen	Geeco	9.4% (chlorine)	SL	02392
2 Fungo	Dax		SL	H138.
3 Mascot Moss Killer	Rigby Taylor	340 g/l	SL	02439
4 Panacide M	BDH	30% w/w	SL	01536
5 Super Mosstox	RP Environ.	340 g/l	SL	02046

Uses

Mosses, algae in TURF, PATHS AND DRIVES, HARD SURFACES [1, 2]. Mosses in TURF, PATHS AND DRIVES, HARD SURFACES [3, 5]. Dollar spot, red thread, fairy rings in TURF [1]. Red thread in TURF [5]. Mosses, liverworts in POT PLANTS [1]. Algae in CAPILLARY BENCHES [1]. Slime bacteria in WATER TANKS AND IRRIGATION LINES [1]. Dry bubble, cobweb, wet bubble, fusarium in MUSHROOMS [4]. Fungus diseases in MUSHROOM BOXES [2].

Notes

Efficacy
• For moss control spray at any time when moss is growing. Supplement spraying with other measures to improve fertility and drainage
• Spray regularly to prevent formation of algal film on capillary beds [1]
• Prevent moss and liverwort growth in pot plants by spraying compost [1]
• Use on inside of empty tanks to prevent growth of slime bacteria, may also be added to water, see label for details [1]
• To control mushroom diseases use on trays, floors etc as part of an environmental hygiene programme [2, 4]

Crop safety/Restrictions
• Must not be applied to growing mushroom crops [2, 4]

Special precautions/Environmental safety
• Harmful to fish. Do not contaminate ponds, waterways or ditches with chemical or used container

Protective clothing/Label precautions
• 21, 29, 53, 63, 65

FOR FULL CONDITIONS OF USE ALWAYS READ THE PRODUCT LABEL

327   dichlorophen + ferrous sulphate
A mosskiller/fertilizer mixture for use on turf

Products					
1 Aitkens Lawn Sand Plus	Aitken	0.3:10% w/w	GR	04542
2 SHL Lawn Sand Plus	Sinclair	0.3:10% w/w	GR	04439

Uses   Mosses in AMENITY TURF.

Notes

Efficacy
* Apply from late spring to early autumn when soil moist but not when grass wet or damp with dew
* Heavy infestations may need a repeat treatment

Crop safety/Restrictions
* Use only on established turf. Do not treat newly sown grass or freshly laid turf for the first year
* Do not apply during drought or freezing conditions or when rain imminent
* Avoid walking on treated areas until it has rained or turf has been watered
* Do not mow for 3-4 d before or after application

Special precautions/Environmental safety
* Harmful to fish. Do not contaminate ponds, waterways or ditches with chemical or used container

Protective clothing/Label precautions
* 22, 29, 53, 63

---

228   dichlorophen + 4-indol-3-ylbutyricacid + 1-naphthylacetic acid
A plant growth regulator for promoting rooting of cuttings

Products				
Synergol	Hortichem	0.25% w/w:5:5 g/lSL	04594

Uses   Rooting of cuttings in ORNAMENTALS.

Notes

Efficacy
* Dip base of cuttings into diluted concentrate immediately before planting
* Suitable for hardwood and softwood cuttings
* See label for details of concentration and timing for different species

---

229   1,3-dichloropropene
A volatile fumigant nematicide for use in soil

Products				
Telone II	DowElanco	94% w/w	VP	02097

Uses	Potato cyst nematode, free-living nematodes in POTATOES. Stem and bulb nematodes in NARCISSI. Stem nematodes in STRAWBERRIES. Nematodes, virus vectors in RASPBERRIES, STRAWBERRIES, HOPS.

Notes	**Efficacy**

- Before treatment soil should be in friable seed bed condition above 5°C, with adequate moisture and all crop remains decomposed. Tilling to 30 cm improves results
- Treat loams and clay-loams when relatively dry, do not use on heavy clays or soils with many large stones
- Apply with sub-surface injector combined with soil-sealing roller or, in hops, with a hollow tined injector. See label for details of suitable machinery
- Leave soil undisturbed for 21 d after treatment. Wet or cold soil needs longer exposure
- For control of potato-cyst nematode contact firm's representative, check nematode level and use in integrated control programme
- For use in hops apply in May or Jun following the year of grubbing and leave as long as possible before replanting
- Do not use water to clean out apparatus nor use aluminium or magnesium alloy containers which may corrode

Crop safety/Restrictions
- Do not drill or plant until odour of fumigant is eliminated
- Do not use fertilizer containing ammonium salts after treatment. Use only nitrate nitrogen fertilizers until crop well established and soil temperature above 18°C

Special precautions/Environmental safety
- Toxic if swallowed. Harmful in contact with skin and by inhalation
- Irritating to eyes, skin and respiratory system
- Flammable

Protective clothing/Label precautions
- A, C, M
- 4c, 5a, 5b, 6a, 6b, 6c, 12c, 14, 16, 18, 22, 26, 27, 28, 29, 36, 37, 54, 64, 65, 70, 79

230   dichlorprop

A translocated phenoxy herbicide for use in cereals
See also *bentazone + dichlorprop*
bromoxynil + dichlorprop + ioxynil
bromoxynil + dichlorprop + ioxynil + MCPA
clopyralid + dichlorprop + MCPA
2,4-D + dichlorprop + MCPA + mecoprop
dicamba + dichlorprop + MCPA

Products	1 Campbell's Redipon	MTM Agrochem.	500 g/l	SL	00419
	2 MSS 2,4-DP	Mirfield	500 g/l	SL	01394

Uses	Annual dicotyledons, black bindweed, redshank, perennial dicotyledons in BARLEY, WHEAT, OATS [2]. Annual dicotyledons, black bindweed, redshank, perennial dicotyledons in BARLEY, WHEAT, WINTER OATS [1].

FOR FULL CONDITIONS OF USE ALWAYS READ THE PRODUCT LABEL

Efficacy
- Best results achieved by application to young seedling weeds in good growing conditions in a strongly competing crop
- Effectiveness may be reduced by rain within 12 h of spraying [1], 6 h [2], by very cold weather or drought conditions

Crop safety/Restrictions
- Spray winter crops in spring from fully tillered stage to before first node detectable (GS 29-30), spring crops from 1-leaf unfolded to before first node detectable (GS 11-30)
- Spray crops undersown with grass as soon as possible after grass begins to tiller
- Do not spray rye, crops undersown with legumes, lucerne, peas or beans
- Do not roll or harrow within 7 d before or after treatment
- Avoid spray drifting onto nearby susceptible crops

Special precautions/Environmental safety
- Harmful in contact with skin and if swallowed
- Keep livestock out of treated areas until foliage of any poisonous weeds such as ragwort has died and become unpalatable

Protective clothing/Label precautions
- 5a, 5c, 18, 21, 29, 36, 37, 43, 54, 63, 66, 70, 78

231   dichlorprop + ferrous sulphate + MCPA
A herbicide/fertilizer combination for moss and weed control in turf

Products	SHL Turf Feed and Weed + Mosskiller	Sinclair	0.474:20:0.366% w/w	DP	04438

Uses   Annual dicotyledons, perennial dicotyledons, mosses in AMENITY TURF.

Notes

Efficacy
- Apply from late spring to early autumn when grass in active growth and soil moist
- A repeat treatment may be needed after 4-6 wk to control perennial weeds or if moss regrows

Crop safety/Restrictions
- Use only on established turf. Do not treat newly sown grass or freshly laid turf for the first year
- Do not apply during drought or freezing conditions or when rain imminent
- Avoid walking on treated areas until it has rained or turf has been watered
- Do not mow for 3-4 d before or after application
- Do not use first 4 mowings after treatment for composting

Special precautions/Environmental safety
- Harmful to fish. Do not contaminate ponds, waterways or ditches with chemical or used container

Protective clothing/Label precautions
- 22, 29, 53, 63

232 dichlorprop + MCPA

A translocated herbicide for use in cereals and turf

Products

1 Campbell's Redipon Extra	MTM Agrochem.	350:150 g/l	SL	00420
2 Farmon 2,4-DP+MCPA	Farm Protection	358:177 g/l	SL	00829
3 Hemoxone	ICI	392:210 g/l	SL	01043
4 MSS 2,4-DP+MCPA	Mirfield	500 g/l (total)	SL	01396
5 Seritox 50	RP	330:150 g/l	SL	01923
6 Seritox Turf	RP Environ	333:167 g/l	SL	03571
7 SHL Turf Feed and Weed	Sinclair	0.474:0.366 g/l	DP	04437

Uses

Annual dicotyledons, hemp-nettle, black bindweed, redshank, perennial dicotyledons in WHEAT, BARLEY, OATS [1-5]. Annual dicotyledons, perennial dicotyledons, buttercups, clovers, daisies, dandelions in AMENITY GRASS [6, 7].

Notes

Efficacy
- Best results achieved by application to young seedling weeds in active growth
- Do not spray during cold weather, if rain or frost expected, if crop wet or in drought
- Use on turf from Apr-Sep. Avoid close mowing for 3-4 d before or after treatment [6, 7]

Crop safety/Restrictions
- Apply to winter crops in spring from fully tillered, leaf-sheath erect to before first node detectable (GS 29-30)
- Apply to spring wheat from 3-leaf stage (GS 13) [4], 5-leaf stage (GS 15) [1, 2, 3, 4], to spring barley from 3-leaf stage (GS 13) [2, 4, 5], 5-leaf stage (GS 15) [1, 3], to spring oats from 1-leaf stage (GS 11) [3], 2-leaf stage (GS 12) [1], 3-leaf stage (GS 13) [2, 4, 5] but before first node detectable (GS 30)
- Do not use on undersown crops
- Do not spray crops suffering from herbicide damage or stress
- Do not roll or harrow for 7 d before or after spraying
- Avoid drift of spray onto nearby susceptible crops
- Do not use on turf in year of sowing [6, 7]

Special precautions/Environmental safety
- Harmful if swallowed and in contact with skin [1, 3-6]
- Harmful if swallowed [2]
- Keep livestock out of treated areas until foliage of any poisonous weeds such as ragwort has died and become unpalatable
- Harmful to fish. Do not contaminate ponds, waterways or ditches with chemical or used container

Protective clothing/Label precautions
- A, C, H [5, 6]
- 5a, 5c [1, 3-6]; 5c [2]; 18, 21, 28, 29, 36, 37, 43, 63, 66, 70, 78 [1-7]; 53 [4]; 54 [1-3, 5-7]

FOR FULL CONDITIONS OF USE ALWAYS READ THE PRODUCT LABEL

233 dichlorprop + mecoprop
A translocated herbicide for use in cereals

Products Campbell's CMPP/DP MTM Agrochem. 520:130 g/l SL 00415

Uses Annual dicotyledons, perennial dicotyledons in WHEAT, BARLEY, OATS.

Notes **Efficacy**
- Best results on young seedling weeds growing actively in vigorous crop
- Effectiveness may be reduced by rain within 12 h or by cold or drought conditions

Crop safety/Restrictions
- Apply to winter crops in spring from fully tillered leaf sheath erect stage to before first node detectable (GS 29-30)
- Apply to spring crops from 1-leaf unfolded to before first node detectable (GS 11-30)
- Do not use on undersown cereals
- Do not roll or harrow within 7 d before or after spraying
- Avoid drift of spray onto nearby susceptible crops

Special precautions/Environmental safety
- Harmful in contact with skin and if swallowed. Irritating to eyes
- Keep livestock out of treated areas until foliage of any poisonous weeds such as ragwort has died and become unpalatable

Protective clothing/Label precautions
- 5a, 5c, 6a, 18, 21, 29, 36, 37, 43, 54, 63, 66, 70, 78

234 dichlorvos
A contact and fumigant organophosphorus insecticide

Products 1 Darmycel Dichlorvos Darmycel 500 g/l EC 02420
2 Nuvan 500 EC Ciba-Geigy 500 g/l EC 03861

Uses Mushroom flies in MUSHROOM HOUSES [1]. Flies, mosquitoes, beetles, poultry ectoparasites in POULTRY HOUSES [2]. Western flower thrips in NON-EDIBLE GLASSHOUSE ORNAMENTALS, GLASSHOUSE VEGETABLE CROPS *(off-label)* [2].

Notes **Efficacy**
- Spray walls, roof, floor and ventilators of mushroom houses. Avoid spraying mushroom beds. Use as an aerosol spray or wet spray [1]
- For use in poultry houses apply as surface spray, cold fog or mist using suitable applicator. Direct away from poultry. See label for details [2]

Special precautions/Environmental safety
- A chemical subject to the Poisons Rules 1982 and Poisons Act 1972
- This product contains an organophosphorus compound. Do not use if under medical advice not to work with such compounds
- Toxic in contact with skin, by inhalation or if swallowed

- Flammable
- Keep unprotected persons out of treated areas for at least 12 h
- Harmful to game, wild birds and animals
- Dangerous to bees. Do not apply at flowering stage. Keep down flowering weeds
- Dangerous to fish. Do not contaminate ponds, waterways or ditches with chemical or used container

Protective clothing/Label precautions
- A, C, H
- 1, 4a, 4b, 4c, 12c, 14, 16, 18, 21, 25, 28, 29, 35, 36, 37, 38, 46, 48, 52, 64, 66, 70, 79

Harvest interval
- 1 d

Approval
- May be applied as a ULV spray. See label for details. See introductory notes on ULV application [1, 2]
- Off-label Approval for use on non-edible glasshouse ornamentals and vegetable crops to Feb 1991 (OLA 0657/89) [2]. See OLA for various restrictions which apply

---

235 dichlorvos + propoxur
A general purpose contact and residual insecticide

Products	Blattanex Residual Spray	Bayer	0.5:2%	AE	H0462

Uses General insect control in AGRICULTURAL PREMISES.

Notes **Efficacy**
- Apply as spot treatment directed to areas known to harbour pests or as space spray
- Spray into cracks and crevices to flush out hidden pests

Crop safety/Restrictions
- Do not allow spray to contact food or working surfaces on which food may be placed

---

236 diclofop-methyl
A translocated phenoxypropionic herbicide for grass weed control

Products	Hoegrass	Hoechst	378 g/l	EC	01063

Uses Annual grasses, wild oats, volunteer oats, awned canary grass, blackgrass, ryegrass, Yorkshire fog, rough meadow grass in WHEAT, DURUM WHEAT, BARLEY, WINTER RYE, TRITICALE, SUGAR BEET, RED BEET, BROAD BEANS, FIELD BEANS, CABBAGES, BRUSSELS SPROUTS, BROCCOLI, CARROTS, CELERY,

---

FOR FULL CONDITIONS OF USE ALWAYS READ THE PRODUCT LABEL

DWARF BEANS, LETTUCE, LINSEED, LUCERNE, LUPINS, MUSTARD, OILSEED RAPE, ONIONS, PARSNIPS, PEAS, POTATOES.

Notes

Efficacy
- Best results applied in autumn, winter and spring to seedling grasses up to 3-4 leaf stage. See label for details of rates and timing. Annual meadow-grass not controlled
- Do not spray if foliage wet or covered in ice. Effectiveness reduced by dry conditions
- Spray is rainfast within 1 h of application
- Do not spray after a sudden drop in temperature or a period of warm days/cold nights
- Recommended for tank mixture with a range of herbicides, growth regulators and other pesticides. See label for details
- Do not mix with hormone herbicides

Crop safety/Restrictions
- Apply to wheat, rye and triticale at any time after crop emergence
- Apply to winter barley from emergence to no later than leaf sheath erect (GS 07-30), to spring barley from emergence to 4-fully expanded leaves and 2-tillers (GS 07-22)
- Apply to broad-leaved crops from 100% emergence (cotyledons fully expanded), to onions from 1-true leaf onwards (not at crook stage)
- In Scotland do not treat spring cereals, winter cereals may only be treated until the end of Feb and only at the low rate
- Do not use on cereal crops undersown with grasses
- Do not roll or harrow cereals within 7 d of spraying
- Do not spray crops under stress, suffering from drought or waterlogging or if soil compacted

Special precautions/Environmental safety
- Flammable
- Harmful to fish. Do not contaminate ponds, waterways or ditches with chemical or used container

Protective clothing/Label precautions
- A, C
- 12c, 21, 29, 35, 53, 63, 66

Withholding period
- Keep livestock out of treated areas for at least 7 d

Harvest interval
- 6 wk

237 dicloran

A protectant nitroaniline fungicide used as a glasshouse fumigant

Products   Fumite Dicloran Smoke   Hunt   14 g   FU   00930

Uses   Botrytis, rhizoctonia in TOMATOES, PROTECTED LETTUCE, CUCUMBERS, GLASSHOUSE FLOWERS.

Notes   **Efficacy**
- Treat at first sign of disease and at 14 d intervals if necessary

- For best results fumigate in late afternoon or evening when temperature is at least 16°C. Keep house closed overnight or for at least 4 h. Do not fumigate in bright sunshine, windy conditions or when temperature too high
- Water plants, paths and straw used as mulches several h before fumigating. Ensure that foliage is dry before treatment

Crop safety/Restrictions
- Do not fumigate young seedlings or plants being hardened off
- Treat only crops showing strong growth. Do not treat crops which are dry at the roots
- Cut all open carnation or chrysanthemum blooms before fumigating
- Test treat small numbers of new or unusual plants before committing whole batches

Special precautions/Environmental safety
- Irritating to eyes and respiratory system
- Ventilate glasshouse thoroughly before re-entering
- Harmful to fish. Do not contaminate ponds, waterways or ditches with chemical or used container

Protective clothing/Label precautions
- 6a, 6c, 18, 28, 29, 35, 36, 53, 63, 67

Harvest interval
- cucumbers, tomatoes 2 d; lettuce 14 d

238 dicofol

A non-systemic acaricide for use on various horticultural crops

Products					
1 Fumite Dicofol Smoke	Hunt	5 or 10 g	FU	03030
2 Kelthane	Rohm & Haas	18.5% w/w	EC	01131

Uses

Red spider mites in GLASSHOUSE CROPS, CUCUMBERS, TOMATOES, PROTECTED LETTUCE, ORNAMENTALS [1]. Red spider mites in HOPS, APPLES, STRAWBERRIES, CUCUMBERS, TOMATOES [2]. Tarsonemid mites in STRAWBERRIES [2].

Notes

Efficacy
- Spray apples when winter eggs hatched and summer eggs laid and repeat 3 wk later if necessary. Spray other crops at any time before stated harvest interval
- Strawberries may be sprayed after picking where necessary
- Apply fumigant treatment at first sign of infestation and repeat every 14 d
- Do not fumigate in bright sunshine, when foliage wet or roots dry
- Resistance has developed in some areas, in which case sprays are unlikely to give satisfactory control, especially after second or subsequent application

Crop safety/Restrictions
- Do not fumigate young seedlings or plants which are being hardened off
- Consult supplier before treating crops other than those listed on label [1]
- Do not spray apples within 4 wk of petal fall

FOR FULL CONDITIONS OF USE ALWAYS READ THE PRODUCT LABEL

- Do not spray cucumber or tomato seedlings before mid-May or in bright sunshine

Special precautions/Environmental safety
- Harmful in contact with skin and if swallowed [2]
- Irritating to eyes and respiratory system [1]
- Flammable [2]

Protective clothing/Label precautions
- A [2]
- 6a, 6c, 18, 28, 29, 35, 36, 37, 67 [1]; 5a, 5c, 12c, 14, 18, 21, 25, 28, 29, 35, 36, 37, 54, 63, 65, 70, 78 [2]

Harvest interval
- edible crops 2 d [1]; tomatoes, cucumbers 2 d [2]; apples, strawberries 7 d [2]; hops 3 wk [2]

239   dicofol + tetradifon
A contact acaricide for fruit and glasshouse crops

Products	1 Childion	Hortichem	166:58.7 g/l	EC	03821
	2 Childion	ICI	170:62.5 g/l	EC	00486

Uses

Bryobia mites in APPLES, PEARS. Red spider mites in APPLES, PEARS, HOPS, STRAWBERRIES, BLACKCURRANTS, GLASSHOUSE CROPS. Leaf and bud mite in APPLES, PEARS. Tarsonemid mites in STRAWBERRIES, GLASSHOUSE CROPS. Broad mite in GLASSHOUSE CROPS.

Notes

Efficacy
- Apply to hops and glasshouse crops as soon as mites appear. Other crops should be sprayed post blossom, strawberries and blackcurrants being treated only if mites appear. Repeat as necessary in all cases

Crop safety/Restrictions
- Do not apply to apples or pears until 3 wk after petal fall, to strawberries during blossoming or to young plants or cucumbers in bright sunshine or before mid-May
- Treat small numbers of new glasshouse subjects before committing whole batches
- Do not apply to cissus, dahlias, ficus, gloxinias, impatiens, kalanchoes, primulas or stephanotis

Special precautions/Environmental safety
- Harmful in contact with skin or if swallowed. Irritating to eyes and skin
- Flammable

Protective clothing/Label precautions
- A
- 5a, 5c, 6a, 6b, 12c, 18, 21, 29, 36, 37, 54, 63, 65, 70, 78

Harvest interval
- cucumbers, tomatoes 2 d; apples, blackcurrants, pears, strawberries 7 d; hops 21 d

240   dienochlor

A specific acaricide for use on ornamentals

Products   Pentac Aquaflow   DowElanco   480 g/l   EC   04238

Uses   Glasshouse red spider mite in ORNAMENTALS.

Notes   **Efficacy**
- Spray roses when pest first seen and repeat 7 d later
- Spray ornamentals when pest first seen. One application normally sufficient but when problem severe repeat 7 d later
- Add wetter/spreader and spray surfaces to point of run-off

Crop safety/Restrictions
- Do not use on rare or unusual plants without first testing on small scale

Special precautions/Environmental safety
- Irritating to eyes and skin
- Dangerous to fish. Do not contaminate ponds, waterways or ditches with chemical or used container
- Do not use on edible crops

Protective clothing/Label precautions
- A, C
- 6a, 6b, 18, 21, 28, 29, 34, 36, 37, 52, 63, 66, 70

241   difenacoum

An anticoagulant rodenticide

Products

1 Killgerm Ratak Cut Wheat Rat Bait	Killgerm	0.005% w/w	RB	04448
2 Killgerm Ratak Rat Pellets	Killgerm	0.005% w/w	RB	04446
3 Killgerm Ratak Whole Wheat Rat Bait	Killgerm	0.005% w/w	RB	04447
4 Neosorexa Concentrate	Sorex	0.1% w/w	CB	01475
5 Neosorexa Liquid Concentrate	Sorex	0.25% w/w	CB	04640
6 Neosorexa Ready to Use Rat and Mouse Bait	Sorex	0.005%	RB	01474
7 Neosorexa Throw Packs	Sorex	0.005%	RB	01474
8 Ratak	ICI Professional	0.005%	RB	02586

Uses   Mice, rats in FARM BUILDINGS [1-8].

FOR FULL CONDITIONS OF USE ALWAYS READ THE PRODUCT LABEL

Efficacy
- Chemical is a chronic poison and rodents need to feed several times before accumulating a lethal dose. Effective against rats and mice resistant to other commonly used anticoagulants
- Apply ready-to-use baits or concentrate diluted with suitable base (whole grain, oatmeal etç) in baiting programme
- Lay small baits about 1 m apart throughout infested areas for mice, larger baits for rats near holes and along runs
- Cover baits by placing in bait boxes, drain pipes or under boards
- Inspect bait sites frequently and top up as long as there is evidence of feeding

Special precautions/Environmental safety
- Cover bait to prevent access by children, animals or birds

Protective clothing/Label precautions
- 25, 29, 63, 67, 81, 82, 83, 84

242 difenzoquat

A post-emergence herbicide for control of wild oats in cereals

Products

1	Avenge 2	Cyanamid	150 g/l	SL	03241
2	Avenge 2	Schering	150 g/l	SL	04470

Uses

Wild oats in BARLEY, WHEAT, DURUM WHEAT, MAIZE, RYE, TRITICALE, RYEGRASS SEED CROPS.

Notes

Efficacy
- Autumn and winter spraying controls wild oats from 2-leaf stage to mid-tillering, spring treatment to end of tillering
- Dose rate varies with season and growth-stage of weed. See label for details
- On barley in spring treatment provides control of powdery mildew in addition to wild oats
- Do not apply if rain expected within 6 h
- Recommended for tank mixture with a range of herbicides, growth regulators and other pesticides. See label for details. Other products can be applied after 24 h

Crop safety/Restrictions
- Apply to crops from 2-leaf stage to 50% of plants with 3 nodes detectable (GS 12-33) but before flag leaf visible (GS 37)
- Treatment of wheat, durum wheat and triticale limited to named cultivars. See label for details
- Apply to crops undersown with ryegrass and clover either before emergence or after grass has reached 3-leaf stage
- Spray ryegrass seed crops after 3-leaf stage, maize as soon as weeds at susceptible stage
- Do not spray crops suffering stress from waterlogging, drought or other factors
- Temporary yellowing may follow application, especially under extremes of temperature, but is normally outgrown rapidly

Special precautions/Environmental safety
- Harmful if swallowed and in contact with skin
- Irritating to eyes and skin. Risk of serious damage to eyes

- Harmful to fish. Do not contaminate ponds, waterways or ditches with chemical or used container

Protective clothing/Label precautions
- A, C
- 5a, 5c, 6a, 6b, 9, 18, 21, 26, 27, 28, 29, 36, 37, 41, 53, 63, 65, 70, 78

Withholding period
- Harmful to livestock. Keep all livestock out of treated areas for at least 6 wk

Approval
- Approved for aerial application on wheat, barley, rye, maize, herbage seed crops. See introductory notes

243 diflubenzuron

A selective, persistent, contact and stomach poison insecticide

Products   Dimilin WP   ICI   25% w/w   WP   03810

Uses   Cabbage white butterfly, cabbage moth, diamond-back moth in BRASSICAS.
Caterpillars, codling moth, tortrix moths, winter moth, leaf miners, sawflies, earwigs, rust mite in APPLES, PEARS. Pear sucker in PEARS. Plum fruit moth, winter moth, tortrix moths in PLUMS. Sawflies, winter moth in BLACKCURRANTS. Caterpillars in AMENITY TREES AND SHRUBS, ORNAMENTALS, TOMATOES, PEPPERS, GLASSHOUSE CROPS. Caterpillars, browntail moth, pine looper, winter moth in FORESTRY. Sciarid flies in MUSHROOMS.

Notes   **Efficacy**
- Acts by disrupting chitin synthesis and preventing hatching of eggs. Most active on young caterpillars and most effective control achieved by spraying as eggs start to hatch
- Dose and timing of spray treatments vary with pest and crop. See label for details
- Addition of wetter recommended for use on brassicas and pears
- Negligible effect on Phytoseiulus or Encarsia used for biological control or on bees
- Apply as case mixing treatment on mushrooms or as post-casing drench

Crop safety/Restrictions
- Consult supplier regarding range of ornamentals which can be treated and check for varietal differences on a small sample in first instance
- Do not use as a compost drench or incorporated treatment on ornamental crops

Protective clothing/Label precautions
- 29, 35, 54, 63, 67, 70

Harvest interval
- tomatoes, peppers 24 h; apples, pears, plums, blackcurrants, brassicas 14 d

FOR FULL CONDITIONS OF USE ALWAYS READ THE PRODUCT LABEL

244 diflufenican

A shoot absorbed herbicide available only in mixtures

245 diflufenican + isoproturon

A contact and residual herbicide for use in winter cereals

Products

1	Javelin	RP	62.5:500 g/l	SC	03696
2	Javelin Gold	RP	20:500 g/l	SC	04783
3	Panther	RP	50:500 g/l	SC	03773

Uses Annual dicotyledons, annual grasses, blackgrass, wild oats in WINTER WHEAT, WINTER BARLEY [1]. Annual dicotyledons, annual meadow grass in WINTER WHEAT, WINTER BARLEY, TRITICALE, RYE [2].

Notes **Efficacy**
* Best results by application to fine, firm seedbed moist at or after application
* Weeds controlled from before emergence to 6 true leaf stage
* Do not use on soils with more than 10% organic matter, nor on waterlogged soil
* Any trash, ash or burnt straw should be buried during seedbed preparation

Crop safety/Restrictions
* Spray winter wheat pre- or early post-emergence (from 1st expanded leaf stage - GS11-onwards) up to end of Feb, winter barley only early post-emergence [1]
* Spray winter wheat and barley pre- or post-emergence before crop reaches second node detectable stage (GS 32) [2, 3]
* Use only pre-emergence treatments on triticale and rye [2, 3]
* Drill crop to normal depth (2.5 cm) and ensure seed well covered
* Do not use on other cereals, broadcast or undersown crops or crops to be undersown
* Do not use on sands, very light, very stony or gravelly soils
* Do not spray crops suffering from stress, frost, deficiency, pest or disease attack
* Do not harrow after application nor roll autumn-treated crops until spring
* See label for details of time and land preparation needed before sowing other crops (12 wk for most crops). In the event of crop failure winter wheat can be redrilled without ploughing, other crops only after ploughing - see label for details

Special precautions/Environmental safety
* Harmful to fish. Do not contaminate ponds, waterways and ditches with chemical or used container

Protective clothing/Label precautions
* 30, 53, 60, 63, 66

246 diflufenican + trifluralin

A contact and residual herbicide for use in winter cereals

Products Ardent   Embetec   40:400 g/l   SC   04248

Annual dicotyledons, annual meadow grass in WINTER WHEAT, WINTER BARLEY.

Notes **Efficacy**
- Apply as a pre-crop emergence spray, not after end of Nov
- Weeds controlled pre- or post-emergence up to 4 leaf stage (2 leaf for annual meadow grass)
- Best results achieved on fine, firm, moist seedbeds
- Do not use on soils with more than 10% organic matter nor on waterlogged soil
- Any residues from straw burning should be buried or dispersed prior to drilling

Crop safety/Restrictions
- Do not use on other cereals, nor on broadcast or undersown crops
- Ensure crop is evenly drilled and seed well covered
- Do not harrow after application or roll until spring
- Do not use on sands or very light, very stony or gravelly soils
- See label for details of time and land preparation needed before sowing other crops and for crops which may be planted in the event of crop failure

Special precautions/Environmental safety
- Irritating to skin and eyes
- Harmful to fish. Do not contaminate ponds, waterways or ditches with chemical or used container

Protective clothing/Label precautions
- A, C
- 6a, 6b, 18, 29, 36, 37, 53, 60, 63, 66, 70

247 dikegulac
A growth regulator for use on hedges and ornamentals

Products Atrinal   RP Environ.   200 g/l   SL   00170

Uses Growth retardation in HEDGES, WOODY ORNAMENTALS. Increasing branching in ORNAMENTALS, AZALEAS, BEGONIAS, FUCHSIAS, KALANCHOES, VERBENA.

Notes **Efficacy**
- For controlling hedge growth apply when in full leaf and growing actively, normally May-Jun. Cut just before spraying but leave a good leaf coverage
- Spray on a mild, calm day when no rain expected
- Recommended concentration varies with species. See label for details
- For promoting side branching in ornamentals mainly applied from Mar to Sep when plants actively growing. Details of treatment recommended vary with species. See label

Crop safety/Restrictions
- Only spray healthy hedges at least 3 yr old
- Do not spray more than once in a season

FOR FULL CONDITIONS OF USE ALWAYS READ THE PRODUCT LABEL

● Slight, temporary yellowing of shoot tips may occur a few weeks after treatment

Protective clothing/Label precautions
● 29, 54, 63, 65

---

248   di-1-p-menthene

An antitranspirant and coating agent available only in mixtures
See also chlormequat + di-1-p-menthene
          dalapon + di-1-p-menthene

---

249   dimethoate

A contact and systemic organophosphorus insecticide and acaricide
See also chlorpyrifos + dimethoate

Products	1 Ashlade Dimethoate	Ashlade	400 g/l	EC	04814
	2 Atlas Dimethoate 40	Atlas	400 g/l	EC	03044
	3 BASF Dimethoate 40	BASF	400 g/l	EC	00199
	4 Brabant Dimethoate	Bos	400 g/l	EC	00310
	5 Campbell's Dimethoate 40	MTM Agrochem.	400 g/l	EC	00398
	6 Chiltern Dimethoate 40	Chiltern	400 g/l	EC	03201
	7 Clifton Dimethoate 40	Clifton	400 g/l	EC	03037
	8 Croptex Dimethoate	Hortichem	400 g/l	EC	02824
	9 Portman Dimethoate 40	Portman	400 g/l	EC	01527
	10 Power Dimethoate	Power	400 g/l	EC	03125
	11 Quad Dimethoate 40	Quadrangle	400 g/l	EC	01667
	12 Tripart Dimethoate 40	Tripart	400 g/l	EC	02199
	13 Turbair Systemic Insecticide	PBI	10 g/l	UL	02250
	14 Unicrop Dimethoate 40	Unicrop	400 g/l	EC	02265

Uses   Aphids in CEREALS [2-7, 9-13]. Aphids in WHEAT, BARLEY [1]. Wheat bulb fly in
WHEAT [2, 3, 5-7, 9-12, 14]. Aphids in HERBAGE SEED CROPS [1-3, 7, 14].
Aphids in BROAD BEANS, FRENCH BEANS, RUNNER BEANS [3, 4, 6-14].
Aphids in FIELD BEANS [5]. Aphids, mangold fly in BEET CROPS [1-7, 9-12, 14].
Aphids in SUGAR BEET SEED CROPS, MANGOLD SEED CROPS [3, 5, 7, 10].
Cabbage aphid in LEAF BRASSICAS [1-14]. Aphids in CARROTS [1-14]. Aphids in
PARSNIPS, CELERY [13]. Aphids, pea midge, thrips in PEAS [1-14]. Aphids in
POTATOES [1-7, 9-14]. Aphids in LETTUCE, PROTECTED LETTUCE [3, 5, 7, 13].
Aphids, sawflies, suckers, capsids, bryobia mites, red spider mites in APPLES, PEARS
[1-9, 11-14]. Aphids, red spider mites, sawflies in PLUMS [1-9, 11-14]. Aphids, cherry
fruit moth, red spider mites in CHERRIES [4-6, 8, 9, 11-13]. Aphids in PEACHES
[13]. Aphids, red spider mites in CURRANTS [4-6, 8, 9, 11, 12]. Aphids, red spider
mites in BLACKCURRANTS [1-3, 7, 13, 14]. Aphids, capsids, red spider mites in
GOOSEBERRIES [3, 6-9, 11-13]. Aphids, capsids, red spider mites, leafhoppers in
CANE FRUIT [4-9, 11-13]. Aphids in RASPBERRIES [3, 7]. Aphids, red spider mites
in STRAWBERRIES [1-9, 11-14]. Aphids, red spider mites in HOPS [3, 7, 14]. Aphids,
leaf miners, red spider mites in ORNAMENTALS, FLOWERS, TREES AND
SHRUBS [3-9, 11-14]. Aphids, leaf miners in ROSES, CARNATIONS [5, 6, 8, 9, 11,
12, 14]. Aphids, red spider mites in TOMATOES [3, 5-9, 11, 12]. Aphids in
GLASSHOUSE CROPS [13]. Thrips in CELERY, LEEKS, MARROWS, SALAD

ONIONS, SWEETCORN *(off-label)* [2-8, 9-12, 14]. Aphids, capsids, red spider mites, thrips in GRAPEVINES *(off-label)* [2-8, 9-12, 14].

Notes **Efficacy**
- Chemical has quick knock-down effect and systemic activity lasts for up to 14 d
- With a number of crops products differ in ranges of pests listed as controlled. Uses section above provides summary. See labels for details
- For most pests apply when pest first seen and repeat 2-3 wk later or as necessary. Timing and number of sprays varies with crop and pest. See label for details
- Best results achieved when crop growing vigorously. Systemic activity reduced when crops suffering from drought or other stress
- In hot weather apply in early morning or late evening
- Do not tank mix with alkaline materials. Do not mix Turbair formulation with any other product. See label for recommended tank-mixes
- Populations of aphids and mites resistant to organophosphorus compounds have developed in some areas, especially in beet crops
- Consult processor before spraying crops grown for processing

Crop safety/Restrictions
- Do not treat chrysanthemums or ornamental prunus [2-9, 11, 12, 14]. Do not treat peaches or plants in flower [2, 4-9, 11, 12, 14]
- Do not treat Chinese cabbage, ornamental prunus, chrysanthemums or indoor roses [13]
- Test for varietal susceptibility on all unusual plants or new cultivars
- In protected lettuce needing treatment between Oct and Feb do not use more than one application per crop

Special precautions/Environmental safety
- This product contains an anticholinesterase organophosphorus compound. Do not use if under medical advice not to work with such compounds
- Harmful in contact with skin or if swallowed
- Harmful to game, wild birds and animals. Bury all spillages
- Dangerous to bees. Do not apply at flowering time except as directed on peas. Keep down flowering weeds in all crops
- Harmful to fish. Do not contaminate ponds, waterways or ditches with chemical or used container

Protective clothing/Label precautions
- A, C
- 5a, 5c, 14, 16, 18, 21, 25, 28, 29, 35, 36, 37, 41, 46, 49, 53, 60, 64, 66, 70, 78 [3-10, 14]; 5a, 5b, 14, 18, 29, 35, 36, 37, 41, 46, 48, 53, 63, 65, 70, 78 [2, 11, 12]; 5a, 5b, 14, 21, 28, 29, 35, 36, 37, 41, 46, 48, 53, 70, 78 [13]

Withholding period
- Harmful to livestock. Keep all livestock out of treated areas for at least 7 d

Harvest interval
- protected lettuce 4 wk;other crops 7d

FOR FULL CONDITIONS OF USE ALWAYS READ THE PRODUCT LABEL

- Approved for aerial application on cereals, peas, ware potatoes, sugar beet, brassicas, carrots [2]; cereals, peas, ware potatoes, beet crops, brassicas, beans, herbage seed crops [3]; cereals, peas, ware potatoes, beet crops, carrots, soft fruit, tree nurseries [4]; cereals, peas, ware potatoes, sugar beet [5]; cereals, peas, ware potatoes, sugar beet, beans [6, 9, 11]; cereals, peas, ware potatoes, beet crops, brassicas, carrots, beans, herbage seed crops [7]; beet crops, carrots, peas, beans, ware potatoes [10]; cereals, peas, ware potatoes, sugar beet, beans [12]; cereals, peas, ware potatoes, sugar beet, herbage seed crops [13]; cereals, peas, ware potatoes, beet crops, herbage seed crops [14]. See introductory notes
- May be applied with ULV equipment. See label for details. See introductory notes [13]
- Off label approval to Feb 1991 for use in celery, leeks, marrows, salad onions, sweetcorn, grapevines (OLA 0144, 0163/88) [2]; (OLA 0143, 0162/88) [3]; (OLA 0135, 0142, 0154, 0161/88) [4]; (OLA 0141, 0160/88) [5]; (OLA 0139, 0158/88) [6]; (OLA 0138, 0157/88) [7]; (OLA 0136, 0155/88) [8]; (OLA 0129, 0150/88) [9]; (OLA 0128, 0149/88) [10]; (OLA 0127. 0148/88) [11]; (OLA 0125, 0146/88) [12]; (OLA 0124, 0145/88) [14]

250 dinocap

A protectant dinitrophenol powdery mildew fungicide

Products	1 Campbell's Dinocap 50% Miscible	MTM Agrochem.	500 g/l (tech)	EC	00400
	2 Karathane Liquid	Rohm & Haas	350 g/l	EC	01126

Uses

Powdery mildew in APPLES, GOOSEBERRIES, BLACKCURRANTS, STRAWBERRIES, HOPS, CUCUMBERS, CARNATIONS, CHRYSANTHEMUMS, ROSES [1]. Powdery mildew in APPLES, GOOSEBERRIES, RASPBERRIES, STRAWBERRIES, HOPS, CHRYSANTHEMUMS, ROSES [2].

Notes

Efficacy
- Spray at 7-14 d intervals to maintain protective film
- Product must be applied to roses before disease becomes established
- Regular use on apples suppresses red spider mites and rust mites
- Addition of wetting agent recommended for high volume application [1]

Crop safety/Restrictions
- When applied to apples during blossom period may cause spotting but no adverse effect on pollination or fruit set. Do not apply to Golden Delicious during blossom period
- Do not apply when temperature is above 24°C. Do not apply with white oils
- Certain chrysanthemum cultivars may be susceptible
- May cause petal spotting on white roses

Special precautions/Environmental safety
- Harmful by inhalation, if swallowed and in contact with skin [1]
- Irritating to eyes, skin and respiratory system [1], to eyes and skin [2]
- May cause sensitization by skin contact [2]
- Flammable
- Dangerous to fish. Do not contaminate ponds, waterways or ditches with chemical or used container

• A, C, M, [1]; A, C, J, M [2]
• 5a, 5b, 5c, 6a, 6b, 6c, 14, 18, 21, 25, 28, 29, 35, 36, 37, 52, 63, 66, 70, 78 [1]; 6a, 6b, 10a, 12c, 16, 18, 23, 25, 26, 27, 28, 29, 35, 36, 37, 52, 63, 65, 78 [2]

Harvest interval
• outdoor crops 7 d, protected crops 2 d [1]; apples, strawberries 7d, blackcurrants, gooseberries, raspberries, hops 21d [2]

251   dinoseb

A dinitro herbicide and desiccant approvals for use and storage of which were revoked in 1988

252   diphenamid

A pre-emergence amide herbicide for use in horticultural crops

Products	Enide 50W	ICI	50% w/w	WP	00800

Uses   Annual dicotyledons, annual grasses in DWARF BEANS, RUNNER BEANS, COURGETTES, MARROWS, TOMATOES, STRAWBERRIES, BULBS, HARDY ORNAMENTAL NURSERY STOCK, FOREST NURSERIES.

Notes   **Efficacy**
• Apply prior to weed germination. Best results achieved by application to firm, moist, weed-free seedbed
• Do not use on soils with more than 10% organic matter
• In dry weather irrigate (10-15 cm repeated after 7-10 d) before weeds germinate
• Application rate varies with soil type and crop species. See label for details
• May be applied to a wide range of container-grown ornamentals (see label for list) for control of weeds including annual meadow grass and hairy bitter-cress
• Recommendations for tank-mixing with chlorthal-dimethyl on various crops and with lenacil on strawberries

Crop safety/Restrictions
• Apply immediately after drilling or planting before weeds germinate
• Apply to established strawberries before flowering or after harvest (not during flowering/harvest period), immediately after removal of existing weeds. Do not use on runner beds
• Do not apply post-emergence in forest nurseries until 2 wk after full emergence when first needles fully extended
• Autumn sown cereals or grass should not follow treated crops. Allow 3 mth before drilling or planting other crops (12 mth for lettuce)

FOR FULL CONDITIONS OF USE ALWAYS READ THE PRODUCT LABEL

Special precautions/Environmental safety

• Harmful to fish. Do not contaminate ponds, waterways or ditches with chemical or used container

Protective clothing/Label precautions

• 21, 25, 29, 53, 63, 65, 70

253   diquat

A non-residual bipyridyl contact herbicide and crop desiccant
See also amitrole + diquat + paraquat + simazine

Products	1 Midstream	ICI Professional	10%	GW	01348
	2 Power Diquat	Power	200 g/l	SL	05002
	3 Reglone	ICI	200 g/l	SL	04444

Uses

Pre-harvest desiccation in POTATOES, OILSEED RAPE, PEAS, LUPINS, CLOVERS, FIELD BEANS, LAID BARLEY AND OATS, LINSEED, EVENING PRIMROSE [3]. Chemical stripping in HOPS [3]. Pre-harvest desiccation in POTATOES, OILSEED RAPE, DRYING PEAS, FIELD BEANS [2]. Annual dicotyledons in ROW CROPS [3]. Aquatic weeds in AREAS OF WATER [1, 3]. Pre-harvest desiccation in NAVY BEANS, WHITE MUSTARD, BROWN MUSTARD, SUNFLOWERS, BORAGE, SOYA BEANS *(off-label)* [3]. Annual dicotyledons in NAVY BEANS, ORNAMENTALS, GRAPEVINES, ASPARAGUS, SAGE *(off-label)* [3].

Notes

Efficacy

• Acts rapidly on green parts of plants and rainfast in 15 min
• Best results achieved by spraying in bright light and low humidity conditions
• Add Agral wetter for improved weed control, not on aquatic weeds [3]
• Apply to potatoes when tubers the desired size, to other crops when mature or approaching maturity. See label for details of timing and of periods to be left before harvesting potatoes and for timing of pea and bean desiccation sprays
• Combining of linseed can normally begin 10-20 d after spraying
• Apply as hop stripping treatment when shoots have reached top wire
• Apply to floating and submerged aquatic weeds in still or slow moving water [1, 3]
• Do not spray in muddy water [1, 3]

Crop safety/Restrictions

• Up to 2 treatments per crop may be used on potatoes and evening primrose, only one treatment on other crops
• Do not apply haulm destruction treatment when soil dry. Tubers may be damaged if spray applied during or shortly after dry period. See label for details of maximum allowable soil-moisture deficit and varietal drought resistance scores
• Apply to laid barley and oats to be used only for stock feed, only on peas for harvesting dry, on field beans for pigeon and animal feed only
• Do not add wetters to desiccant sprays for potatoes or processing peas. Agral wetter may be added to spray for use on peas to be used as animal fodder [3]
• For weed control in crops apply as overall spray before crop emergence or as interrow spray in row crops. Keep off crop foliage
• Do not apply through mist-blower or in windy conditions

Special precautions/Environmental safety
- Harmful if swallowed. Irritating to eyes and skin
- Do not feed treated straw or haulm to livestock within 4 d of spraying [2, 3]
- Do not use treated water for human consumption within 24 h or for overhead irrigation within 10 d of treatment [1, 3]

Protective clothing/Label precautions
- A, C
- 5c, 6a, 6b, 17, 18, 21, 28, 29, 36, 37, 41, 44, 54, 55, 56, 63, 65, 70, 78

Withholding period
- Harmful to livestock. Keep all livestock out of treated areas for at least 24 h [2, 3]

Approval
- Approved for aquatic weed control. See introductory notes on use of herbicides in or near water [1, 3]
- Off-label Approval to Mar 1991 for use on navy beans (OLA 0246/88) [3]; to May 1992 for use on white and brown mustard (OLA 0461/89), sunflowers (OLA 0462/89), ornamentals (OLA 0463/89), borage (OLA 0469/89); to June 1992 for use on grapevines (OLA 0536/89), asparagus pre-planting (OLA 0537), soya beans (OLA 0538/89), sage (OLA 0539/89) [3]

---

254   diquat + paraquat
A non-residual bipyridyl contact herbicide and crop desiccant

Products

1 Farmon PDQ	Farm Protection	80:120 g/l	SL	02886
2 Parable	ICI	100:100 g/l	SL	03805

Uses

Annual dicotyledons, annual grasses, volunteer cereals, perennial non-rhizomatous grasses in FIELD CROPS *(autumn stubble, pre-planting/sowing, stubble burning, sward destruction/direct drilling)* [1, 2]. Annual dicotyledons, annual grasses, volunteer cereals, perennial non-rhizomatous grasses in POTATOES, ROW CROPS, FRUIT CROPS, STRAWBERRIES, HOPS, FORESTRY, NON-CROP AREAS [1, 2]. Annual dicotyledons, annual grasses, perennial non-rhizomatous grasses in BULBS, HARDY ORNAMENTALS [2]. Chemical stripping in HOPS [2]. Spawn control in RASPBERRIES *(off-label)* [1, 2].

Notes

Efficacy
- Apply to young emerged weeds less than 15 cm high, annual grasses must have at least 2 leaves when sprayed. Best results when foliage dry and weather dull and humid
- Addition of approved non-ionic wetter recommended for control of certain species and with low dose rates. See label for details
- Allow at least 4 d between spraying and cultivation, as long a period as possible after spraying creeping perennial grasses
- For chemical stripping apply in July or after hops have reached top wire. Do not use on hops under drought conditions
- Chemical rapidly inactivated in moist soil, activity reduced in dirty or muddy water
- Spray is rainfast in 10 min

FOR FULL CONDITIONS OF USE ALWAYS READ THE PRODUCT LABEL

- For stubble burning fire 7-10 d after spraying

Crop safety/Restrictions
- Apply up to just before sown crops emerge or just before planting
- On sandy or immature peat soils and on forest nursery seedbeds allow 3 d between spraying or planting
- Where trash or dying weeds are left on surface allow at least 3 d before planting or crop emergence
- In potatoes spray earlies up to 10% emergence, maincrop to 40% emergence, provided plants are less than 15 cm high. Do not use post-emergence on potatoes from diseased or small tubers or under very hot, dry conditions
- Use guarded no-drift sprayers to kill interrow weeds and strawberry runners
- Apply to fruit crops as a directed spray, preferably in dormant season
- If spraying bulbs at end of season ensure all crop foliage is detached from bulbs. Do not use on very sandy soils

Special precautions/Environmental safety
- A chemical subject to the Poisons Rules 1982 and Poisons Act 1972
- Toxic if swallowed. Paraquat can kill if swallowed
- Harmful in contact with skin. Irritating to eyes and skin
- Do not put in a food or drinks container. Keep out of reach of children
- Not to be used by amateur gardeners
- Paraquat can be harmful to hares, where possible spray stubbles early in the day

Protective clothing/Label precautions
- A, C
- 4c, 5a, 6a, 6b, 17, 18, 21, 23, 24, 28, 29, 36, 37, 41, 47, 54, 64, 65, 70, 79

Withholding period
- Harmful to livestock. Keep all livestock out of treated areas for at least 24 h

Approval
- Off-label approval to Mar 1991 for spawn control in raspberries (OLA 0267/88) [1], (OLA 0268/88) [2]

255   diquat + paraquat + simazine
A total herbicide with contact and residual activity

Products   Soltair   ICI Professional   2.5:2.5:5% w/w   SG   03601

Uses   Annual dicotyledons, annual grasses, perennial dicotyledons, perennial grasses in ESTABLISHED TREES AND SHRUBS, HEDGES. Total vegetation control in NON-CROP AREAS.

Notes   **Efficacy**
- Apply to emerged weeds at any time of year. Do not disturb soil after treatment
- Light rain shortly after treatment does not reduce effectiveness
- Do not use on soils with very high content of organic matter
- Deep rooted perennials may need retreatment

Crop safety/Restrictions
- May be applied up to mature stems with no green tissue but avoid contact with leaves or green stems of desirable trees and shrubs
- Desirable species may be damaged if heavy rain falls soon after application on light sand or gravel soils. See label for list of sensitive species
- Some newly planted shrubs growing on light, sandy soils may be damaged
- Do not treat slopes where runoff might cause damage to lawns, flowerbeds etc.

Protective clothing/Label precautions
- A, C
- 14, 22, 28, 29, 41, 54, 64, 65

Withholding period
- Harmful to livestock. Keep all livestock out of treated areas for at least 24 h

256 disulfoton

A systemic organophosphorus aphicide and insecticide
See also chlorpyrifos + disulfoton

Products

1 Campbell's Disulfoton FE 10	MTM Agrochem.	10% w/w	GR	00402
2 Campbell's Disulfoton P 10	MTM Agrochem.	10% w/w	GR	00403
3 Disyston FE-10	Bayer	10% w/w	GR	00714
4 Disyston P-10	Bayer	10% w/w	GR	00715

Uses

Aphids in BRUSSELS SPROUTS, CABBAGES, CAULIFLOWERS, POTATOES [1-4]. Aphids in BROCCOLI, PARSNIPS [1, 2]. Aphids in SUGAR BEET, FODDER BEET, MANGOLDS [2]. Aphids in SUGAR BEET [4]. Aphids in CARROTS [1-3]. Carrot fly in CARROTS, PARSNIPS [1-4]. Aphids in CELERY [3]. Carrot fly in CELERY [3, 4]. Aphids in FRENCH BEANS, RUNNER BEANS [3, 4]. Aphids in BROAD BEANS, FIELD BEANS [2, 4]. Aphids, pea midge, thrips in PEAS [1]. Aphids in MARROWS, PARSLEY [3, 4]. Aphids in STRAWBERRIES [4]. Large bulb fly in NARCISSI *(off-label)* [4]. Carrot willow aphid in OUTDOOR HERBS, PROTECTED HERBS *(off-label)* [3, 4]. Aphids in PROTECTED MARROWS *(off-label)* [3].

Notes

Efficacy
- Granules based on fullers earth (FE) are applied at drilling or planting those based on pumice (P) can also be used as foliar treatments
- For effective results granules applied at drilling or planting should be incorporated in soil. See label for details of suitable application machinery
- One application is normally sufficient, a split application may be used on carrots
- Persistence may be reduced in fen peat soils

Crop safety/Restrictions
- On potatoes treat only second early and maincrop varieties
- Consult processor before treating crops grown for processing

---

FOR FULL CONDITIONS OF USE ALWAYS READ THE PRODUCT LABEL

Special precautions/Environmental safety
- This product contains an organophosphorus compound. Do not use if under medical advice not to work with such compounds
- Toxic in contact with skin, by inhalation or if swallowed
- Dangerous to game, wild birds and animals
- Applied correctly treatment will not harm bees
- Dangerous to fish. Do not contaminate ponds, ditches or waterways with chemical or used container

Protective clothing/Label precautions
- A, B, C, C or D + E, H, J, K, M
- 1, 4a, 4b, 4c, 16, 18, 22, 23, 25, 28, 29, 35, 36, 37, 45, 52, 64, 67, 70, 79

Withholding period
- Dangerous to livestock. Keep all livestock out of treated areas for at least 6 wk. Bury or remove spillages

Harvest interval
- 6 wk

Approval
- Approved for aerial application on brassicas, broad beans [2]; brassicas, beans, sugar beet, carrots [4]. See introductory notes. Consult firm for details
- Off-label Approval to Sep 1992 for use on outdoor narcissus (OLA 0749/89) [4]; to Jan 1993 for use on outdoor and protected herbs (0076/90) [4], (0077/90) [3]; for use on protected marrows (0078/90) [3]

---

257 disulfoton + quinalphos
A systemic and contact organophosphorus insecticide for brassicas

Products	Knave	Hortichem	5:2.5% w/w	GR	02534

Uses

Aphids, cabbage root fly in BROCCOLI, BRUSSELS SPROUTS, CABBAGES, CALABRESE, CAULIFLOWERS.

Notes

Efficacy
- Apply as band treatment with suitable granule applicator at drilling using 'bow-wave' technique. See label for details
- Apply as a sub-surface band to crops transplanted after mid-Apr using a 'Leeds' coulter set so that roots are planted below treated layer of soil
- Product can be used on all mineral and organic soils

Crop safety/Restrictions
- Consult processor before using on crops for processing

Special precautions/Environmental safety
- A chemical subject to the Poisons Rules 1982 and Poisons Act 1972
- This product contains an organophosphorus compound. Do not use if under medical advice not to work with such compounds
- Toxic in contact with skin, by inhalation or if swallowed
- Dangerous to game, wild birds and animals, bury spillages

- Extremely dangerous to fish. Do not contaminate ponds, waterways or ditches with chemical or used container

Protective clothing/Label precautions
- A, B, C, H or L, M
- 1, 4a, 4b, 4c, 14, 16, 18, 22, 25, 28, 29, 35, 36, 37, 45, 51, 64, 67, 70, 79

Harvest interval
- 6 wk

258   dithianon
A protectant and eradicant quinone fungicide for scab control

Products	1 Delan-Col	ICI	600 g/l	SC	00662
	2 Dithianon Flowable	Shell	750 g/l	SC	04065

Uses   Scab in APPLES, PEARS.

Notes   **Efficacy**
- Apply as protective spray at bud-burst and repeat every 10-14 d until danger of scab infection ceases
- Application at high rate within 48 h of a Mills period prevents new infection
- Spray programme also reduces summer infection with apple canker

Crop safety/Restrictions
- Do not use on Golden Delicious apples after green cluster
- Do not mix with lime sulphur or highly alkaline products [2]

Special precautions/Environmental safety
- Harmful if swallowed. Irritating to eyes and skin

Protective clothing/Label precautions
- 5c, 6a, 6b, 18, 21, 28, 29, 36, 37, 54, 60, 63, 66, 70, 78

Harvest interval
- zero

FOR FULL CONDITIONS OF USE ALWAYS READ THE PRODUCT LABEL

diuron

A residual urea herbicide for non-crop areas and woody crops

See also amitrole + atrazine + diuron
           amitrole + bromacil + diuron
           amitrole + 2,4-D + diuron
           amitrole + 2,4-D + diuron + simazine
           amitrole + diuron
           atrazine + bromacil + diuron
           bromacil + diuron
           chlorpropham + diuron
           chlorpropham + diuron + propham
           2,4-D + dalapon + diuron

Products

1	Campbell's Diuron 80	MTM Agrochem.	80% w/w	WP	04708
2	Chipman Diuron Flowable	Chipman	500 g/l	SC	04132
3	Diuron 80	Staveley	80% w/w	WP	00730
4	Diuron 50 FL	Staveley	500 g/l	SC	02814
5	Hoechst Diuron	Promark	80% w/w	WP	01056
6	Karmex	Du Pont	80% w/w	WP	01128
7	MSS Diuron 50FL	Mirfield	500 g/l	SC	04786
8	MSS Diuron 80WP	Mirfield	80% w/w	WP	04785
9	Promark Diuron	Promark	80% w/w	WP	03906
10	Unicrop Flowable Diuron	Unicrop	500 g/l	SC	02270

Uses

Total vegetation control in NON-CROP AREAS [1-5, 7-10]. Annual dicotyledons, annual grasses in APPLES, PEARS [5, 6, 9, 10]. Annual dicotyledons, annual grasses in ASPARAGUS, BLACKCURRANTS, GOOSEBERRIES [5, 9]. Annual dicotyledons, annual grasses in ESTABLISHED TREES AND SHRUBS, WOODY NURSERY STOCK [3].

Notes

Efficacy
- Best results when applied to moist soil and rain falls soon afterwards
- Length of residual activity may be reduced on heavy or highly organic soils
- Apply at low rates for selective, high rates for non-selective control
- Selective rates must be applied to weed-free soil and activity persists for 2-3 mth
- Application for total vegetation control may be at any time of year, best results obtained in late winter to early spring
- Application to frozen ground not recommended

Crop safety/Restrictions
- Apply to weed-free soil in apple and pear orchards established for at least 1 yr (4 yr [5, 9]) during Feb-Mar
- Apply to weed-free soil in blackcurrants and gooseberries established for at least 1 yr in early spring, or as split application in Oct-Nov and before bud-burst
- Apply to established asparagus before crop emergence
- Do not use on sands, very light or gravelly soils or where less than 1% organic matter
- Do not apply to areas intended for replanting during next 12 mth (2 yr for vegetables)
- Do not apply non-selective rates on or near desirable plants where chemical may be washed into contact with roots

Special precautions/Environmental safety
- Irritating to eyes, skin and respiratory system

• Harmful in contact with skin and if swallowed [1]

Protective clothing/Label precautions
• A, C [2, 4, 5, 6, 9, 10]; A, D, E [1]
• 5a, 5c, 78 [1]; 6a, 6b, 6c, 16, 18, 22, 28, 29, 36, 37, 53, 63, 67 [1-9]

Harvest interval
• apples, pears, blackcurrants, gooseberries, asparagus 60 d [5, 9]

260   diuron + paraquat
A total herbicide with contact and residual activity

Products					
	Dexuron	Chipman	300:100 g/l	SC	00689

Uses   Annual dicotyledons, annual grasses, perennial dicotyledons, perennial grasses in ESTABLISHED TREES AND SHRUBS, WOODY NURSERY STOCK. Total vegetation control in NON-CROP AREAS, PATHS.

Notes   **Efficacy**
• Apply to emerged weeds at any time of year. Best results achieved by application in spring or early summer
• Effectiveness not reduced by rain soon after treatment

Crop safety/Restrictions
• Avoid contact of spray with green bark, buds or foliage of desirable trees or shrubs
• In nurseries use low dose rate as inter-row spray, not more than one spray per year

Special precautions/Environmental safety
• A chemical subject to the Poisons Rules 1982 and Poisons Act 1972
• Toxic if swallowed
• Harmful in contact with skin. Irritating to eyes and skin
• Keep livestock out of treated areas for at least 2 wk and until foliage of any poisonous weeds such as ragwort has died and become unpalatable
• Harmful to fish. Do not contaminate ponds, waterways or ditches with chemical or used container

Protective clothing/Label precautions
• A, C
• 4c, 5a, 6a, 6b, 14, 16, 18, 21, 28, 29, 36, 43, 53, 64, 66, 70, 79

261   diuron + simazine
A residual herbicide mixture for non-crop areas and woody crops

Products					
	Simfix Granules	RP Environ.	4.12:2.88% w/w	GR	04425

FOR FULL CONDITIONS OF USE ALWAYS READ THE PRODUCT LABEL

Annual dicotyledons, annual grasses in ROSES, ORNAMENTAL TREES AND SHRUBS. Total vegetation control in NON-CROP AREAS.

Efficacy
- Best results achieved when applied to fine, firm, moist soil, free of established weeds from Oct to Apr
- Effect of treatment reduced in soils with high content of organic matter or carbon

Crop safety/Restrictions
- Use as selective application on shrubs, roses and conifers established for at least 1 yr
- Avoid application to green bark, buds or foliage
- Do not apply around Acers, Betula, Deutzia, Euonymus, Forsythia, Fraxinus, Ligustrum, Lonicera, Philadelphus, Prunus, Senecio, Spirea or Viburnum
- When used for total vegetation control do not plant treated areas for 1.5 - 3.5 yr, depending on dose
- Heavy rain after application on a slope may cause surface run-off

Special precautions/Environmental safety
- Harmful to fish. Do not contaminate ponds, waterways or ditches with chemical or used container

Protective clothing/Label precautions
- 29, 34, 36, 37, 53, 63, 65, 70

262 DNOC

A dinitro insecticide approvals for supply of which were revoked in December 1988 and for use and storage in December 1989

263 dodecylbenzyl trimethylammonium chloride

A quaternary ammonium algicide for use in glasshouses

Products   Country Fresh Disinfectant   Dimex   150 g/l   SL   04443

Uses   Algae in CAPILLARY BENCHES, SAND BEDS. Prevention of rooting through in POT PLANTS.

Notes   **Efficacy**
- Apply to beds and benches before standing out and repeat half-way through season

Crop safety/Restrictions
- Pots should be removed from beds and benches before spraying or spray should be directed at base of pot only

Special precautions/Environmental safety
- Irritating to eyes and skin
- Harmful to fish. Do not contaminate ponds, waterways or ditches with chemical or used container

Protective clothing/Label precautions
• A, C, H
• 6a, 6b, 53

264 dodemorph

A systemic morpholine fungicide for powdery mildew control

Products F238   BASF   400 g/l   EC   0020€

Uses Powdery mildew in ROSES. Powdery mildew in HARDY ORNAMENTAL NURSERY STOCK, NURSERY FRUIT TREES AND BUSHES *(off-label)*.

Notes **Efficacy**
• Spray roses every 10-14 d during mildew period or every 7 d and at increased dose if cleaning up established infection or if disease pressure high
• Add Citowett when treating rose varieties which are difficult to wet
• Tank-mix with Bavistin/Bavistin DF/Bavistin FL for blackspot control
• Product has negligible effect on Phytoseiulus spp being used to control red spider mites

Crop safety/Restrictions
• Do not use on seedling roses
• Do not apply to roses under hot, sunny conditions, particularly roses under glass, but spray early in the morning or during the evening
• Increase the humidity some hours before spraying roses under glass

Special precautions/Environmental safety
• Harmful in contact with skin. Irritating to eyes, skin and respiratory system
• Flammable
• Harmful to fish. Do not contaminate ponds, waterways or ditches with chemical or used container

Protective clothing/Label precautions
• A, C
• 5a, 6a, 6b, 6c, 12c, 16, 18, 21, 25, 28, 29, 36, 37, 53, 60, 63, 66, 78

Approval
• Off-label Approval to April 1992 for use on ornamental nursery stock and nursery fruit trees and bushes (OLA 0362/89)

265 dodine

An aliphatic nitrogen protectant and eradicant fungicide

Products
1 Dodine FL	Truchem	450 g/l	SC	00737
2 Dodine WP	Truchem	65% w/w	WP	00738
3 Radspor FL	Truchem	450 g/l	SC	01685

FOR FULL CONDITIONS OF USE ALWAYS READ THE PRODUCT LABEL

| 4 Radspor 65 WP | Truchem | 65% w/w | WP | 01684 |

Uses Scab in APPLES, PEARS [1-4]. Currant leaf spot in BLACKCURRANTS, GOOSEBERRIES [1-4].

Notes **Efficacy**
- Apply protective spray on apples and pears at bud-burst and at 10-14 d intervals until late Jun to early Jul
- Apply post-infection spray within 36 h of rain responsible for initiating infection. Where scab already present spray prevents production of spores
- Apply to blackcurrants immediately after flowering, repeat every 10-14 d to within 1 mth of harvest and once or twice post-harvest

Crop safety/Restrictions
- Do not apply in very cold weather (under 5°C) or under slow drying conditions to pears or dessert apples during bloom or immediately after petal fall
- Do not spray Conference pears [2, 4]. Do not mix with lime sulphur or tetradifon

Special precautions/Environmental safety
- Harmful if swallowed [1, 3]
- Irritating to skin, eyes and respiratory system [1-4]

Protective clothing/Label precautions
- A, C
- 5c [1, 3]; 6a, 6b, 6c, 18, 22, 29, 36, 37, 54, 63, 65, 70, 78 [1-4]

Harvest interval
- blackcurrants 1 mth [1]

266   drazoxolon
An oxazole fungicide seed treatment also active on powdery mildew

Products | Mil-Col 30 | ICI | 300 g/l | LI | 01352 |

Uses Damping off, fusarium in FIELD BEANS, BROAD BEANS, DWARF BEANS, NAVY BEANS, RUNNER BEANS, PEAS, GRASS SEED. Powdery mildew in HARDY ORNAMENTAL NURSERY STOCK *(off-label)*. Root rot in EDIBLE PODDED PEAS *(off-label)*.

Notes **Efficacy**
- Apply to seed as liquid seed treatment using Plantector seed treater

Special precautions/Environmental safety
- Use of this product as a seed treatment is permitted only in premises registered under the Factories Act
- Harmful if swallowed
- This substance is poisonous. Inhalation of vapour, mist, spray or dust may have harmful consequences. It may also be dangerous to let it come into contact with skin or clothing
- Treated seed not to be used as food or feed. Do not re-use sack
- Dangerous to game, wild birds and animals. Bury spillages

- Harmful to fish. Do not contaminate ponds, waterways or ditches with chemical or used container

Protective clothing/Label precautions
- A, C, H
- 5c, 14, 16, 18, 21, 25, 28, 29, 35, 36, 37, 40, 45, 53, 60, 63, 65, 70, 72, 73, 74, 75, 76, 77, 78

Withholding period
- Dangerous to livestock. Keep all livestock out of treated areas for at least 4 wk

Approval
- Off-label Approval to April 1991 for use on hardy ornamental nursery stock (OLA 0351/89); to April 1992 for use as seed treatment on edible podded peas (OLA 0352/89)

---

267   endosulfan
A contact and ingested cyclodiene insecticide and acaricide

Products					
1 Thiodan 20	Promark		200 g/l	EC	02122
2 Thiodan 35EC	Promark	`	350 g/l	EC	02123

Uses

Pollen beetles, pod midge, cabbage seed weevil in SPRING OILSEED RAPE, MUSTARD [2]. Pollen beetles in WINTER OILSEED RAPE [2]. Big-bud mite in BLACKCURRANTS [1]. Blackberry mite in BLACKBERRIES [1]. Tarsonemid mites in STRAWBERRIES [1]. Damson-hop aphid in HOPS [2]. Bulb scale mite in NARCISSI [1]. Western flower thrips in NON-EDIBLE GLASSHOUSE ORNAMENTALS *(off-label)* [2]. Colorado beetle in POTATOES *(off-label)* [2]. Insect pests in POT PLANTS, BEDDING PLANTS, GLASSHOUSE CUT FLOWERS *(off-label)* [2].

Notes

Efficacy
- Adjust spray volume to achieve total cover. See label for minimum dilutions
- On blackcurrants apply 3 sprays, at first flower, end of flowering and fruit set [1]
- On blackberries apply 3 sprays at 14 d intervals before flowering [1]
- On strawberries apply immediately after whole crop picked, where a second crop to be picked in autumn within 1 wk of mowing old foliage [1]
- Apply as drench to boxed narcisi a few days after bringing into greenhouse [1]
- On hops apply up to 6 sprays at 10-14 d intervals [2]
- On oilseed rape and mustard apply 1 or 2 sprays. See label for timing [2]
- On glasshouse ornamentals apply up to 8 sprays [2]

Special precautions/Environmental safety
- A chemical subject to the Poisons Rules 1982 and Poisons Act 1972
- Toxic if swallowed. Harmful in contact with skin
- Flammable
- Keep unprotected persons out of treated areas for at least 1 d

---

FOR FULL CONDITIONS OF USE ALWAYS READ THE PRODUCT LABEL

- Hops must not be handled for at least 4 d after spraying unless wearing rubber gloves
- Disbudding must not take place until at least 24 h after treatment
- Harmful to bees. Do not apply at flowering stage. Keep down flowering weeds. To minimize risk when using on oilseed rape and mustard remove hives and warn local beekeepers
- Dangerous to fish. Do not contaminate ponds, waterways or ditches with chemical or used container

Protective clothing/Label precautions
- A, C, H, J, K, L, M
- 4c, 5a, 12c, 14, 16, 18, 21, 25, 28, 29, 35, 36, 37, 38, 40, 50, 52, 64, 66, 70, 79

Withholding period
- Dangerous to livestock. Keep all livestock out of treated areas for at least 3 wk

Harvest interval
- blackcurrants, blackberries, strawberries 6 wk; hops, oilseed rape, mustard 3 wk; ornamental crops 24 h before intended sale

Approval
- Off-label approval to Feb 1991 for use on non-edible glasshouse crops (OLA 1393/88) [2]; to Nov 1991 for use on potatoes for Colorado beetle control only as required by Statutory Notice (OLA 1367/88) [2]; to Jan 1992 for use on pot plants, bedding plants and plants for flower production (OLA 0007/89) [2]

---

268 EPTC

A residual pre-planting thiocarbamate herbicide for use in potatoes

Products Eptam 6E Farm Protection 720 g/l EC 03944

Uses Annual grasses, couch, fat-hen, chickweed, speedwells in POTATOES.

Notes **Efficacy**
- Apply as pre-planting incorporated treatment immediately before planting
- For couch control cultivate thoroughly during 2 wk before spraying to chop up rhizomes and stimulate growth. Land should not have been ploughed deeper than 15 cm
- Incorporation to 15 cm must be done within 15 min of spraying using a suitable rotary cultivator or tandem disc harrow. See label for details
- Do not use on soils with more than 10% organic matter, nor on cold, wet soils

Crop safety/Restrictions
- Plant potatoes immediately after or within 2 wk of incorporation. Planting on flat recommended with final ridging 3-4 wk later
- Any crop may be sown or planted after treated crop has been harvested

Special precautions/Environmental safety
- Harmful if swallowed. Irritating to skin and eyes
- Harmful to fish. Do not contaminate ponds, waterways or ditches with chemical or used container

Protective clothing/Label precautions
- A, C

269 ethirimol

A pyrimidine fungicide available only in mixtures

270 ethirimol + flutriafol + thiabendazole

A systemic, non-mercurial fungicide seed treatment for barley

Products					
	1 Ferrax	Bayer	400:30:10 g/l	FS	02827
	2 Ferrax	ICI	400:30:10 g/l	FS	02827
	3 Ferrax	DowElanco	400:30:10 g/l	FS	03885

Uses

Loose smut, covered smut, seedling blight and foot rot, leaf stripe, net blotch, powdery mildew, leaf blotch, rhynchosporium, yellow rust, brown rust in BARLEY.

Notes

Efficacy
• Apply with suitable seed treatment machinery. See label for details
• Controls early attacks of seed, soil and air-borne diseases. Additional treatment may be required later
• Recalibrate drill for treated seed
• Do not use treated seed on soils with more than 20% organic matter
• Disease control may be reduced under dry conditions
• May be mixed with gamma-HCH for wireworm control provided both seed dressings applied together. Do not apply to seed already treated with another seed treatment

Crop safety/Restrictions
• Do not apply to seed with more than 16% moisture
• Emergence of seed may be delayed, especially under poor germination conditions
• Treated seed should be stored in cool, well ventilated conditions and drilled as soon as possible. Test germination if stored until next season

Special precautions/Environmental safety
• Irritating to eyes and skin
• Treated seed not to be used as food or feed. Do not re-use sack
• Harmful to game, wild birds and animals. Bury spillages
• Harmful to fish. Do not contaminate ponds, waterways or ditches with chemical or used container

Protective clothing/Label precautions
• A, C
• 6a, 6b, 14, 18, 21, 28, 36, 37, 46, 53, 63, 65, 70, 72, 73, 74, 75, 76, 77, 78

FOR FULL CONDITIONS OF USE ALWAYS READ THE PRODUCT LABEL

71   ethofumesate

A herbicide for grass weed control in various crops

See also bromoxynil + ethofumesate + ioxynil
       chloridazon + ethofumesate

ses

Annual dicotyledons, annual meadow grass, blackgrass in SUGAR BEET, FODDER
BEET, MANGOLDS, RED BEET [1, 2]. Annual grasses, blackgrass, volunteer cereals,
chickweed, cleavers in GRASS SEED CROPS, LEYS, ESTABLISHED GRASSLAND
[1, 2]. Annual grasses, chickweed, cleavers in STRAWBERRIES [1]. Annual grasses,
chickweed, cleavers in HORSERADISH *(off-label)* [1].

otes

Efficacy
* May be applied pre- or post-emergence of crop or weeds
* Apply in beet crops in tank mixes with other pre- or post-emergence herbicides.
  Recommendations vary for different mixtures. See label for details
* In grass crops apply to moist soil as soon as possible after sowing or post-emergence
  when crop in active growth, normally mid-Oct to mid-Dec. See label for details
* Volunteer cereals not well controlled pre-emergence, weed grasses should be sprayed
  before fully tillered
* Do not use on soils with more than 10% organic matter
* Grass crops may be sprayed during rain or when wet. Not recommended in very dry
  conditions or prolonged frost
* Do not graze or cut grass for 14 d after, or roll less than 7 d before or after spraying

Crop safety/Restrictions
* Safe timing on beet crops varies with other ingredient of tank mix. See label for details
* May be used in Italian, hybrid and perennial ryegrass, timothy, cocksfoot, meadow
  fescue and tall fescue. Apply pre-emergence to autumn-sown leys, post-emergence after
  2 to 3-leaf stage. See label for details
* Do not use on swards reseeded without ploughing
* Clovers will be killed or severely checked
* In strawberries apply only to Cambridge Favourite established at least one season from
  Oct to Dec. Do not spray on soils classed as sands nor in periods of prolonged frost [1]
* Any crop may be sown 3 mth after application of mixtures in beet crops following
  ploughing, 5 mth after application in grass crops

Special precautions/Environmental safety
* Flammable
* Harmful to fish. Do not contaminate ponds, waterways or ditches with chemical or used
  container

Protective clothing/Label precautions
* 12c, 21, 28, 29, 53, 60, 63, 66

Approval
* Off-label approval to June 1993 for use on horseradish (OLA 0369/90) [1]

272 ethofumesate + phenmedipham
A contact and residual herbicide for beet crops

Products Betanal Tandem Schering 100:80 g/l EC 0385

Uses Annual dicotyledons in SUGAR BEET, FODDER BEET, MANGOLDS.

Notes **Efficacy**
* Best results achieved by repeat applications to cotyledon stage weeds
* Apply up to 3 treatments at 7-10 d intervals on all soil types
* Do not spray wet foliage or if rain imminent
* Spray must be applied low volume. See label for details
* Pre-emergence use recommended in combination with post-emergence treatment

Crop safety/Restrictions
* Apply to sugar beet with at least 2 fully expanded true leaves
* At temperatures above 21°C spray after 5 pm
* Do not apply immediately after frost or if frost expected
* Do not spray crops under stress from wind damage, nutrient deficiency, pest or disease attack etc

Special precautions/Environmental safety
* Irritating to eyes, skin and respiratory system
* Flammable

Protective clothing/Label precautions
* 6a, 6b, 6c, 12c, 18, 21, 28, 29, 36, 37, 53, 58, 60, 63, 66, 70, 78

273 ethoprophos
An organophosphorous nematicide and insecticide

Products Mocap 10G RP 10% w/w GR 04935

Uses Potato cyst nematode, wireworms in POTATOES.

Notes **Efficacy**
* Broadcast shortly before or during final soil preparation with air flow fertilizer spreader or other suitable machine and incorporate immediately to 10-15 cm. See label for details
* Treatment can be applied on all soil types. Control of pests reduced on organic soils
* Effectiveness dependent on soil moisture. Drought after application may reduce control of pests

FOR FULL CONDITIONS OF USE ALWAYS READ THE PRODUCT LABEL

Special precautions/Environmental safety
- This product contains an organophosphorous compound. Do not use if under medical advice not to work with such compounds
- Harmful by inhalation and in contact with skin
- Dangerous to game, wild birds and animals
- Dangerous to fish. Do not contaminate ponds, waterways or ditches with chemical or used container

Protective clothing/Label precautions
- A, B, C, C or D + E, H, K, M
- 1, 5a, 5b, 14, 16, 18, 25, 28, 29, 36, 37, 40, 45, 52, 63, 67, 70, 78

Withholding period
- Dangerous to livestock. Keep all livestock out of treated areas for at least 2 wk. Bury or remove spillages

Harvest interval
- 10 wk

---

274   etridiazole

A protective thiazole fungicide for incorporation into soil or compost

Products   Aaterra WP   ICI   35% w/w   WP   03795

Uses   Damping off in VEGETABLE SEEDLINGS, VEGETABLE TRANSPLANTS, TOMATOES, PEPPERS, CUCUMBERS. Phytophthora in CONTAINER-GROWN STOCK, HARDY ORNAMENTAL NURSERY STOCK. Phytophthora, pythium in TULIPS. Red core in STRAWBERRIES *(Scotland only)*. Root diseases in TOMATOES GROWN ON ROCK WOOL *(off-label)*.

Notes   **Efficacy**
- Best results obtained when incorporated thoroughly into soil or compost. Drench application also recommended, especially for strawberries
- Do not apply to wet soil
- Treat compost as soon as possible before use

Crop safety/Restrictions
- After drenching wash spray residue from crop foliage
- Do not use on escallonia, pyracantha, gloxinia spp., pansies or lettuces
- Germination of lettuce in previously treated soil may be impaired
- Do not drench seedlings until well established
- When treating compost for blocking reduce dose by 50%
- Test on small numbers of plants in advance when treating subjects of unknown susceptibility or using compost with more than 20% inert material
- Apply only to spring planted strawberries established for at least 3 mth

Special precautions/Environmental safety
- Irritating to eyes and skin

Protective clothing/Label precautions
- A, C or E + F
- 6a, 6b, 14, 16, 18, 22, 28, 29, 36, 37, 54, 63, 67, 70

Harvest interval
● tomatoes, peppers, cucumbers, mustard and cress 3 d; strawberries 6 mth

Approval
● Off-label Approval to Feb 1991 for use on tomatoes grown on rock wool (OLA 0165/88)

275 etrimfos
A contact organophosphorus insecticide for stored grain crops

Products					
	1 Satisfar	Nickerson	525 g/l	EC	04180
	2 Satisfar Dust	Nickerson	2% w/w	DP	04085

Uses Grain storage pests, grain beetles, grain weevils, grain storage mites in STORED GRAIN, STORED OILSEED RAPE, GRAIN STORES.

Notes **Efficacy**
● Spray internal surfaces of clean store with knapsack or motorized sprayer
● Allow sufficient time for pests to emerge from hiding places and make contact with chemical before grain is stored
● Apply as admixture treatment to grain as it enters store using a suitable stored grain sprayer or automatic seed treater if using dust
● Controls malathion-resistant beetles and gamma-HCH-resistant mites

Crop safety/Restrictions
● Cool grain to below 15°C before treatment and storage. Moisture content of grain should not exceed 16%

Special precautions/Environmental safety
● This product contains an organophosphorus compound. Do not use if under medical advice not to work with such compounds
● Harmful to game, wild birds and animals
● Harmful to livestock. Keep all livestock out of treated areas for at least 7 d
● Dangerous to fish. Do not contaminate ponds, waterways or ditches with chemical or used container

Protective clothing/Label precautions
● A, C
● 22, 28, 29, 41, 46, 52, 63, 67

276 fatty acids
A soap concentrate insecticide and moss killer

Products					
	1 Koppert Moss Killer	Koppert		SL	03568
	2 Savona	Koppert	50.5% w/w	SL	03569

FOR FULL CONDITIONS OF USE ALWAYS READ THE PRODUCT LABEL

Mosses in TURF [1]. Whitefly, mealy bugs, scale insects, aphids, woolly aphid, thrips in BRUSSELS SPROUTS, CABBAGES, BEANS, PEAS, LETTUCE, FRUIT TREES, WOODY ORNAMENTALS, AUBERGINES, CUCUMBERS, PEPPERS, PUMPKINS, TOMATOES, POT PLANTS [2].

Efficacy
- Use only soft or rain water for diluting spray
- For turf use apply when moss in active growth
- For glasshouse use apply when insects first seen and repeat as necessary
- To control whitefly spray when required and use biological control after 12 h

Crop safety/Restrictions
- Do not use on new transplants, newly rooted cuttings or plants under stress

Special precautions/Environmental safety
- Harmful to fish. Do not contaminate ponds, waterways or ditches with chemical or used container

Protective clothing/Label precautions
- 29, 53, 63, 65

Harvest interval
- zero

277   fenarimol

A systemic, curative and protective pyrimidine fungicide

Products   Rubigan   DowElanco   120 g/l   SC   02926

Uses   American gooseberry mildew in BLACKCURRANTS, GOOSEBERRIES. Powdery mildew in APPLES, RASPBERRIES, STRAWBERRIES, GRAPEVINES, ORNAMENTALS, ROSES, CUCUMBERS. Scab in APPLES, PEARS. Powdery mildew in PROTECTED TOMATOES *(off-label)*.

Notes   **Efficacy**
- Recommended spray interval varies from 7-14 d depending on crop and climatic conditions. See label for timing details
- Efficient coverage and short spray intervals essential, especially for scab control
- Mildew spray programme also controls low to moderate levels of blackcurrant leaf spot
- Spray strawberry runner beds and vine nurseries regularly throughout season
- May be used on cucumbers where Phytoseiulus and Encarsia are being used for biological control

Crop safety/Restrictions
- Do not use on trees that are under stress from drought, severe pest damage, mineral deficiency or poor soil conditions
- Test treat small numbers of ornamentals of unknown susceptibility in advance

Special precautions/Environmental safety
- Dangerous to fish. Do not contaminate ponds, waterways or ditches with chemical or used container

Protective clothing/Label precautions
• A, C
• 21, 25, 28, 29, 35, 37, 52, 63, 66

Harvest interval
• tomatoes 2 d; other crops 14 d

Approval
• Off-label approval to Jan 1993 for use on protected tomatoes (OLA 0305/90)

278   fenbutatin oxide

A selective contact and ingested organotin acaricide

Products   Torque   ICI   50% w/w   WP   03303

Uses   Red spider mites, straw mites in PROTECTED CROPS, CUCUMBERS, TOMATOES, ORNAMENTALS, STRAWBERRIES.

Notes

Efficacy
• Active on larvae and adult mites. Spray may take 7-10 d to effect complete kill but mites cease feeding and crop damage stops almost immediately
• Apply as soon as mites first appear on glasshouse crops and repeat as necessary, normally every 10 d or more
• On tunnel-grown strawberries apply when mites first appear, usually before flowering starts, and repeat 10-14 d later. A post-harvest spray is also recommended
• Not harmful to bees or to Encarsia or Phytoseiulus being used for biological control

Crop safety/Restrictions
• Do not add wetters or mix with anything other than water
• Do not use within 28 d of applying white petroleum oil or within 7 d of any other pesticide. Do not apply to crops which are under stress for any reason
• On subjects of unknown susceptibility test treat on a small number of plants in advance

Special precautions/Environmental safety
• Irritating to eyes, skin and respiratory system
• Extremely dangerous to fish. Do not contaminate ponds, waterways or ditches with chemical or used container

Protective clothing/Label precautions
• A, C, D, H, M
• 6a, 6b, 6c, 18, 21, 28, 29, 36, 37, 51, 63, 70, 78

Harvest interval
• glasshouse cucumber, tomatoes, ornamentals 3 d; tunnel-grown strawberries 7 d

---

FOR FULL CONDITIONS OF USE ALWAYS READ THE PRODUCT LABEL

fenitrothion

A broad spectrum, contact organophosphorus insecticide

Products					
1 Cooper Fenitrothion WP	Wellcome	25% w/w	WP	H2366
2 Dicofen	PBI	500 g/l	EC	00693
3 Micromite	Micro-Bio	500 g/l	EC	H2639
4 Unicrop Fenitrothion 50	Unicrop	500 g/l	EC	02267

Uses

Leatherjackets, saddle gall midge in CEREALS [2, 4]. Wheat-blossom midges in WHEAT [2]. Thrips in BARLEY, WHEAT [2]. Frit fly in MAIZE [2, 4]. Frit fly in SWEETCORN [2]. Flour beetles, grain weevils, grain beetles in GRAIN STORES [1, 2]. Aphids, pea and bean weevils, pea moth, pea midge, thrips in PEAS [2, 4]. Aphids, apple blossom weevil, capsids, codling moth, sawflies, suckers, tortrix moths, winter moth in APPLES [2, 4]. Common green capsid in PEARS [4]. Aphids, caterpillars in PEARS, PLUMS [2]. Aphids, sawflies, winter moth in PLUMS, CHERRIES [4]. Capsids, sawflies in BLACKCURRANTS, GOOSEBERRIES [2, 4]. Raspberry beetle, raspberry cane midge in RASPBERRIES, BLACKBERRIES [4]. Raspberry beetle, raspberry cane midge in RASPBERRIES [2]. Raspberry cane midge in LOGANBERRIES [4]. Tortrix moths in STRAWBERRIES [2, 4]. Poultry house pests in POULTRY HOUSES [3]. Thrips in LEEKS, SALAD ONIONS, SWEETCORN *(off-label)* [2].

Notes

Efficacy
- Number and timing of sprays vary with disease and crop. See label for details
- Apply as bran bait for leatherjacket control in cereals
- For cane midge add suitable spreader and apply to lower 60 cm of young canes
- For control of grain store pests apply to all surfaces of empty stores before filling with grain. Remove dust and debris before applying [1, 2]
- For use in poultry houses apply as band spray after depopulation [3]

Crop safety/Restrictions
- Do not mix with magnesium sulphate or highly alkaline materials [2]
- On raspberries may cause slight yellowing of leaves of some varieties which closely resembles that caused by certain viruses [2]

Special precautions/Environmental safety
- This product contains an organophosphorus compound. Do not use if under medical advice not to work with such compounds
- Harmful in contact with skin or if swallowed [2-4]. Irritating to eyes and skin [2, 4]
- Flammable
- Harmful to game, wild birds and animals
- Harmful to fish. Do not contaminate ponds, waterways or ditches with chemical or used container
- Dangerous to bees. Do not apply at flowering stage. Keep down flowering weeds

Protective clothing/Label precautions
- A, C
- 1, 5a, 5c, 12c, 14, 18, 21, 25, 26, 27, 28, 29, 35, 36, 37, 41, 46, 48, 53, 63, 65, 70, 78 [1-4]; 6a, 6b [2, 4]

Withholding period
- Harmful to livestock. Keep all livestock out of treated areas for at least 7 d

Harvest interval
• raspberries 7 d; other crops 2 wk

Approval
• Approved for aerial application on cereals, peas. See introductory notes [2]
• Off-label approval to Feb 1991 for use on leeks, salad onions, sweetcorn (OLA 0166/88) [2]

280   fenitrothion + permethrin + resmethrin
An organophosphate/pyrethroid insecticide mixture for grain stores

Products	Turbair Grain Store Insecticide	PBI	10:20:2 g/l	UL	02238

Uses   Flour beetles, grain beetles, grain weevils, grain moths in GRAIN STORES.

Notes   **Efficacy**
• Apply as a surface spray in empty stores using a suitable fan-assisted ULV sprayer
• Clean store thoroughly before applying
• Combines knock-down effect with up to 5 mth residual activity
• Should not be mixed with other sprays

Special precautions/Environmental safety
• This product contains an organophosphorus compound. Do not use if under medical advice not to work with such compounds
• Irritating to eyes, skin and respiratory system
• Highly flammable
• Extremely dangerous to fish. Do not contaminate ponds, waterways or ditches with chemical or used container

Protective clothing/Label precautions
• A, C
• 1, 6a, 6b, 6c, 12b, 18, 21, 28, 29, 36, 37, 51, 63, 65, 70

Approval
• Product formulated for application as a ULV spray. See label for details. See introductory notes on ULV application

281   fenoxaprop-ethyl
A phenoxypropionic herbicide for grass weed control in wheat

Products	Cheetah R	Hoechst	60 g/l	EW	04932

Uses   Blackgrass, wild oats, rough meadow grass in WINTER WHEAT, SPRING WHEAT.

FOR FULL CONDITIONS OF USE ALWAYS READ THE PRODUCT LABEL

Efficacy
- Apply post-weed emergence from 2-leaf to flag leaf ligule visible stage of weeds
- Susceptible weeds stop growing within 3 d of spraying and die usually within 2-4 wk
- Spray is rainfast 1 h after application
- Do not spray onto wet or iced foliage

Crop safety/Restrictions
- Do not apply after flag leaf visible stage of crop (GS 39)
- Do not mix with products containing bifenox or hormone herbicides
- Do not roll or harrow within 1 wk of spraying
- Do not spray crops under stress, crops suffering from drought, waterlogging or nutrient deficiency or those which have been grazed or if soil compacted

Special precautions/Environmental safety
- Not harmful or irritant
- Dangerous to fish. Do not contaminate ponds, waterways or ditches with chemical or used container

Protective clothing/Label precautions
- A, C
- 52 (no further information)

282 fenpropathrin
A contact and ingested pyrethroid acaricide and insecticide

Products Meothrin   Shell   100 g/l   EC   04993

Uses Red spider mites, caterpillars in APPLES, HOPS. Two-spotted spider mite in STRAWBERRIES. Two-spotted spider mite, damson-hop aphid in HOPS. Two-spotted spider mite in ROSES.

Notes **Efficacy**
- Acts on motile stages of mites and gives rapid kill
- Apply on apples as post-blossom spray and repeat 3-4 wk later if necessary
- Apply on hops when hatch of spring eggs complete in May and repeat if necessary
- Apply on roses when pest first seen and repeat as necessary
- Apply on strawberries as a pre-flowering spray and repeat after harvest if necessary

Crop safety/Restrictions
- Do not apply more than 2 sprays per crop on apples and hops, 1 pre-flowering and 1 post harvest for strawberries

Special precautions/Environmental safety
- Toxic in contact with skin or if swallowed. Irritating to eyes and skin
- Flammable
- Extremely dangerous to bees. Do not apply at flowering stage, keep down flowering weeds
- Extremely dangerous to fish. Do not contaminate ponds, waterways or ditches with chemical or used container

- Do not operate air-assisted sprayers within 80 m of surface water or ditches, other wheeled sprayers within 6 m, hand-held sprayers within 2 m and direct spray away from water

Protective clothing/Label precautions
- A, C, H, J
- 4a, 4c, 6a, 6b, 12c, 14, 16, 18, 21, 28, 29, 35, 36, 37, 48, 51, 64, 65, 70, 79

Harvest interval
- apples, hops 7d; strawberries latest application pre-flowering

283 fenpropidin
A systemic, curative and protectant piperidine fungicide

Products Patrol   ICI   750 g/l   EC   03317

Uses Powdery mildew, brown rust, yellow rust in BARLEY, WHEAT. Leaf blotch in BARLEY. Leaf spot, glume blotch in WHEAT.

Notes

Efficacy
- Best results obtained when applied at early stage of disease development. See label for details of recommended timing alone and in mixtures
- Disease control enhanced by vapour-phase activity. Control can persist for 4-6 wk
- Apply a maximum of 3 sprays on winter crops, 2 sprays on spring crops
- Alternate with triazole fungicides to discourage build-up of resistance

Crop safety/Restrictions
- May be applied up to and including ear emergence complete (GS 59)

Special precautions/Environmental safety
- Harmful in contact with skin and if swallowed. Irritating to skin
- Risk of serious damage to eyes
- Dangerous to fish. Do not contaminate ponds, waterways or ditches with chemical or used container

Protective clothing/Label precautions
- A, C
- 5a, 5c, 6b, 9, 14, 16, 18, 21, 29, 35, 36, 37, 52, 63, 66, 70, 78

Harvest interval
- cereals 5 wk

FOR FULL CONDITIONS OF USE ALWAYS READ THE PRODUCT LABEL

fenpropimorph

A contact and systemic morpholine fungicide
See also chlorothalonil + fenpropimorph

Products

1 Corbel	BASF	750 g/l	EC	00578
2 Mistral	RP	750 g/l	EC	04582
3 Power Fenpropimorph 750	Power	750 g/l	EC	04610
4 Power Task	Power	750 g/l	EC	04943

Uses

Mildew in WHEAT, BARLEY, OATS [1-4]. Brown rust, yellow rust in WHEAT, BARLEY [1-4]. Brown rust, yellow rust in TRITICALE [1, 2]. Rhynchosporium in BARLEY [1-4]. Mildew in RYE [1]. Rust in FIELD BEANS [1-4]. Ring spot, light leaf spot, alternaria in BRUSSELS SPROUTS, CABBAGES [1]. Rust in LEEKS [1, 2].

Notes

Efficacy
- On cereals spray at start of disease attack. See label for recommended tank mixes
- On field beans a second application may be needed after 2-3 wk
- On Brussels sprouts and cabbages apply in tank mix with Bavistin F at start of disease attack [1]
- Product rainfast after 2 h [1]

Crop safety/Restrictions
- Use up to 2 applications on spring-sown cereals and beans, up to 3 on winter-sown cereals, up to 5 on Brussels sprouts and cabbages
- On leeks up to 6 applications may be made at 14-21 d intervals
- Scorch may occur if applied during frosty weather or in high temperatures [1]

Special precautions/Environmental safety
- Harmful by inhalation. Irritating to eyes, skin and respiratory system
- Dangerous to fish. Do not contaminate ponds, waterways or ditches with chemical or used container

Protective clothing/Label precautions
- A, C
- 5b, 6a, 6b, 6c, 18, 21, 26, 27, 28, 29, 36, 37, 52, 63, 66, 70, 78

Harvest interval
- Brussels sprouts, cabbages 2 wk;leeks 3 wk; cereals, field beans 5 wk

Approval
- Provisional Approval for aerial application on restricted area of cereals and field beans [1, 2]. Contact manufacturer for area allocation

285 fenpropimorph + gamma-HCH + thiram

A combined fungicide and insecticide seed treatment

Products

Lindex-Plus FS	DowElanco	43:545:73 g/l	FS	03934

Uses

Flea beetles in OILSEED RAPE. Alternaria, canker, damping off in OILSEED RAPE.

Efficacy
- Do not treat seed with a moisture content exceeding 9%
- Keep treated seed in a cool, dry place, sow as soon as possible after treatment and do not store for more than 6 mth

Crop safety/Restrictions
- Do not mix with other treatments
- Control may be reduced if strains of fungi tolerant to either fungicide develop
- Treatment may reduce germination, especially if seed is grown, harvested or stored under adverse conditions

Special precautions/Environmental safety
- Toxic if swallowed. Irritating to skin, eyes and respiratory system
- Dangerous to fish. Do not contaminate ponds, waterways or ditches with chemical or used container

Protective clothing/Label precautions
- A, C
- 4c, 6a, 6b, 6c, 18, 21, 29, 36, 37, 52, 60, 61, 63, 65, 70, 79

286   fenpropimorph + prochloraz
A fungicide mixture for late season disease control in cereals

Products   Sprint   Schering   375:225 g/l   EC   03986

Uses   Powdery mildew, yellow rust, brown rust in WINTER WHEAT, SPRING WHEAT, BARLEY. Septoria in WINTER WHEAT, SPRING WHEAT. Net blotch, rhynchosporium in BARLEY.

Notes   **Efficacy**
- In wheat, if disease present, spray as soon as ligule of flag leaf visible (GS 39). If no disease, delay until first signs appear or use as protectant treatment before flowering
- In barley spray when disease appears on new growth or as protective spray when flag leaf ligule just visible (GS 39)

Crop safety/Restrictions
- Do not apply more than twice to any crop. Do not spray later than full ear emergence (GS 59) or less than 5 wk before harvest
- Do not apply to crops suffering from stress

Special precautions/Environmental safety
- Harmful if swallowed
- Flammable
- Dangerous to fish. Do not contaminate ponds, waterways or ditches with chemical or used container

Protective clothing/Label precautions
- A, C

FOR FULL CONDITIONS OF USE ALWAYS READ THE PRODUCT LABEL

287   fentin acetate
An organotin fungicide available only in mixtures

288   fentin acetate + maneb
A curative and protectant fungicide for use in potatoes

Products	1 Brestan 60	Hoechst	54:16% w/w	WP	00325
	2 Hytin	Agrichem	6.77:72% w/w	WP	01101
	3 Trimastan	Pennwalt	11:33% w/w	WP	04181

Uses   Blight in POTATOES [1-3].

Notes   **Efficacy**
- Spray before haulm meets across the rows or at blight warning
- Repeat at 7-14 d intervals up to a maximum of 6 sprays per crop
- Do not spray if rain imminent

Special precautions/Environmental safety
- A chemical subject to the Poisons Rules 1982 and the Poisons Act 1972
- Harmful if swallowed. Irritating to eyes, skin and respiratory system
- Harmful to fish. Do not contaminate ponds, waterways or ditches with chemical or used container

Protective clothing/Label precautions
- A, C, H, J, M
- 5c, 6a, 6b, 6c, 14, 16, 18, 22, 28, 29, 35, 36, 37, 41, 53, 64, 67, 70, 78

Withholding period
- Harmful to livestock. Keep all livestock out of treated areas for at least 1 wk

Harvest interval
- zero

Approval
- Approved for aerial application on potatoes. See introductory notes (1-3)

289   fentin hydroxide
A curative and protectant organotin fungicide

Products	1 Ashlade Flotin	Ashlade	625 g/l	SC	03535
	2 Chiltern Super-Tin 4L	Chiltern	480 g/l	SC	02995
	3 Du-Ter 50	ICI	47.5% w/w	WP	00778
	4 Farmatin 560	Farm Protection	560 g/l	SC	04595
	5 Farmatin FL	Farm Protection	500 g/l	SC	03560
	6 Quadrangle Super-Tin 4L	Quadrangle	480 g/l	SC	03842

Uses	Blight in POTATOES [1-6]. Ramularia leaf spot, phoma leaf spot in SUGAR BEET SEED CROPS [3].

Notes	**Efficacy**

- Apply to potatoes before haulm meets across rows or on receipt of blight warning and repeat at 7-14 d intervals throughout season. Late sprays protect against tuber blight
- Best results on potatoes achieved by using for 2 final sprays following programme with maneb + zinc [2] or cymoxanil + mancozeb [5]
- Apply to sugar beet seed crops in May and repeat at 14-21 d intervals for maximum of 3 applications. Make final application as late in season as possible [3]

Crop safety/Restrictions
- Slight adverse effects sometimes occur on young potato foliage [4, 5]
- Do not mix with oil-based insecticides [3]

Special precautions/Environmental safety
- A chemical subject to the Poisons Rules 1982 and the Poisons Act 1972
- Toxic if swallowed [1], harmful if swallowed [2, 6], harmful in contact with skin, by inhalation or if swallowed [3-5]
- Irritating to eyes and skin [1], to eyes, skin and respiratory system [2, 6]
- Beet crops must not be fed to animals after the seed has been harvested [3]
- Harmful to fish. Do not contaminate ponds, waterways or ditches with chemical or used container

Protective clothing/Label precautions
- A, C, H, J, M [1, 2, 4-6]; A, C, H [3]
- 4c, 6a, 6b [1]; 5c, 6a, 6b, 6c, [2,6]; 5a, 5b, 5c [3-5]; 14, 16, 18, 21, 25, 26, 28, 29, 36, 37, 41, 53, 64, 66, 70, 78 [1-6]

Withholding period
- Harmful to livestock. Keep all livestock out of treated areas for at least 1 wk

Harvest interval
- potatoes zero [3], 1 wk [4, 5]

Approval
- Approved for aerial application on potatoes. See introductory notes [1-6]

290 fentin hydroxide + maneb + zinc
A curative and protectant fungicide for use in potatoes

Products	Chiltern Tinman	Chiltern	60:480:- g/l	SC	03120

Uses	Blight in POTATOES.

Notes	**Efficacy**

- Spray just before haulm meets across rows, normally Jun to early Jul, or earlier if blight warning and repeat every 10-14 d, every 7-10 d under severe blight conditions

FOR FULL CONDITIONS OF USE ALWAYS READ THE PRODUCT LABEL

Special precautions/Environmental safety
- A chemical subject to the Poisons Rules 1982 and Poisons Act 1972
- Harmful if swallowed. Irritating to skin, eyes and respiratory system
- Harmful to fish. Do not contaminate ponds, waterways or ditches with chemical or used container

Protective clothing/Label precautions
- A, C, H, J, M
- 5c, 6a, 6b, 6c, 14, 16, 18, 21, 25, 26, 28, 29, 35, 36, 41, 53, 64, 66, 70, 78

Withholding period
- Harmful to livestock. Keep all livestock out of treated areas for at least 7 d

Harvest interval
- potatoes 7 d

Approval
- Approved for aerial application on potatoes. See introductory notes

---

291   fentin hydroxide + metoxuron
A combined fungicide and haulm desiccant for potatoes

Products   Endspray   PBI   6.7:66.7% w/w   WP   02799

Uses   Blight in POTATOES. Haulm destruction in POTATOES.

Notes   **Efficacy**
- Spray at start of senescence and leave 21-28 d before harvest to obtain maximum haulm kill, proper skin set and tuber protection
- Do not use in crops with heavy infestation of late-germinating weeds

Crop safety/Restrictions
- Unsuitable for crops grown specifically for seed or where haulm is to be destroyed when still actively growing

Special precautions/Environmental safety
- A chemical subject to the Poisons Rules 1982 and the Poisons Act 1972
- Irritating to eyes, skin and respiratory system
- Harmful to fish. Do not contaminate ponds, waterways or ditches with chemical or used container

Protective clothing/Label precautions
- A, C, H
- 6a, 6b, 6c, 16, 18, 22, 28, 29, 35, 36, 37, 41, 53, 64, 67, 70, 78

Withholding period
- Harmful to livestock. Keep all animals out of treated areas for at least 7 d

Harvest interval
- potatoes 2 wk

292   fenuron

A urea herbicide available only in mixtures

See also chloridazon + chlorpropham + fenuron + propham
chloridazon + fenuron + propham
chlorpropham + cresylic acid + fenuron
chlorpropham + fenuron
chlorpropham + fenuron + propham

293   fenvalerate

A contact pyrethroid insecticide for foliar application

Products   Sumicidin   Shell   100 g/l   EC   02568

Uses   Aphids in CEREALS. Cabbage stem flea beetle in OILSEED RAPE. Seed weevil, pod midge, cabbage aphid, pollen beetles in FLOWERING OILSEED RAPE. Aphids, caterpillars in LEAF BRASSICAS. Weevils in FIELD BEANS. Aphids, pea moth, pea and bean weevils in PEAS. Aphids, apple sucker, caterpillars, codling moth, fruitlet mining tortrix, fruit tree tortrix moth, summer-fruit tortrix moth in APPLES. Aphids, caterpillars, pear sucker in PEARS. Damson-hop aphid in HOPS.

Notes   **Efficacy**
- Combines rapid knock-down effect with long lasting control and is highly effective against aphids and caterpillars. Ensure foliage well covered by spray
- Spray for aphid control in winter cereals in high-risk situations in mid-Oct and repeat at end Oct/early Nov, in other situations only use end Oct/early Nov spray
- Dose, timing and number of sprays vary with crop and pest. See label for details
- Use of pheromone trapping records advised to improve accuracy of timing of post-blossom sprays on apples
- Effective against organophosphorus-resistant strains of green aphid in sugar beet and provides suppression of virus yellows
- For damson-hop aphid control on hops consult suppliers for latest information

Crop safety/Restrictions
- Treatment presents minimal hazard to bees but on flowering oilseed rape spray in evening, early morning or in dull weather as a precaution
- Apply up to 6 sprays per crop on vegetable brassicas, 3 per crop on peas, 2 per crop on field beans, winter oilseed rape, wheat and barley

Special precautions/Environmental safety
- Harmful in contact with skin or if swallowed. Irritating to eyes and skin
- Flammable
- Harmful to bees. Do not apply at flowering stage except as directed in oilseed rape and peas. Keep down flowering weeds in all crops
- Extremely dangerous to fish. Do not contaminate ponds, waterways or ditches with chemical or used container

FOR FULL CONDITIONS OF USE ALWAYS READ THE PRODUCT LABEL

Protective clothing/Label precautions
- A, C
- 5a, 5c, 6a, 6b, 12c, 16, 18, 21, 28, 29, 35, 36, 37, 48, 51, 63, 66, 70, 78

Harvest interval
- zero

Approval
- Provisional Approval for aerial application to limited areas of winter wheat, winter barley, winter and spring oilseed rape and peas. Contact supplier for allocation before spraying

---

294   ferbam

A dithiocarbamate fungicide available only in mixtures

---

295   ferbam + maneb + zineb

A protectant dithiocarbamate complex fungicide

Products	Trimanzone	Pennwalt	10:65:10% w/w	WP	04291

Uses

Brown rust, leaf blotch, net blotch, powdery mildew in BARLEY. Brown rust, leaf blotch, glume blotch, powdery mildew in WHEAT. Alternaria in OILSEED RAPE. Blight in POTATOES. Clubroot in BRASSICAS. Rust, leek leaf blotch, white tip in LEEKS. Downy mildew in ONIONS. Fire in TULIPS.

Notes

Efficacy
- Apply before first disease symptoms occur and repeat at 10-14 d intervals
- For use in cereals normally tank-mixed with other fungicides
- Spray potatoes when plants meet in the row and repeat regularly through season
- Apply to soil before planting to reduce incidence of clubroot on brassicas
- For tulip fire rogue out plants showing primary infection before spraying

Crop safety/Restrictions
- Do not use on lettuce

Special precautions/Environmental safety
- Irritating to eyes, skin and respiratory system

Protective clothing/Label precautions
- A
- 6a, 6b, 6c, 18, 22, 29, 35, 36, 37, 54, 63, 66

Harvest interval
- 14 d

Approval
- Approved for aerial application on cereals, oilseed rape, Brussels sprouts, potatoes. See introductory notes

296 ferrous sulphate

A herbicide/fertilizer combination for moss control in turf

See also *chloroxuron + ferrous sulphate*
dichlorophen + ferrous sulphate
dichlorprop + ferrous sulphate + MCPA

Products

1	Elliott's ETB Mosskiller	Elliott	43% w/w	GR	04764
2	Elliott's Lawn Sand	Elliott	17% w/w	SA	04860
3	Elliott's Mosskiller	Elliott	45% w/w	GR	04909
4	Greenmaster Autumn	Fisons	6% w/w	GR	03211
5	Greenmaster Mosskiller	Fisons	8.9% w/w	GR	00881
6	Green-Up Mossfree	Synchemicals	95% w/w	SP	03270
7	Hart Lawn Sand	Maxwell Hart	17% w/w	SA	04861
8	Hart Mosskiller	Maxwell Hart	45% w/w	GR	04930
9	Maxicrop Mosskiller & Conditioner	Maxicrop	16.4% w/w	SL	04635
10	Taylors Lawn Sand	Rigby Taylor	7.35% w/w	SA	04451
11	Vitax Micro Gran 2	Vitax	20% w/w	MG	04541
12	Vitax Turf Tonic	Vitax	15% w/w	GR	04354
13	Walkover Mosskiller	Walkover	16.4% w/w	SL	04662

Uses

Mosses in TURF [1-13]. Algae in TURF [6].

Notes

Efficacy

- Apply Autumn treatment from Sep onwards but not when heavy rain expected or in frosty weather [4]
- Apply from Mar to Sep except during drought or when soil frozen [5]
- Apply in autumn or spring when turf moist and moss actively growing [6]
- For best results apply when light showers expected, mow 3 d before treatment, do not mow or walk on treated area until well watered and water after 2 d if no rain
- Rake out dead moss thoroughly 7-14 d after treatment
- Fertilizer component of Autumn treatment encourages root growth, that of mosskiller formulations promotes tillering

Crop safety/Restrictions

- If spilt on paving, concrete, clothes etc brush off immediately to avoid discolouration

Special precautions/Environmental safety

- Harmful to fish. Do not contaminate ponds, waterways or ditches with chemical or used container [4, 5]

Protective clothing/Label precautions

- 29, 63, 67 [1-13]; 53 [4, 5]; 54 [1-3, 6-13]

FOR FULL CONDITIONS OF USE ALWAYS READ THE PRODUCT LABEL

297   flamprop-M-isopropyl

A translocated, post-emergence arylalanine wild-oat herbicide

Products

	1 Commando	Shell	200 g/l	EC	00564
	2 Gunner	Quadrangle	200 g/l	EC	04547
	3 Power Flame	Power	200 g/l	EC	04892
	4 Power Flamprop	Power	200 g/l	EC	04837

Uses   Wild oats in WHEAT, BARLEY, DURUM WHEAT, RYE, TRITICALE [1, 2]. Wild oats in WHEAT, BARLEY, RYE, TRITICALE [3, 4].

Notes   **Efficacy**
- Must be applied when crop and weeds growing actively under conditions of warm, moist days and warm nights
- Timing determined mainly by crop growth stage. Best control at later stages of wild oats but not after weeds visible above crop
- Best results achieved before 3rd node detectable stage (GS 33) in barley, or 4th node detectable (GS 34) in wheat [1]
- Good spray coverage and retention essential. Spray rainfast after 2 h
- Do not treat thin, open crops
- See label for recommended tank mixes and sequential treatments

Crop safety/Restrictions
- Latest time of application for winter and spring wheat before flag leaf opening (GS 47) [1, 2], before 4th node detectable (GS 34) [3, 4], for winter and spring barley before first awns visible (GS 49) [1, 2], before 3rd node detectable (GS 33) [3, 4], for rye and triticale before 3rd node detectable (GS 33) [1-4]
- May be used on crops undersown with ryegrass and clover [1], not recommended on undersown crops [2-4]
- Do not use on crops under stress or during periods of high temperature
- Do not roll or harrow within 7 d of spraying
- Do not mix with any herbicide other than fluroxypyr

Special precautions/Environmental safety
- Irritating to eyes and skin
- Flammable
- Do not use straw from barley treated after stage GS 33 or wheat treated after stage GS 34 as feed or bedding for animals [1, 2]
- Harmful to fish. Do not contaminate ponds, waterways or ditches with chemical or used container

Protective clothing/Label precautions
- A, C
- 6a, 6b, 12c, 14, 16, 18, 21, 28, 29, 36, 37, 53, 60, 63, 66, 70

Approval
- Approved for aerial application on wheat, barley, rye, triticale. See introductory notes [1]

371

298   fluazifop-P-butyl

A phenoxypropionic grass weed herbicide for broadleaved crops

Products	Fusilade 5	ICI	125 g/l	EC	02833

Uses   Annual grasses, perennial grasses, wild oats, volunteer cereals in SUGAR BEET, FODDER BEET, WINTER OILSEED RAPE, CARROTS, ONIONS, PEAS, BLACKCURRANTS, GOOSEBERRIES, RASPBERRIES, STRAWBERRIES. Annual grasses, perennial grasses, wild oats, volunteer cereals in KALE, SWEDES *(stockfeed only)*. Annual grasses, perennial grasses in PARSNIPS *(off-label)*. Annual grasses, blackgrass, wild oats, volunteer cereals in RED FESCUE SEED CROPS, CLOVERS, TREFOIL, SAINFOIN, LUCERNE *(off-label)*.

Notes   **Efficacy**
- Best results achieved by application when weed growth active under warm conditions with adequate soil moisture. Agral wetter must always be added to spray
- Spray weeds from 2-expanded leaf stage to fully tillered, couch from 4 leaves when majority of shoots have emerged, with a second application if necessary
- Control of perennial grasses may be reduced under dry conditions. Do not cultivate for 2 wk after spraying couch
- Annual meadow grass is not controlled
- May also be used to remove grass cover crops

Crop safety/Restrictions
- Apply to sugar and fodder beet from 1-true leaf to 50% ground cover, but not later than 8 wk before harvest
- Apply to winter oilseed rape from 1-true leaf to established plant, but not after 31 Dec
- Apply in fruit crops after harvest but before flowering. For other crops see label for details of timing
- Before using on onions or peas use crystal violet test to check that leaf wax is sufficient
- Off-label uses on legumes refer to seed crops only
- Do not sow cereals for at least 8 wk after application of high rate, 2 wk after low rate
- Do not apply through CDA sprayer or from air

Special precautions/Environmental safety
- Irritating to eyes and skin
- Harmful to fish. Do not contaminate ponds, waterways or ditches with chemical or used container

Protective clothing/Label precautions
- A, C, H
- 6a, 6b, 18, 21, 29, 36, 37, 53, 63, 66, 70, 78

Harvest interval
- sugar beet, fodder beet carrots 8 wk; onions 4 wk; fruit crops not between flowering and harvest

FOR FULL CONDITIONS OF USE ALWAYS READ THE PRODUCT LABEL

299 fluroxypyr

A post-emergence phenoxy herbicide for use in cereals and grassland

See also *bromoxynil + fluroxypyr*
       bromoxynil + fluroxypyr + ioxynil
       clopyralid + fluroxypyr + ioxynil
       cyanazine + fluroxypyr

Products	Starane 2	DowElanco	200 g/l	EC	03193

Uses

Annual dicotyledons, chickweed, cleavers, black bindweed, hemp-nettle, forget-me-not, docks, volunteer potatoes in WHEAT, BARLEY, OATS, RYE, WINTER TRITICALE, DURUM WHEAT, GRASSLAND.

Notes

Efficacy
- Best results achieved under good growing conditions in a strongly competing crop
- Apply in autumn or spring. Autumn treatment usually needs repeat spring spray
- A number of tank mixtures with HBN and other herbicides are recommended for use in autumn and spring to extend range of species controlled. See label for details
- Spray is rainfast in 1 h
- Do not spray if frost imminent

Crop safety/Restrictions
- Apply to winter wheat and winter barley up to boot swollen stage (GS 45), spring wheat and spring barley to flag leaf ligule visible (GS 39), other cereals to first node detectable (GS 31)
- Timing varies in tank mixtures. See label for details
- Do not use on crops undersown with clovers or other legumes
- Crops undersown with grass may be sprayed provided grasses are tillering
- Do not treat crops suffering stress caused by any factor
- Do not roll or harrow for 7 d before or after treatment

Special precautions/Environmental safety
- Flammable
- Keep livestock out of treated areas for at least 3 d and until the foliage of any poisonous weeds such as ragwort has died and become unpalatable
- Harmful to fish. Do not contaminate ponds, waterways or ditches with chemical or used container

Protective clothing/Label precautions
- 12c, 21, 28, 29, 43, 53, 63, 66

300   flusilazole

A systemic, protectant and curative conazole fungicide for cereals

See also carbendazim + flusilazole

Products	Sanction	Du Pont	400 g/l	EC	04773

Uses   Eyespot, septoria, yellow rust, brown rust, powdery mildew in WINTER WHEAT, WINTER BARLEY, SPRING BARLEY. Rhynchosporium, net blotch in WINTER BARLEY, SPRING BARLEY.

Notes

Efficacy
- Use as a routine preventative spray or when disease first develops
- Best control of eyespot achieved by spraying between leaf-sheath erect and second node detectable stages (GS 30-32)
- Product active against both mbc sensitive and mbc resistant strains of eyespot
- Rain occurring within 2 h after application may reduce effectiveness
- See label for recommended tank-mixes

Crop safety/Restrictions
- May be applied at any time up to grain watery ripe stage (GS 71)
- Do not apply more than 3 sprays of any product containing flusilazole to any crop of winter wheat or more than 2 sprays per crop of winter or spring barley. Do not use high rate more than once in any crops
- Do not apply to crops under stress or during frosty weather

Special precautions/Environmental safety
- Harmful if swallowed
- Flammable
- Dangerous to fish. Do not contaminate ponds, waterways or ditches with chemical or used container

Protective clothing/Label precautions
- A, C
- 5c, 12c, 18, 24, 28, 29, 36, 37, 52, 63, 66, 70, 78

301   flutriafol

A triazole fungicide available only in mixtures

See also carbendazim + flutriafol
         chlorothalonil + flutriafol
         ethirimol + flutriafol + thiabendazole

FOR FULL CONDITIONS OF USE ALWAYS READ THE PRODUCT LABEL

fonofos

An organophosphorus insecticide for soil and seed treatment

1 Cudgel	ICI	433 g/l	CS	04349
2 Dyfonate 10 G	Farm Protection	10% w/w	MG	03945
3 Dyfonate MS	Farm Protection	550 g/l	CS	04336
4 Fonofos Seed Treatment	ICI	433 g/l	CS	00018

Frit fly in WINTER WHEAT, WINTER BARLEY [2, 3, 4]. Wheat bulb fly, wireworms in WINTER WHEAT [2, 4]. Wheat bulb fly, wireworms in WINTER BARLEY [2]. Wheat bulb fly, yellow cereal fly in WINTER CEREALS [3]. Frit fly in RYEGRASS [4]. Cabbage stem flea beetle, cabbage root fly in WINTER OILSEED RAPE [2]. Cabbage root fly in LEAF BRASSICAS, ROOT BRASSICAS [1]. Sciarid flies, vine weevil in ORNAMENTALS [1]. Sciarid flies, vine weevil in HARDY ORNAMENTAL NURSERY STOCK, NURSERY FRUIT TREES AND BUSHES *(off-label)* [1]. Cabbage root fly in CHINESE CABBAGE *(off-label)* [1]. Cabbage root fly in CALABRESE, ROSCOFF CAULIFLOWER *(off-label)* [2].

Efficacy
- Apply granules in winter cereals as overall treatment during final seedbed preparation or at drilling with incorporation to 25-50 mm on day of application [2]Apply spray treatment in cereals at drilling and incorporate or at any time from Jan until egg hatch, depending on pest
- Apply seed treatment with suitable liquid seed treatment machinery [4]
- Apply granules in oilseed rape 4-6 wk after adult beetles first seen, usually mid-Oct or Nov/Dec depending on sowing date, or at drilling as for cereals [2]
- May be used in brassicas as pre- or post-planting treatment
- For cabbage root fly control apply as surface band, sub-surface band, spot treatment, by bow-wave technique or incorporated in peat blocks. See label for details
- Do not tank-mix with any other product [3]
- On ornamentals spray as hygiene treatment, incorporate into compost or drench compost or soil [1]

Crop safety/Restrictions
- Do not drill carrots or parsnips within 12 mth of application to soil
- Do not treat seed with more than 16% moisture
- Treated seed may be stored for up to 1 mth in cool, well ventilated place. Drill seed not more than 40 mm deep
- With drench treatment rinse off foliage immediately, do not drench over foliage under hot, dry conditions or on plants in open flower [1]

Special precautions/Environmental safety
- A chemical subject to the Poisons Rules 1982 and Poisons Act 1972
- This product contains an organophosphorus compound. Do not use if under medical advice not to work with such compounds
- Toxic [4]; harmful in contact with skin, by inhalation or if swallowed [2]
- Harmful in contact with skin [1, 3]
- Dangerous to livestock. Keep all livestock out of treated areas for at least 6 wk. Bury or remove spillages
- Dangerous to game, wild birds and animals [1-4]
- Extremely dangerous [1], dangerous [2-4], to fish. Do not contaminate ponds, waterways or ditches with chemical or used container

- Treated seed not to be used as food or feed. Do not re-use sack. Do not apply treated seed aerially [4]

Protective clothing/Label precautions
- A, C, H, J, K, L, M [1]; A, B, C, C or D + E, H, M [2]; A, C, H, J, K, M [3]; A, C H, K, M [4]
- 1, 14, 16, 18, 22, 25, 28, 29, 36, 37, 45, 64, 65, 70, 78 [1-4]; 5a, 26, 27, 51 [1]; 5a, 5b 5c, 52 [2]; 5a, 26, 27, 52 [3]; 4, 40, 52 [4]

Harvest interval
- 6 wk

Approval
- Off-label Approval to Apr 1992 for use as drench on hardy ornamental nursery stock and nursery fruit trees and bushes (OLA 0374/89) [1]; to Sep 1992 for use as drench or incorporated in compost on Chinese cabbage (OLA 0748/89) [1]; to Mar 1993 for use on calabrese and Roscoff cauliflower (0200/90) [2]

303   formaldehyde

A fumigant fungicide for glasshouses and soil sterilization
See also gamma-HCH + formaldehyde

Products					
1 Dyna-Form	Fargro	35% w/w	HN	00781
2 Formaldehyde	Monro	38% w/w	HN	04504
3 Formaldehyde Solution 40%	BH & B	40% w/w	SL	00919

Uses

Soil-borne diseases in GLASSHOUSE CROPS [1, 2, 3]. Fungus diseases in GLASSHOUSE STRUCTURES AND SURROUNDS [1].

Notes

Efficacy
- Apply as drenching spray for soil sterilization. Keep windows and doors open during application and, after closing, do not re-enter until fumes have cleared
- To disinfect empty glasshouses and mushroom houses apply undiluted through Dyna-Fog thermal fogging machine or through high volume sprayer. Apply from door or opening in glass when surfaces dry [1]

Crop safety/Restrictions
- After fumigation allow at least 3 d before entering structure or planting a crop
- After soil treatment allow 3 wk for light soils, 6 wk for heavy soils before planting
- Do not plant until pungent smell has disappeared from soil, normally about 4 wk
- Do not use near growing plants

Special precautions/Environmental safety
- Toxic if swallowed. Harmful in contact with skin or by inhalation
- Maximum exposure limits apply to this chemical. See HSE Approved Code of Practice for the Control of Substances Hazardous to Health
- Do not enter building while fogging

FOR FULL CONDITIONS OF USE ALWAYS READ THE PRODUCT LABEL

● A, D, E, J, L, M [1]; A [2, 3]
● 4c, 5a, 5c, 21, 24, 28, 29, 36, 37, 54, 60, 63, 65, 79

304 formaldehyde + gamma-HCH
A fumigant insecticide and fungicide for poultry houses

Products	Microgen Plus	Micro-Bio	90% w/w	FU	01340

Uses Poultry diseases in POULTRY HOUSES. Poultry house pests in POULTRY HOUSES.

Notes

Efficacy
● Fumigate by applying in Microgen generators
● Most effective in humid conditions above 18°C. Fumigate after washing down and before introduction of litter
● Close all openings throughout fumigation operation

Special precautions/Environmental safety
● Toxic if swallowed. Harmful in contact with skin or by inhalation
● Do not enter building while fumigating

Protective clothing/Label precautions
● A, D, E, J, L, M
● 4c, 5a, 5c, 21, 28, 29, 54, 63, 65

305 fosamine-ammonium
A contact phosphonate herbicide for woody weed control

Products					
1 Krenite	Du Pont	480 g/l	SL	01165
2 Krenite	Selectokil	480 g/l	SL	01165

Uses Woody weeds in NON-CROP AREAS, FORESTRY, WATERSIDE AREAS.

Notes

Efficacy
● Apply as overall spray to foliage during Aug-Oct when growth ceased but before leaves have started to change colour. Effects develop in following spring
● Thorough coverage of leaves and stems needed for effective control
● Addition of non-ionic wetter recommended. Rain within 24 h may reduce effectiveness
● Most deciduous species controlled or suppressed but evergreens resistant
● Little effect on underlying herbaceous vegetation

Crop safety/Restrictions
● Do not use in conifer plantations

Protective clothing/Label precautions
● 21, 29, 54, 60, 63, 65

• May be used alongside river, canal and reservoir banks but should not be sprayed into water. See introductory notes on use of herbicides in or near water

306 fosetyl-aluminium

A systemic phosphonate fungicide for various horticultural crops
See also captan + fosetyl-aluminium + thiabendazole

Products	Aliette	Embetec	80% w/w	WP	02484

Uses Collar rot, crown rot in APPLES. Downy mildew in BROAD BEANS, HOPS, PROTECTED LETTUCE. Phytophthora in HARDY ORNAMENTAL NURSERY STOCK, GLASSHOUSE CROPS, CAPILLARY BENCHES. Red core in STRAWBERRIES. Downy mildew, ascochyta in VINING PEAS, DRIED PEAS *(off-label)*. Downy mildew, damping off in BRASSICAS *(off-label)*. Downy mildew in LETTUCE *(off-label)*. Red core in STRAWBERRIES *(spring treatment, off-label)*.

Notes **Efficacy**
• Spray young orchards for crown rot protection after blossom and repeat after 3-6 wk. Apply as paste to bark of apples to control collar rot
• Apply to broad beans at early flowering and 14 d later
• Spray autumn-planted strawberries 2-3 wk after planting or use dip treatment at planting. Spray established crops in late summer/early autumn after picking
• Apply to hops as early season band spray and as foliar spray every 10-14 d
• May be used incorporated in blocking compost for protected lettuce
• Apply as drench to rooted cuttings of hardy nursery stock after first potting and repeat monthly. Up to 6 applications may be needed
• See label for details of application to capillary benches and list of tolerant plants

Crop safety/Restrictions
• When using on strawberries do not mix with any other product
• Apply up to 6 foliar sprays per year on hops, up to 2 basal sprays per year
• Use in protected lettuce only from Sep to Apr
• Do not apply to ornamentals in mixture with nutrient solutions
• Apply test treatment on lettuce cultivars or ornamentals of unknown tolerance before committing whole batches

Special precautions/Environmental safety
• Irritating to eyes, skin and respiratory system

Protective clothing/Label precautions
• A, C, H
• 6a, 6b, 6c, 18, 22, 27, 35, 36, 37, 54, 63, 67, 78

Harvest interval
• Hops 4 d; broad beans 17 d; apples 14 d, applied as bark paste 5 mth

FOR FULL CONDITIONS OF USE ALWAYS READ THE PRODUCT LABEL

• Off-label approval to Feb 1991 for use on peas (OLA 0169/88), apples, strawberries and brassicas (OLA 0170/88); to Sep 1991 for use on lettuce (OLA 0982/88)

307 fuberidazole
An MBC fungicide available only in mixtures

308 fuberidazole + imazalil + triadimenol
A systemic fungicide seed dressing for barley

Products	Baytan IM	Bayer	3:3.3:25% w/w	DS	00226

Uses Brown foot rot and ear blight, covered smut, leaf stripe, loose smut, net blotch, mildew, rhynchosporium, yellow rust, brown rust in WINTER BARLEY.

Notes **Efficacy**
• Apply treatment through suitable seed treatment machinery
• Calibrate drill for treated seed and drill at 2.5-4 cm into firm, well prepared seed bed
• Also reduces snow rot, seedling blight and fusarium foot rot

Crop safety/Restrictions
• Treatment may accentuate effects of adverse seedbed conditions on emergence
• Do not use on seed with more than 16% moisture content or on sprouted, cracked or skinned seed. Test germination of all seed batches to ensure suitability for treatment
• Treated seed should be drilled in the season of purchase
• Only compatible with Gammasan 30 and New Kotol

Special precautions/Environmental safety
• Treated seed not to be used as food or feed. Do not re-use sack
• Dangerous to fish. Do not contaminate ponds, waterways or ditches with chemical or used container

Protective clothing/Label precautions
• A
• 21, 29, 52, 60, 63, 67, 71, 73, 74, 76

309 fuberidazole + triadimenol
A broad spectrum systemic fungicide seed treatment for cereals

Products					
Products	1 Baytan	Bayer	3:25% w/w	DS	00225
	2 Baytan	ICI	3:25% w/w	DS	03946
	3 Baytan Flowable	Bayer	22.5:187.5 g/l	FS	02593
	4 Baytan Flowable	ICI	22.5:187.5 g/l	FS	03845

Uses Brown foot rot and ear blight, covered smut, loose smut, mildew, rhynchosporium, yellow rust, brown rust in BARLEY [1-4]. Leaf stripe in SPRING BARLEY [1-4]. Brown foot

rot and ear blight, bunt, loose smut, septoria, mildew, yellow rust, brown rust in
WHEAT [1-4]. Brown foot rot and ear blight, loose smut, pyrenophora leaf spot, mildew
in OATS [1-4]. Brown foot rot and ear blight, stripe smut in RYE [1-4]. Brown foot rot
and ear blight in TRITICALE [1, 3]. Blue mould in WHEAT, TRITICALE [1, 3].

Notes   **Efficacy**
● Apply through suitable seed treatment machinery. See label for details
● Calibrate drill for treated seed and drill at 2.5-4 cm into firm, well prepared seedbed
● In addition to seed-borne diseases early attacks of various foliar, air-borne diseases are
controlled or suppressed. See label for details
● Product may be used in conjunction with Gammasan 30 or Wireworm FS and applied
to seed already treated with New Kotol. No other combinations recommended [1, 3]

Crop safety/Restrictions
● Do not drill treated winter wheat or rye seed after end of Nov. Seed rate should be
increased as drilling season progresses
● Do not use on seed with moisture content above 16%, on sprouted, cracked or skinned
seed or on seed already treated with another seed treatment
● Store treated seed in cool, dry, well-ventilated store and drill as soon as possible,
preferably in season of purchase
● Treatment may accentuate effects of adverse seedbed conditions on crop emergence

Special precautions/Environmental safety
● Treated seed not to be used as food or feed. Do not re-use sack
● Dangerous to fish. Do not contaminate ponds, waterways or ditches with chemical or
used container

Protective clothing/Label precautions
● A
● 14, 29, 52, 60, 63, 70, 72, 73, 74, 76

310   furalaxyl
A protective and curative acylalanine fungicide for ornamentals

Products   Fongarid 25WP   Ciba-Geigy   25% w/w   WP   03595

Uses   Phytophthora, pythium in HARDY ORNAMENTAL NURSERY STOCK, POT
PLANTS. Damping off in BEDDING PLANTS.

Notes   **Efficacy**
● Apply by incorporation into compost or as a drench to obtain at least 12 wk protection
● Apply drench within 3 d of seeding or planting out as a protective treatment or as soon
as first signs of root disease appear as curative treatment
● Do not apply to field or border soil

FOR FULL CONDITIONS OF USE ALWAYS READ THE PRODUCT LABEL

Crop safety/Restrictions
- Manufacturer's Guide lists 86 genera which have been treated successfully. With others, treat a few plants before committing whole batches

Special precautions/Environmental safety
- Irritating to eyes and skin
- Harmful to fish. Do not contaminate ponds, waterways or ditches with chemical or used container

Protective clothing/Label precautions
- A, C
- 6a, 6b, 14, 18, 22, 28, 29, 36, 37, 53, 63, 66, 70

311 gamma-HCH

A contact, ingested and fumigant organochlorine insecticide
See also aldicarb + gamma-HCH
*   captan + gamma-HCH*
*   carboxin + gamma-HCH + thiram*
*   fenpropimorph + gamma-HCH + thiram*

Products					
1	Ashlade Gamma HCH	Ashlade	560 g/l	SC	02780
2	Atlas Steward	Atlas	560 g/l	SC	03062
3	Fumite Lindane 10	Hunt	21.1 g a.i.	FU	00933
4	Fumite Lindane 40	Hunt	84.4 g a.i.	FU	00934
5	Fumite Lindane Pellets	Hunt	2.7 g or 7.6 g a.i.	FU	00937
6	Gamma-Col	ICI	800 g/l	SC	00964
7	Gamma-Col Turf	ICI Professional	800 g/l	SC	00964
8	Gamma-HCH Dust	Avon	0.5% w/w	DP	04392
9	Gammasan 30	ICI	30% w/w	FS	00969
10	Lindane Flowable	PBI	800 g/l	SC	02610
11	Murfume Grain Store Smoke	DowElanco	84.4 g a.i.	FU	01425
12	Murfume Lindane Smoke	DowElanco	20.5 g a.i.	FU	01426
13	New Kotol	Embetec	12.5% w/w	FS	01487
14	Power Lindane 80	Power	800 g/l	SC	02720
15	Unicrop Leatherjacket Pellets	Unicrop	1.8% w/w	GB	02272
16	Wireworm FS Seed Treatment	DowElanco	100 g/l	FS	03963

Uses

Cutworms, leatherjackets, wireworms in CEREALS, GRASSLAND [1, 2, 6, 10, 14]. Leatherjackets in CEREALS, SUGAR BEET, GRASSLAND [15]. Wireworms in CEREALS [8, 9, 13, 16]. Grain beetles, grain weevils, grain storage pests in STORED GRAIN [3, 4, 11]. Cutworms, leatherjackets, flea beetles, millipedes, pygmy mangold beetle, springtails, symphylids, wireworms in SUGAR BEET [1, 2, 6, 8, 10, 14]. Cabbage stem flea beetle, cabbage stem weevil, pollen beetles, weevils in OILSEED RAPE, BRASSICAS [1, 2, 6, 8, 14]. Cabbage stem flea beetle, cabbage stem weevil, cutworms, leatherjackets, wireworms in BRASSICAS [6]. Flea beetles in BRASSICAS [6, 10]. Cabbage stem flea beetle, pollen beetles, cabbage stem weevil, seed weevil in OILSEED RAPE, BRASSICA SEED CROPS [10]. Soil pests in PEAS, BEANS [8]. Chafer grubs, leatherjackets, millipedes, wireworms in TURF [7]. Aphids, capsids, leafhoppers, leaf miners in VEGETABLES, ORNAMENTALS [2]. Aphids, cutworms, leatherjackets in LETTUCE [6, 8]. Sawflies, suckers in APPLES, PEARS [1, 2, 6].

Apple blossom weevil in APPLES, PEARS [6]. Raspberry cane midge in CANE FRUIT [1, 2, 6]. Raspberry cane midge in RASPBERRIES, LOGANBERRIES [6]. Chafer grubs, leatherjackets, wireworms in STRAWBERRIES [2, 6]. Soil pests in HOPS [8]. Adelgids, aphids, capsids, leafhoppers, leaf miners, rhododendron bug, chrysanthemum midge, symphylids, springtails in ORNAMENTALS [6]. Adelgids, aphids, leafhoppers, rhododendron bug in TREES AND SHRUBS [2]. Adelgids, large pine weevil in FORESTRY PLANTATIONS [10]. Clay-coloured weevil, cutworms, poplar leaf beetles, strawberry root weevil in FOREST NURSERY BEDS [10]. Soil pests in FOREST NURSERY BEDS [8]. Ambrosia beetle, elm bark beetle in CUT LOGS [10]. Great spruce bark beetle in CUT TIMBER [10]. Ants, aphids, capsids, earwigs, leaf miners, mushroom flies, thrips, whitefly, woodlice in GLASSHOUSE CROPS [1, 5, 12]. Aphids, capsids, leafhoppers, leaf miners, springtails, symphylids in TOMATOES [2, 6]. Springtails, thrips in TOMATOES [2, 8]. Springtails, symphylids, thrips in CUCUMBERS [1, 6].

Notes

Efficacy
- Apply as foliar spray, soil spray with incorporation, soil drench, dip, bran-based bait, fumigant or seed treatment as appropriate
- Method of application dose, timing and number of applications vary with formulation, crop and pest. See labels for details

Crop safety/Restrictions
- Do not use if potatoes or carrots are to be planted within 18 mth
- Do not apply to potatoes, carrots, cucurbits, blackcurrants or soft fruit after flowering
- In spring and summer apply in early morning or late evening to reduce hazard to bees
- Do not fumigate newly pricked-out seedlings until root action has started again. Advisable to cut blooms before fumigating. Do not fumigate rare or unusual plants without first testing on small scale

Special precautions/Environmental safety
- Toxic or harmful in contact with skin, by inhalation or if swallowed. Precautionary statement varies with product, see label for details
- Irritating to eyes, and respiratory system [5, 11, 12, 14]
- Dangerous to bees. Do not apply at flowering stage. Keep down flowering weeds [1, 2, 6, 7, 8, 10, 14]
- Dangerous or harmful to fish. Do not contaminate ponds, waterways or ditches with chemical or used container
- Treated seed not to be used as food or feed. Do not re-use sack [9, 13, 16]
- Do not handle treated seed without protective gloves [9, 13, 16]

Protective clothing/Label precautions
- 4c, 5a, 5c, 6a, 6c, 14, 18, 22, 28, 29, 35, 36, 37, 41, 48, 52, 53, 60, 63, 66, 67, 70, 78 [1-16]; 73, 74, 75, 76 [9, 13, 16]

Withholding period
- Harmful to livestock. Keep all livestock out of treated areas for at least 14 d [2, 6, 8, 10, 14, 15]

Harvest interval
- 2 d [5, 12]; 14 d [1, 2, 6, 8, 10, 14]

FOR FULL CONDITIONS OF USE ALWAYS READ THE PRODUCT LABEL

312   gamma-HCH + phenylmercury acetate
An insecticide and fungicide seed dressing for cereals

Products   Mergamma 30   ICI   30:2% w/w   DS   01327

Uses   Stinking smut in WHEAT. Covered smut, loose smut, pyrenophora leaf spot in OATS. Covered smut, leaf stripe, net blotch in BARLEY. Stripe smut, stinking smut in RYE. Wireworms in CEREALS.

Notes   **Efficacy**
- Mix thoroughly with seed in automatic seed treater or use rotary drum or shake in closed container. Do not mix with shovel on open floor
- Treat seed at least one day, but preferably several days, before sowing

Crop safety/Restrictions
- Do not treat seed with moisture content above 16%
- Do not use on seed already treated with an organomercury seed dressing
- Store treated seed in cool, draught-free place for not more than 6 mth and do not allow moisture content to rise above 16%
- Stock carried over to following season should be tested for germination
- Treatment may lower germination capacity, especially if seed not of good quality

Special precautions/Environmental safety
- Harmful in contact with skin, by inhalation or if swallowed
- Organomercury compounds can cause rashes or blisters on skin
- Harmful to fish. Do not contaminate ponds, waterways or ditches with chemical or used container
- Treated seed not to be used as food or feed. Do not re-use sack
- Special precautions needed in event of spillage or leakage. See label for details

Protective clothing/Label precautions
- A
- 5a, 5b, 5c, 18, 23, 24, 28, 29, 36, 37, 53, 63, 65, 70, 72, 73, 74, 76, 78

---

313   gamma-HCH + resmethrin/tetramethrin
A contact and residual insecticide for fly control

Products   Dairy Flyspray   Deosan   0.06:0.1% w/w   AL   00667

Uses   Flies in LIVESTOCK HOUSES.

Notes   **Efficacy**
- Apply undiluted as fine fog or spray for knockdown effect, as coarse spray on surfaces for residual effect
- Spray frequently giving particular attention to entrances of milking parlours and sheds

Special precautions/Environmental safety
- Irritating to eyes, skin and respiratory system

- Extinguish all naked flames when applying
- Do not apply directly to foodstuffs, remove exposed water and milk and collect eggs before applying. Protect feed, milking machinery and containers from contamination
- Extremely dangerous to fish. Do not contaminate ponds, waterways or ditches with chemical or used container

Protective clothing/Label precautions
- A, D, E
- 6a, 6b, 6c, 18, 21, 24, 28, 29, 36, 37, 51, 63, 80

Approval
- May be applied with ULV equipment. See label for details. See introductory notes

314 gamma-HCH + tecnazene
An insecticide and fungicide fumigant for glasshouse crops

Products	Fumite Tecnalin Smoke	Hunt	6.7:26.6 g	FU	04134

Uses
Aphids, ants, capsids, caterpillars, earwigs, leafhoppers, mealy bugs, sawflies, springtails, thrips, whitefly, woodlice in PROTECTED LETTUCE, TOMATOES, CHRYSANTHEMUMS, PROTECTED FLOWERS. Botrytis in PROTECTED LETTUCE, TOMATOES, CHRYSANTHEMUMS, PROTECTED FLOWERS.

Notes

Efficacy
- Start treatment at first sign of disease or infection and repeat as required at 2-3 wk intervals, on lettuce not more frequently than every 3 wk
- For best results fumigate in the late afternoon or evening when temperature is at least 16°C. Keep house closed overnight or for at least 4 h. Do not fumigate in bright sunshine, windy conditions or when temperature too high
- Water plants several hours before fumigating. Damp down paths and straw used as mulch. Ensure foliage dry before treatment

Crop safety/Restrictions
- Do not treat lettuce until 4 wk after planting or tomatoes until the third truss has set
- Do not use on cucurbits, schizanthus and some orchids or on young seedlings or plants being hardened off
- Consult manufacturer before treating ornamental or other crops not listed on the label

Special precautions/Environmental safety
- Irritating to eyes and respiratory system
- Ventilate glasshouse thoroughly before re-entering
- Harmful to fish. Do not contaminate ponds, waterways or ditches with chemical or used container

Protective clothing/Label precautions
- 6a, 6b, 18, 28, 29, 35, 36, 37, 53, 63, 67

FOR FULL CONDITIONS OF USE ALWAYS READ THE PRODUCT LABEL

384

315 gamma-HCH + thiabendazole + thiram
An insecticide and fungicide seed dressing for brassica crops

Products	Hysede FL	Agrichem	400:120:140 g/l	FS	02863

Uses Cabbage root fly, flea beetles, millipedes in OILSEED RAPE, CABBAGES, BRUSSELS SPROUTS, BROCCOLI, SWEDES, TURNIPS. Canker, damping off in OILSEED RAPE, BRUSSELS SPROUTS, BROCCOLI, SWEDES, TURNIPS.

Notes

Efficacy
● Apply with automated seed dresser if treating large quantities of seed. For smaller quantities use batch-type equipment or other forms of closed container and ensure even coverage by not treating batches of more than 25 kg
● May be applied diluted with water, in which case treated seed may require drying

Special precautions/Environmental safety
● Harmful in contact with skin or if swallowed
● Irritating to eyes, skin and respiratory system
● Treated seed not to be used as food or feed. Do not re-use sack
● Harmful to fish. Do not contaminate ponds, waterways or ditches with chemical or used container

Protective clothing/Label precautions
● A
● 5a, 5c, 6a, 6b, 6c, 18, 22, 28, 29, 36, 37, 53, 60, 63, 65, 70, 72, 73, 74, 76, 77, 78

316 gamma-HCH + thiophanate-methyl
An insecticide and lumbricide for use on turf

Products	Castaway Plus	RP Environ.	60:500 g/l	SC	03740

Uses Earthworms, leatherjackets in TURF.

Notes

Efficacy
● Apply by spraying in spring or autumn. Drenching is not required
● Do not mow for 2 d after spraying. If mown beforehand leave clippings
● Do not spray during drought or if ground frozen
● Effectiveness is not impaired by rain or irrigation immediately after application
● Do not mix with any other product

Special precautions/Environmental safety
● Irritating to eyes and skin
● Dangerous to bees. Do not apply at flowering stage. Keep down flowering weeds

- Harmful to fish. Do not contaminate ponds, waterways or ditches with chemical or used container

Protective clothing/Label precautions
- P
- 6a, 6b, 14, 16, 18, 21, 28, 29, 36, 37, 48, 53, 63, 66, 70

317 gamma-HCH + thiram
An insecticide and fungicide seed treatment for brassicas

Products	Hydra-guard	Agrichem	615:230 g/l	FS	03278

Uses Damping off in BRASSICAS, OILSEED RAPE, SWEDES. Flea beetles in BRASSICAS, OILSEED RAPE, SWEDES.

Notes
Efficacy
- Apply with suitable seed treatment equipment undiluted or mixed with up to equal volume of water. Carry out dilution immediately before using
- After use of diluted material seed may require drying

Crop safety/Restrictions
- For use only by agricultural contractors

Special precautions/Environmental safety
- Toxic if swallowed and in contact with skin
- Irritating to skin, eyes and respiratory system
- Treated seed not to be used as food or feed. Do not re-use sack
- Dangerous to fish. Do not contaminate ponds, waterways or ditches with chemical or used container

Protective clothing/Label precautions
- A, C
- 4a, 4c, 6a, 6b, 6c, 18, 21, 28, 29, 36, 37, 52, 63, 65, 72, 73, 74, 75, 77, 78

318 gibberellins
A plant growth regulator for use in apples, pears etc

Products					
1 Berelex	ICI	1 g/tablet	TB	00231
2 Regulex	ICI	10 g/l	SL	03010

Uses Reducing fruit russeting in APPLES [2]. Increasing fruit set in PEARS [1]. Increasing germination in CELERY, NOTHOFAGUS [2]. Increasing yield in CELERY, RHUBARB [1].

FOR FULL CONDITIONS OF USE ALWAYS READ THE PRODUCT LABEL

Efficacy
- Apply to apples at completion of petal fall and repeat 3 times at 10 d intervals. Good coverage of fruitlets is essential for successful results. Best results achieved by spraying under humid, slow drying conditions [2]
- Useful in pears when blossom sparse, setting conditions poor or where frost has killed many flowers. Apply as single or split application. See label for details. Resulting fruit may be seedless [1]
- May be used on pears if 80% of flowers frosted but not effective on very severely frosted blossom. Conference pear responds well, Beurre Hardy only in some seasons. Young trees generally less responsive [1]
- Apply to seed as a soak treatment, 24 h for nothofagus, 48 h at 5°C for celery, and sow immediately
- Apply to celery 3 wk before harvest to increase head size, to rhubarb crowns on transfer to forcing shed or as drench at first signs of growth in field [1]

Crop safety/Restrictions
- Good results achieved on Cox's Orange Pippin, Discovery, Golden Delicious and Karmijn apples. For other cultivars test on a small number of trees [2]
- Do not apply to pears after petal fall [1]

Protective clothing/Label precautions
- 29, 54, 60, 61, 63, 65, 66, 70

---

319   glyphosate
A translocated non-residual phosphonoglycine herbicide

Products

1 Barclay Gallup	Barclay	360 g/l	SL	05020
2 FAL Glyphosate	Fine	480 g/l	SL	04353
3 Mascot Sonic	Rigby Taylor	288 g/l	SL	03376
4 Muster	ICI	240 g/l	SL	03549
5 Portman Glider	Portman	480 g/l	SL	04695
6 Portman Glyphosate 360	Portman	360 g/l	SL	04699
7 Power Glyphosate 360	Power	360 g/l	SL	04714
8 Roundup	Monsanto	360 g/l	SL	01828
9 Roundup	Schering	360 g/l	SL	03947
10 Roundup Four 80	Monsanto	480 g/l	SL	03176
11 Roundup Pro	Chipman	360 g/l	SL	04146
12 Spasor	RP Environ.	360 g/l	SL	03436
13 Spasor CDA	RP Environ.	174 g/l	EW	04866
14 Sting	Monsanto	180 g/l	SL	02789
15 Sting CT	Monsanto	165 g/l	SL	04754
16 Stirrup	Chipman	144 g/l	EC	04174

Uses

Annual dicotyledons, annual grasses, volunteer cereals, perennial grasses, couch, perennial dicotyledons in FIELD CROPS *(autumn stubble, pre-planting/sowing)* [1, 2, 4-10]. Annual dicotyledons, annual grasses, volunteer cereals, perennial grasses, couch, perennial ryegrass, rough meadow grass, perennial dicotyledons in FIELD CROPS *(sward destruction/direct drilling, minimum cultivation)* [1, 3, 4, 6-10]. Annual weeds in FIELD CROPS *(autumn stubble, pre-planting/sowing)* [14, 15]. Annual weeds, perennial weeds, couch in WHEAT, BARLEY, OATS *(pre-harvest)* [1, 2, 4-10]. Pre-harvest desiccation in WHEAT, BARLEY, OATS [1, 2, 4-6, 8-10]. Annual weeds, perennial weeds, couch in OILSEED RAPE, FIELD BEANS, PEAS *(pre-harvest)* [1, 2, 4, 6, 8-10]. Annual weeds, perennial weeds, couch in MUSTARD *(pre-harvest)* [4, 8-10]. Annual weeds, perennial weeds, couch in LINSEED *(pre-harvest)* [8, 9]. Volunteer potatoes in FIELD CROPS *(autumn stubble)* [1, 5-10]. Annual weeds, perennial weeds in GRASSLAND *(pre-cut/graze)* [1, 2, 4, 6-10]. Annual weeds, destruction of short term leys in GRASSLAND [1, 4, 7-10, 14, 15]. Annual weeds, perennial weeds in SETASIDE [8, 9]. Annual weeds, perennial weeds in APPLES, PEARS, PLUMS, CHERRIES [1, 2, 4, 6, 8, 9]. Annual weeds, perennial weeds in DAMSONS [1, 4, 8, 9]. Sucker control in APPLES, PEARS, PLUMS, CHERRIES [1, 2, 4, 8, 9]. Sucker control in DAMSONS [1, 4, 8, 9]. Annual weeds, perennial weeds in WOODY ORNAMENTALS, FENCELINES, MOWING MARGINS [11, 13, 16]. Annual weeds, perennial weeds in AMENITY TREES AND SHRUBS *(pre-planting)* [11, 13, 16]. Wild oats in CEREALS *(wiper glove)* [8, 9]. Weed beet in SUGAR BEET *(wiper application)* [8, 9]. Perennial dicotyledons in GRASSLAND, ORCHARDS, FORESTRY *(wiper application)* [8, 9]. Annual weeds, perennial weeds in AMENITY GRASS *(wiper application)* [12]. Perennial grasses, reeds, rushes, sedges, waterlilies in AQUATIC SITUATIONS [3, 8, 9, 11, 12]. Total vegetation control in NON-CROP AREAS, FIELD BOUNDARIES [1-4, 6, 8, 9, 11-13]. Annual weeds, couch, perennial grasses, perennial dicotyledons, rushes, stinging nettle, woody weeds, rhododendrons, bracken, brambles, heather in FORESTRY [1, 2, 6, 8, 9]. Chemical thinning in FORESTRY [2, 6, 8, 9]. Annual weeds, perennial weeds in GRAPEVINES, TREE NUTS *(off-label)* [8, 9].

Notes

Efficacy
- Perennial grasses must have adequate leaf area (5 leaves 10-15 cm long for couch) and be growing actively for effective control
- Perennial dicotyledons most susceptible when at or near flowering and actively growing
- Volunteer potatoes and polygonums are not controlled by harvest aid rates
- Do not cultivate before spraying perennials nor cultivate, graze or apply lime, fertilizer etc for 7 d afterwards (14 d for direct drilling after sward destruction)
- Cultivation or drilling can be done 24 h after spraying annuals
- After pre-cut/graze treatments do not graze or cut for silage or hay for 5 d
- With pre-harvest treatment of cereals mean moisture content must be below 30%
- A rainfree period of at least 6 h (preferably 24 h) must follow spraying [1-14, 16]
- A rainfree period of at least 1 h (preferably 3 h) must follow spraying for control of volunteer cereals and annual grasses in stubbles; at least 3 h (preferably 6 h) for all other weeds [15]
- Do not spray weeds stressed by drought or frost
- Fruit tree suckers best treated in late spring
- Wiper application best on weeds at or near flowering
- Chemical thinning treatment can be applied as stump spray or by injection
- Do not tank mix with any product other than recommended surfactants
- For rhododendron control apply to stumps or regrowth. Addition of adjuvant Mixture B recommended for application to foliage by knapsack sprayer [8, 9]
- Apply only through special applicator [13, 16]

Crop safety/Restrictions
- Do not use pre-harvest treatment on undersown crops, on cereals for seed certification or on other crops to be used as seed
- Decaying remains of plants killed by spraying must be dispersed before direct drilling

FOR FULL CONDITIONS OF USE ALWAYS READ THE PRODUCT LABEL

- Do not use treated straw as mulch or growing medium for horticultural crops. It may be used for animal feed or bedding
- For use in orchards, grapevines and tree nuts care must be taken to avoid contact with trees. Do not use in orchards established less than 2 yr and keep off low-lying branches
- Treat apples and pears after harvest but before green cluster, plums, cherries and damsons after harvest but before white bud [8, 9]
- Do not spray root suckers in orchards in late summer or autumn
- Do not use under glass or polythene as damage to crops may result
- Take extreme care to avoid drift onto other crops
- With wiper application weeds should be at least 10 cm taller than crop
- Certain conifers may be sprayed overall in dormant season. See label for details [8, 9]
- Use a tree guard when spraying in established forestry plantations

Special precautions/Environmental safety
- Irritating to eyes and skin [1-13, 15, 16]. Irritating to eyes. Harmful if swallowed [14]
- Dangerous to fish [4, 14], harmful to fish [1-3, 5-13, 15, 16]. Do not contaminate ponds, waterways or ditches with chemical or used container
- Maximum permitted concentration in treated water 0.2 ppm. [3, 8, 9, 11, 12]
- Do not mix, store or apply in galvanized or unlined mild steel containers or spray tanks
- Do not leave spray in spray tanks for long periods and make sure tanks are well vented
- Take extreme care to avoid drift
- Treated poisonous plants must be removed before grazing or conserving [7-10]
- For field edge treatment direct spray away from hedge bottoms

Protective clothing/Label precautions
- A, C [1-3, 5, 7, 10, 14]; A, H, M [13]; A, C, H, M [4, 6]; A, C, M, N [11, 12, 16]; A, C, F, H, M, N [8, 9]
- 5c, 6a, 9, 78 [14]; 6a, 6b [3-12]; 14, 18, 21, 28, 29, 36, 37, 53, 60, 63, 70 [1-14]; 55 [3, 8, 9, 11, 12]

Harvest interval
- grassland 5d; wheat, barley, oats, field beans, combining peas, linseed 7d; mustards 8d; oilseed rape 14 d; stubbles 2-5 d (depending on dose) -see label for details

Approval
- Approved for aquatic weed control. See introductory notes on use of herbicides in or near water [3, 8, 9, 11, 12]
- May be applied through CDA equipment. See label for details. See introductory notes on ULV application [1, 8-13, 16]
- Off-label Approval to Feb 1991 for use on grapevines and tree nuts (OLA 0173, 0174/88) [8]; (OLA 0175, 0177/88) [9]

---

320   glyphosate + simazine
A translocated and residual total herbicide

Products

1 Mascot Ultrasonic	Rigby Taylor	100:280 g/l	SC	03546
2 Rival	Chipman	100:280 g/l	SC	03377

Uses

Total vegetation control in NON-CROP AREAS [1, 2]. Annual weeds, perennial grasses, couch in TREES AND SHRUBS, WOODY NURSERY STOCK [1, 2]. Annual weeds, perennial grasses in CONIFER PLANTATIONS [2].

Efficacy
  • Apply to fully emerged weeds with actively growing green leaves
  • Spray perennial dicotyledons at or near flowering but before onset of senescence
  • Spray couch when shoots have 10-15 cm new leaf growth
  • For forestry use apply to active weed growth using a directed spray or tree guard. Tree must be established for at least 1 yr [2]
  • Do not cultivate before spraying. Do not mix with other chemicals
  • A rainfree period of at least 6 h (preferably 24 h) must follow spraying
  • Do not spray weeds suffering from drought or in frosty weather

Crop safety/Restrictions
  • Apply as carefully directed spray among woody ornamentals established for at least 1 yr. See label for list of species which should not be sprayed
  • Do not apply on slopes where heavy rain is likely to cause wash onto desirable plants

Special precautions/Environmental safety
  • Irritating to eyes and skin
  • Take extreme care to avoid drift
  • Harmful to fish. Do not contaminate ponds, waterways or ditches with chemical or used container
  • Do not mix, store or apply in galvanized or unlined mild steel containers or spray tanks
  • Do not leave spray mixtures in tank for long periods and ensure tanks are well vented

Protective clothing/Label precautions
  • A, C, H, M
  • 6a, 6b, 14, 18, 21, 28, 29, 36, 37, 53, 60, 63, 66

Approval
  • May be applied through CDA equipment. See label for details. See introductory notes on ULV application [1, 2]

321 guazatine
A guanidine fungicide seed dressing for wheat

Products	Rappor	DowElanco	300 g/l	LS	04900

Uses Bunt, glume blotch in WHEAT.

Notes **Efficacy**
  • Apply with conventional seed treatment machinery
  • After treating, bag seed immediately and keep in dry, draught-free store
  • Treatment also reduces brown foot rot and seedling blights

Crop safety/Restrictions
  • Do not treat grain with moisture content above 16% and do not allow moisture content of treated seed to exceed 16%
  • Do not apply to cracked, split or sprouting seed

FOR FULL CONDITIONS OF USE ALWAYS READ THE PRODUCT LABEL

- Do not store treated seed for more than 6 mth
- Treatment may lower germination capacity, particularly if seed grown, harvested or stored under adverse conditions

Special precautions/Environmental safety
- Harmful if swallowed. Irritating to eyes
- Harmful to fish. Do not contaminate ponds, waterways or ditches with chemical or used container

Protective clothing/Label precautions
- A, C
- 5c, 6a, 18, 21, 36, 37, 53, 63, 65, 70, 72, 73, 76, 78

322   guazatine + imazalil
A fungicide seed treatment for barley and oats

Products	Rappor Plus	DowElanco	300:25 g/l	LS	04899

Uses   Leaf stripe, net blotch, foot rot in BARLEY. Pyrenophora leaf spot in OATS.

Notes

Efficacy
- Apply with conventional seed treatment machinery
- After treating, bag seed immediately and keep in dry, draught-free store
- Treatment also reduces fusarium foot rot and seedling blights

Crop safety/Restrictions
- Do not treat grain with moisture content above 16% and do not allow moisture content of treated seed to exceed 16%
- Do not apply to cracked, split or sprouting seed
- Do not store treated seed for more than 6 mth
- Treatment may lower germination capacity, particularly if seed grown, harvested or stored under adverse conditions

Special precautions/Environmental safety
- Harmful if swallowed. Irritating to eyes
- Harmful to fish. Do not contaminate ponds, waterways or ditches with chemical or used container

Protective clothing/Label precautions
- A, C
- 5c, 6a, 18, 21, 36, 37, 53, 63, 65, 70, 72, 73, 76, 78

323   heptenophos
A contact, systemic and fumigant organophosphorus insecticide
See also deltamethrin + heptenophos

Products	Hostaquick	Hoechst	550 g/l	EC	01079

Uses	Aphids in CEREALS, BRASSICAS, BROAD BEANS, FIELD BEANS, FRENCH BEANS, RUNNER BEANS, CELERY, COURGETTES, MARROWS, LETTUCE, PEAS, APPLES, PEARS, BLACKCURRANTS, GOOSEBERRIES, RASPBERRIES, STRAWBERRIES, PEPPERS, TOMATOES, CUCUMBERS, GLASSHOUSE POT PLANTS. Woolly aphid in APPLES, PEARS. Leafhoppers in TOMATOES, POT PLANTS. Thrips, leafhoppers in CUCUMBERS. American serpentine leaf miner in BEANS, PEAS, BEET CROPS, BRASSICAS, LEEKS, LETTUCE, POTATOES, RHUBARB, ARTICHOKES, ASPARAGUS, CELERY, CHICORY, CUCURBITS, ENDIVES, FENNEL, SPINACH, AUBERGINES, PEPPERS, ORNAMENTALS *(off label)*.

Notes

Efficacy
- Spray when pests first seen and repeat as necessary. Knock-down effect is rapid
- Use sufficient water for good spray coverage. On glasshouse crops apply at high volume
- May be used in integrated control programmes with Phytoseiulus or Encarsia in glasshouses. Allow at least 4 d after treatment before introducing any new beneficial insects

Crop safety/Restrictions
- In glasshouses do not apply in early morning following a low night temperature nor in excessively high daytime temperatures. Do not spray crops at temperatures above 30°C
- Do not apply through a fogging or ULV machine in glasshouse crops

Special precautions/Environmental safety
- This product contains an organophosphorus compound. Do not use if under medical advice not to work with such compounds
- Toxic if swallowed, harmful in contact with skin
- Flammable
- Dangerous to bees. Do not apply at flowering stage except as directed on peas. Keep down flowering weeds in all crops
- Harmful to fish. Do not contaminate ponds, waterways or ditches with chemical or used container

Protective clothing/Label precautions
- A, C
- 1, 4c, 5a, 12c, 14, 16, 18, 21, 25, 28, 29, 35, 36, 37, 48, 53, 60, 63, 66, 70, 79

Harvest interval
- 1 d

Approval
- Approved for aerial application on cereals, peas, brassicas, beans. See introductory notes. Consult firm for details
- Off-label Approval to Dec 1992 for use in wide range of vegetable crops and non-edible ornamentals (OLA 1058/89)

FOR FULL CONDITIONS OF USE ALWAYS READ THE PRODUCT LABEL

324   hexazinone

A contact and residual triazinone herbicide for use in forestry

Products	1 Velpar Liquid	Du Pont	240 g/l	SL	02288
	2 Velpar Liquid	Selectokil	240 g/l	SL	02288

Uses Annual weeds, perennial grasses, perennial dicotyledons, woody weeds in CONIFER PLANTATIONS.

Notes **Efficacy**
- Apply as 1 m strip with trees in centre of strip or by spot-gun to circles round trees
- Chemical is active via foliage and soil. Adequate soil moisture increases soil uptake
- Best control of grasses and response of trees will follow spring application prior to flushing. Apply to perennial dicotyledons when in active growth
- Best results on woody weeds achieved by application to foliage of actively growing plants up to 1 m high. Heathers and evergreens are not adequately controlled

Crop safety/Restrictions
- Apply to conifers planted out for at least 1 yr, on Sitka and Norway spruce from early Mar to bud-burst or from after cessation of flushing and growth hardened to end of Aug, on pines from end Mar to end Aug
- Do not use on or allow drift onto larches, deciduous or other desirable species
- Do not use on trees suffering from pests, diseases, winter injury or other stress
- Do not treat bare soil or soils with less than 1% organic matter

Special precautions/Environmental safety
- Irritant. Risk of serious damage to eyes
- Flammable
- Do not use within 250 m of water courses

Protective clothing/Label precautions
- A, C, H, J, M
- 6, 9, 18, 21, 27, 29, 36, 37, 54, 60, 63, 65

---

325   8-hydroxyquinoline sulphate

A protectant fungicide and bactericide for soil-borne diseases

Products	Cryptonol	Fargro	140 g/l	LI	03126

Uses Soil-borne diseases in ORNAMENTALS.

Notes **Efficacy**
- Use as spray or soil drench pre-emergence or post-planting for protection or post-emergence for control of infection
- Repeat application at 7-14 d intervals. See label for details of rates and timing
- Where damping off known to be a problem drench soil 1 wk before sowing
- May also be used for disinfecting tools, seedboxes etc and for washing down premises
- Use plastic containers for storage, dipping or spraying

Crop safety/Restrictions
* Do not treat during seedling germination
* Do not spray gloxinias, begonias or petunias during cotyledon stage
* With new cultivars check for varietal susceptibility

Special precautions/Environmental safety
* Irritating to eyes, skin and respiratory system

Protective clothing/Label precautions
* A, C
* 6a, 6b, 6c, 14, 18, 21, 25, 28, 29, 34, 36, 37, 54, 63, 65

326   hymexazol
A systemic oxazole fungicide for pelleting sugar beet seed

Products   Tachigaren 70 WP   Sumitomo   70% w/w   WP   02649

Uses   Black leg, damping off in SUGAR BEET.

Notes   **Efficacy**
* Incorporate into pelleted sugar beet seed using suitable seed pelleting machinery

Special precautions/Environmental safety
* Irritating to eyes, skin and respiratory system
* Wear protective gloves when handling treated seed
* Treated seed not to be used as food or feed
* Harmful to fish. Do not contaminate ponds, waterways or ditches with chemical or used container

Protective clothing/Label precautions
* A, C, F
* 6a, 6b, 6c, 18, 29, 36, 37, 53, 63, 67, 72, 73, 74, 76

327   imazalil
A systemic and protectant imidazole fungicide
See also carboxin + imazalil + thiabendazole
*   fuberidazole + imazalil + triadimenol*
*   guazatine + imazalil*

Products					
1 Fungaflor	Hortichem	200 g/l	EC	03452
2 Fungaflor Smoke	Hortichem	15% w/w	FU	03599

FOR FULL CONDITIONS OF USE ALWAYS READ THE PRODUCT LABEL

Powdery mildew in CUCUMBERS, MARROWS, COURGETTES, ROSES, ORNAMENTALS [1]. Powdery mildew in CUCUMBERS [2]. Powdery mildew in GLASSHOUSE ORNAMENTALS, GLASSHOUSE ROSES [1, 2].

Efficacy
- Treat cucurbits before or as soon as disease appears and repeat every 10-14 d or every 7 d if infection pressure great or with susceptible cultivars [1, 2]

Crop safety/Restrictions
- Do not spray in full, bright sunshine. When spraying in the evening the spray should dry before nightfall. May cause damage if open flowers are sprayed [1]
- Do not use on rose cultivar Dr A.J. Verhage [1]
- With ornamentals of unknown tolerance test on a few plants in first instance [1]

Special precautions/Environmental safety
- Irritating to eyes and skin [1], to eyes [2]
- Flammable [1], highly flammable [2]
- Keep unprotected persons out of treated areas for at least 2 h [2]
- Harmful to bees. Do not apply at flowering stage [1]
- Harmful to fish. Do not contaminate ponds, waterways or ditches with chemical or used container [1]

Protective clothing/Label precautions
- A, C [1]
- 6a, 6b, 12c, 21, 28, 29, 35, 50, 53, 63, 67, 70 [1]; 6a, 12b, 18, 35, 36, 37, 38, 63, 67 [2]

Harvest interval
- cucumbers 1 d

328   imazalil + thiabendazole
A fungicide for treatment of potatoes at planting

Products   Seedtect   MSD Agvet   0.5:2% w/w   WP

Uses   Silver scurf, skin spot, black scurf and stem canker in POTATOES.

Notes   **Efficacy**
- Apply directly to potatoes in planter hopper. Ensure all tuber surfaces evenly treated
- Seed should only be treated after chitting has started

Crop safety/Restrictions
- Do not mix with any other product
- Delayed emergence noted under certain conditions but with no final effect on yield

Special precautions/Environmental safety
- Irritating to respiratory system
- Dangerous to fish. Do not contaminate ponds, waterways or ditches with chemical or used container
- Handle with care and mix only in a closed container

329 imazamethabenz-methyl

A post-emergence grass weed herbicide for use in winter cereals

Products	Dagger	Cyanamid	300 g/l	SC	03737

Uses

Wild oats, blackgrass, onion couch, loose silky bent, charlock, volunteer oilseed rape in WINTER WHEAT, WINTER BARLEY.

Notes

Efficacy
• Has contact and residual activity. Wild oats controlled from pre-emergence to 4 leaves + 3 tillers stage, charlock and volunteer oilseed rape to 10 cm. Good activity on onion couch up to 7.5 cm
• Best results achieved when applied to fine, firm, clod-free seedbed when soil moist
• Do not use on soils with more than 10% organic matter
• Effects of autumn/winter treatment normally persist to control spring flushes of weed
• Tank mixture with isoproturon recommended for control of tillered blackgrass, with pendimethalin for general grass and dicotyledon control

Crop safety/Restrictions
• Apply from 2-fully expanded leaf stage of crop to leaf sheath erect (GS 12-30)
• Do not use on durum wheat
• Do not use on soils where surface water is likely to accumulate
• In case of crop failure land must be ploughed to 15 cm and re-drilled in spring. Sugar beet, oilseed rape or other brassicas should not be sown in these circumstances

Special precautions/Environmental safety
• Irritating to eyes

Protective clothing/Label precautions
• A, C
• 6a, 18, 21, 29, 35, 37, 54, 60, 63, 66, 70, 78

330 imazamethabenz-methyl + isoproturon

A contact and residual grass weed herbicide for winter cereals

Products					
1 Pinnacle	Cyanamid	300:553 g/l	SC	04103
2 Pinnacle 400	Cyanamid	100:300 g/l	SC	04687

FOR FULL CONDITIONS OF USE ALWAYS READ THE PRODUCT LABEL

Uses	Blackgrass, wild oats, annual grasses, annual dicotyledons in WINTER WHEAT, WINTER BARLEY.

Notes	**Efficacy**

- Apply in autumn or spring pre- or post-emergence of weeds
- Best results from post-emergence treatment, wild oats and meadow grass to early tillering stage, dicotyledons to young plant stage, cleavers to 1 whorl
- Blackgrass controlled to late tillering at higher dose
- Residual activity reduced on soils with more than 10% organic matter
- Weed control reduced by prolonged dry conditions
- Do not roll or harrow after application
- Disperse and bury trash and ash from straw burning during seedbed preparation

Crop safety/Restrictions

- Apply from 1 leaf unfolded stage of crop to leaf sheath erect (GS 11-30)
- Do not use on durum wheats
- Do not use on spring varieties drilled in autumn or winter [2]
- Do not use where surface water may accumulate or if frost imminent
- Do not spray crops suffering stress from any cause
- Early sown crops may be damaged by application during rapid growth in autumn
- In case of crop failure plough to 15 cm and re-drill with specified crops in spring. Do not re-drill with sugar beet, oilseed rape or other brassicas

Special precautions/Environmental safety

- Irritating to skin and eyes
- Harmful to fish. Do not contaminate ponds, waterways or ditches with chemical or used container

Protective clothing/Label precautions

- A, C
- 6a, 6b, 18, 21, 28, 29, 36, 37, 53, 63, 66, 70

331   imazapyr

A non-selective translocated and residual imidazolinone herbicide
See also atrazine + imazapyr

Products	1 Arsenal	Chipman	250 g/l	SL	02904
	2 Arsenal	Cyanamid	250 g/l	SL	04064
	3 Arsenal 50	Chipman	50 g/l	SL	03774
	4 Arsenal 50	Cyanamid	50 g/l	SL	04070

Uses	Total vegetation control, bracken in NON-CROP AREAS.

Notes	**Efficacy**

- Chemical is absorbed through roots and foliage, kills underground storage organs and gives long term residual control
- May be applied before weed emergence but gives best results from application at any time of year when weeds are growing actively
- Kill of plants is slow, complete kill not occurring for several weeks

Crop safety/Restrictions
- Avoid drift onto desirable plants
- Do not apply to soil which may later be used to grow desirable plants
- Do not apply where roots of desirable plants may extend
- Not recommended for bracken control on hillsides or agricultural land

Special precautions/Environmental safety
- Irritating to eyes
- Not to be used on food crops

Protective clothing/Label precautions
- A, C, M
- 6a, 14, 18, 21, 28, 36, 37, 54, 63, 66

Approval
- May be applied through CDA equipment. See label for details. See introductory notes

---

332   indol-3-ylacetic acid
A plant growth regulator for promoting rooting of cuttings

Products					
	1 Rhizopon A Powder	Fargro	0.5% w/w	DP	01797
	2 Rhizopon A Tablets	Fargro	50 mg ai	TB	01799

Uses   Rooting of cuttings in ORNAMENTALS.

Notes   **Efficacy**
- Apply by dipping base of cuttings into powder or dissolved tablets
- See label for details of concentrations recommended for different species

---

333   4-indol-3-ylbutyric acid
A plant growth regulator promoting the rooting of cuttings
See also dichlorophen + 4-indol-3-yl butyric acid + 1-naphthylacetic acid

Products					
	1 Chryzoplus	Fargro	0.8% w/w	DP	00510
	2 Chryzopon	Fargro	0.1% w/w	DP	00509
	3 Chryzosan	Fargro	0.6% w/w	DP	00511
	4 Chryzotek	Fargro	0.4% w/w	DP	00512
	5 Rhizopon AA Powder	Fargro	0.1% w/w	DP	01798
	6 Rhizopon AA Tablets	Fargro	50 mg ai	TB	01800
	7 Seradix 1	Embetec	0.1% w/w	DP	04680
	8 Seradix 2	Embetec	0.3% w/w	DP	04681
	9 Seradix 3	Embetec	0.8% w/w	DP	04682

Uses   Rooting of cuttings in ORNAMENTALS.

FOR FULL CONDITIONS OF USE ALWAYS READ THE PRODUCT LABEL

Efficacy
- Dip base of cuttings into powder or dissolved tablets immediately before planting
- Powders or solutions of different concentration are required for different types of cutting. Lowest concentration for softwood [7], intermediate for semi-ripe [8], highest for hardwood [9]
- See label for details of concentration and timing recommended for different species

Crop safety/Restrictions
- Use of too strong a powder or solution may cause injury to cuttings

---

334   iodofenphos

A contact and ingested organophosphorus insecticide
See also benomyl + iodofenphos + metalaxyl

Products

1 Elocril 50 WP	Ciba-Geigy	50% w/w	WP	00787
2 Nuvanol N 500 FW	Ciba-Geigy	39.1% w/w	SC	H2168

Uses

Caterpillars in BRUSSELS SPROUTS, CABBAGES [1]. Cabbage root fly in BRUSSELS SPROUTS [1]. Onion fly in ONIONS. Ants, fleas, cockroaches, crickets, silverfish, flies, bugs in AGRICULTURAL PREMISES [2]. Grain storage pests in GRAIN STORES, AGRICULTURAL PREMISES [2]. Flies, crickets in REFUSE TIPS [2]. Lesser mealworm, hide beetle, ectoparasites in POULTRY HOUSES [2].

Notes

Efficacy
- For caterpillar control in brassicas apply as soon as pest first seen in crop [1]
- To reduce attack of cabbage root fly maggots in Brussels sprout buttons spray 28 d prior to anticipated harvest date and repeat 10 d later. Apply with inter-row pendant lance equipment to obtain maximum cover of lower buttons [1]
- For onion fly control treat seed with a seed coating compound containing product. Seed may be over-coated with benomyl but not thiram [1]
- Do not mix with strongly alkaline materials
- Used as indoor spray treatment pests controlled for 2-3 mth [2]
- May be mixed with approved disinfectant for terminal disinfection of poultry houses [2]

Special precautions/Environmental safety
- This product contains an organophosphorus compound. Do not use if under medical advice not to work with such compounds
- Irritating to eyes and skin
- Do not apply to surfaces on which food or feed is stored, prepared or eaten [2]
- Remove or cover all foodstuffs before application. Protect food preparing equipment and eating utensils from contamination [2]
- Remove exposed milk and collect eggs before application. Protect milk machinery and containers from contamination [2]
- Do not apply directly to livestock [2]
- Dangerous to bees. Do not apply at flowering stage. Keep down flowering weeds
- Dangerous to fish. Do not contaminate ponds, waterways or ditches with chemical or used container

Protective clothing/Label precautions
- A, C

Harvest interval
• Brussels sprouts 7 d; cabbage 14 d

335 iodophor
See also nonylphenoxypoly(ethyleneoxy)ethanol-iodine complex

336 ioxynil
A contact acting HBN herbicide for use in turf and onions
See also benazolin + bromoxynil + ioxynil
benazolin + bromoxynil + ioxynil + mecoprop
bromoxynil + chlorsulfuron + ioxynil
bromoxynil + dichlorprop + ioxynil
bromoxynil + dichlorprop + ioxynil + MCPA
bromoxynil + ethofumesate + ioxynil
bromoxynil + fluroxypyr + ioxynil
bromoxynil + ioxynil
bromoxynil + ioxynil + isoproturon
bromoxynil + ioxynil + isoproturon + mecoprop
bromoxynil + ioxynil + mecoprop
clopyralid + fluroxypyr + ioxynil
2,4-D + dicamba + ioxynil

Products	1 Actrilawn 10	RP Environ.	100 g/l	SL	04431
	2 Totril	Embetec	225 g/l	EC	02813

Uses Annual dicotyledons in ONIONS, LEEKS, SHALLOTS, GARLIC [2]. Annual dicotyledons in NEWLY SOWN TURF, TURF [1].

Notes **Efficacy**
• Best results on seedling to 4-leaf stage weeds in active growth during mild weather
• In newly sown turf apply after first flush of weed seedlings, in spring normally 4 wk after sowing
• May be used on established turf from May to Sep under suitable conditions. Do not mow within 7 d of treatment

Crop safety/Restrictions
• Apply to sown onion crops as soon as possible after plants have 3 true leaves or to transplanted crops when established
• Apply to newly sown turf after the 2-leaf stage of grasses
• Do not use on crested dogstail

Special precautions/Environmental safety
• Harmful in contact with skin or if swallowed [2], if swallowed [1]

FOR FULL CONDITIONS OF USE ALWAYS READ THE PRODUCT LABEL

- Irritating to eyes and skin
- Do not apply with hand-held equipment or at concentrations higher than those recommended
- Keep livestock out of treated areas for at least 6 wk after treatment and until foliage of any poisonous weeds such as ragwort has died and become unpalatable
- Harmful to fish. Do not contaminate ponds, waterways or ditches with chemical or used container

Protective clothing/Label precautions
- A, C
- 5a, 5c, 6a, 6b, 18, 21, 25, 28, 29, 36, 37, 43, 53, 60, 63, 66, 70, 78

Harvest interval
- onions 12 wk; leeks 14 wk [1]

---

337 ioxynil + isoproturon + mecoprop
A contact and residual post-emergence herbicide for use in cereals

Products					
1 Musketeer	Hoechst	50:250:180 g/l	SC	01461
2 Post-Kite	Schering	50:250:180 g/l	SC	03832

Uses

Annual dicotyledons, chickweed, cleavers, speedwells, annual grasses, annual meadow grass, blackgrass in WINTER WHEAT, WINTER BARLEY, SPRING WHEAT, SPRING BARLEY [2, 3]. Annual dicotyledons, chickweed, cleavers, speedwells, annual grasses, annual meadow grass, blackgrass in WINTER WHEAT, WINTER BARLEY [1].

Notes

Efficacy
- Best results achieved by application to young weed seedlings growing actively when rain is not expected for 6 h. For effective blackgrass control adequate soil moisture and rain within 3 wk after spraying are required
- Grass weeds should have at least 1 main leaf unfolded
- Residual activity reduced on soils with more than 10% organic matter and on stony or gravelly soils if heavy rain falls soon after application
- Trash and ash from straw burning should be dispersed and buried before spraying

Crop safety/Restrictions
- Apply to crops from 2-leaves fully expanded to first node detectable (GS 12-31)
- Do not treat durum wheat, undersown crops or those due to be undersown
- Do not spray crops under stress from waterlogging, drought, disease or pest attack
- Early sown crops (eg Sep) may be prone to damage if spraying precedes or coincides with rapid growth [1, 2]
- Do not roll within 7 d before or after spraying
- Do not harrow within 7 d before or at any time after spraying
- On free draining stony or gravelly soils crop may be damaged by heavy rain soon after application

Special precautions/Environmental safety
- Irritating to eyes and skin
- Do not apply with knapsack sprayer or at concentrations higher than those recommended

- Harmful to bees. Do not apply at flowering stage. Keep down flowering weeds
- Harmful to fish. Do not contaminate ponds, waterways or ditches with chemical or used container

Protective clothing/Label precautions
- A, C
- 6a, 6b, 16, 18, 22, 25, 28, 29, 36, 37, 43, 50, 53, 63, 65

Approval
- May be applied through CDA equipment. See label for details. See introductory notes on ULV application [2]

---

338   ioxynil + mecoprop
A contact and translocated herbicide for use in amenity turf

Products					
1 Iotox	RP Environ.	72:214 g/l	SL	04583
2 Synox	Synchemicals	75:225 g/l	SL	02080

Uses

Annual dicotyledons, perennial dicotyledons, speedwells, pearlwort, parsley piert in TURF, AMENITY GRASS.

Notes

Efficacy
- Apply at any time from Apr to Oct on warm, dry day when soil moist and weeds growing actively
- For speedwell control apply in early spring before flower heads appear, usually by end of Apr, and repeat in autumn
- Use of a tank mixture with liquid fertilizer recommended or feed turf later to encourage grass to fill in as weeds die

Crop safety/Restrictions
- Apply to established turf which has been regularly close mown
- Do not apply during drought. Do not mow for 3-4 d before or after treatment

Special precautions/Environmental safety
- Harmful in contact with skin and if swallowed. Irritating to eyes
- Do not apply with hand-held equipment or at concentrations higher than those recommended
- Keep livestock out of treated areas for at least 6 wk and until foliage of any poisonous weeds such as ragwort has died and become unpalatable
- Harmful to fish. Do not contaminate ponds, waterways or ditches with chemical or used container

Protective clothing/Label precautions
- A, C
- 5a, 5c, 6a, 18, 21, 25, 28, 29, 36, 37, 43, 53, 60, 63, 66, 70, 78

FOR FULL CONDITIONS OF USE ALWAYS READ THE PRODUCT LABEL

339   iprodione
A protectant dicarboximide fungicide with some eradicant activity

Products

1 CDA Rovral	RP Environ.	200 g/l	UL	04679
2 Rovral Dust	Hortichem	1.25% w/w	DP	03836
3 Rovral Flo	RP	250 g/l	SC	04526
4 Rovral Granules	RP Environ.	10% w/w	GR	02956
5 Rovral Green	RP Environ.	250 g/l	SC	01835
6 Rovral WP	Embetec	50% w/w	WP	03575
7 Turbair Rovral	PBI		UL	02248

Uses

Glume blotch in WINTER WHEAT [3]. Net blotch in BARLEY [3]. Chocolate spot in FIELD BEANS [3]. Alternaria, botrytis in OILSEED RAPE, BRASSICAS, STUBBLE TURNIPS, MUSTARD, KALE SEED CROPS, BRASSICA SEED CROPS [3]. Alternaria, botrytis in STORED CABBAGES [6]. Alternaria in BRASSICA SEED CROPS, CARNATIONS [6]. Sclerotinia stem rot in OILSEED RAPE [3]. Alternaria in OILSEED RAPE, BRASSICAS, FLOWER SEEDS *(seed treatment)* [6]. Botrytis in PEAS [3]. Black scurf and stem canker in SEED POTATOES [6]. Brown patch, dollar spot, fusarium patch, melting out, red thread, grey snow mould in TURF, AMENITY GRASS [1, 4, 5]. Botrytis in LETTUCE, TOMATOES, CHRYSANTHEMUMS [7]. Botrytis in LETTUCE, PROTECTED LETTUCE, RASPBERRIES, STRAWBERRIES, POT PLANTS, CUCUMBERS, TOMATOES [6]. Botrytis in PROTECTED LETTUCE [2]. Leaf rot, neck rot in ONIONS, SALAD ONIONS [3]. Botrytis in STRAWBERRIES [3]. Botrytis in FREESIAS [2]. Stemphylium in ASPARAGUS *(off-label)* [3, 6]. Botrytis, sclerotinia in FRENCH BEANS, VINING PEAS, DRIED PEAS, MANGE-TOUT PEAS *(off-label)* [3, 6]. Botrytis in TREE NUTS *(off-label)* [3, 6]. Chocolate spot in BROAD BEANS *(off-label)* [3]. Botrytis, sclerotinia in LETTUCE, RADICCHIO *(off-label)* [3]. Fungus diseases in LINSEED *(off-label)* [3]. Fungus diseases in EVENING PRIMROSE *(off-label)* [3, 6]. Botrytis in HARDY ORNAMENTAL NURSERY STOCK, NURSERY FRUIT TREES AND BUSHES *(off-label)* [3, 6]. Alternaria in WHITE TURNIPS *(off-label)* [3]. Botrytis, fusarium in ORNAMENTAL BULBS *(off-label)* [3]. Grey mould, powdery mildew in CUCURBITS *(off-label)* [3]. Botrytis in LETTUCE, ORNAMENTALS, AUBERGINES, PEPPERS, TOMATOES *(off-label)* [3]. Botrytis in FLOWER CROPS, BEDDING PLANTS *(off-label)* [6]. Botrytis in GLASSHOUSE CUCURBITS, PEPPERS, AUBERGINES, POT PLANTS, BEDDING PLANTS UNDER PROTECTION, FLOWER CROPS UNDER PROTECTION *(off-label)* [7]. Fungus diseases in BORAGE *(off-label)* [[6]].

Notes

Efficacy
- Many diseases require a programme of 2 or more sprays at intervals of 2-4 wk
- Number and timing of sprays vary with disease and crop. See label for details
- Apply turf treatments after mowing and do not mow for 24 h [3, 4]
- Treatment harmless to Phytoseiulus and Encarsia being used for integrated pest control

Crop safety/Restrictions
- Do not treat oats [3]
- Use a maximum of 3 sprays on oilseed rape and brassica seed crops [3]
- Only 3 applications permitted on protected lettuce in winter [2, 6, 7]Do not dust lettuce under stress from lack of moisture, cold draughts etc [2]
- See label for pot plants showing good tolerance. Check other species before applying on a large scale [6]. Check safety on new chrysanthemum cultivars [7]
- Use a drench to control Rhizoctonia on bedding plants and cabbage storage diseases [6]

- Some flower seeds may be soaked in a solution of product and sown or packed after drying. See label for details [6]

Special precautions/Environmental safety
- Irritating to eyes and skin [1, 3-5], to eyes [6]
- Harmful to fish. Do not contaminate ponds, waterways or ditches with chemical or used container

Protective clothing/Label precautions
- A, C, H, M [1]; F [2]; A, C [3,5]
- 6a, 6b [1, 3-5]; 6a [6]; 18, 29, 36, 37, 53, 63 [1, 3-5]; 21, 28, 65 [1]; 28, 29, 53, 63, 67 [2]; 21, 35, 66 [3]; 67, 70 [4]; 16, 21, 66, 70 [5]; 35, 53, 63, 67, 70 [6]; 28, 35, 53, 63, 65 [7]

Withholding period
- Stored cabbage 2 mth

Harvest interval
- strawberries, raspberries 1 d; tomatoes 1 d [6], 2 d [7]; cucumbers 2d; lettuce, onions, peppers 7d; brassicas, brassica seed crops, oilseed rape, peas, beans 21 d; winter lettuce 4 wk

Approval
- Approved for aerial application on oilseed rape, peas [3]. See introductory notes
- Approved for ULV application [7]. See label for details. See introductory notes
- Off-label Approval to Mar 1991 for use on asparagus, french beans, vining, dried and mange-tout peas and tree nuts (OLA 1394/88) [3], (OLA 1209/88) [6]; to June 1992 for use on linseed (OLA 0569/89) [3], broad beans (OLA 0570/89) [3], lettuce and radicchio (OLA 0571/89) [3], evening primrose (OLA 0572/89) [3]; to July 1992 for use on hardy ornamental nursery stock and nursery fruit trees and bushes (OLA 0613/89) [3], white turnips (OLA 0674/89) [3]; to Sep 1992 for use on lettuce, ornamentals, ornamental bulbs, aubergines, cucurbits, peppers, tomatoes (OLA 0755/89) [3]; to Jan 1993 for use on flower crops and bedding plants (OLA 0067/90) [6], on protected cucurbits, peppers, aubergines, pot plants, bedding plants and flower crops (OLA 0068/90) [7], on evening primrose (OLA 0073/90) [6], on hardy ornamental nursery stock and nursery fruit trees and bushes (OLA 0074/90) [6], on borage (OLA 0075/90) [6]

---

340   iprodione + metalaxyl + thiabendazole
A fungicide polymer seed coating for carrots

Products	Polycote Prime, ingredients 1 & 2	Seedcote	50:45:24% w/w	DS	04075, 04076

Uses   Damping off, alternaria blight, black rot in CARROTS.

Notes   **Efficacy**
- Mix active ingredients with polymer supplied according to label instructions and apply to seed in Polycote Seed Coater machine

---

FOR FULL CONDITIONS OF USE ALWAYS READ THE PRODUCT LABEL

Crop safety/Restrictions
- Seed treatment may alter flow characteristics of seed. Recalibrate drill before drilling
- Do not use on seed of low germination vigour and viability

Special precautions/Environmental safety
- Dangerous to fish. Do not contaminate ponds, waterways or ditches with chemical or used container

Protective clothing/Label precautions
- 18, 22, 29, 36, 37, 52, 63, 65

---

341   iprodione + thiophanate-methyl
A protectant and systemic fungicide for various field crops

| **Products** | Compass | RP | 167:167 g/l | LI | 04580 |

Uses Glume blotch, botrytis, sooty moulds in WINTER WHEAT, SPRING WHEAT. Rhynchosporium, net blotch in WINTER BARLEY, SPRING BARLEY. Sclerotinia stem rot, light leaf spot, alternaria, grey mould in WINTER OILSEED RAPE. Chocolate spot in FIELD BEANS.

Notes   **Efficacy**
- Recommended timing of sprays varies with crop and disease, see label for details
- Eyespot in cereals can be suppressed by treatment where there is no resistance to MBC fungicides
- Other fungicides may be needed in control programmes for certain diseases, see label for details

Crop safety/Restrictions
- Do not apply to oilseed rape after end of flowering (GS 5,0)
- Do not use on oats
- A maximum of 2 treatments may be applied to any crop in one season

Special precautions/Environmental safety
- Harmful to fish. Do not contaminate ponds, waterways or ditches with chemical or used container

Protective clothing/Label precautions
- 29, 35, 53, 63, 66

Harvest interval
- oilseed rape, field beans 3 wk;cereals, do not apply after grain watery-ripe stage (GS 71)

Approval
- Provisional Approval for aerial application to a limited area of winter oilseed rape, winter and spring wheat, barley and field beans. Contact manufacturer for allocation

342   isoproturon

A pre- or post-emergence urea herbicide for use in cereals

See also   *bifenox + isoproturon*
           bifenox + isoproturon + mecoprop
           bromoxynil + ioxynil + isoproturon
           bromoxynil + ioxynil + isoproturon + mecoprop
           cyanazine + isoproturon
           diflufenican + isoproturon
           imazamethabenz-methyl + isoproturon
           ioxynil + isoproturon + mecoprop

Products

1	Arelon	Hoechst	553 g/l	SC	04544
2	Arelon WDG	Hoechst	82.5% w/w	WG	04494
3	Chiltern IPU	Chiltern	500 g/l	SC	04304
4	Hytane 500 FW	Ciba-Geigy	500 g/l	SC	01098
5	Portman Isotop	Portman	500 g/l	SC	03434
6	Power Isoproturon	Power	500 g/l	SC	03922
7	Power Swing	Power	500 g/l	SC	04434
8	Sabre	Schering	553 g/l	SC	04148
9	Sabre WDG	Schering	80% w/w	WG	04719
10	Tolkan 500	FCC	500 g/l	SC	04141
11	Tolkan Liquid	RP	500 g/l	SC	04562

Uses

Annual grasses, blackgrass, rough meadow grass, wild oats, annual dicotyledons in WINTER WHEAT, SPRING WHEAT, WINTER BARLEY, WINTER RYE, TRITICALE [1, 2, 3, 8, 9]. Annual grasses, blackgrass, rough meadow grass, wild oats, annual dicotyledons in WINTER WHEAT, AUTUMN SOWN SPRING WHEAT, WINTER BARLEY, WINTER RYE, TRITICALE [4]. Annual grasses, blackgrass, rough meadow grass, wild oats, annual dicotyledons in WINTER WHEAT, WINTER BARLEY [5, 6]. Annual grasses, blackgrass, rough meadow grass, wild oats, annual dicotyledons in WINTER WHEAT, SPRING WHEAT, AUTUMN SOWN SPRING WHEAT, WINTER BARLEY, SPRING BARLEY [7]. Annual grasses, blackgrass, rough meadow grass, wild oats, annual dicotyledons in WINTER WHEAT, WINTER BARLEY, WINTER RYE, TRITICALE [10]. Annual grasses, blackgrass, rough meadow grass, wild oats, annual dicotyledons in WINTER WHEAT, SPRING WHEAT, AUTUMN SOWN SPRING WHEAT, WINTER BARLEY, SPRING BARLEY, WINTER RYE, TRITICALE [11].

Notes

Efficacy

- May be applied in autumn or spring as a pre- or post-weed emergence treatment. Best results normally achieved by early post-emergence treatment
- See label for details of rates and timings for different weed problems and tank mixes
- Effectiveness may be reduced in seasons of above average rainfall, when heavy rain falls shortly after application or by prolonged dry weather
- Residual activity reduced on soils with more than 10% organic matter. Only use on such soils in spring
- Any trash or ash from straw burning must be buried and dispersed before spraying

---

FOR FULL CONDITIONS OF USE ALWAYS READ THE PRODUCT LABEL

Crop safety/Restrictions
- Recommended timing of treatment varies depending on crop to be treated, method of sowing, season of application, weeds to be controlled and product being used. See label for details. Do not use on waterlogged or very cloddy soils, if frost imminent or after onset of frosty weather
- Crop damage may occur on free draining, stony or gravelly soils if heavy rain falls soon after spraying. Early sown crops may be damaged if spraying precedes or coincides with a period of rapid growth
- Do not roll for 1 wk before or after treatment or harrow for 1 wk before or any time after
- Take care to avoid overlapping of spray swaths where autumn sown spring wheat is treated post-emergence in spring [4]

Special precautions/Environmental safety
- Irritating to skin and eyes [5]. Irritating to skin [6, 8, 11]
- Harmful to fish. Do not contaminate ponds, waterways or ditches with chemical or used container [3]

Protective clothing/Label precautions
- 6a [6, 8, 11]; 6a, 6b [5]; 29, 54, 60, 63, 66, [1]; 21, 29, 54, 63, 67 [2]; 14, 29, 53, 60, 63, 66 [3]; 18, 21, 29, 36, 37, 54, 63, 66, 70 [4, 5, 6, 8]; 17 [8]; 28 [4, 6, 8, 11]; 54 [4, 5, 6, 8, 11]; 60 [11]

Approval
- Approved for aerial application on cereals [1-4, 8]. See introductory notes.
- May be applied through CDA equipment. See label for details. See introductory notes [4, 10, 11]

---

343   isoproturon + isoxaben
A residual pre- and early post-emergence herbicide for cereals

Products

1 Fanfare 469FW	Ciba-Geigy	450:19 g/l	SC	04020
2 Ipso	DowElanco	450:19 g/l	SC	04089

Uses

Annual grasses, blackgrass, annual dicotyledons in WINTER WHEAT, AUTUMN SOWN SPRING WHEAT, WINTER BARLEY, WINTER RYE, TRITICALE.

Notes

Efficacy
- Apply in autumn or spring pre- or early post-emergence of weeds. Best results normally achieved early post-emergence
- See label for details of rates and timings for different weed problems
- Prolonged dry weather after application reduces effectiveness
- Residual activity reduced on soils with more than 10% organic matter

Crop safety/Restrictions
- May be used pre- or post-emergence of winter wheat or barley up to 31 Mar or 2nd node detectable (GS 32), whichever comes first
- On autumn sown spring wheat use as post-emergence spray in spring
- On winter rye and triticale use only as a pre-emergence treatment on named cultivars
- Do not use on waterlogged or very cloddy soils. Crop damage may occur on free draining, stony or gravelly soils if heavy rain falls soon after spraying

- Broadcast crops should be treated post-emergence
- Do not use on durum wheat or on crops undersown or about to be undersown
- Do not treat if frost imminent or after onset of frosty weather
- Early sown crops may be damaged if spraying precedes or coincides with a period of rapid growth
- Do not roll for 1 wk before or after treatment or harrow for 1 wk before or at any time after
- Following crops of beet, oilseed rape and other brassicas are sensitive to isoxaben if direct drilled or minimally cultivated. Land must be ploughed to at least 20 cm before drilling subsequent crops

Special precautions/Environmental safety
- Harmful to fish. Do not contaminate ponds, waterways or ditches with chemical or used container

Protective clothing/Label precautions
- A
- 21, 28, 29, 53, 60, 63, 65

344   isoproturon + metsulfuron-methyl
A contact and residual herbicide for use in cereals

Products	Oracle	Du Pont	50:20% w/w	KK	04027

Uses   Annual grasses, blackgrass, annual dicotyledons in WINTER WHEAT, SPRING WHEAT , WINTER BARLEY.

Notes   **Efficacy**
- Treatment most effective when weeds small and growing actively
- Blackgrass should be treated as early as possible after emergence and before tillering. Useful control can be obtained up to jointing stage of grass if growing actively in moist soil
- Residual activity is reduced on soils with more than 10% organic matter

Crop safety/Restrictions
- Apply after 1 Jan from 2-leaf stage of wheat (GS 12) or 3-leaf of barley (GS 13)
- Do not treat Igri barley before leaf sheath erect (GS 30)
- Do not use on durum wheat or any undersown cereal
- Do not apply to crops under stress from drought, waterlogging, cold or any other factor
- Do not treat if frost imminent or after onset of frosty weather
- Do not spray within 7 d of rolling
- Do not spray more than once in any crop or on crops previously treated with Ally, Glean TP or Harmony M
- Sow only cereals, oilseed rape, field beans or grass in same year after treatment. In the event of crop failure sow only wheat within 3 mth

---

FOR FULL CONDITIONS OF USE ALWAYS READ THE PRODUCT LABEL

- Take extreme care to avoid drift onto nearby broad-leaved crops, land intended for cropping, desirable trees or other plants. Do not drain or flush out spraying equipment on land planted with or to be planted with trees or crops other than cereals
- Use specified procedure for cleaning out spraying equipment after use

Protective clothing/Label precautions
- A
- 16, 18, 22, 28, 29, 36, 37, 54, 63, 66, 70

545 isoproturon + pendimethalin
A contact and residual herbicide for use in winter cereals

Products

1 Encore	Cyanamid	125:250 g/l	SC	04737
2 Trump	Cyanamid	236:236 g/l	SC	03687

Uses

Annual grasses, blackgrass, wild oats, annual dicotyledons in WINTER WHEAT, WINTER BARLEY, WINTER RYE, TRITICALE.

Notes

Efficacy
- Annual grasses controlled from pre-emergence to 3-leaf stage, extended to 3-tiller stage with blackgrass by tank mixing with isoproturon. Best results on wild oats post-emergence in autumn
- Annual dicotyledons controlled pre-emergence and up to 4-leaf stage [1], up to 8-leaf stage [2]. See label for details
- Best results achieved by application to fine, firm seedbed with adequate soil moisture. Trash, ash or straw should have been incorporated evenly
- Contact activity is reduced by rain within 6 h of application
- Activity may be reduced on soils with more than 6% organic matter or ash. Do not use on soils with more than 10% organic matter
- Do not roll or harrow after application

Crop safety/Restrictions
- Apply to crops from immediately post-drilling to early tillering stage (GS 21-22)
- On winter rye and triticale use pre-emergence only on named cultivars
- Do not use on durum wheat, spring cereals or spring cultivars drilled in autumn
- Do not use pre-emergence unless crop seed covered with at least 32 mm settled soil. May be applied to shallowly drilled crops after crop has emerged
- Do not use on crops suffering stress from disease, drought, waterlogging, poor seedbed conditions or other causes or apply post-emergence when frost imminent
- Do not undersow treated crops
- In the event of autumn crop failure spring wheat, spring barley, maize, potatoes, beans or peas may be grown following ploughing to at least 15 cm

Special precautions/Environmental safety
- Dangerous to fish. Do not contaminate ponds, waterways or ditches with chemical or used container

Protective clothing/Label precautions
- A [2]; A, C [1]
- 21, 25, 28, 29, 52, 66, 60

346   isoproturon + trifluralin

A residual pre- and early post-emergence herbicide for cereals

Products	Autumn Kite	Schering	300:200 g/l	EC	0383C

Uses

Annual grasses, blackgrass, annual dicotyledons in WINTER WHEAT, WINTER BARLEY.

Notes

Efficacy
- Provides contact and residual control. Best results achieved by application to dicotyledons pre-emergence or up to 2-leaf stage, to blackgrass up to 2-3 tillers
- Pre-emergence treatment best on moist soils, post-emergence when leaves dry, weeds in active growth and rain not expected for 2 h. Early pre-emergence treatment may need to be followed by post-emergence spray
- Effectiveness may be reduced by prolonged dry or sunny weather after application
- Do not use on soils with more than 10% organic matter
- If used in conjunction with minimum cultivation ensure that all trash and burnt straw is removed, buried or dispersed before spraying
- Do not harrow treated crops

Crop safety/Restrictions
- Apply to drilled crops from after drilling to 4-leaf and 2 tillers stage [GS 22] and to broadcast crops after 3-leaf stage (GS 13)
- Do not use on durum wheat or crops to be undersown
- Do not use on sands and do not incorporate in soil. On very stony, gravelly or other free draining soils crops may be damaged if heavy rain falls soon after treatment
- Do not spray crops stressed by frost, waterlogging, deficiency or pest attack
- Do not roll after treatment until following spring
- In case of crop failure only sow carrots, peas or sunflowers within 5 mth and plough to at least 15 cm. Do not sow sugar beet in spring following treatment

Special precautions/Environmental safety
- Irritating to skin and eyes
- Flammable
- Harmful to fish. Do not contaminate ponds, waterways or ditches with chemical or used container

Protective clothing/Label precautions
- A, C
- 6a, 6b, 12c, 18, 21, 25, 28, 29, 36, 37, 53, 60, 63, 66, 70

FOR FULL CONDITIONS OF USE ALWAYS READ THE PRODUCT LABEL

347   isoxaben

A soil-acting amide herbicide for use in cereals, grass and fruit
See also isoproturon + isoxaben

Products

1 Flexidor	DowElanco	500 g/l	SC	03481
2 Knot Out	Synchemicals	125 g/l	SC	04701
3 Tripart Ratio	Tripart	125 g/l	SC	04659

Uses

Annual dicotyledons in WHEAT, BARLEY, OATS, WINTER RYE, DURUM WHEAT, TRITICALE, NEWLY SOWN LEYS, GRASS SEED CROPS, AMENITY GRASS, APPLES, PEARS, PLUMS, BLACKCURRANTS, GOOSEBERRIES, RASPBERRIES [1]. Annual dicotyledons in AMENITY GRASS [2]. Annual dicotyledons in WINTER WHEAT, WINTER BARLEY, WINTER OATS, DURUM WHEAT, RYE, TRITICALE, NEWLY SOWN LEYS, GRASS SEED CROPS [3].

Notes

Efficacy
- When used alone apply pre-weed emergence. Best results by spraying as soon as possible after drilling into well-prepared, seedbed when 5-10 mm rain falls within 14 d
- Effectiveness is reduced in dry conditions. Weed seeds germinating at depth are not controlled
- Activity reduced on soils with more than 10% organic matter. Do not use on peaty soils
- Various tank mixtures are recommended for early post-weed emergence treatment (especially for grass weeds). See label for details
- Best results on turf achieved by applying to firm, moist seedbed within 2 d of sowing. Avoid disturbing soil surface after application [2]

Crop safety/Restrictions
- Apply pre- or post-crop emergence in cereals from Sep to 31 Mar or before second node detectable (GS 32), whichever comes first
- May be used on sand, light, medium and heavy soils [1]
- Treated crops may be undersown, but not with clover
- See label for details of crops which may be sown in spring in the event of failure of a treated crop

Protective clothing/Label precautions
- A, C
- 54, 60, 63, 66, 70

Withholding period
- Dangerous to livestock. Keep all livestock out of treated grass for at least 50 d

---

348   isoxaben + methabenzthiazuron

A soil acting herbicide for use in winter cereals

Products

Glytex	Bayer	3.4:70% w/w	WP	04230

Uses

Annual dicotyledons, annual meadow grass, rough meadow grass in WINTER WHEAT, WINTER BARLEY, WINTER OATS, WINTER RYE, WINTER TRITICALE.

Efficacy
- Best applied pre-weed emergence but may be used up to 1-2 leaf stage of weeds. Apply within 3 d of drilling early crops
- Apply to fine, firm seedbed free of ash and surface trash. Consolidate loose seedbeds by rolling before application
- Control may be reduced on soils with more than 4% organic matter. Do not use where more than 10% organic matter
- Continuous and effective agitation essential during mixing and spraying
- Do not disturb soil surface after application

Crop safety/Restrictions
- Apply to winter wheat pre- or post-emergence before end of Feb
- Apply to winter barley pre-emergence before end Dec (not after mid-Oct on sands in E Anglia). Soil, varietal and regional restrictions apply to post-emergence treatment on barley. See label for details
- Apply to winter oats, rye and triticale pre-emergence before end of Dec (not after mid-Oct on sands in E. Anglia)
- Do not tank-mix with other products or apply other products within 7 d on barley
- Cereals, grass leys, potatoes or maize may be planted following a treated crop using normal cultivation, other crops after 10-12 mth following ploughing to 20 cm
- In the event of crop failure cultivate treated soil and sow only cereals or grass

Special precautions/Environmental safety
- Harmful to fish. Do not contaminate ponds, waterways or ditches with chemical or used container

Protective clothing/Label precautions
- 21, 28, 29, 53, 63, 67

349   lambda-cyhalothrin
A quick-acting contact and ingested pyrethroid insecticide

Products   Hallmark   ICI   50 g/l   EC   04466

Uses   Aphids, apple sucker, apple leaf midge in APPLES. Pear sucker in PEARS. Damson-hop aphid, red spider mites in HOPS. Caterpillars in BRUSSELS SPROUTS, CABBAGES, CAULIFLOWERS, BROCCOLI, CALABRESE. Pea and bean weevils, pea moth in PEAS.

Notes   **Efficacy**
- Apply at first signs of pest attack or as otherwise recommended and repeat after 3-4 wk (10-14 d for pea moth) if necessary
- Add Agral wetter to spray for use on brassicas

Special precautions/Environmental safety
- Harmful in contact with skin or if swallowed. Irritating to eyes and skin
- Extremely dangerous to bees. Do not apply at flowering stage (except as directed on peas). Keep down flowering weeds

FOR FULL CONDITIONS OF USE ALWAYS READ THE PRODUCT LABEL

- Extremely dangerous to fish. Do not contaminate ponds, waterways or ditches with chemical or used container
- Do not allow spray from vehicle mounted sprayers to fall within 6 m of surface waters or ditches, or from hand-held sprayers within 2 m, direct spray away from water

Protective clothing/Label precautions
- A, C, H, J, K, L, M
- 5a, 5c, 6a, 6b, 12c, 14, 18, 21, 29, 36, 37, 49, 51, 64, 66, 70, 78

Harvest interval
- zero

350   lenacil

A residual, soil-acting uracil herbicide for beet and other crops
See also chloridazon + lenacil

Products					
1 Ashlade Lenacil	Ashlade	80% w/w	WP	03538
2 Venzar	Du Pont	80% w/w	WP	02293
3 Vizor	Farm Protection	440 g/l	SC	02315

Uses

Annual dicotyledons, annual meadow grass in BEET CROPS, STRAWBERRIES, BLACKCURRANTS, REDCURRANTS, GOOSEBERRIES, RASPBERRIES, BLACKBERRIES, LOGANBERRIES, MINT, DAFFODILS, TULIPS, ROSES, DAHLIAS, MICHAELMAS DAISY, CHRYSANTHEMUM MAXIMUM, ESTABLISHED WOODY ORNAMENTALS, HERBACEOUS PERENNIALS [1, 2].
Annual dicotyledons, annual meadow grass in BEET CROPS, STRAWBERRIES [3].
Annual dicotyledons, annual grasses in SPINACH *(off-label)* [1, 2, 3].

Notes

Efficacy
- Weeds controlled as germinating not after emergence. Rain or irrigation necessary after spraying to activate chemical. Effectiveness may be reduced by dry conditions
- May be used in fruit and ornamentals on soils with less than 10% organic matter, in beet crops on more highly organic soils with incorporation

Crop safety/Restrictions
- Apply to beet crops pre-drilling incorporated, pre- or post-emergence
- Recommended alone as pre-drilling incorporated treatment on soils with more than 15% organic matter, pre-emergence on mineral soils
- Recommended in tank mixture with other beet herbicides. See label for details
- Do not use on soils lighter than loamy sand or heavier than sandy-clay loam, nor on stony or gravelly soils
- Heavy rain after application may cause damage to beet and maiden strawberrie
- Strawberries may be treated immediately after planting (including cold-stored runners), as maidens, established crops or runner beds. Products differ in varieties recommended for treatment. See label for details. Rainfall and soil type may influence safety of treatment
- Apply to bush or cane fruit in late winter to early spring. Plant cuttings at least 15 cm deep and firm soil well
- Apply to bulbs before or shortly after emergence, to ornamentals before weed growth starts in spring
- Do not spray when crops are flowering or fruiting

413

- Do not use more than 2 applications in 1 yr or apply other residual herbicides for at least 3 mth
- Succeeding crops should not be planted or sown for at least 4 mth after treatment following ploughing to at least 15 cm

Special precautions/Environmental safety
- Irritating to eyes, skin and respiratory system
- Harmful to fish. Do not contaminate ponds, waterways or ditches with chemical or used container

Protective clothing/Label precautions
- A, C
- 6a, 6b, 6c, 18, 21, 28, 29, 36, 37, 53, 63, 65, 67

Approval
- Off-label Approval to Feb 1991 for use on spinach (OLA 0179, 0180, 0181/88) [1, 2, 3]

351 lenacil + phenmedipham
A contact and residual herbicide for use in sugar beet

Products	DUK-880	Du Pont	440:114 g/l	KL	04121

Uses Annual dicotyledons in SUGAR BEET.

Notes **Efficacy**
- Apply at any stage of crop when weeds in cotyledon stage
- Germinating weeds controlled by root uptake for several weeks
- Best results achieved under warm, moist conditions on a fine, firm seedbed
- Treatment may be repeated up to a maximum of 3 applications on later weed flushes
- Residual activity may be reduced on highly organic soils or under very dry conditions
- Rain falling 1 h after application does not reduce activity

Crop safety/Restrictions
- Do not apply when temperature above or likely to exceed 21 °C on day of spraying or under conditions of high light intensity
- Do not spray any crop under stress from drought, waterlogging, cold, wind damage or any other cause
- Heavy rain after application may reduce stand of crop particularly in very hot weather. In severe cases yield may be reduced
- Do not apply more than 3 treatments per crop
- Do not sow or plant any crop within 4 mth of treatment. In case of crop failure only sow or plant beet crops, strawberries or other tolerant horticultural crop within 4 mth

Special precautions/Environmental safety
- Harmful to fish. Do not contaminate ponds, waterways or ditches with chemical or used container

FOR FULL CONDITIONS OF USE ALWAYS READ THE PRODUCT LABEL

352   linuron

A contact and residual urea herbicide for various field crops
See also chlorpropham + linuron
        2,4-DB + linuron + MCPA

Products

1 Afalon	Hoechst	450 g/l	SC	04665
2 Ashlade Linuron FL	Ashlade	480 g/l	SC	03439
3 Ashlade Linuron 42FL	Ashlade	421 g/l	SC	04696
4 Atlas Linuron	Atlas	370 g/l	LI	03054
5 Campbell's Linuron 45% Flowable	MTM Agrochem.	450 g/l	SC	00408
6 Du Pont Linuron 50	Du Pont	50% w/w	WP	01209
7 Du Pont Linuron 4L	Du Pont	480 g/l	SC	01207
8 Linuron 450 FL	FCC	450 g/l	SC	02709
9 Linuron Flowable	PBI	450 g/l	SC	02965
10 Liquid Linuron	PBI	133 g/l	EC	01556
11 Liquid Linuron 15	Farm Protection	150 g/l	EC	01216
12 Rotalin	Farm Protection	300 g/l	SC	01827

Uses

Annual dicotyledons, black bindweed, chickweed, fat-hen, redshank, corn marigold, annual meadow grass in SPRING CEREALS, POTATOES, NEWLY SOWN LEYS, CARROTS, PARSNIPS, PARSLEY, CELERY [3, 6, 7, 12]. Annual dicotyledons, black bindweed, chickweed, fat-hen, redshank, corn marigold, annual meadow grass in SPRING CEREALS, POTATOES, CARROTS, PARSNIPS, PARSLEY, CELERY [4, 5]. Annual dicotyledons, black bindweed, chickweed, redshank, annual meadow grass in WINTER WHEAT, WINTER BARLEY [4]. Annual dicotyledons, black bindweed, chickweed, fat-hen, redshank, corn marigold, annual meadow grass in SPRING CEREALS, POTATOES, NEWLY SOWN LEYS, CARROTS, PARSLEY, CELERY [2]. Annual dicotyledons, black bindweed, chickweed, fat-hen, redshank, corn marigold, annual meadow grass in SPRING CEREALS, POTATOES, CARROTS, PARSNIPS, CELERY [11]. Annual dicotyledons, black bindweed, chickweed, fat-hen, redshank, annual meadow grass in POTATOES, CARROTS, PARSNIPS, PARSLEY, CELERY [1]. Annual dicotyledons, black bindweed, chickweed, fat-hen, redshank, annual meadow grass in POTATOES, CARROTS, PARSNIPS, CELERY [9, 10]. Annual dicotyledons, black bindweed, chickweed, fat-hen, redshank, annual meadow grass in POTATOES, CARROTS, PARSNIPS, PARSLEY [8]. Annual weeds in EVENING PRIMROSE, HORSERADISH *(off-label)* [10, 11]. Annual weeds in HERBS, HORSERADISH *(off-label)* [6]. Annual weeds in LINSEED, HERBS, POT PLANTS, BEDDING PLANTS, FLOWERS *(off-label)* [12]. Annual weeds in ORNAMENTALS *(off-label)* [11]. Annual weeds in HERBS, CELERIAC, ONIONS, LEEKS, GARLIC *(off-label)* [6, 9].

Notes

Efficacy
• Many weeds controlled pre-emergence or post-emergence to 2-3 leaf stage, some (annual meadow grass, mayweed) only susceptible pre-emergence. See label for details
• Best results achieved by application to firm, moist soil of fine tilth
• Little residual effect on soils with more than 10% organic matter

- For chickweed control in newly sown grass leys in autumn apply post-crop emergence. Clovers will be severely checked [3, 6, 7, 12]

Crop safety/Restrictions
- Drill spring cereals at least 3 cm deep and apply pre-emergence of crop or weeds
- Do not use on undersown cereals or crops grown on sands or very light soils or soils heavier than sandy clay loam or with more than 10% organic matter
- Apply to potatoes well earthed up to a rounded ridge pre-crop emergence and do not cultivate after spraying
- Apply within 4 d of drilling carrots or post-emergence after first rough leaf stage.
- Do not apply to emerged crops of carrots, parsnips or parsley under stress
- Recommendations for parsnips, parsley and celery vary. See label for details
- Poor conditions at drilling or planting may be followed by crop damage
- Potatoes, carrots and parsnips may be planted at any time after application. Lettuce should not be grown within 12 mth of treatment. Transplanted brassicas may be grown from 3 mth after treatment

Special precautions/Environmental safety
- Irritating to eyes, skin and respiratory system [2, 4, 5, 10-12]
- Flammable [10, 11]
- Harmful to fish. Do not contaminate ponds, waterways or ditches with chemical or used container [2, 3, 8-10]

Protective clothing/Label precautions
- A, C [6, 7]; A, C, K, M [10]
- 6a, 6b, 6c, [2, 4, 5, 9-12]; 12c [10, 11]; 18, 21, 28, 29, 36, 37, 63, 65 or 66 [2-11]; 21, 29, 63, 66 [1, 12]; 53 [2, 3, 8-10]; 54 [1, 4-7, 11]

Withholding period
- Do not graze crops within 4 wk after treatment [12]. Do not allow stock to graze treated grass for 5 mth after application [6, 7]

Harvest interval
- Herbs 7d [12]

Approval
- Approval for aerial application on potatoes, carrots, parsnips, celery, parsley [1]
- Off-label approval to Apr 1992 for use on various herbs (OLA 0342/89) and horseradish (OLA 0376/89) [6]; to Apr 1992 for use on evening primrose (OLA 0378/89), to May 1992 for use on horseradish (OLA 0435/89) [10]; to Apr 1992 for use on horseradish (OLA 0377/89), to May 1992 for use on evening primrose (OLA 0470/89) and ornamentals (0473/89) [11]; to May 1992 for use on linseed (OLA 0456/89) [12]; to Jan 1993 for use on herbs, celeriac, onions, leeks, garlic (OLA 0056/90) [6], (OLA 0054, 0070/90) [9], on pot plants, bedding plants, flower crops, herbs (OLA 0055/90) [12]

FOR FULL CONDITIONS OF USE ALWAYS READ THE PRODUCT LABEL

353   linuron + terbutryn

A contact and residual herbicide for potatoes

Products	Tempo	Farm Protection	150:150 g/l	SC	02736

Uses   Annual dicotyledons, fumitory, annual meadow grass in POTATOES.

Notes

Efficacy
- Weeds controlled pre- to early post-emergence. Apply before seedlings have passed cotyledon stage
- Higher dose rate recommended on soils with more than 10% organic matter but residual activity limited on such soils
- Rain soon after application is essential to carry chemical into root zone of germinating weeds to provide optimum residual effect

Crop safety/Restrictions
- Apply after planting crop but before 10% emergence of crop
- Do not sow lettuce in same season following treatment. Other crops may be sown or planted 3 mth after treatment (4 mth after treatment where prolonged drought)

Special precautions/Environmental safety
- Irritating to eyes, skin and respiratory system
- Harmful to fish. Do not contaminate ponds, waterways or ditches with chemical or used container

Protective clothing/Label precautions
- 6a, 6b, 6c, 21, 28, 29, 53, 60, 63, 66

---

354   linuron + trietazine

A contact and residual herbicide for potatoes

Products	Bronox	Schering	24:24% w/w	WP	03864

Uses   Annual dicotyledons, annual grasses in POTATOES. Annual dicotyledons, annual grasses in HARDY ORNAMENTAL NURSERY STOCK, NURSERY FRUIT TREES AND BUSHES *(off-label)*.

Notes

Efficacy
- Weeds controlled from pre-emergence to cotyledon stage. Best results achieved by application to shallow, rounded ridges as soon as possible after planting
- Application soon after planting gives control until harvest.
- Effectiveness is reduced under very dry conditions
- Do not use on soil with more than 10% organic matter
- If irrigation is used it should not be heavy enough to disturb soil surface
- Do not use on potatoes that have been re-ridged unless previous weed growth killed

Crop safety/Restrictions
- Apply before 10% crop emergence
- Do not use on soils classed as sands or on soils with a very open texture

- If heavy rain falls soon after application some yellowing of foliage may occur from which crop normally recovers completely
- Do not plant brassicas within 14 wk after application or other crops within 10 wk and cultivate to 15 cm before sowing or planting

Special precautions/Environmental safety
- Harmful if swallowed. Irritating to skin and eyes
- Flammable
- Harmful to fish. Do not contaminate ponds, waterways or ditches with chemical or used container

Protective clothing/Label precautions
- 5c, 6a, 6b, 12c, 21, 28, 29, 53, 60, 63, 65

Approval
- Off-label approval to Apr 1992 for use on hardy ornamental nursery stock and nursery fruit trees and bushes (OLA 0343/89)

355 linuron + trietazine + trifluralin
A pre-emergence herbicide for use in winter cereals

Products	Pre-Empt	Schering	46:54:208 g/l	EC	03825

Uses Annual dicotyledons, annual grasses in WINTER WHEAT, WINTER BARLEY.

Notes

Efficacy
- Gives residual control of weeds germinating in surface layers of soil
- Best results by application to fine, firm, moist, clod-free seedbed within 5 d of drilling
- Effectiveness reduced if prolonged dry or sunny weather after spraying
- Do not use on soils with more than 10% organic matter
- A suitable post-emergence treatment may be needed in spring after a mild, wet winter or where barley has been sown early
- All surface trash or burnt straw must be removed, buried or dispersed before spraying

Crop safety/Restrictions
- Apply in autumn sown crops as soon as possible after drilling
- Do not incorporate into soil or apply after emergence
- Crop seed must be covered by at least 2.5 cm of settled soil. Special care needed with broadcast crops. Any slits caused by direct-drilling must be closed by cultivation or rolling before spraying
- Do not use on durum wheat or crops to be undersown
- On free-draining stony or gravelly soils there is risk of crop damage especially if heavy rain falls soon after treatment
- Do not harrow after treatment or roll until following spring

Special precautions/Environmental safety
- Irritating to eyes, skin and respiratory system

FOR FULL CONDITIONS OF USE ALWAYS READ THE PRODUCT LABEL

- Harmful to fish. Do not contaminate ponds, waterways or ditches with chemical or used container

Protective clothing/Label precautions
- 6a, 6b, 6c, 18, 21, 25, 28, 29, 36, 37, 53, 60, 63, 66, 70

Approval
- May be applied through CDA equipment. See label for details. See introductory notes on ULV application

---

356 linuron + trifluralin

A residual pre-emergence herbicide for use in winter cereals

Products					
1 Ashlade Flint	Ashlade	120:240 g/l	EC	04471
2 Atlas Janus	Atlas	127.5:255 g/l	LI	03085
3 Campbell's Solo	MTM Agrochem.	125:240 g/l	EC	03702
4 Campbell's Trifluron	MTM Agrochem.	250:480 g/l	KL	02682
5 Chandor	DowElanco	120:240 g/l	EC	03051
6 Linnet	PBI	106:192 g/l	EC	01555
7 Marksman	FCC	250:480 g/l	KL	01290
				01291
8 Onslaught	Quadrangle	160:320 g/l	SC	02548
9 Trifarmon FL	Farm Protection	160:320 g/l	SC	02870

Uses

Annual dicotyledons, annual grasses, annual meadow grass, rough meadow grass, perennial ryegrass in WINTER WHEAT, WINTER BARLEY [1-4, 6, 7]. Annual dicotyledons, annual grasses, annual meadow grass, rough meadow grass, perennial ryegrass in WINTER WHEAT, WINTER BARLEY, TRITICALE [5, 8, 9]. Annual weeds in LINSEED *(off-label)* [5, 6].

Notes   **Efficacy**
- Effective against weeds germinating near soil surface. Best results achieved by application to fine, firm, moist seedbed free of clods, crop residues or established weeds
- Effectiveness reduced by long dry period after application or on waterlogged soil
- Do not use on peaty soils or where organic matter exceeds 10% (8%[6])
- With autumn application residual effects normally last until spring but further herbicide treatment may be needed on thin or backward crops
- With loose seedbeds on lighter soils results improved by rolling after drilling

Crop safety/Restrictions
- Apply without incorporation as soon as possible after drilling and before crop emergence, within 3 d on early drilled crops
- Do not treat durum wheat or undersown crops
- Crop seed must be well covered, minimum depth specified varies from 12 to 30 mm
- Do not use on soils classed as sands. Do not harrow after treatment
- Only spring barley or spring wheat may be sown within 6 mth of application

Special precautions/Environmental safety
- Harmful if swallowed [2]. Irritating to eyes, skin and respiratory system [1-9]
- Harmful to fish. Do not contaminate ponds, waterways or ditches with chemical or used container

Protective clothing/Label precautions
• A, C [2, 5, 6]; A, C, H [3]
• 5c [2]; 6a, 6b, 6c, 14, 18, 21, 25, 28, 29, 36, 37, 53, 60, 61, 63, 66, 70[1-11]

Approval
• Off-label Approval to Apr 1992 (OLA 0363/89) [5], to Jan 1993 (OLA 0052/90) [6], for use on linseed

357 malathion

A broad-spectrum, contact, organophosphorus insecticide and acaricide

Products Malathion 60   Farm Protection   600 g/l   LI   01242

Uses Aphids in POTATOES, CARROTS, CELERY, PARSNIPS, PEAS, FIELD BEANS, BROAD BEANS, FRENCH BEANS, RUNNER BEANS, RED BEET, KALE, SWEDES, TURNIPS, LETTUCE, APPLES, PEARS, CHERRIES, PLUMS, DAMSONS, PEACHES, APRICOTS, NECTARINES, CURRANTS, CANE FRUIT, STRAWBERRIES, FLOWERS. Pollen beetles, cabbage stem weevil, cabbage seed weevil in OILSEED RAPE, MUSTARD. Celery fly in CARROTS, CELERY, PARSNIPS. Thrips in ONIONS, LEEKS, FLOWERS. Bryobia mites, leafhoppers, suckers in APPLES, PEARS. Red spider mites in APPLES, PEARS, CHERRIES, PLUMS, DAMSONS, GOOSEBERRIES. Codling moth in APPLES. Sawflies in GOOSEBERRIES. Leafhoppers, raspberry beetle in RASPBERRIES. Scale insects in ROSES. Aphids, thrips, whitefly, mealy bugs, scale insects, leafhoppers in GLASSHOUSE FLOWERS, PROTECTED LETTUCE, TOMATOES, GLASSHOUSE ROSES. Sciarid flies in MUSHROOMS.

Notes **Efficacy**
• Spray when pest first seen and repeat as necessary, usually at 7-14 d intervals
• Number and timing of sprays vary with crop and pest. See label for details
• Repeat spraying routinely for scale insect and whitefly control in glasshouses
• Apply as drench to control sciarid larvae in mushroom casing
• Strains of aphids and red spider mite resistant to organophosphorus compounds have developed in some areas

Crop safety/Restrictions
• Do not use on antirrhinums, crassula, ferns, fuchsia, gerbera, petunias, pileus, sweet peas or zinnias
• Pick mushrooms hard before treatment and do not harvest for 4 d afterwards

Special precautions/Environmental safety
• This product contains an organophosphorus compound. Do not use if under medical advice not to work with such compounds
• Harmful to bees. Do not apply at flowering stage. Keep down flowering weeds
• Harmful to fish. Do not contaminate ponds, waterways or ditches with chemical or used container

---

FOR FULL CONDITIONS OF USE ALWAYS READ THE PRODUCT LABEL

Protective clothing/Label precautions
- 1, 21, 28, 29, 35, 50, 53, 63, 65

Harvest interval
- 1 d; to avoid possibility of taint 4 d; crops for processing 7 d

Approval
- Approved for aerial application on arable field crops except oilseed rape. See introductory notes

358   maleic hydrazide

A plant growth regulator suppressing grass and bud growth
See also dicamba + maleic hydrazide + MCPA

Products

1 Bos MH	Bos	180 g/l	SL	03589
2 Burtolin	RP Environ.	185 g/l	SL	00355
3 Chiltern Fazor	Chiltern	80% w/w	SP	03314
4 Mazide 25	Synchemicals	250 g/l	SL	02067
5 MSS MH 18	Mirfield	180 g/l	SL	03065
6 Regulox K	RP Environ.	250 g/l	SL	01716
7 Royal Slo-Gro	Uniroyal	217 g/l	SL	01837

Uses

Growth suppression in AMENITY GRASS [1, 4-7]. Growth suppression in GRASS NEAR WATER [1, 6]. Sprout suppression in ONIONS [1, 3, 4, 5]. Sprout suppression in STORED POTATOES [3]. Volunteer suppression in POTATOES [3]. Sucker inhibition in AMENITY TREES AND SHRUBS [2, 4]. Growth retardation in HEDGES [4].

Notes

Efficacy
- Apply to grass at any time of year when growth active, best when growth starting in Apr-May and repeated when growth recommences
- Uniform coverage and 8 h dry weather necessary for effective results
- Mow 3-5 d before and 2-3 wk after spraying. Need for mowing reduced for up to 6 wk
- Apply to onions at 50% necking stage to prevent sprouting in store
- Apply to potatoes at least 3 wk before haulm destruction. Accurate timing essential. See label for details
- To control tree suckers wet trunks thoroughly, especially pruned and basal bud areas
- Spray hawthorn hedges in full leaf, privet 7 d after cutting, in Apr-May

Crop safety/Restrictions
- Do not apply to grass in drought or when suffering from pest, disease or herbicide damage. Do not treat fine turf or grass seeded less than 8 mth previously
- May be applied to grass along water courses but not to water surface [1, 6]
- Treated onions may be stored until Mar but must then be removed to avoid browning
- Only use on potatoes of good keeping quality, not on seed, first earlies or crops for seed
- Avoid drift onto nearby vegetables, flowers or other garden plants

Special precautions/Environmental safety
- Only apply to grass not to be used for grazing
- Do not use treated water for irrigation purposes within 3 wk of treatment or until concentration in water falls below 0.02 ppm [1, 6]

- Maximum permitted concentration in water 2 ppm [1, 6]
- Harmful to fish. Do not contaminate ponds, waterways or ditches with chemical or use container [1]

Protective clothing/Label precautions
- 21, 29, 60, 63, 65, 66, 70 [1-6]; 53 [1]; 54 [2-6]; 55, 56, [1, 6]

Harvest interval
- onions 1 wk; potatoes 3 wk

Approval
- Approved for aquatic weed control. See introductory notes on use of herbicides near water [1, 6]

---

359 mancozeb

A protective dithiocarbamate fungicide for potatoes and other crops

See also benalaxyl + mancozeb
carbendazim + mancozeb
cymoxanil + mancozeb

Products					
	1 Ashlade Mancozeb FL	Ashlade	410 g/l	SC	03208
	2 Dithane 945	PBI	80% w/w	WP	04017
	3 Dithane 945	Rohm & Haas	80% w/w	WP	04017
	4 Dithane Dry Flowable	PBI	75% w/w	WG	04255
	5 Dithane Dry Flowable	Rohm & Haas	75% w/w	WG	04251
	6 Karamate Dry Flo	Rohm & Haas	75% w/w	WG	04252
	7 Karamate N	Rohm & Haas	80% w/w	WP	01125
	8 Manzate 200	Du Pont	80% w/w	WP	01281
	9 Penncozeb	Shell	80% w/w	WP	02716
	10 Penncozeb Water Soluble Pack	Shell	80% w/w	SP	04609
	11 Portman Mancozeb 80	Portman	80% w/w	WP	01528
	12 Unicrop Flowable Mancozeb	Unicrop	410 g/l	SC	04700
	13 Unicrop Mancozeb	Unicrop	80% w/w	WP	02273

Uses Septoria diseases, rust, sooty moulds, mildew, rhynchosporium in WHEAT, BARLEY, OATS, RYE, TRITICALE [1, 12]. Septoria diseases, rust, sooty moulds, net blotch in WHEAT, BARLEY [2-5]. Septoria diseases, rust, sooty moulds, net blotch, rhynchosporium in WHEAT, BARLEY [9, 10]. Septoria diseases, rust, sooty moulds, mildew in CEREALS [11]. Downy mildew in WINTER OILSEED RAPE [2-5]. Blight in POTATOES [1-5, 8-13]. Downy mildew in LETTUCE, PROTECTED LETTUCE [6, 7]. Scab in APPLES, PEARS [6, 7]. Scab in APPLES [9, 10]. Currant leaf spot in BLACKCURRANTS, GOOSEBERRIES [6, 7]. Currant leaf spot in BLACKCURRANTS [11]. Rust in ROSES, CARNATIONS, GERANIUMS [6, 7]. Black spot in ROSES [6, 7]. Ray blight in CHRYSANTHEMUMS [6]. Fire in TULIPS [6]. Downy mildew in GRAPEVINES *(off-label)* [7].

FOR FULL CONDITIONS OF USE ALWAYS READ THE PRODUCT LABEL

Efficacy
- Optimum timing in cereals depends on disease situation in the crop. See label for details
- Use in tank-mixture with other cereal fungicides to improve disease control [2-5, 9, 10]
- Apply to potatoes before haulm meets across rows (usually mid-Jun) or at earlier blight warning, and repeat every 10-14 d depending on conditions
- May be used on potatoes up to desiccation of haulm
- Number and timing of sprays vary with crop and disease. See label for details
- On oilseed rape apply as soon as disease develops between cotyledon and 5-leaf stage (GS 1,0-1,5)

Crop safety/Restrictions
- May be applied to cereals from 4-leaf stage (GS 14) to before grain milky ripe (GS 73). Statements vary, see label for details
- Treat oilseed rape before end of Dec
- On protected lettuce only 2 post-planting applications of mancozeb or of any combination of products containing EBDC fungicides (mancozeb, maneb, thiram, zineb) either as a spray or a dust are permitted within 2 wk of planting out and none thereafter

Special precautions/Environmental safety
- Irritating to respiratory system. May cause sensitization by skin contact
- Harmful to fish. Do not contaminate ponds, waterways or ditches with chemical or used container [1,-7, 12, 13]

Protective clothing/Label precautions
- A
- 6c, 10a, 18, 21, 28, 29, 36, 37, 63, 67, 70, [1-13]; 53 [1-7, 12, 13]; 54 [8-11]

Harvest interval
- potatoes 7 d; cereals, apples, pears 28 d; lettuce 21 d; grapevines 30 d; blackcurrants, gooseberries 1 mth

Approval
- Approved for aerial application on potatoes [1-5, 8, 9, 12, 13]; on cereals [11, 12]. See introductory notes
- Off-label Approval to Feb 1991 for use on grapevines (OLA 0182/88) [7]

360   mancozeb + metalaxyl
A systemic and protectant fungicide for potatoes and other crops

Products					
1	Fubol 58WP	Ciba-Geigy	48:10% w/w	WP	00927
2	Fubol 75WP	Ciba-Geigy	67.5:7.5% w/w	WP	03462

Uses

Blight in POTATOES [2]. Cavity spot in CARROTS, PARSNIPS [1]. Phytophthora fruit rot in APPLES *(off-label)* [2]. White blister in CABBAGES, CAULIFLOWERS, BROCCOLI, CALABRESE, KALE *(off-label)* [2]. White tip in LEEKS *(off-label)* [2]. Downy mildew in ONIONS *(off-label)* [2]. Damping off in WATERCRESS *(off-label)* [1]. Root rot in RASPBERRIES *(off-label)* [1].

Efficacy

- Apply as protectant spray on potatoes immediately risk of blight in district or as crops begin to meet in (not across) rows and repeat every 10-21 d according to blight risk (every 10-14 d on irrigated crops) [2]
- Use for first part of spray programme to mid-Aug. Later use a protectant fungicide (preferably tin-based) up to complete haulm destruction or harvest [2]
- Do not treat potato crops showing active blight infection
- Best results on carrots and parsnips achieved by application to damp soil. Do not use where these crops have been grown on same site during either of 2 previous years [1]
- Do not apply on potatoes for 2-3 h before rainfall or when raining

Crop safety/Restrictions

- On potatoes apply up to 5 sprays on one crop. Do not use acylalanine products after final Fubol spray [2]
- On carrots and parsnips apply single spray within 6 wk of drilling, on field and broad beans and oilseed rape up to 2 sprays, on Brussels sprouts up to 3 sprays, on outdoor lettuce up to 5 sprays, protected lettuce up to 8 sprays [1]

Special precautions/Environmental safety

- Irritating to skin, eyes and respiratory system
- Harmful to fish. Do not contaminate ponds, waterways or ditches with chemical or used container

Protective clothing/Label precautions

- A
- 6a, 6b, 6c, 18, 21, 29, 35, 36, 37, 53, 63, 67

Harvest interval

- Early and maincrop potatoes 1 d; Brussels sprouts, outdoor lettuce, onions 14 d; field and broad beans, apples, protected lettuce 4 wk; carrots 8 wk; parsnips 16 wk

Approval

- Approved for aerial application on potatoes. See introductory notes [2]
- Off-label Approval to Feb 1991 for use on apples, cabbages, cauliflowers, broccoli, calabrese, kale, leeks and onions (OLA 0185/88) [2]; to Jul 1992 for use on protected seedling watercress beds (OLA 0604/89) [1]; to Sep 1993 for use as drench on raspberries (OLA 0553/90) [1]

361 mancozeb + oxadixyl

A protectant and systemic fungicide for potato blight control

Products					
1 Recoil	Bayer	56:10% w/w	WP	02781
2 Recoil	Schering	56:10% w/w	WP	04038

Uses Blight in POTATOES.

FOR FULL CONDITIONS OF USE ALWAYS READ THE PRODUCT LABEL

Efficacy
- Systemic component (oxadixyl) most effective on young growing foliage and efficacy declines with onset of senescence. Apply as protectant before crop infected
- Spray immediately risk of blight occurs in district or before crop meets in row
- Apply up to 5 treatments at 10-21 d intervals according to degree of blight risk
- Do not apply after end of Aug. Follow up with a protectant treatment (preferably tin-based) until haulm destruction
- Do not apply within 2-3 h of rainfall or irrigation
- Do not treat crops showing active blight infection

Crop safety/Restrictions
- Do not use more than 5 applications of acylalanine based products in one season

Special precautions/Environmental safety
- Irritating to eyes, skin and respiratory system
- May cause sensitization by skin contact

Protective clothing/Label precautions
- A
- 6a, 6b, 6c, 10a, 14, 18, 23, 24, 28, 29, 36, 37, 54, 63, 67, 70

Harvest interval
- zero

Approval
- Approved for aerial application on potatoes. See introductory notes. Consult firm for details [1, 2]

362   maneb

A protectant dithiocarbamate fungicide
See also carbendazim + maneb
carbendazim + maneb + sulphur
carbendazim + maneb + tridemorph
copper oxychloride + maneb + sulphur
fentin acetate + maneb
fentin hydroxide + maneb + zinc
ferbam + maneb + zineb

Products

1	Ashlade Maneb Flowable	Ashlade	480 g/l	SC	02911
2	Campbell's X-Spor SC	MTM Agrochem.	480 g/l	SC	03252
3	Clifton Maneb	Clifton	80% w/w	WP	03033
4	Headland Spirit	WBC Technology	480 g/l	SC	04548
5	Maneb 80	Rohm & Haas	80% w/w	WP	01276
6	Manzate	Du Pont	80% w/w	WP	02436
7	Manzate Flowable	Du Pont	480 g/l	SC	04055
8	Power Maneb 80	Power	80% w/w	WP	03236
9	Trimangol 80	Pennwalt	80% w/w	WP	04294
10	Unicrop Flowable Maneb	Unicrop	480 g/l	SC	04323
11	Unicrop Maneb	Unicrop	80% w/w	WP	02274

Uses

Blight in POTATOES [1-11]. Brown rust, yellow rust, septoria, sooty moulds, eyespot, net blotch in WINTER WHEAT [1-10]. Rhynchosporium, brown rust, yellow rust in BARLEY [1, 2, 4-7, 9, 10]. Alternaria, downy mildew in OILSEED RAPE,

BRASSICAS [1, 7, 10]. Cercospora leaf spot in SUGAR BEET [1, 7, 10]. Downy mildew in LETTUCE, PROTECTED LETTUCE [9]. Blight in TOMATOES [3, 9, 11]. Leaf mould in TOMATOES [3]. Black spot in ROSES [3]. Fire in TULIPS [9]. Powdery mildew in CELERY, MARROWS *(off-label)* [9]. Didymella in TOMATOES *(off-label)* [5, 6, 8, 9, 11].

Notes

Efficacy
- Apply to potatoes before blight infection occurs, at blight warning or before haulms meet in row and repeat every 10-14 d
- Most effective on cereals as a preventative treatment before disease established, an early application at about first node stage (GS 31), a late application after flag leaf emergence (GS 37) and during ear emergence before watery ripe stage (GS 71) [6]
- Also recommended in cereals as tank mix with carbendazim, captafol or tridemorph products. Timing and number of sprays vary with crop and other ingredient of mixture. See label for details
- Apply to tomatoes at first sign of disease and repeat every 7-14 d

Crop safety/Restrictions
- Do not apply if frost or rain expected, if crop wet or suffering from drought or physical or chemical stress
- Apply up to 2 sprays on cereals but not after grain watery-ripe stage (GS 71) [6]
- Do not apply more than 5 treatments in any crop [10]
- Do not feed treated sugar beet tops to livestock [7]
- On protected lettuce only 2 post-planting applications of maneb or of any combination of products containing EBDC fungicides (mancozeb, maneb, thiram or zineb) either as a spray or a dust are permitted within 2 wk of planting out and none thereafter

Special precautions/Environmental safety
- Harmful if swallowed [2, 4]. Irritating to skin, eyes and respiratory system [1, 2, 4], to respiratory system [3, 5, 6, 9, 11]
- May cause sensitization by skin contact [5]

Protective clothing/Label precautions
- A [1, 3, 5, 7-11]; A, C [2, 4, 6]
- 5c, 6a, 6b, 6c, 16, 18, 21, 29, 35, 36, 37, 54, 63, 67, 70 [1-11]

Harvest interval
- Protected lettuce 3wk; outdoor celery 14 d; outdoor crops 7 d; protected crops 2 d

Approval
- Approved for aerial application on potatoes [2-6, 9, 11]; winter cereals [2, 4, 6]; winter wheat [3, 11]; spring barley [2]. See introductory notes. Consult firm for details
- Off-label Approval to Feb 1991 for use on outdoor celery and marrows (OLA 0184/88) [9]; to Mar 1991 for use on tomatoes (OLA 0250/8) [2]; (OLA 0252/88) [3]; (OLA 0254, 0258/88) [5]; (OLA 0255/88) [6]; (OLA 0257/88) [8]; (OLA 0260/88) [9]; (OLA 0262/88) [11]

FOR FULL CONDITIONS OF USE ALWAYS READ THE PRODUCT LABEL

363   maneb + zinc

A protectant fungicide for use in potatoes and other crops

Products

	1 Chiltern Manex	Chiltern	480:- g/l	SC	01279
	2 Quadrangle Manex	Quadrangle	480:- g/l	SC	03406
	3 Vassgro Manex	Vass	480:- g/l	SC	03405

Uses

Glume blotch, septoria, powdery mildew, rust, sooty moulds in CEREALS [1, 2]. Brown foot rot and ear blight, sooty moulds, rust, septoria in WHEAT [3]. Brown foot rot and ear blight, rhynchosporium, rust in BARLEY [3]. Cercospora leaf spot in SUGAR BEET [1-3]. Downy mildew, alternaria in OILSEED RAPE, BRASSICAS [1-3]. Blight in POTATOES [1-3]. Scab in APPLES [1, 2]. Currant leaf spot in BLACKCURRANTS [1, 2]. Black spot, rust in ROSES [1, 2]. Fire in TULIPS [1, 2]. Blight, leaf mould in TOMATOES [1, 2].

Notes

Efficacy
- Apply spray before or at first appearance of disease and repeat at 10-14 d intervals (7-10 d on potatoes under severe blight conditions)
- Number and timing of sprays vary with crop and disease. See label for details
- Optimum time for treating cereals is boot stage (GS 45), repeat spray if necessary

Crop safety/Restrictions
- Spray cereals from last leaf just visible to ear three-quarters emerged (GS 20-57)
- Apply a maximum of 5 sprays in sugar beet

Special precautions/Environmental safety
- Irritating to eyes, skin and respiratory system
- Do not feed treated sugar beet tops to livestock
- Do not graze livestock in treated orchards

Protective clothing/Label precautions
- A, C
- 6a, 6b, 6c, 18, 21, 29, 35, 36, 37, 54, 63, 66, 70 [1-3]

Harvest interval
- protected crops 2 d; field crops 7 d

Approval
- Approved for aerial application on cereals. See introductory notes [1-3]

364   maneb + zinc oxide

A protectant fungicide for use on potatoes

Products

| | Mazin | Unicrop | 80:2.5% w/w | WP | 01309 |

Uses

Blight in POTATOES. Powdery scab in POTATOES.

Efficacy
* Apply as spray for blight control before disease appears or at blight warming and repeat at 10-14 d intervals. Three, 4 or more sprays may be needed through season
* May also be used as dust application for reduction in tuber-borne powdery scab (not in situations of high disease pressure)

Special precautions/Environmental safety
* Irritating to eyes, skin and respiratory system
* Keep away from fire and sparks
* Use contents of opened container as soon as possible and do not store opened containers until following season. If product becomes wet effectiveness may be reduced and inflammable vapour produced

Protective clothing/Label precautions
* A, C, F
* 6a, 6b, 6c, 18, 22, 28, 29, 35, 36, 37, 54, 67

Harvest interval
* 7 d

Approval
* Approved for aerial application on potatoes. See introductory notes

---

365   manganese zinc ethylenebisdithiocarbamate complex
A protectant dithiocarbamate fungicide for potatoes and cereals

Products   Power Manplex   Power   80% w/w   WP   02954

Uses   Blight in POTATOES. Septoria, sooty moulds, powdery mildew, rust in WINTER WHEAT.

Notes   **Efficacy**
* Apply to potatoes after first blight warning or when plants start to meet along rows and repeat every 10-14 d. Do not apply if blight already present
* Apply to cereal crops at onset of disease, with second application if necessary. Recommended for tank mixing with carbendazim in wheat and with tridemorph in barley. See label for details

Crop safety/Restrictions
* Do not spray cereal crops after full ear emergence (GS 59)

Special precautions/Environmental safety
* Irritating to eyes, skin and respiratory system

Protective clothing/Label precautions
* A
* 6a, 6b, 6c, 18, 21, 29, 36, 37, 54, 63, 67

FOR FULL CONDITIONS OF USE ALWAYS READ THE PRODUCT LABEL

366   manganese zinc ethylenebisdithiocarbamate + ofurace

A contact and systemic protectant fungicide mixture

Products   Patafol Plus   ICI   67:5.8% w/w   WP   02808

Uses   Blight in POTATOES. Downy mildew in WINTER OILSEED RAPE.

Notes   **Efficacy**
- Apply to potatoes at first blight warning or just before foliage meets in row and repeat at 10-14 d intervals, not after end of Aug
- Apply before crop infested, not to crops already showing blight
- Do not apply to wet foliage, if rain likely in 2-3 h or after shoot growth stops
- Organotin products recommended as follow-up treatment up to haulm destruction
- Apply to oilseed rape at seedling to young plant stage as soon as infection visible in crop but before end of Nov

Crop safety/Restrictions
- Maximum number of treatments, potatoes 5, oilseed rape 1

Special precautions/Environmental safety
- Irritating to eyes, skin and respiratory system
- Harmful to fish. Do not contaminate ponds, waterways or ditches with chemical or used container

Protective clothing/Label precautions
- A, C
- 6a, 6b, 6c, 18, 23, 28, 29, 35, 36, 37, 53, 63, 67, 70

Harvest interval
- 7 d

Approval
- Approved for aerial application on potatoes and oilseed rape. See introductory notes

367 MCPA

A translocated phenoxyacetic herbicide for cereals and grassland

See also benazolin + 2,4-DB + MCPA
bentazone + MCPA + MCPB
bromoxynil + dichlorprop + ioxynil + MCPA
clopyralid + dichlorprop + MCPA
2,4-D + dichlorprop + MCPA + mecoprop
2,4-DB + linuron + MCPA
2,4-DB + MCPA
dicamba + dichlorprop + MCPA
dicamba + maleic hydrazide + MCPA
dicamba + MCPA + mecoprop
dichlorprop + ferrous sulphate + MCPA
dichlorprop + MCPA

Products

1 Agrichem MCPA-25	Agrichem	250 g/l	SL	00050
2 Agrichem MCPA-50	Agrichem	500 g/l	SL	04097
3 Agricorn 500	FCC	476 g/l	SL	00055
4 Agritox 50	RP	500 g/l	SL	02588
5 Agroxone 50	ICI	475 g/l	SL	03629
6 Atlas MCPA	Atlas	480 g/l	SL	03055
7 BASF MCPA Amine 50	BASF	500 g/l	SL	00209
8 BH MCPA 75	RP Environ.	750 g/l	SL	03363
9 Campbell's MCPA 25	MTM Agrochem.	250 g/l	SL	00409
10 Campbell's MCPA 50	MTM Agrochem.	500 g/l	SL	00381
11 Chafer MCPA 675	BritAg	675 g/l	SL	02815
12 Farmon MCPA 50	Farm Protection	475 g/l	SL	00842
13 MSS MCPA 50	Mirfield	500 g/l	SL	01412
14 Phenoxylene 50	Schering	500 g/l	SL	03854
15 Power MCPA	Power	250 g/l	SL	02790
16 Quad MCPA 50%	Quadrangle	500 g/l	SL	01669
17 Star MCPA	Star	500 g/l	SL	03157

Uses

Annual dicotyledons, perennial dicotyledons, charlock, hemp-nettle, fat-hen, wild radish in WHEAT, BARLEY, OATS, WINTER RYE [1-4, 6, 7, 9-16]. Annual dicotyledons, perennial dicotyledons, charlock, hemp-nettle, fat-hen, wild radish in WHEAT, BARLEY, OATS [16, 17]. Annual dicotyledons, perennial dicotyledons, charlock, hemp-nettle, fat-hen, wild radish in CEREALS [5]. Annual dicotyledons, perennial dicotyledons, charlock, hemp-nettle, fat-hen, wild radish in SPRING RYE [2, 15]. Annual dicotyledons, perennial dicotyledons, buttercups, dandelions, docks, thistles in ESTABLISHED GRASSLAND [1-7, 9-16]. Annual dicotyledons, perennial dicotyledons in YOUNG LEYS [11, 14, 15]. Annual dicotyledons, perennial dicotyledons in GRASS SEED CROPS [4, 11, 14, 15]. Annual dicotyledons, perennial dicotyledons in LINSEED [4]. Annual dicotyledons, perennial dicotyledons in ASPARAGUS [14]. Annual dicotyledons, perennial dicotyledons, creeping thistle, daisies, docks in AMENITY GRASS, ROAD VERGES [8].

FOR FULL CONDITIONS OF USE ALWAYS READ THE PRODUCT LABEL

Efficacy

- Best results achieved by application to weeds in seedling to young plant stage under good growing conditions when crop growing actively
- Spray perennial weeds in grassland before flowering. Most susceptible growth stage varies between species. See label for details
- Do not spray during cold weather, drought, if rain or frost expected or if crop wet

Crop safety/Restrictions

- Apply to winter cereals in spring from fully tillered, leaf sheath erect stage to before first node detectable (GS 30)
- Apply to spring barley and wheat from 5-leaves unfolded (GS 15), to oats from 1-leaf unfolded (GS 11) to before first node detectable (GS 30)
- Apply to cereals undersown with grass after grass has 2-3 leaves unfolded
- Do not use on cereals before undersowing
- Do not roll, harrow or graze for a few days before or after spraying, see label
- Recommendations for crops undersown with legumes vary. Red clover may withstand low doses after 2-trifoliate leaf stage, especially if shielded by taller weeds but white clover is more sensitive. See label for details
- Apply to linseed as soon as possible after 5 cm high and before 25 cm [4]
- Do not use on grassland where clovers are an important part of the sward
- Apply to grass seed crops from 2-3 leaf stage to 5 wk before head emergence
- On asparagus apply to established plants only after cutting finished and avoid fern [14]
- Do not use on any crop suffering from stress or herbicide damage
- Do not direct drill brassicas or legumes within 6 wk of spraying grassland
- Avoid spray drift onto nearby susceptible crops

Special precautions/Environmental safety

- Harmful if swallowed [1, 3-8, 10-14, 16, 17]. Harmful in contact with skin [8, 11, 14, 17]
- Risk of serious damage to eyes [7]
- Harmful to fish. Do not contaminate ponds, waterways or ditches with chemical or used container [2, 3, 7, 15]

Protective clothing/Label precautions

- A, C
- 5a [8, 11, 14, 17]; 5c [1, 3-8, 10-14, 16, 17]; 8 [7]; 18, 21, 29, 36, 37, 43, 60, 63, 65, 66, 70, 78 [1-17]; 53 [2, 3, 7, 15]; 54 [1, 4-6, 8-14, 16, 17]

Approval

- May be applied through CDA equipment. See label for details. See introductory notes on ULV application [7]

368   MCPA + MCPB
A translocated herbicide for undersown cereals and grassland

Products					
	1 Campbell's Bellmac Plus	MTM Agrochem.	38:262 g/l	SL	00385
	2 Farmon MCPB Plus	Farm Protection	41:244 g/l	SL	00843
	3 MSS MCPB + MCPA	Mirfield	25:275 g/l	SL	01413
	4 Trifolex-Tra	Shell	34:216 g/l	SL	02175
	5 Tropotox Plus	RP	37.5:262.5 g/l	SL	02224

Uses	Annual dicotyledons, perennial dicotyledons in CEREALS, UNDERSOWN CEREALS, DREDGE CORN CONTAINING PEAS, DIRECT-SOWN SEEDLING CLOVERS, CLOVER SEED CROPS, PERMANENT PASTURE, ESTABLISHED LEYS, SAINFOIN [5]. Annual dicotyledons, perennial dicotyledons in CEREALS, UNDERSOWN CEREALS, LEYS, ESTABLISHED GRASSLAND, SAINFOIN [3]. Annual dicotyledons, perennial dicotyledons in UNDERSOWN CEREALS, LEYS, ESTABLISHED GRASSLAND, SAINFOIN [2]. Annual dicotyledons, perennial dicotyledons in UNDERSOWN CEREALS, LEYS, ESTABLISHED GRASSLAND, PEAS [4]. Annual dicotyledons, perennial dicotyledons in UNDERSOWN CEREALS, LEYS, SAINFOIN [1].

Notes

Efficacy
- Best results achieved by application to weeds in seedling to young plant stage under good growing conditions when crop growing actively. Spray perennials when adequate leaf surface before flowering. Retreatment often needed in following year
- Spray in leys and sainfoin before crop provides cover for weeds
- Rain, cold or drought may reduce effectiveness

Crop safety/Restrictions
- Apply to cereals from 2-expanded leaf stage to before jointing (GS 12-30) and, where undersown, after 1-trifoliate leaf stage of clover
- Apply to direct-sown seedling clover after 1-trifoliate leaf stage
- Apply to mature white clover for fodder at any stage. Do not spray red clover after flower stalk has begun to form
- Do not spray clovers for seed [2, 3, 4], not in first year [5]
- Apply to sainfoin after first compound leaf stage [1, 2, 5], second compound leaf [3]
- Do not spray peas [1, 2, 3]. Only spray peas in tank mix with Fortrol from 4-leaf stage to before buds visible (GS 104-201) [4]. Only spray peas in dredge-corn, apply when peas have 3-6 leaves (GS 103-106) [5]
- Do not apply prior to sowing any crop [1-4]
- Do not roll or harrow for a few days before or after spraying [1-4]
- Avoid spray drift onto nearby sensitive crops

Special precautions/Environmental safety
- Keep livestock out of treated areas until foliage of any poisonous weeds such as ragwort has died and become unpalatable
- Harmful to fish. Do not contaminate ponds, waterways or ditches with chemical or used container

Protective clothing/Label precautions
- 18, 21, 29, 36, 37, 43, 53, 60, 63, 65, 66, 70, 78

369 MCPA + mecoprop
A translocated selective herbicide for amenity and roadside grass

Products	Greenmaster Extra	Fisons	0.49:0.57% w/w	GR	03130

FOR FULL CONDITIONS OF USE ALWAYS READ THE PRODUCT LABEL

Annual dicotyledons, perennial dicotyledons in AMENITY GRASS [1]. Annual dicotyledons, perennial dicotyledons in ROAD VERGES, NON-CROP AREAS [2].

Efficacy
- Apply from Apr to Sep, when weeds growing actively and have large leaf area for absorption of chemical
- Granular formulation contains NPK fertilizer to encourage grass growth
- Do not apply when heavy rain expected or during prolonged drought
- Do not mow for 2-3 d before or after treatment
- Irrigate after 1-2 d unless rain has fallen

Crop safety/Restrictions
- Avoid drift of spray onto nearby crops, shrubs or flower beds. Do not use near water
- Compost for not less than 6 mth before using first 4 mowings as compost or mulch
- Do not apply to newly sown or turfed areas for at least 6 mth or reseed bare patches for 8 wk

Special precautions/Environmental safety
- Keep livestock out of treated areas until foliage of any poisonous weeds such as ragwort has died and become unpalatable

Protective clothing/Label precautions
- 29, 43, 63, 67

370 MCPB

A translocated phenoxybutyric herbicide for use in peas etc.
See also bentazone + MCPA + MCPB
bentazone + MCPB
MCPA + MCPB

Products

1	Campbell's Bellmac Straight	MTM Agrochem.	400 g/l	SL	00386
2	Tropotox	RH	400 g/l	SL	03496

Uses
Annual dicotyledons, perennial dicotyledons in UNDERSOWN CEREALS, LEYS, PEAS, WHITE CLOVER SEED CROPS, BLACKCURRANTS [1]. Annual dicotyledons, perennial dicotyledons in PEAS, WHITE CLOVER SEED CROPS, BLACKCURRANTS [2]. Annual dicotyledons, perennial dicotyledons in EDIBLE PODDED PEAS *(off-label)* [1].

Notes
Efficacy
- Best results achieved by spraying young seedling weeds in good growing conditions
- Best results on perennials by spraying before flowering
- Effectiveness may be reduced by rain within 12 h, by very cold or dry conditions

Crop safety/Restrictions
- Apply to undersown cereals from 2-leaves unfolded to first node detectable of cereal (GS 12-31) [1], and after first trifoliate leaf stage of clover [1, 2]
- Red clover seedlings may be temporarily damaged but later growth is normal
- Apply to white clover seed crops in Mar to early Apr, not after mid-May, and allow 3 wk before cutting and closing up for seed

- Apply to peas from 3-6 leaf stage but before flower buds detectable (GS 103-201). See label for details of cultivars which can be treated and requirements for leaf-wax testing
- Do not use on beans, vetches or established red clover
- Apply to blackcurrants in Aug after harvest and after shoot growth ceased but while bindweed still growing actively; direct spray onto weed as far as possible
- Do not roll or harrow for 7 d before or after treatment
- Avoid drift onto nearby sensitive crops

Special precautions/Environmental safety
- Harmful in contact with skin and if swallowed [1], in contact with skin [2]
- Keep livestock out of treated areas until foliage of any poisonous weeds such as ragwort has died and become unpalatable
- Harmful to fish. Do not contaminate ponds, waterways or ditches with chemical or used container

Protective clothing/Label precautions
- 5a, 5c [1]; 5a [2]; 18, 21, 29, 36, 37, 43, 53, 60, 63, 66, 70, 78

Approval
- Off-label Approval to Feb 1991 for use on edible podded peas (OLA 0186/88) [1]

371 mecoprop

A translocated phenoxy herbicide for cereals and grassland
See also benazolin + bromoxynil + ioxynil + mecoprop
bifenox + isoproturon + mecoprop
bromoxynil + ioxynil + isoproturon + mecoprop
bromoxynil + ioxynil + mecoprop
clopyralid + mecoprop
cyanazine + mecoprop
2,4-D + dicamba + mecoprop
2,4-D + dichlorprop + MCPA + mecoprop
2,4-D + mecoprop
2,4-DB + mecoprop
dicamba + MCPA + mecoprop
dicamba + mecoprop
dicamba + mecoprop + triclopyr
dichlorprop + mecoprop
ioxynil + isoproturon + mecoprop
ioxynil + mecoprop
MCPA + mecoprop

Products					
1 Atlas CMPP	Atlas	570 g/l	SL	03050
2 BH CMPP Extra	RP Environ.	570 g/l	SL	02854
3 Campbell's CMPP	MTM Agrochem.	570 g/l	SL	02918
4 Chafer CMPP Super	BritAg	570 g/l	SL	00475
5 Cleanacres CMPP 60	Cleanacres	600 g/l	SL	02483
6 Clenecorn	FCC	570 g/l	SL	00542
7 Clifton CMPP Amine 60	Clifton	600 g/l	SL	03020
8 Clovotox	RP Environ.	300 g/l	SL	00547

FOR FULL CONDITIONS OF USE ALWAYS READ THE PRODUCT LABEL

	9 Compitox Extra	RP	570 g/l	SL	00566
	10 Farmon CMPP	Farm Protection	570 g/l	SL	00835
	11 Headland Charge	WBC Technology	570 g/l	SL	04495
	12 Herrifex DS	Bayer	570 g/l	SL	01047
	13 Hymec	Agrichem	600 g/l	SL	01091
	14 Iso-Cornox 57	Schering	570 g/l	SL	03868
	15 Mascot Cloverkiller	Rigby Taylor	300 g/l	SL	02438
	16 Methoxone	ICI	570 g/l	SL	01334
	17 MSS CMPP	Mirfield	600 g/l	SL	01405
	18 MSS CMPP Amine 60	Mirfield	600 g/l	SL	02446
	19 Power CMPP	Power	570 g/l	SL	02616
	20 Quad CMPP 60	Quadrangle	570 g/l	SL	01666
	21 Star CMPP	Star	570 g/l	SL	03132

Uses

Annual dicotyledons, perennial dicotyledons, chickweed, cleavers in WHEAT, BARLEY, OATS [7, 13, 17, 18, 20]. Annual dicotyledons, perennial dicotyledons, chickweed, cleavers in WHEAT, BARLEY, OATS, YOUNG LEYS [3, 6, 7, 9, 11]. Annual dicotyledons, perennial dicotyledons, chickweed, cleavers in WHEAT, BARLEY, OATS, ESTABLISHED GRASSLAND [1, 16, 19, 21]. Annual dicotyledons, perennial dicotyledons, chickweed, cleavers in WHEAT, BARLEY, OATS, WINTER TRITICALE, YOUNG LEYS [10]. Annual dicotyledons, perennial dicotyledons, chickweed, cleavers in WHEAT, BARLEY, OATS, YOUNG LEYS, GRASS SEED CROPS [4]. Annual dicotyledons, perennial dicotyledons, chickweed, cleavers in WHEAT, BARLEY, OATS, YOUNG LEYS, ESTABLISHED GRASSLAND, GRASS SEED CROPS [14]. Annual dicotyledons, perennial dicotyledons, chickweed, cleavers in WHEAT, BARLEY, OATS, YOUNG LEYS, GRASS SEED CROPS, ORCHARD GRASS, TURF [12]. Annual dicotyledons, perennial dicotyledons, clovers, chickweed in TURF [2, 8, 15].

Notes

Efficacy
- Best results achieved by application to weeds in seedling to young plant stage growing actively in a strongly competing crop
- Do not spray when rain imminent or when growth hard from cold or drought. Rainfall within 8-12 h may reduce effectiveness
- For dock control in grassland cut in mid-summer and spray regrowth in Aug-Sep
- Spray turf weeds in spring or early summer when in active growth. More resistant perennials may need repeat treatment after 4-6 wk

Crop safety/Restrictions
- Range of timings for winter cereals include from 1-4 leaves fully expanded (GS 11-14) to start of tillering or second node detectable (GS 21-32) until 31 Dec, from leaf sheath erect to first or second node detectable (GS 30-31 or 32) in spring. See label
- Range of timings for spring cereals include from 1-2 leaves fully expanded (GS 11-12) to first or second node detectable (GS 31-32)
- Do not spray rye, crops undersown with legumes or crops about to be undersown
- If hard frosts occur within 3-4 wk of application to barley of low vigour on light soils crop may be scorched or stunted
- Do not roll or harrow within a few days before or after treatment
- Apply to grasses after 3-fully expanded leaves and 1 tiller stage
- Clovers present in sprayed grass will be severely damaged
- Do not direct-drill brassicas or legumes within 6 wk after application
- Do not use first 4 mowings for mulching. Compost for at least 6 mth before use
- Avoid drift onto nearby fruit trees or other susceptible crops

* A, C
* Harmful in contact with skin and if swallowed. Irritating to eyes and skin
* Keep livestock out of treated areas until foliage of any poisonous weeds such as ragwort has died and become unpalatable

Protective clothing/Label precautions
* 5a, 5c, 6a, 6b, 18, 21, 29, 36, 37, 43, 54, 63, 66, 70, 78

---

372 mecoprop-P

A translocated phenoxy herbicide for cereals and grassland

Products					
Products	1 Astix	RP	600 g/l	SL	04472
	2 Duplosan New System CMPP	BASF	600 g/l	SL	04481

Uses

Annual dicotyledons, perennial dicotyledons, chickweed, cleavers in WHEAT, BARLEY, OATS, YOUNG LEYS [1, 2]. Annual dicotyledons, perennial dicotyledons, chickweed, cleavers in ESTABLISHED GRASSLAND, GRASS SEED CROPS [2].

Notes

Efficacy
* Best results achieved by application to weeds in seedling to young plant stage growing actively in a strongly competing crop
* Do not spray during cold weather, periods of drought, if rain or frost expected or if crop wet

Crop safety/Restrictions
* Spray winter cereals in spring from leaf sheath erect stage to first node detectable (GS 30-31) or up to second node detectable (GS 32) if necessary, spring cereals from first fully expanded leaf to before first node detectable (GS 11-30)
* Do not apply to winter cereals from start of tillering to leaf sheath erect (GS 21-30)
* Spray cereals undersown with grass after grass starts to tiller
* Do not spray cereals undersown with clovers or legumes or to be undersown with legumes or grasses
* Spray newly sown grass leys when grasses have at least 3 fully expanded leaves and have begun to tiller. Any clovers will be damaged
* Do not spray grass seed crops within 5 wk of seed head emergence
* Do not spray crops suffering from herbicide damage or physical stress
* Do not roll or harrow for 7 d before or after treatment
* Avoid drift onto beans, beet, brassicas, fruit and other sensitive crops

Special precautions/Environmental safety
* Harmful if swallowed and in contact with skin. Irritating to eyes
* Keep livestock out of treated areas until foliage of any poisonous weeds, such as ragwort, has died and become unpalatable
* Harmful to fish. Do not contaminate ponds, waterways or ditches with chemical or used container

---

FOR FULL CONDITIONS OF USE ALWAYS READ THE PRODUCT LABEL

373   mefluidide

A plant growth regulator for suppression of grass growth

Products					
1 Echo	ICI Professional	240 g/l	SL	04744
2 Embark	Gordon	240 g/l	SL	02513
3 Mowchem	RP Environ.	240 g/l	SL	02787

Uses   Growth suppression in AMENITY GRASS, ROADSIDE GRASS.

Notes   **Efficacy**
- Apply when grass growth active in Apr/May and repeat when growth re-commences
- Grass growth suppressed for up to 8 wk
- Mow (not below 2 cm) 2-3 d before or 5-10 d after spraying
- Only apply to dry foliage but do not use in drought
- A rainfree period of 8 h is needed after spraying
- May be tank-mixed with suitable 2,4-D or mecoprop product for control of dicotyledons
- Delay application for 3-4 wk after application of high-nitrogen fertilizer

Crop safety/Restrictions
- Up to 2 applications per year may be made
- Do not use on fine turf areas or on turf established for less than 4 mth
- Do not use on grass suffering from herbicide damage
- Treated turf may appear less dense and be temporarily discoloured but normal colour returns in 3-4 wk
- Do not reseed within 2 wk after application

Special precautions/Environmental safety
- Do not allow animals to graze treated areas

Protective clothing/Label precautions
- 29, 54, 63, 66, 70

374   mephosfolan

A systemic organophosphorus insecticide for use on hops

Products					
Cytro-Lane	Cyanamid	250 g/l	EC	00626

Uses   Damson-hop aphid in HOPS.

Notes   **Efficacy**
- Apply as accurately measured drench to lower part of bines and centre of hop plant at ground level in mid- to late May after most bines have reached 1.5-2 m. A second application may be made before mid-Jul. See label for details of suitable applicators

- Base of hop plants should be free of weeds and dead leaves
- Bines emerging away from main stock or near poles may require separate treatment
- Systemic uptake may be delayed by very dry conditions
- Product may be used in integrated control programmes. Predator insects not harmed

Crop safety/Restrictions
- Do not apply as a spray
- Total dose per season should not exceed 8 l/1000 plants

Special precautions/Environmental safety
- A chemical subject to the Poisons Act 1972 and Poisons Rules 1982
- This product contains an organophosphorus compound. Do not use if under medical advice not to work with such compounds
- Very toxic in contact with skin, by inhalation and if swallowed. Irritating to skin
- Keep unprotected persons out of treated areas for at least 1 d
- Dangerous to game, wild birds and animals
- Harmful to bees. Keep down flowering weeds
- Dangerous to fish. Do not contaminate ponds, waterways or ditches with chemical or used container

Protective clothing/Label precautions
- A, C, H + K or L, M
- 1, 3a, 3b, 3c, 6b, 14, 16, 18, 21, 25, 28, 29, 35, 36, 37, 38, 41, 45, 50, 52, 64, 65, 70, 79

Withholding period
- Harmful to livestock. Keep out of treated areas for at least 10 d

Harvest interval
- 4 wk

375 mepiquat chloride
A plant growth regulator available only in mixtures
See also 2-chloroethylphosphonic acid + mepiquat chloride

376 mercuric oxide
An inorganic mercury fungicide paint for canker control

Products	Kankerex	Unicrop	3% w/w	PA	01123

Uses   Canker in APPLES, PEARS.

Notes   **Efficacy**
- Cut back canker to sound wood and apply to clean wounds as thin, continuous coat
- Apply only after harvest and before bud burst

FOR FULL CONDITIONS OF USE ALWAYS READ THE PRODUCT LABEL

- In hot weather and on dry wounds use a more liberal application. If wounds wet, wait until dry before treating

Crop safety/Restrictions
- Do not apply to grafting cuts, green wood, very young trees or to buds

Special precautions/Environmental safety
- Harmful if swallowed
- Harmful to fish. Do not contaminate ponds, waterways or ditches with chemical or used container

Protective clothing/Label precautions
- 5c, 18, 25, 29, 36, 37, 53, 60, 63, 65, 70, 78

---

377 mercurous chloride

A soil applied inorganic mercury fungicide for brassicas and onions

Products					
	1 Calomel	Hortichem	100%	TC	03177
	2 F S Calomel 4% Dust	Ford Smith	4% w/w	DP	03296
	3 F S Mercurous Chloride	Ford Smith	100%	TC	01326

Uses

Clubroot in BRASSICAS [1-3]. White rot in ONIONS [3].

Notes

Efficacy
- For white rot control pellet with onion seed (using methyl cellulose or other suitable sticker) and dry before storage
- For club root control apply as a pre-planting dust in seed beds, mixed with sand post-planting or as a suspension dip. See label for details
- Do not use metal containers to mix or store mercurous chloride, avoid contact with alkaline materials

Crop safety/Restrictions
- Treatment of compost for propagation of Brussels sprouts is not recommended
- Plants in treated compost grow less vigorously and take longer to reach planting out stage though yield should not be affected
- Do not mix chlorpyrifos into treated compost

Special precautions/Environmental safety
- Harmful if swallowed
- Harmful to fish. Do not contaminate ponds, waterways or ditches with chemical or used container

Protective clothing/Label precautions
- A
- 5c, 22, 25, 28, 29, 36, 37, 53, 63, 67, 70, 78

378 metalaxyl

A phenylamide (acylalanine) fungicide available only in mixtures
See also benomyl + iodofenphos + metalaxyl
carbendazim + metalaxyl
chlorothalonil + metalaxyl
copper oxychloride + metalaxyl
iprodione + metalaxyl + thiabendazole
mancozeb + metalaxyl

379 metalaxyl + thiabendazole

A protective fungicide seed treatment for grass seed

Products   Apron T 69WS   Ciba-Geigy   45:24% w/w   FS   02661

Uses   Seed borne diseases in GRASS SEED.

Notes   **Efficacy**
• Apply through continuous flow seed treaters which should be calibrated before use
• Check calibration of seed drill with treated seed before drilling

Crop safety/Restrictions
• Ensure moisture content of treated seed satisfactory and store in a dry place

Special precautions/Environmental safety
• Irritating to eyes, skin and respiratory system
• Harmful to fish. Do not contaminate ponds waterways or ditches with chemical or used container

Protective clothing/Label precautions
• A, E
• 6a, 6b, 6c, 22, 28, 29, 37, 53, 63, 65, 70, 72, 73, 74, 76

380 metalaxyl + thiabendazole + thiram

A protective fungicide seed treatment for peas and beans

Products   Apron Combi 453FS   Ciba-Geigy   233:120:100 g/l   FS   04110

Uses   Ascochyta, downy mildew, damping off in VINING PEAS, DRIED PEAS, MANGE-TOUT PEAS, BEANS, DWARF BEANS, RUNNER BEANS.

Notes   **Efficacy**
• Apply through continuous flow seed treaters which should be calibrated before use
• Check calibration of seed drill with treated seed before drilling

FOR FULL CONDITIONS OF USE ALWAYS READ THE PRODUCT LABEL

Crop safety/Restrictions
• Ensure moisture content of treated seed satisfactory and store in a dry place

Special precautions/Environmental safety
• Harmful if swallowed. Irritating to eyes
• Treated seed not to be used as food or feed. Do not re-use sack
• Harmful to fish. Do not contaminate ponds, waterways or ditches with chemical or used container

Protective clothing/Label precautions
• A, C
• 5c, 6a, 18, 22, 25, 29, 36, 37, 53, 63, 65, 72, 73, 74, 75, 76, 78

---

381 metalaxyl + thiram
A systemic and protectant fungicide for use in lettuce

Products   Favour 600FW   Ciba-Geigy   100:500 g/l   SC   04000

Uses   Downy mildew in LETTUCE, PROTECTED LETTUCE.

Notes   **Efficacy**
• Apply from 100% emergence and repeat at 14 d intervals
• Resistant strains of lettuce downy mildew may develop against which product may be ineffective. See label for details

Crop safety/Restrictions
• On protected lettuce only 2 post-planting applications of thiram or any combination of products containing thiram or other EBDC compounds (mancozeb, maneb, zineb) either as a spray or a dust are permitted within 2 wk of planting out and none thereafter. On winter lettuce, if only thiram based products are being used, 3 applications may be made within 3 wk of planting out and none thereafter
• On outdoor lettuce up to 5 applications per crop may be made
• Do not use as a soil block treatment

Special precautions/Environmental safety
• Harmful if swallowed. Irritating to eyes, skin and respiratory system
• Harmful to fish. Do not contaminate ponds, waterways or ditches with chemical or used container

Protective clothing/Label precautions
• A, C
• 5a, 6a, 6b, 6c, 18, 21, 29, 35, 36, 37, 53, 63, 66, 70, 78

Harvest interval
• protected lettuce 3 wk; outdoor lettuce 2 wk

382　metaldehyde

A molluscicide bait for controlling slugs and snails

Products

1 Chiltern Metaldehyde Slug Killer Mini Pellets	Chiltern	6% w/w	PT	00491
2 Doff Metaldehyde Slug Killer Mini Pellets	Doff Portland	6% w/w	PT	00741
3 Doff Metaldehyde Slug Killer Mini Pellets with Animal Repellent	Doff Portland	6% w/w	PT	04898
4 Farmon Mini Slug Pellets	Farm Protection	6% w/w	PT	03913
5 Gastratox 6G Slug Pellets	Truchem	6% w/w	PT	04066
6 Helarion	Fisons	5% w/w	PT	02520
7 Mifaslug	FCC	6% w/w	PT	01349
8 HM Metaldehyde Mini Slug Pellets	Doff Portland	6% w/w	PT	04901
9 MSS Metaldehyde Slug Killer Pellets	Doff Portland	6% w/w	PT	04896
10 PBI Slug Pellets	PBI	6% w/w	PT	02611
11 Power Metaldehyde 6P	Power	6% w/w	PT	02619
12 Quad Mini Slug Pellets	Quadrangle	6% w/w	PT	01670
13 Slug Destroyer	Schering	6% w/w	PT	03919
14 Tripart Mini Slug Pellets	Tripart	6% w/w	PT	02207
15 Unicrop 6% Mini Slug Pellets	Unicrop	6% w/w	PT	02275
16 Vassgro Mini Slug Pellets	Vass	6% w/w	PT	03579

Uses

Slugs, snails in FIELD CROPS, VEGETABLES, FRUIT CROPS, ORNAMENTALS, GLASSHOUSE CROPS [1-16].

Notes

Efficacy
- Apply pellets by hand, fiddle drill, fertilizer distributor, by air or in admixture with seed. See labels for rates and timings.
- Best results achieved by application during mild, damp weather when slugs and snails most active. May be applied in standing crops
- Do not apply when rain imminent or water glasshouse crops within 4 d of application
- To prevent slug build up apply at end of season to brassicas and other leafy crops
- To reduce tuber damage in potatoes apply twice in Jul and Aug

Special precautions/Environmental safety
- Keep poultry out of treated areas for at least 7 d
- Dangerous to game, wild birds and animals
- Harmful to fish. Do not contaminate ponds, waterways or ditches with chemical or used container

Protective clothing/Label precautions
- 29, 37, 42, 45, 53, 63, 66, 67, 70 [1-16]

FOR FULL CONDITIONS OF USE ALWAYS READ THE PRODUCT LABEL

• Approved for aerial application on all crops. See introductory notes [1, 2, 4, 7, 10, 12-16]

---

383   metamitron

A contact and residual triazinone herbicide for use in beet crops

Products	1 Goltix WG	Bayer	70% w/w	WG	02430
	2 Power Countdown	Power	70% w/w	WG	04953

Uses   Annual dicotyledons, annual grasses, fat-hen, annual meadow grass in SUGAR BEET, FODDER BEET, RED BEET, MANGOLDS.

Notes   **Efficacy**
• Low dose programme (LDP). Apply post-weed emergence as a fine spray, including adjuvant oil, timing treatment according to weed emergence and size. See label for details and for recommended tank mixes and sequential treatments. On mineral soils the LDP should be preceeded by pre-drilling or pre-emergence treatment [1]
• Use up to 5 LDP sprays on mineral soils (up to 3 where also used pre-emergence) and up to 6 sprays on organic soils [1]
• Traditional application. Apply pre-drilling before final cultivation with incorporation to 8-10 cm, pre-crop emergence at or soon after drilling into firm, moist seedbed or to emerged weeds from cotyledon to first true leaf stage
• On emerged weeds at or beyond 2-leaf stage addition of adjuvant oil advised
• Up to 3 post-emergence sprays may be used on soils with over 10% organic matter
• For control of wild oats and certain other weeds tank mixes with other herbicides or sequential treatments are recommended. See label for details

Crop safety/Restrictions
• Using traditional method post-crop emergence on mineral soils do not apply before first true leaves have reached 1 cm long
• Crop tolerance may be reduced by stress caused by growing conditions, effects of pests, disease or other pesticides, nutrient deficiency etc
• Only sugar beet, fodder beet or mangolds may be drilled within 4 mth after treatment. Winter cereals may be sown in same season after ploughing, provided 16 wk passed since last treatment

Protective clothing/Label precautions
• 29, 54, 63, 70

---

384   metam-sodium

A liquid dithiocarbamate sterilant for glasshouse and nursery soils

Products	1 Campbell's Metham Sodium	MTM Agrochem.	400 g/l	SL	00412
	2 Sistan	Unicrop	380 g/l	SL	01957

Uses   Soil-borne diseases in GLASSHOUSE CROPS, TOMATOES, ORNAMENTALS, CHRYSANTHEMUMS, CARNATIONS [1, 2]. Soil insects, nematodes in

GLASSHOUSE CROPS, TOMATOES, ORNAMENTALS, CHRYSANTHEMUMS, CARNATIONS [1, 2]. Weed seeds in GLASSHOUSE CROPS, TOMATOES, ORNAMENTALS, CHRYSANTHEMUMS, CARNATIONS [1, 2]. Soil-borne diseases in OUTDOOR TOMATOES *(Jersey)* [1]. Soil insects, nematodes in OUTDOOR TOMATOES *(Jersey)* [1]. Weed seeds in OUTDOOR TOMATOES *(Jersey)* [1]. Dutch elm disease in ELM TREES *(off-label)* [2].

Notes	

Efficacy
- Product acts by breaking down in contact with soil to release methyl isothiocyanate
- Apply to glasshouse soils as a drench, inject undiluted to 20 cm at 30 cm intervals and seal with water or apply to surface, rotavate and seal
- After treatment allow sufficient time (several weeks) for residues to dissipate and aerate soil by forking. Time varies with season. See label for details
- Apply to outdoor soils by drenching or sub-surface application
- Apply when soil temperatures exceed 7°C, preferably above 10°C, between 1 Apr and 31 Oct. Soil must be of fine tilth and adequate moisture content
- Product may also be used by mixing into potting soils
- Avoid using in equipment incorporating natural rubber parts
- Use in control programme against dutch elm disease to sever root grafts which can transmit disease from tree to tree. See OLA notice for details [2]

Crop safety/Restrictions
- No plants must be present during treatment. Do not treat glasshouses within 2 m of growing crops. Do not plant until soil is entirely free of fumes
- Cress test advised to check for absence of residues. See label for details

Special precautions/Environmental safety
- Harmful in contact with skin and if swallowed
- Irritating to eyes, skin and respiratory system

Protective clothing/Label precautions
- A, C, D, H, M
- 5a, 5c, 6a, 6b, 6c, 14, 16, 18, 21, 28, 29, 36, 37, 54, 63, 65, 78 [1, 2]; see label for additional precautions

Approval
- Off-label Approval to Apr 1993 for use around elm trees infected with Dutch elm disease (OLA 0297/90) [2]

385 metazachlor

A residual anilide herbicide for use in brassica and ornamental crops

Products					
1	Butisan S	BASF	500 g/l	SC	00357
2	Power Metazachlor 50	Power	500 g/l	SC	04346
3	Power Track	Power	500 g/l	SC	04954

Uses	

Annual dicotyledons, triazine resistant groundsel, blackgrass, annual meadow grass in WINTER OILSEED RAPE, SPRING OILSEED RAPE, SWEDES, TURNIPS,

---

FOR FULL CONDITIONS OF USE ALWAYS READ THE PRODUCT LABEL

CABBAGES, BRUSSELS SPROUTS, CAULIFLOWERS, CALABRESE,
BROCCOLI, ORNAMENTALS, HARDY ORNAMENTAL NURSERY STOCK [1].
Annual dicotyledons, blackgrass, annual meadow grass in WINTER OILSEED RAPE
[2, 3].

Efficacy
- Activity is dependent on root uptake. For pre-emergence use apply to firm, moist, clod-free seedbed
- Some weeds (chickweed, mayweeds, blackgrass etc) susceptible up to 2- or 4-leaf stage. Moderate control of cleavers achieved provided weeds not emerged and adequate soil moisture present
- Split pre- and post-emergence treatments recommended for certain weeds in winter oilseed rape on light and/or stony soils
- Effectiveness is reduced on soils with more than 10% organic matter
- Various tank-mixtures recommended. See label for details

Crop safety/Restrictions
- On winter oilseed rape may be applied pre-emergence from drilling until seed chits, post-emergence after fully expanded cotyledon stage (GS 1,0) or by split dose technique depending on soil and weeds. See label for details. Do not use later than the end of Jan
- On spring oilseed rape, swedes, turnips and transplanted brassicas recommended as a pre-emergence sequential treatment following trifluralin
- On spring oilseed rape may also be used pre-weed-emergence from cotyledon to 6-leaf stage of crop (GS 1,0-1,6)
- With pre-emergence treatment ensure seed covered by 15 mm of well consolidated soil. Harrow across slits of direct-drilled crops
- Ensure brassica transplants have roots well covered and are well established
- In ornamentals and hardy nursery stock apply after plants established and hardened off as a directed spray or, on some subjects, as an overall spray. See label for list of tolerant subjects. Do not treat plants in containers [1]
- Do not use on sands, very light or poorly drained soils
- Do not treat protected crops or spray overall on ornamentals with soft foliage
- Do not spray crops suffering from wilting, pest or disease
- Do not spray broadcast crops or if a period of heavy rain forecast
- Any crop can follow normally harvested treated oilseed rape. See label for details of crops which may be planted in event of crop failure

Special precautions/Environmental safety
- Harmful if swallowed. Irritating to eyes and skin
- Keep livestock out of treated areas until foliage of any poisonous weeds such as ragwort has died and become unpalatable [1]
- Harmful to fish. Do not contaminate ponds, waterways or ditches with chemical or used container

Protective clothing/Label precautions
- A, C [2,3]; A, C, H, M [1]
- 5c, 6a, 6b, 18, 21, 28, 29, 36, 37, 53, 63, 66, 70, 78 [1, 2]; 43 [1]

Harvest interval
- transplanted and direct-drilled brassicas, spring oilseed rape 6 wk

386   methabenzthiazuron

A contact and residual urea herbicide for cereals and ryegrass
See also isoxaben + methabenzthiazuron

Products   Tribunil   Bayer   70% w/w   WP   0216⁹

Uses   Annual dicotyledons, annual meadow grass, rough meadow grass in WINTER WHEAT, WINTER BARLEY, SPRING BARLEY, WINTER OATS, WINTER RYE, WINTER TRITICALE, DURUM WHEAT, RYEGRASS LEYS, RYEGRASS SEED CROPS. Blackgrass in WINTER WHEAT, WINTER BARLEY. Annual dicotyledons, blackgrass, annual meadow grass in RED FESCUE SEED CROPS *(off-label)*.

Notes   **Efficacy**
* Apply pre-weed emergence on fine, firm seedbed or post-emergence to 1-2 true leaf stage of weeds, blackgrass (at higher rates) not beyond 2-leaf stage or after end Nov
* Do not use lower rates on soils with more than 10% organic matter
* Do not use higher rate for blackgrass control on soils with more than 4% organic matter or where seedbed preparation has not been preceded by ploughing

Crop safety/Restrictions
* Apply in winter wheat pre- or post-emergence up to 6 wk after drilling, provided weeds in correct stage, in oats, rye, triticale and durum wheat pre-emergence only
* Autumn sown spring wheat may be treated as for winter wheat
* Do not use on oats, rye or triticale after mid-Oct on sands in E. Anglia
* Apply in winter barley pre- or post-emergence (to end Oct and lower rates only) to named cultivars in specified localities. See label for details
* Only use pre-emergence in spring barley drilled to at least 2.5 cm and not on sands
* Apply pre-emergence in perennial ryegrass drilled before mid-Oct; not on sands
* Do not use on crops to be undersown with clover
* Use lower rate on direct drilled crops and roll before spraying where harrowed or disced after drilling
* Do not graze treated crops until at least 2 wk after spraying

Special precautions/Environmental safety
* Keep livestock out of treated areas for at least 14 d after application
* Harmful to fish. Do not contaminate ponds, waterways or ditches with chemical or used container

Protective clothing/Label precautions
* 21, 28, 29, 43, 53, 63, 67

Approval
* Approved for aerial application on cereals, grass leys. See introductory notes
* May be applied through CDA equipment. See label for details. See introductory notes on ULV application
* Off-label Approval to Jul 1991 for use on red fescue seed crops (OLA 0622/88)

FOR FULL CONDITIONS OF USE ALWAYS READ THE PRODUCT LABEL

methacrifos

An organophosphorus insecticide and acaricide for stored cereals

Damfin 950EC   Ciba-Geigy   950 g/l   EC   00648

Flour beetles, flour moths, grain beetles, grain storage mites, grain weevils in STORED GRAIN, GRAIN STORES, GRAIN TRANSPORT VEHICLES.

Efficacy
- Apply as spray to cleaned surfaces before grain is loaded into store, taking special care to treat cracks and crevices. Make up only sufficient for each day's use
- Apply to clean, dry grain as admixture treatment using a suitable grain sprayer as grain travels along conveyor auger system
- Product is cleared by Brewers' Society for use on malting barley before storage

Crop safety/Restrictions
- Only use on stored wheat, oats and barley

Special precautions/Environmental safety
- This product contains an organophosphorus compound. Do not use if under medical advice not to work with such compounds
- Harmful if swallowed. Irritating to eyes and skin
- Harmful to game, wild birds and animals
- Harmful to fish. Do not contaminate ponds, waterways or ditches with chemical or used container

Protective clothing/Label precautions
- A, C, H, M
- 1, 5c, 6a, 6b, 14, 16, 18, 25, 28, 29, 36, 37, 53, 63, 65, 70, 78

388 methiocarb

A stomach acting carbamate molluscicide and insecticide

Products
1 Club	ICI	4% w/w	PT	03800
2 Draza	Bayer	4% w/w	PT	00765

Uses Slugs, snails in FIELD CROPS, VEGETABLES, BLACKCURRANTS, STRAWBERRIES [1, 2]. Leatherjackets in CEREALS *(reduction)* [1, 2]. Strawberry seed beetle in STRAWBERRIES [1, 2]. Slugs, snails in PROTECTED CROPS *(off-label)* [2].

Notes **Efficacy**
- Use as a surface, overall application when pests active (normally mild, damp weather), pre-drilling or post-emergence. May also be used on cereals and ryegrass as admixture with seed [1, 2]
- Also reduces populations of cutworms and millipedes
- Best on potatoes in late Jul to Aug, on peas before start of pod formation
- Apply to strawberries before strawing down, to blackcurrants at grape stage to prevent snails contaminating crop

- See label for details of suitable application equipment

Crop safety/Restrictions
- Do not allow pellets to lodge in crops as this may affect marketability
- Drill cereal seed admixed with Club or Draza within 3 mth of mixing

Special precautions/Environmental safety
- Harmful if swallowed
- This product contains an anticholinesterase compound. Do not use if under medical advice not to work with such compounds
- Keep poultry out of treated areas for at least 7 d
- Harmful to game, wild birds and animals. Keep away from animals
- Harmful to fish. Do not contaminate ponds, waterways or ditches with chemical or used container

Protective clothing/Label precautions
- A
- 2, 5c, 18, 29, 35, 36, 37, 42, 46, 53, 63, 67, 70, 78

Harvest interval
- 7 d

Approval
- Approved for aerial application on all crops. See introductory notes [1, 2]
- Off-label Approval to Feb 1992 for use on all protected crops (OLA 0120/89) [2]

389 methomyl

A carbamate insecticide for fly control in animal houses

Products	Sorex Golden Fly Bait	Sorex	1% w/w	GR	02731

Uses   Flies in LIVESTOCK HOUSES.

Notes   **Efficacy**
- Product contains attractant and is intended to be used as fly bait, killing by ingestion
- For paint-on application mix with water and apply with paint brush to surfaces where flies congregate
- Granules may also be scattered in dry areas where flies congregate or sprinkled onto pasted boards for hanging in suitable areas
- Repeat application as necessary

Crop safety/Restrictions
- Only apply out of reach of livestock

Special precautions/Environmental safety
- An anti-cholinesterase compound. Do not use if under medical advice not to work with such compounds
- Apply only in positions inaccessible to livestock

FOR FULL CONDITIONS OF USE ALWAYS READ THE PRODUCT LABEL

- Protect exposed water, feed and milk from contamination
- Prevent access to bait by children and domestic animals, especially cats, dogs and pigs
- Harmful to birds
- Dangerous to fish. Do not contaminate ponds, waterways or ditches with chemical or used container

Protective clothing/Label precautions
- A, H
- 2, 25, 29, 36, 52, 63, 65, 81, 82

390 2-methoxyethylmercury acetate
An organomercury fungicide treatment for cereal seeds

| Products | Panogen M | Embetec | 31.8 g/l | LS | 01541 |

Uses Bunt in WHEAT. Covered smut, leaf stripe, net blotch in BARLEY. Covered smut, leaf spot, loose smut in OATS. Stripe smut in RYE.

Notes

Efficacy
- For best results apply through specialised seed treaters. See label for details
- Some strains of leaf spot of oats are resistant to mercury based seed treatments

Crop safety/Restrictions
- Do not treat seed with moisture content above 16% or seed already treated with an organomercury or dual-purpose powder seed treatment
- Treatment may reduce germination capacity, particularly if seed not of good quality
- Store treated seed in cool, dry, airy place for up to 6 mth
- Incorporation of red dye in product reduces chance of accidental feeding of treated seed to stock

Special precautions/Environmental safety
- When used in premises subject to the Factories Act, H.M. Inspector of Factories should be consulted on necessary precautions
- Harmful in contact with skin, by inhalation or if swallowed
- Flammable
- Organomercury compounds can cause rashes or blisters on the skin
- Treated seed not to be used as food or feed. Do not re-use sack
- Harmful to game, wild birds and animals. Bury spillages
- Do not sow treated seed from aircraft
- Special precautions needed in case of spillage. See label for details
- Special precautions needed in disposing of empty container. See label for details

Protective clothing/Label precautions
- A, C, H
- 5a, 5b, 5c, 12c, 18, 21, 29, 36, 37, 54, 63, 68, 72, 73, 74, 76, 78

391 methyl bromide with amyl acetate
A highly toxic soil fumigant

Products	Fumyl-O-Gas	Jones	99.7%	GA	04833

Uses

Soil-borne diseases in FIELD CROPS, GLASSHOUSE CROPS, MUSHROOM COMPOST. Nematodes, soil insects in FIELD CROPS, GLASSHOUSE CROPS, MUSHROOM COMPOST. Weed seeds in FIELD CROPS, GLASSHOUSE CROPS.

Notes

Efficacy
* Chemical is released from containers as a highly volatile, highly toxic gas and must be released under gas-proof sheeting
* Amyl acetate is added as a warning odourant
* Soil must be clear of trash and cultivated to at least 40 cm before treatment
* Gas must be confined in soil for 48-96 h

Crop safety/Restrictions
* Special equipment is needed for application and the chemical may only be used by professional operators trained in its use and familiar with the precautionary measures to be observed. Refer to HSE Guidance Note CS 12 before applying
* Crops may be planted 3-21 d after removal of gas-proof sheeting. For some crops a leaching irrigation is essential. See firms' literature for details
* Only one application may be made per year

Special precautions/Environmental safety
* A chemical subject to the Poisons Act 1972 and Poisons Rules 1982
* Product not for retail sale
* Very toxic by inhalation. Causes burns. Risk of serious damage to health by prolonged exposure
* Wear suitable respiratory equipment during fumigation and while removing sheeting
* Do not wear gloves or rubber boots
* Ventilate treated areas thoroughly when gas has cleared
* Keep unprotected persons and livestock out of treated areas for at least 24 h
* Dangerous to game, wild birds, animals and bees
* Harmful to fish. Do not contaminate ponds, waterways or ditches with chemical or used container

Protective clothing/Label precautions
* D
* 3b, 8, 9, 11, 16, 18, 23, 24, 25, 26, 27, 29, 36, 37, 38, 41, 45, 48, 53, 61, 64, 78

392 methyl bromide with chloropicrin
A highly toxic fumigant for treatment of soil and stored products

Products					
1 Methyl Bromide 98	Bromine & Chem.	980:20 g/l	GA	01335
2 Sobrom BM 98	Jones	980:20 g/l	GA	04508

FOR FULL CONDITIONS OF USE ALWAYS READ THE PRODUCT LABEL

Soil-borne diseases in FIELD CROPS, GLASSHOUSE CROPS. Nematodes, soil insects in FIELD CROPS, GLASSHOUSE CROPS. Weed seeds in FIELD CROPS, GLASSHOUSE CROPS. Grain storage pests in GRAIN STORES.

Efficacy
- Chemical is released from containers as a highly volatile, highly toxic gas and for soil fumigation must be released under a plastic tarpaulin
- Chloropicrin is added as a warning odourant tear gas
- Soil must be clear of trash and cultivated to at least 40 cm before treatment
- Gas must be confined in soil for 48-96 h

Crop safety/Restrictions
- Special equipment is needed for application and the chemical may only be used by professional operators trained in its use and familiar with the precautionary measures to be observed. Refer to HSE Guidance Note CS 12 before applying. Crops may be planted 3-21 d after removal of tarpaulin. For some crops a leaching irrigation is essential. See firms literature for details
- Only one application may be made per year

Special precautions/Environmental safety
- A chemical subject to the Poisons Act 1972 and Poisons Rules 1982
- Product not for retail sale
- Very toxic by inhalation. Causes burns. Risk of serious damage to eyes
- Danger of serious damage to health by prolonged exposure
- Wear suitable respiratory equipment during fumigation and while removing sheeting
- Do not wear gloves or rubber boots
- Ventilate treated areas thoroughly when gas has cleared
- Keep unprotected persons and livestock out of treated areas for at least 24 h
- Dangerous to game, wild birds, animals and bees
- Harmful to fish. Do not contaminate ponds, waterways or ditches with chemical or used container

Protective clothing/Label precautions
- D
- 3b, 8, 9, 11, 16, 18, 23, 24, 25, 26, 27, 29, 36, 37, 38, 41, 45, 48, 53, 61, 64, 78

---

393   metoxuron

A contact and residual urea herbicide for cereals, carrots etc
See also fentin hydroxide + metoxuron

Products					
1 Dosaflo	Farm Protection	500 g/l	SC	00752
2 Dosaflo	ICI	500 g/l	SC	00753

Uses

Annual dicotyledons, annual grasses, blackgrass, barren brome in WINTER WHEAT, DURUM WHEAT, WINTER BARLEY, TRITICALE [1, 2]. Annual dicotyledons, annual grasses, mayweeds in CARROTS [1, 2]. Annual dicotyledons, annual grasses, mayweeds in PARSNIPS *(off-label)* [1, 2].

Notes

Efficacy
- Best results achieved by application to weeds in seedling to young plant stage

- Best control of blackgrass from seedling stage to tillering in period Feb-May
- Best control of barren brome in autumn or winter from 1-3 leaf stage
- Effects on blackgrass reduced on soils of high organic matter or low pH

Crop safety/Restrictions
- Apply to named cereal cultivars from 2-leaf unfolded stage to before first node detectable (GS 12-30). See label for lists of resistant and susceptible cultivars
- Apply to carrots after 2-true leaf stage. Do not spray when soil very dry or wet
- Do not spray cereals on sands or very light soils, or carrots on soils with more than 80% sand or less than 1% organic matter
- Do not apply during prolonged frosty weather, when frost or heavy rain imminent or ground waterlogged
- If treatment repeated in spring at least 6 wk must elapse after autumn/winter spray
- Do not roll for 7 d before or after spraying
- Do not cultivate after spraying or harrow or cultivate within 14 d beforehand
- Do not spray crops checked by pests, wind, frost or waterlogging until recovered
- No crop should be sown within 6 wk of treatment

Protective clothing/Label precautions
- 21, 28, 29, 54, 60, 63, 65, 70

Approval
- Approved for aerial application on winter wheat, winter barley. See introductory notes
- Off-label Approval to June 1992 for use on parsnips (OLA 0517/89, 0515/89) [1, 2]

---

394   metoxuron + simazine
A contact and residual post-emergence herbicide for winter cereals

Products	Atlas Hermes	Atlas	480:30 g/l	SC	03040

Uses   Annual dicotyledons, annual grasses, blackgrass in WINTER WHEAT, WINTER BARLEY, TRITICALE.

Notes

Efficacy
- Best results achieved by application to weeds in seedling to young plant stage
- Blackgrass controlled from 1-main leaf to 5-tiller stage, barren brome in 1-3 leaf stage
- For effective control of blackgrass adequate soil moisture at spraying and rain within 3 wk are required. Control reduced on soils of high organic matter or low pH
- Do not use on soils with more than 10% organic matter

Crop safety/Restrictions
- Apply to named cereal cultivars as soon as possible after crop has 2 main leaves unfolded but before first node detectable (GS 12-31). See label for lists of susceptible and resistant cultivars
- Do not spray undersown crops. Do not spray during frost
- On sands, very light or stony soils heavy rain after application may check crop

---

FOR FULL CONDITIONS OF USE ALWAYS READ THE PRODUCT LABEL

- Do not roll within 7 d of spraying, harrow for 7 d before or any time after, graze before or after, or apply other herbicides within 7 d before or after spraying

Protective clothing/Label precautions
- 21, 28, 29, 54, 60, 63, 66

---

395   metribuzin

A contact and residual triazinone herbicide for use in potatoes

Products					
1 Lexone 70DF	Du Pont	70% w/w	WG	04991
2 Power Metribuzin 70	Power	70% w/w	WG	04674
3 Sencorex WG	Bayer	70% w/w	WG	01918

Uses

Annual dicotyledons, annual grasses in POTATOES.

Notes

Efficacy
- Best results achieved by application to weeds in cotyledon to 1-leaf stage
- Apply to moist soil with well-rounded ridges and few clods
- Activity reduced by dry conditions and on soils with high organic matter content
- Do not cultivate after treatment
- On fen and moss soils pre-planting incorporation to 10-15 cm gives increased activity
- Effective control using a programme of reduced doses is made possible by using a spray of smaller droplets, thus improving retention. See label for details

Crop safety/Restrictions
- On first earlies apply pre-emergence only, on second earlies pre-emergence or post-emergence on named varieties, on maincrop pre-emergence (except for Maris Piper and Sante on sands or very light soils) or post-emergence before longest shoots reach 15 cm (except for certain named varieties). See label for details
- On stony or gravelly soils there is risk of crop damage, especially if heavy rain falls soon after application
- When days are hot and sunny delay spraying until evening
- Some varieties may be sensitive to post-emergence treatment if crop under stress
- Ryegrass, cereals or winter beans may be sown in same season provided at least 16 wk elapsed after treatment and ground ploughed to 15 cm and thoroughly cultivated soon after harvest. Other crops may be sown normally in spring of next year
- Do not drill lettuce or radish in season following treatment

Protective clothing/Label precautions
- 21, 25, 28, 29, 54, 63, 67

---

396   metsulfuron-methyl

A contact and residual sulphonyl-urea herbicide for use in cereals
See also isoproturon + metsulfuron-methyl

Products					
Ally	Du Pont	20% w/w	WG	02977

Uses	Annual dicotyledons, chickweed, mayweeds in WHEAT, BARLEY, OATS, DURUM WHEAT, TRITICALE.

Notes

Efficacy
- Best results achieved on small, actively growing weeds up to 2-true leaf stage
- Commonly used in tank-mixture with other cereal herbicides to improve control of resistant dicotyledons (cleavers, fumitory, ivy-leaved speedwell), larger weeds and grasses. See label for recommended mixtures
- Always add Ally to tank first

Crop safety/Restrictions
- Product must be applied only after 1 Jan
- Apply to wheat and oats from 2-leaf (GS 12), to barley, durum wheat and triticale from 3-leaf stage (GS 13) until flag-leaf fully emerged (GS 39). Do not spray Igri barley before leaf sheath erect stage (GS 30)
- Recommendations for durum wheat and triticale apply to product alone
- Do not use on broadcast crops or crops undersown with grass or legumes
- Do not use on any crop suffering stress from drought, waterlogging, frost, deficiency, pest or disease attack or apply within 7 d of rolling
- Do not apply more than once or to crops already treated with Glean TP or Harmony M
- Take extreme care to avoid drift onto broad-leaved crops or contamination of land planted or to be planted with any crops other than cereals
- Use recommended procedure to clean out spraying equipment
- Only cereals, oilseed rape, field beans or grass may be sown in same calendar year after treatment
- In the event of crop failure sow only wheat within 3 mth after treatment

Protective clothing/Label precautions
- 18, 21, 28, 29, 36, 37, 54, 63, 67

397 metsulfuron-methyl + thifensulfuron-methyl
A contact residual and translocated herbicide for use in cereals

Products	Harmony M	Du Pont	68:7% w/w	WG	03990

Uses	Annual dicotyledons, cleavers, speedwells, polygonums, field pansy in WINTER WHEAT, SPRING WHEAT, SPRING BARLEY.

Notes

Efficacy
- Best results by application to small, actively growing weeds up to 6-true leaf stage
- Effectiveness may be reduced if heavy rain occurs within 4 h of application or if soil conditions very dry
- Tank-mixture with mecoprop or reduced rate of fluroxypyr improves control of cleavers and other problem weeds
- When used in tank-mixes product must be added to spray tank first and fully dispersed before adding other product

FOR FULL CONDITIONS OF USE ALWAYS READ THE PRODUCT LABEL

Crop safety/Restrictions
- Apply in spring to crops from 3-leaf stage to flag-leaf fully emerged (GS 13-39)
- Do not use on durum wheat or winter barley or on any crop suffering stress from drought, waterlogging, frost, deficiency, pest or disease attack
- Do not apply more than once or to crops previously treated with Ally or Glean TP
- Do not apply within 7 d of rolling
- Take extreme care to avoid drift onto broad-leaved crops or contamination of land planted or to be planted with any crops other than cereals
- Use recommended procedure to clean out spraying equipment
- Only cereals, oilseed rape, field beans or grass may be sown in same calendar year after treatment
- In the event of crop failure sow only wheat within 3 mth after treatment

Protective clothing/Label precautions
- 18, 21, 28, 29, 36, 37, 54, 63, 67

398 monolinuron

A residual urea herbicide for potatoes, leeks and french beans

Products   Arresin   Hoechst   200 g/l   EC   00118

Uses   Annual dicotyledons, annual grasses, annual meadow grass, polygonums, fat-hen in POTATOES, LEEKS, FRENCH BEANS.

Notes   **Efficacy**
- Best results achieved by application on moist, firm, clod-free seedbeds pre-weed emergence or on seedlings up to 2-3 leaf stage
- Light rain after application can improve control
- In potatoes or french beans on fen or peat soils use post-weed emergence but before emergence of crop

Crop safety/Restrictions
- Apply in potatoes and french beans as pre-emergence spray
- May be used in tank-mixture with paraquat on potatoes up to 20% emergence (10% emergence on earlies). Do not apply alone post-emergence
- On early sown french beans apply at least 5 d pre-emergence
- Some bean cultivars may be sensitive on light soils, with application close to emergence with heavy rain after treatment or in unevenly drilled crop. See label for list of cultivars
- Apply to transplanted or direct-sown leeks after established and 18 cm high
- Only apply pre-emergence to crops grown under polythene
- Do not use on very light soils
- Do not sow or plant other crops within 2-3 mth, depending on rate applied. Lettuce must not be sown in same season

Special precautions/Environmental safety
- Irritating to skin, eyes and respiratory system
- Flammable

Protective clothing/Label precautions
- A
- 6a, 6b, 6c, 12c, 16, 18, 21, 28, 29, 37, 54, 63, 66, 70

• Approved for aerial application on potatoes, dwarf beans and leeks. See introductory notes

399 monolinuron + paraquat
A contact and residual herbicide for potatoes and french beans

Products					
	1 Gramonol 5	Hoechst	154:110 g/l	SC	00994
	2 Gramonol Five	ICI	154:110 g/l	SC	00995

Uses Annual dicotyledons, annual grasses, annual meadow grass, blackgrass, wild oats in POTATOES, FRENCH BEANS.

Notes **Efficacy**
• Intended for use where some weeds have emerged and further germination expected
• Best results by application to fine, firm seedbeds with weeds up to 2-3 leaf stage
• Residual activity assisted by moist soil conditions and light rain after application
• Residual action greatest on light soils, very little on highly organic soils

Crop safety/Restrictions
• Apply to maincrop potatoes before 20% emergence, to earlies before 10% emergence
• Do not use if any potato plants are 15 cm or more high
• Apply to certain cultivars of french bean after sowing but at least 3 d before crop emergence. See label for details of cultivars which may be treated
• Do not use on weak or unhealthy crops
• Crops may be sensitive on light, sandy soils, with application close to emergence, with heavy rain after treatment or in unevenly drilled crop

Special precautions/Environmental safety
• A chemical subject to the Poisons Rules 1982 and Poisons Act 1972
• Toxic if swallowed
• Harmful in contact with skin. Irritating to eyes, skin and respiratory system

Protective clothing/Label precautions
• A, C
• 4c, 5a, 6a, 6b, 6c, 17, 18, 21, 28, 29, 36, 37, 41, 54, 64, 65, 70, 79

Withholding period
• Harmful to animals, keep all livestock out of treated areas for at least 24 h

400 myclobutanil
A systemic, protectant and curative fungicide for apples

Products					
	1 Systhane 6 Flo	Promark	60 g/l	SC	03921
	2 Systhane 6 Flo	Rohm & Haas	60 g/l	SC	03921

FOR FULL CONDITIONS OF USE ALWAYS READ THE PRODUCT LABEL

3 Systhane 6 W	PBI	6% w/w	WP	04571
4 Systhane 6 W	Rohm & Haas	6% w/w	WP	04570
5 Systhane 40 W	Rohm & Haas	40% w/w	WP	04267

Uses
Scab, powdery mildew in APPLES [1, 2, 5]. Scab in PEARS [1, 2]. Mildew, black spot, rust in ROSES [3, 4].

Notes
Efficacy
- On apples and pears best results achieved when used as part of routine preventive spray programme from bud burst to end of flowering
- Spray at 7-14 d intervals depending on disease pressure and rate applied
- For improved scab control in post-blossom period tank-mix with mancozeb or captan
- Apply alone from mid-Jun for control of secondary mildew on apples and pears
- On roses spray at first signs of disease and repeat every 2 wk. In high risk areas spray when leaves emerge in spring, repeat 1 wk later and then continue normal spraying programme

Crop safety/Restrictions
- Total application for season should not exceed 15 l/ha [1, 2], 2.25 kg/ha [5] (ie 10 sprays at highest recommended rate)

Special precautions/Environmental safety
- Harmful to fish. Do not contaminate ponds, waterways or ditches with chemical or used container

Protective clothing/Label precautions
- A, C, H, M [5]
- 21, 29, 35, 36, 37, 53, 67 [1, 2]; 29, 53, 63, 67 [3, 4]; 29, 35, 36, 37, 53, 67, 70 [5]

Harvest interval
- 14 d

401 nabam

A dithiocarbamate soil fungicide for tomatoes and chrysanthemums

Products
Campbell's Nabam Soil Fungicide   MTM Agrochem.   320 g/l   SL   00413

Uses
Root rot in TOMATOES. Phoma root rot in CHRYSANTHEMUMS.

Notes
Efficacy
- For brown root rot control in tomatoes apply as drench at least 14 d pre-planting in winter, 3 d in summer, or water soil 1 wk post-planting and continue until Jul/Aug
- For phoma control in chrysanthemums drench pre-planting (at least 14 d pre-planting in winter, 10 d in summer) or water soil every 10 d post-planting

Crop safety/Restrictions
- Do not use on fen peat soil or if other soil troubles are checking root action
- Do not use in overhead irrigation systems or apply to foliage

457

Special precautions/Environmental safety
- Harmful if swallowed. Irritating to eyes, skin and respiratory system
- Harmful to fish. Do not contaminate ponds, waterways or ditches with chemical or used container

Protective clothing/Label precautions
- A, C
- 5c, 6a, 6b, 6c, 16, 18, 21, 29, 36, 37, 53, 63, 66, 70, 78

---

402 1-naphthylacetic acid

A plant growth regulator with uses on fruit and ornamentals
See also dichlorophen + 4-indol-3-ylbutyric acid + 1-naphthylacetic acid

Products					
1 Rhizopon B Powder	Fargro	0.1% w/w	SP	01801
2 Rhizopon B Tablets	Fargro	25 mg ai	TB	01802
3 Tipoff	ICI	57.8 g/l	EC	02139

Uses

Sucker inhibition in APPLES, PEARS, PLUMS, CHERRIES, RASPBERRIES, WOODY ORNAMENTALS, ROSES [3]. Control of water shoots in APPLES [3]. Rooting of cuttings in ORNAMENTALS [1, 2].

Notes

Efficacy
- For sucker control apply to shoots 10-15 cm long and repeat in 6-8 wk if necessary. Ensure thorough coverage. Treatment details vary between species - see label
- For suppression of water shoots spray when shoots 7-10 cm long or mix with canker paint and brush on cuts
- See label for details of concentrations recommended for promotion of rooting in cuttings of different species

Crop safety/Restrictions
- When spraying for control of suckers or water shoots avoid fruit bearing parts of trees
- Product not suitable for control of vigour of raspberry canes [3]

Protective clothing/Label precautions
- 21, 28, 29, 54, 63, 67, 70 [1-3]; 60 [3]

---

403 (2-naphthyloxy)acetic acid

A plant growth regulator for setting tomato fruit

Products				
Betapal Concentrate	Synchemicals	16 g/l	SL	00234

Uses

Increasing fruit set in TOMATOES.

FOR FULL CONDITIONS OF USE ALWAYS READ THE PRODUCT LABEL

Efficacy
- Apply with any fine sprayer or syringe when first half dozen flowers are open
- Spray actual trusses in flower and repeat every 2 wk

Crop safety/Restrictions
- Do not spray growing head of tomato plants

404 napropamide

A soil applied amide herbicide for oilseed rape and fruit

Products		Devrinol		Embetec	450 g/l	SC	1124

Uses

Annual grasses, annual dicotyledons, cleavers, groundsel in WINTER OILSEED RAPE, APPLES, PEARS, PLUMS, BLACKCURRANTS, GOOSEBERRIES, RASPBERRIES, STRAWBERRIES, NEWLY PLANTED TREES AND SHRUBS, ESTABLISHED TREES AND SHRUBS, CONTAINER-GROWN TREES AND SHRUBS.

Notes

Efficacy
- Apply to winter oilseed rape as pre-drilling incorporated treatment in tank-mixture with trifluralin. Addition of TCA recommended where volunteer cereals serious
- Incorporate to 25 mm within 30 min. See label for details of recommended procedure
- With minimum cultivation spray onto stubble (allow 24 h after burning) and incorporate as part of surface cultivation
- Apply in fruit crops and woody ornamentals as pre-weed emergence treatment in Nov-Feb, when rainfall will carry chemical into soil
- Do not use on soils with more than 10% organic matter
- Do not cultivate after treatment

Crop safety/Restrictions
- Apply up to 14 d prior to drilling winter oilseed rape
- Do not use on sands. Do not use on strawberries on very light soil
- Tree and bush fruit must be established for at least 10 mth before spraying
- Apply to named strawberry cultivars established for at least one season or to maiden crops as long as planted carefully and no roots exposed
- Apply immediately after planting strawberries in Nov-Feb. Do not treat before Nov if planted in other months. Do not use on protected crops
- Only oilseed rape, swedes, fodder turnips, brassicas or potatoes should be sown within 12 mth of application
- Many woody ornamentals have been treated without damage but not yellow or golden ornamentals or container-grown alpines. If in doubt treat small area initially to test safety

Special precautions/Environmental safety
- Irritating to skin and eyes
- Harmful to fish. Do not contaminate ponds, waterways or ditches with chemical or used container

Protective clothing/Label precautions
- A, C
- 6a, 6b, 18, 21, 25, 29, 36, 37, 53, 63, 65, 70

405   napropamide + trifluralin
A residual herbicide for oilseed rape

Products	Devrinol T	Embetec	140:140 g/l	EC	03942

Uses

Annual grasses, annual dicotyledons in WINTER OILSEED RAPE. Annual grasses, annual dicotyledons in HARDY ORNAMENTAL NURSERY STOCK, NURSERY FRUIT TREES AND BUSHES *(off-label)*.

Notes

Efficacy
* Apply to soil surface pre-drilling and incorporate into top 25 mm of soil within 30 min. Effectiveness is reduced by incorporating too deeply
* With minimum cultivation spray onto stubble (allow 24 h after burning) and incorporate as part of surface cultivation
* See label for details of recommended incorporation procedure
* Do not use on soils with more than 10% organic matter
* Tank mixture with TCA recommended where volunteer cereals serious

Crop safety/Restrictions
* Apply up to 14 d prior to drilling
* On very light and light soils crop damage may occur under adverse conditions such as dry and/or cold conditions. Do not use on sands
* Only brassicas or potatoes should be sown or planted within 12 mth of application

Special precautions/Environmental safety
* Irritating to skin and eyes
* Harmful to fish. Do not contaminate ponds, waterways or ditches with chemical or used container

Protective clothing/Label precautions
* A, C
* 6a, 6b, 18, 21, 25, 28, 29, 36, 37, 53, 60, 63, 65

Approval
* Off-label approval to Aug 1992 for use on hardy ornamental nursery stock and nursery fruit trees and bushes (OLA 0684/89)

406   nicotine
A general purpose, non-persistent, contact, alkaloid insecticide

Products					
1	Campbell's Nico Soap	MTM Agrochem.	75 g/l	LI	00416
2	Nicotine 40% Shreds	DowElanco	40% w/w	FU	04236

FOR FULL CONDITIONS OF USE ALWAYS READ THE PRODUCT LABEL

| 3 XL-All Insecticide | Synchemicals | 70 g/l | LI | 02369 |
| 4 XL-All Nicotine 95% | Synchemicals | 950 g/l | LI | 02370 |

Uses

Aphids in BROAD BEANS, DWARF BEANS, LEAF BRASSICAS, LETTUCE, APPLES, STRAWBERRIES, CHRYSANTHEMUMS, ROSES, TOMATOES [1]. Aphids, capsids, leafhoppers, leaf miners, sawflies, thrips, woolly aphid in VEGETABLES, FRUIT CROPS, FLOWERS, BULBS, CORMS, GLASSHOUSE CROPS [3, 4]. Potato virus vectors in CHITTING POTATOES [2]. Leaf miners in CELERY, ORNAMENTALS [1]. Sawflies in GOOSEBERRIES [1]. Aphids, capsids, leafhoppers, leaf miners, mealy bugs, millipedes, springtails, thrips in GLASSHOUSE CROPS [2]. Mushroom flies in MUSHROOMS [2]. General insect control in FRENCH BEANS *(off-label)* [3, 4]. American serpentine leaf miner in ARTICHOKES, ASPARAGUS, BEANS, DWARF BEANS, BEET CROPS, LEAF BRASSICAS, ROOT BRASSICAS, CARROTS, CELERIAC, CELERY, FENNEL, PARSNIPS, CHICORY, CUCURBITS, ENDIVES, LEEKS, ONIONS, GARLIC, SHALLOTS, MANGE-TOUT PEAS, POTATOES, RADISHES, RHUBARB, SALSIFY, SPINACH, AUBERGINES, GLASSHOUSE CUCURBITS, PEPPERS *(off-label)* [1, 2]. American serpentine leaf miner in LETTUCE, TOMATOES *(off-label)* [1].

Notes

Efficacy
- Apply as foliar spray, taking care to cover undersides of leaves and repeat as necessary or dip young plants, cuttings or strawberry runners before planting out
- Best results achieved by spraying at air temperatures above 16°C
- Fumigate glasshouse crops and mushrooms at temperatures of at least 16°C. See label for details of recommended fumigation procedure [2]
- Potatoes may be fumigated in chitting houses to control virus-spreading aphids
- May be used in integrated control systems to give partial control of organophosphorus-, organochlorine- and pyrethroid-resistant whitefly

Crop safety/Restrictions
- On plants of unknown sensitivity test first on a small scale

Special precautions/Environmental safety
- Harmful in contact with skin, by inhalation or if swallowed [1, 2, 3]. Toxic in contact with skin, by inhalation or if swallowed [4]
- Keep unprotected persons out of treated glasshouses for at least 12 h
- Wear overall, hood, rubber gloves and respirator if entering glasshouse within 12 h of fumigating [2]
- Highly flammable [2]
- Dangerous to game, wild birds and animals
- Harmful to bees. Do not apply at flowering stage. Keep down flowering weeds
- Dangerous to fish. Do not contaminate ponds, waterways or ditches with chemical or used container

Protective clothing/Label precautions
- A [1, 3]; A, D, H, J, [2]; A, C, H, K, M, [4]
- 5a, 5b, 5c, 14, 16, 18, 21, 23, 25, 27, 28, 29, 35, 36, 37, 38, 41, 45, 50, 52, 63, 65, 66, 67, 70 [1-4]; 64 [2]; 78 [1-3]; 4a, 4b, 4c, 64, 79 [4]

Withholding period
- Dangerous/harmful to livestock. Keep all livestock out of treated areas for at least 12 h

Harvest interval
- 1 d [2]; 2 d [1, 3, 4]

• Off-label Approval to Feb 1991 for use on french beans (OLA 0193, 0194/88) [3, 4]; to May 1992 for use on a wide range of vegetables (OLA 0406/89) [2], (OLA 0407/89) [1]

407 nitrothal-isopropyl

A non-systemic fungicide available only in mixtures

408 nitrothal-isopropyl + zineb-ethylenethiuram disulphide complex

A fungicide for control of powdery mildew

Products	Pallitop	BASF	48:3.2% w/w	WP	01533

Uses   Powdery mildew in APPLES. Powdery mildew in CUCUMBERS *(off-label)*.

Notes   **Efficacy**
• For use on apples apply at green cluster or early pink bud then every 14 d (7-10 d under severe mildew conditions) until cessation of extension growth
• Do not apply if rain expected in near future
• For scab control a specific scab fungicide should be used in programme. See label for details of compatible tank mixtures

Special precautions/Environmental safety
• Irritating to eyes, skin and respiratory system. May cause sensitization by skin contact
• Keep all livestock out of treated areas for at least 2 wk
• Dangerous to fish. Do not contaminate ponds, waterways and ditches with chemical or used container

Protective clothing/Label precautions
• A, C
• 6a, 6b, 6c, 10a, 18, 21, 29, 35, 36, 37, 41, 52, 63, 67, 70

Harvest interval
• 2 wk

Approval
• Off-label Approval to Feb 1991 for use on cucumbers (OLA 0195/88)

FOR FULL CONDITIONS OF USE ALWAYS READ THE PRODUCT LABEL

409   nonylphenoxypoly(ethyleneoxy)ethanol-iodine complex
An organoiodine fungicide available only in mixtures

410   nonylphenoxypoly(ethyleneoxy)ethanol-iodine complex + tecnazene
A protectant fungicide and sprout suppressant for stored potatoes

Products	Bygran F	Dean	1.5:10% w/w	GR	00365

Uses   Sprout suppression in STORED POTATOES. Dry rot, soft rot, silver scurf, skin spot in STORED POTATOES.

Notes   **Efficacy**
- Distribute granules evenly over the face and sides of the clamp, the top of each box or the face of the stack after each load. Follow by a top dressing and final strawing down
- Do not apply to boxes in field
- Potatoes must be clean, dry and free from soil or debris
- Will not control sprouting in tubers which have broken dormancy
- Effective for up to 6-9 mth. Excessive ventilation may reduce period of control

Crop safety/Restrictions
- Air seed for 6 wk before planting. Only sprouted tubers should be planted
- May delay emergence and slightly reduce yield, especially if airing insufficient

Special precautions/Environmental safety
- Harmful to fish. Do not contaminate ponds, waterways or ditches with chemical or used container

Protective clothing/Label precautions
- 29, 35, 53, 63, 67

Harvest interval
- 6 wk

411   nonylphenoxypoly(ethyleneoxy)ethanol-iodine complex + thiabendazole
A systemic and protectant fungicide for control of soil-borne diseases

Products					
1 Byatran	Dean	2.5:2.5% w/w	GR	00362
2 Tubazole	Dean	10:10% w/w	LI	02225
3 Tubazole M	Dean	2.4:2.4% w/w	UL	02470

Uses   Soil-borne diseases in BRASSICAS, TOMATOES [1]. Soil-borne diseases, black scurf and stem canker, dry rot, gangrene, skin spot, silver scurf in STORED POTATOES [1-3].

Efficacy
- Apply granules at planting with suitable applicator attached to planter, broadcast on brassica seedbeds and incorporate lightly to 10 cm or use band treatment on transplanted or direct-drilled crops, apply to glasshouse tomatoes prior to final cultivation before planting out or as band treatment immediately after planting [1]
- Apply Tubazole to potato tubers by thermal fogging apparatus, ULV sprayer or as dip for 5 min. Air dry dipped tubers before storage
- Apply Tubazole M to tubers going into store with ULV applicator. See label for details
- Potato tubers for treatment should be as free of soil and debris as possible

Crop safety/Restrictions
- Do not apply to module-grown tomato plants [1]

Special precautions/Environmental safety
- Harmful to fish. Do not contaminate ponds, waterways or ditches with chemical or used container

Protective clothing/Label precautions
- 29, 35, 53, 63, 67 [1-3]; 21, 28 [2, 3]

Harvest interval
- potatoes 21 d [2, 3]

Approval
- Approved for application with ULV equipment. See introductory notes [2, 3]

---

412 nuarimol

A systemic protectant and eradicant pyrimidine fungicide
See also captan + nuarimol

Products

1 Chemtech Nuarimol	Chemtech	90 g/l	EC	04517
2 Triminol	DowElanco	90 g/l	EC	02467

Uses

Powdery mildew, rhynchosporium, net blotch in BARLEY [1, 2]. Powdery mildew, septoria in WHEAT [1, 2]. Powdery mildew in OATS [1, 2].

Notes

Efficacy
- Apply prior to or at first signs of disease, usually from first node detectable stage (GS 31), with a second application if necessary
- In tank-mixes products complement action of other fungicides against leaf and ear disease complexes. See label for details
- Mildew control may be reduced if rain falls before product has dried on leaves

Crop safety/Restrictions
- Apply 1 treatment per crop for oats and wheat to before flowering (GS 59)
- Apply up to 2 treatments per crop for barley to before boots swollen (GS 45)

Special precautions/Environmental safety
- Irritating to eyes and skin

FOR FULL CONDITIONS OF USE ALWAYS READ THE PRODUCT LABEL

- Flammable
- Harmful to fish. Do not contaminate ponds, waterways or ditches with chemical or used container

Protective clothing/Label precautions
- A, C
- 6a, 6b, 12c, 18, 21, 28, 29, 36, 37, 53, 60, 63, 66

413 octhilinone

An isothiazolone fungicide for treating canker and tree wounds

Products					
	Pancil T	Rohm & Haas	1% w/w	PA	01540

Uses Canker in APPLES, PEARS, TREE FRUIT, WOODY ORNAMENTALS. Silver leaf, pruning wounds in APPLES, PEARS, PLUMS, CHERRIES, TREE FRUIT, WOODY ORNAMENTALS.

Notes
Efficacy
- Treat canker by removing infected shoots, cutting back to healthy tissue and painting evenly over wounds
- Apply to pruning wounds immediately after cutting. Cover all cracks in bark and ensure that coating extends beyond edges of wound
- Best results achieved by application in dry conditions

Crop safety/Restrictions
- Only apply after harvest and before bud burst
- Do not apply to grafting cuts or buds. Do not apply under freezing conditions

Special precautions/Environmental safety
- Irritating to skin, risk of serious damage to eyes
- May cause sensitization by skin contact
- Dangerous to fish. Do not contaminate ponds, waterways or ditches with chemical or used container

Protective clothing/Label precautions
- A, C
- 6b, 9, 10a, 18, 22, 25, 28, 29, 36, 37, 52, 60, 63, 65, 78

414 ofurace

A phenylamide (acylalanine) fungicide available only in mixtures
See also manganese zinc ethylenebisdithiocarbamate + ofurace

415 omethoate

A systemic organophosphorus insecticide and acaricide

Products					
	Folimat	Bayer	575 g/l	EC	00912

Uses	Frit fly in CEREALS, YOUNG LEYS. Yellow cereal fly in WINTER WHEAT, WINTER BARLEY. Wheat bulb fly in WINTER WHEAT. Aphids in SPRING CEREALS. Aphids in PLUMS, DAMSONS, HOPS. Frit fly in GRASS SEED CROPS *(off-label)*.

Notes

Efficacy
- Spray cereals at first signs of damage by wheat bulb fly, frit fly or aphids, at start of egg hatch for control of yellow cereal fly
- For aphid control in plums and damsons spray at cot-split stage in mid-May and again 4 wk later if necessary. Good coverage is essential
- For aphid control in hops apply when first seen, usually late May to early Jun, and repeat at 10-14 d intervals. Good coverage is essential
- Resistant strains of damson-hop aphid have occurred in some hop growing areas

Crop safety/Restrictions
- Treat spring-sown cereals up to 4-leaf stage (GS 14)
- Some leaf yellowing may occur in hops but the crop is not affected

Special precautions/Environmental safety
- A chemical subject to the Poisons Rules 1982 and Poisons Act 1972
- This product contains an organophosphorus compound. Do not use if under medical advice not to work with such compounds
- Toxic in contact with skin or if swallowed
- Flammable
- Harmful to game, wild birds and animals
- Harmful to bees. Do not apply at flowering stage. Keep down flowering weeds
- Harmful to fish. Do not contaminate ponds, waterways or ditches with chemical or used container

Protective clothing/Label precautions
- A, C, H
- 1, 4a, 4c, 12c, 14, 17, 18, 24, 25, 28, 29, 35, 36, 37, 41, 46, 50, 53, 64, 66, 70, 79

Withholding period
- Harmful to livestock. Keep all livestock out of treated areas for at least 7 d

Harvest interval
- hops 14d; cereals 3 wk; plums, damsons 5 wk; grass crops 6 wk

Approval
- Approved for aerial application on winter wheat. See introductory notes
- Off-label Approval to Aug 1992 for use on grass seed crops (OLA 0725/89)

---

416 oxadiazon

A residual and contact herbicide for fruit and ornamental crops
See also carbetamide + oxadiazon

Products	1 Ronstar 2G	Embetec	2% w/w	GR	03723
	2 Ronstar Liquid	Embetec	250 g/l	EC	01822

FOR FULL CONDITIONS OF USE ALWAYS READ THE PRODUCT LABEL

Annual dicotyledons, annual grasses in CONTAINER-GROWN ORNAMENTALS [1]. Annual dicotyledons, cleavers, knotgrass, bindweeds, annual grasses in APPLES, PEARS, BLACKCURRANTS, GOOSEBERRIES, GRAPEVINES, HOPS, WOODY ORNAMENTALS [2].

Notes

Efficacy

* Best results achieved by application pre- or post-weed emergence
* Rain or overhead watering is needed soon after application for effective results
* Pre-emergence activity reduced on soils with more than 10% organic matter and, in these conditions, post-emergence treatment more effective [2]
* Best results on bindweed when first shoots are 10-15 cm long
* Do not cultivate after treatment [2]

Crop safety/Restrictions

* See label for list of ornamental species which may be treated with granules. Treat small numbers of other species to check safety. Do not treat hydrangea or spiraea
* Do not treat container stock under glass or use on plants rooted in media with high sand or non-organic content. Do not apply to plants with wet foliage
* Apply spray to apples and pears from Jan to Jul, avoiding young growth
* Treat bush fruit from Jan to bud-break, avoiding bushes, grapevines in Feb/Mar before start of new growth or in Jun/Jul avoiding foliage
* Treat hops cropped for at least 2 yr in Feb or in Jun/Jul after deleafing
* Treat woody ornamentals from Jan to Jun, avoiding young growth. Do not spray container stock overall
* Do not use more than 8 l/ha in any 12 mth period

Special precautions/Environmental safety

* Irritating to eyes
* Flammable [2]
* Dangerous to fish. Do not contaminate ponds, waterways or ditches with chemical or used container

Protective clothing/Label precautions

* A, C
* 6a, 18, 21, 26, 27, 29, 36, 37, 52, 63, 65, 70 [1, 2]; 12c [2]

---

417 oxadixyl

A phenylamide (acylalanine) fungicide available only in mixtures
See also mancozeb + oxadixyl

---

418 oxamyl

A soil-applied, systemic carbamate nematicide and insecticide

Products

1 Power Blade	Power	10% w/w	GR	04952
2 Vydate 10G	Du Pont	10% w/w	GR	02322

Uses

Potato cyst nematode, free-living nematodes, spraing vectors, aphids in POTATOES [1, 2]. Docking disorder vectors, millipedes, pygmy beetle, mangold fly, aphids in SUGAR BEET [1, 2]. Stem and bulb nematodes in ONIONS [1, 2]. Pea cyst nematode, pea and bean weevils in PEAS [1, 2]. Potato cyst nematode, globodera species, red spider mites, leaf miners, aphids, whitefly in TOMATOES [2]. American serpentine leaf miner, non-indigenous leaf miners in BEETROOT, AUBERGINES, PEPPERS, BROAD BEANS, SOYA BEANS, CUCURBITS, GARLIC, SHALLOTS, NON-EDIBLE ORNAMENTALS *(off-label)* [2].

Notes

Efficacy
- Apply granules with suitable applicator before drilling or planting. See label for details of recommended machines
- In potatoes incorporate thoroughly to 10-15 cm and plant within 3-4 d
- In sugar beet apply in seed furrow at drilling
- In onions apply in seed furrow or use 'bow-wave' technique with higher rates
- In peas apply during final seed-bed preparation and incorporate to 10-15 cm immediately after treatment. Drill within 3-4 d
- For nematode control in tomatoes apply 3-4 d before planting and incorporate. For insect control apply 10-14 d after planting and water in. Repeat 3 wk later
- Use on mineral soils only with tomatoes, on all soil types with other crops

Special precautions/Environmental safety
- This product contains an anticholinesterase compound. Do not use if under medical advice not to work with such compounds
- Harmful by inhalation or if swallowed
- Dangerous to fish. Do not contaminate ponds, waterways or ditches with chemical or used container
- Dangerous to game and wildlife. Cover granules completely. Bury spillages
- Wear protective gloves if handling treated compost or soil within 2 wk after treatment
- Allow at least 12 h, followed by at least 1 h ventilation, before entry of unprotected persons into treated glasshouses

Protective clothing/Label precautions
- A, B, C, C or D, H, K, M
- 2, 5b, 5c, 14, 16, 18, 22, 25, 28, 29, 35, 36, 37, 45, 52, 64, 67, 70, 79

Harvest interval
- tomatoes 2 wk; other crops 3 wk

Approval
- Off-label Approval to Feb 1993 for use on vegetable crops and non-edible ornamentals for control of American serpentine leaf miner and other non-indigenous leaf miners (OLA 0088/90) [2]

419   oxycarboxin
A protectant, eradicant and systemic anilide fungicide

Products					
1 Plantvax 20	Uniroyal	200 g/l	EC	01600
2 Plantvax 75	Fargro	75% w/w	WP	01601
3 Ringmaster	RP Environ.	200 g/l	EC	01805

FOR FULL CONDITIONS OF USE ALWAYS READ THE PRODUCT LABEL

Uses	Yellow rust in WHEAT [1]. Rust in CARNATIONS, GERANIUMS, ROSES, CHRYSANTHEMUMS [2]. Fairy rings in AMENITY GRASS, SPORTS TURF [3].

Notes

Efficacy
- Apply in wheat at first signs of infection. Where heavy infections occur a second application may be made
- Add wetter for first spray on carnations, geraniums and roses if rust already established. May be applied with fogging machine mixed with suitable carrier [2]
- Can be used as a drench on carnations grown in peat bags [2]
- Apply to fairy rings after spiking. On dry soil water before spraying or add wetter. Best results when fungus in active growth [3]

Crop safety/Restrictions
- On chrysanthemums apply alone, not within 2 d of other sprays. Do not apply spraying oil for 14 d before or after. Can scorch chrysanthemum leaves when taken up by roots. Avoid spraying when blooms open [2]

Special precautions/Environmental safety
- Harmful if swallowed. Irritating to eyes, skin and respiratory system [1, 3]
- Flammable [1]
- Not to be used on food crops [3]
- Dangerous to fish. Do not contaminate ponds, waterways or ditches with chemical or used container

Protective clothing/Label precautions
- A, E [1]; A, C, G, J, L, M [2]; A, C [3]
- 14, 21, 28, 29, 36, 37, 52, 63, 67, 70 [2]; 5c, 6a, 6b, 6c, 18, 21, 28, 29, 36, 37, 52, 63, 66, 70, 78 [1, 3]; 34 [3]; 35 [1]

Harvest interval
- 21 d [1]

Approval
- Approved for application with thermal fogging machine. See introductory notes on ULV application [2]

420 oxydemeton-methyl
A contact, fumigant and systemic organophosphorus insecticide

Products	Metasystox R	Bayer	570 g/l	EC	01333

Uses	Aphids in CEREALS, GRASS SEED CROPS, BEET CROPS, POTATOES, BROAD BEANS, FIELD BEANS, FRENCH BEANS, RUNNER BEANS, PEAS, CARROTS, PARSNIPS, CELERY, APPLES, PEARS, PLUMS, PEACHES, APRICOTS, DAMSONS, CHERRIES, CURRANTS, GOOSEBERRIES, RASPBERRIES, STRAWBERRIES, PROTECTED LETTUCE, GLASSHOUSE CROPS, FLOWERS, ROSES. Pea midge in PEAS. Leafhoppers in POTATOES, APPLES, PEARS, STRAWBERRIES. Red spider mites in APPLES, PEARS, PLUMS, DAMSONS, CHERRIES, PEACHES, APRICOTS, TREE NUTS, CURRANTS, GOOSEBERRIES, RASPBERRIES, STRAWBERRIES, CUCUMBERS, TOMATOES, CARNATIONS, ROSES, ORNAMENTALS. Sawflies in APPLES, PEARS, PLUMS, DAMSONS, CHERRIES. Bryobia mites in APPLES, PEARS. Pear

sucker in PEARS. Blackfly in CHERRIES. Blackcurrant leaf midge in BLACKCURRANTS. Insect pests in LUCERNE, GRASS SEED CROPS, HERBS, CELERIAC, FENNEL, MARROWS, COURGETTES, ONIONS, LEEKS, GARLIC, SWEETCORN, RADICCHIO, SPINACH, PROTECTED SPINACH, LETTUCE, BEETROOT, ORNAMENTALS, GRAPEVINES, PROTECTED FRUIT *(off-label)*.

Notes

Efficacy
- With the exception of brassicas, Metasystox R is recommended for the same range of crops as Metasystox 55 (based on demeton-S-methyl) but has a less penetrating odour which may influence choice of product in some situations
- Spray when pests first seen, or beforehand, and repeat as necessary. Number and timing of sprays vary with crop and pest. See label for details
- Treatment of winter cereals advisable if aphids found in late-Sep in areas where serious outbreaks of barley yellow dwarf virus have occurred regularly
- Best results on sugar beet achieved by application in early morning or late evening
- Spray potatoes for seed at 80% emergence whether aphids present or not and repeat 3 times at 14 d intervals
- Apply as spray on ornamentals or as drench if soil damp. Do not apply to dry soil or under alkaline conditions
- Organophosphorus-resistant strains of aphid have developed on sugar beet in certain areas and of red spider mite on certain crops
- Do not mix with any alkaline sprays

Crop safety/Restrictions
- Do not use on any brassica crop
- Do not spray carnations in flower
- On plants of unknown tolerance test first on a small scale

Special precautions/Environmental safety
- A chemical subject to the Poisons Rules 1982 and Poisons Act 1972
- This product contains an organophosphorus compound. Do not use if under medical advice not to work with such compounds
- Toxic if swallowed, harmful in contact with skin
- A minimum interval of 12 h must elapse before entry of unprotected persons into treated buildings
- Harmful to game, wild birds and animals
- Harmful to bees. Do not apply at flowering stage. Keep down flowering weeds
- Harmful to fish. Do not contaminate ponds, waterways or ditches with chemical or used container

Protective clothing/Label precautions
- A, C, H
- 1, 4c, 5a, 14, 16, 18, 23, 24, 25, 28, 29, 35, 36, 37, 41, 46, 50, 53, 64, 66, 70, 79

Withholding period
- Harmful to livestock. Keep all livestock out of treated areas for at least 7 d; treated leys must not be grazed within 6 wk of treatment

Harvest interval
- wheat, barley 2 wk; mangolds, fodder beet for clamping 10 d; mangolds, fodder beet for fodder 3 wk; other crops 3 wk.

FOR FULL CONDITIONS OF USE ALWAYS READ THE PRODUCT LABEL

Approval
- Approved for aerial application on cereals, beet crops, potatoes, carrots, parsnips, celery, beans, peas, top fruit, soft fruit, herbage seed crops, maize, mangolds. See introductory notes
- Off-label Approval to Feb 1993 for use on lucerne (OLA 0155/90), a wide range of outdoor and protected vegetables, fruits, herbs and ornamentals (OLA 0156/90), and grass seed crops (OLA 0157/90)

---

421 paclobutrazol

A plant growth regulator for ornamentals, fruit and grass seed crops

See also dicamba + paclobutrazol

Products					
1 Bonzi	ICI	4 g/l	SC	02155
2 Cultar	ICI	250 g/l	SC	02156

Uses

Stem shortening, increasing flowering in AZALEAS, BEDDING PLANTS, BEGONIAS, CHRYSANTHEMUMS, KALANCHOES, LILIES, ROSES, TULIPS, HARDY ORNAMENTAL NURSERY STOCK [1]. Improving colour in POINSETTIAS [1]. Controlling vigour, increasing fruit set in APPLES, PEARS [2].

Notes

Efficacy
- Chemical is active via both foliage and root uptake
- Apply as spray to produce compact pot plants and container grown nursery stock, and to improve poinsettia bract colour [1]
- Apply as drench to reduce stem length of potted tulips and on some container stock [1]
- Timing is critical and varies with species. See label for details [1]
- Apply to apple and pear trees under good growing conditions as pre-blossom spray (apples only) and post-blossom at 7-14 d intervals [2]
- Timing and dose of orchard treatment vary with species and cultivar. See label [2]

Crop safety/Restrictions
- Do not use on border grown chrysanthemums [1]
- Do not use on trees of low vigour or under stress [2]
- Do not tank mix with or apply on same day as Thinsec [2]
- With 7 d spray programme on fruit trees use up to 6 sprays, with 10 d programme up to 4 sprays with 14 d programme up to 3 sprays [2]
- Do not use on trees from green cluster to 2 wk after full petal fall [2]
- Do not use in underplanted orchards, nor for 3 yr before grubbing [2]

Special precautions/Environmental safety
- Do not use on food crops [1]
- Keep livestock out of treated areas [2]
- Harmful to fish. Do not contaminate ponds, waterways or ditches with chemical or used container

Protective clothing/Label precautions
- A, C [2]
- 29, 53, 60, 63, 66, 70 [1, 2]; 21, 34 [1]; 35, 43 [2]

Harvest interval
• apples, pears 6 wk [2]

422 paraquat

A non-selective, non-residual contact bipyridilium herbicide
See also amitrole + diquat + paraquat + simazine
diquat + paraquat
diquat + paraquat + simazine
diuron + paraquat
monolinuron + paraquat

Products

1 Dextrone X	Chipman	200 g/l	SL	00687
2 Dextrone X	ICI Professional	200 g/l	SL	00687
3 Gramoxone 100	ICI	200 g/l	SL	00997
4 Gramoxone 100	Schering	200 g/l	SL	03867
5 Power Paraquat	Power	200 g/l	SL	04840
6 Scythe	Cyanamid	200 g/l	SL	02455
7 Speedway	ICI Professional	8% w/w	WG	02732

Uses

Annual dicotyledons, annual grasses, volunteer cereals in FIELD CROPS, VEGETABLES, ORNAMENTALS *(pre-emergence)* [1-7]. Annual dicotyledons, annual grasses, volunteer cereals, wild oats, barren brome, creeping bent in FIELD CROPS *(autumn stubble)* [3-6]. General weed growth in CEREALS *(stubble burning)* [3-6]. Annual grasses, perennial ryegrass, rough meadow grass, creeping bent in FIELD CROPS *(sward destruction/direct drilling)* [3-6]. Annual grasses, annual dicotyledons, volunteer cereals, creeping bent in FIELD CROPS *(minimum cultivation)* [3-6]. Annual grasses, annual dicotyledons, volunteer cereals in POTATOES, ROW CROPS, SUGAR BEET [3-6]. Annual dicotyledons, annual grasses, volunteer cereals, wild oats, barren brome, creeping bent in LUCERNE [3, 4]. Annual dicotyledons, annual grasses, perennial ryegrass, rough meadow grass, creeping bent in ORCHARDS, BLACKCURRANTS, GOOSEBERRIES, RASPBERRIES [3-6]. Annual dicotyledons, annual grasses, perennial ryegrass, rough meadow grass, creeping bent in HOPS [3, 4]. Chemical stripping in HOPS [3, 4]. Runner desiccation in STRAWBERRIES [3, 4, 6]. Annual dicotyledons, annual grasses, creeping bent in BULBS [1-4]. Annual dicotyledons, annual grasses, creeping bent in ORNAMENTAL TREES AND SHRUBS [1, 2, 7]. Annual dicotyledons, annual grasses, creeping bent in FORESTRY, FORESTRY TRANSPLANT LINES [1-4]. Annual dicotyledons, annual grasses in FOREST NURSERY BEDS *(stale seedbed)* [1-4]. Firebreak desiccation in FORESTRY [1-4]. Annual dicotyledons, annual grasses, perennial ryegrass, rough meadow grass, creeping bent in NON-CROP AREAS [1-7]. Annual dicotyledons, annual grasses in GRAPEVINES *(off-label)* [1, 3, 4, 6]. Annual dicotyledons, annual grasses in MINT *(off-label)* [3].

Notes

Efficacy
• Effects develop more rapidly under bright conditions but as effective when dull
• Apply at any time when weeds green and preferably before 15 cm high
• Spray is rainfast after 10 min
• Addition of wetter recommended with lower application rates [3-6]

FOR FULL CONDITIONS OF USE ALWAYS READ THE PRODUCT LABEL

- Only use clean water for mixing up spray
- Allow at least 4 h before cultivating, leave overnight if possible
- Apply in autumn to improve stubble burning. To suppress couch spray when shoots have 2 leaves, repeat as necessary and plough after last treatment
- For direct-drilling land should be free of perennial weeds and protection against slugs should be provided. Allow 7-10 d before drilling into sprayed grass
- Best time for use in top fruit is Nov-Apr
- For forestry fire-break use apply in Jul-Aug and fire 7-10 d later

Crop safety/Restrictions
- Chemical is inactivated on contact with soil. Crops can be sown or planted soon after spraying on most soils (24 h [1, 2], 4 h [3, 5], immediately [4], a few hours [6]), after 3 d on sandy or immature peat soils
- Do not use on straw or other artifical growing media
- Apply to lucerne in late Feb/early Mar when crop dormant
- Apply to potatoes up to 10% emergence for earlies and seed crops, up to 40% emergence for maincrop provided plants do not exceed 15 cm high. Do not use under hot, dry conditions
- For interrow use apply with guarded, no-drift sprayer
- Use as a carefully directed spray in blackcurrants, gooseberries, grapevines and other fruit crops, in raspberries only when dormant
- For runner control in strawberries use guarded sprayer and not when flowers or fruit present
- Do not use on hops under drought conditions [3]
- Apply to bulbs pre-emergence (at least 3 d pre-emergence on sandy soils) or at end of season, provided no attached foliage and bulbs well covered (not on sandy soils)
- If using around glasshouses ensure vents and doors closed
- In forestry seedbeds apply up to 3 d before seedling emergence

Special precautions/Environmental safety
- A chemical subject to the Poisons Rules 1982 and Poisons Act 1972
- Toxic if swallowed. Paraquat can kill if swallowed
- Harmful in contact with skin. Irritating to eyes and skin
- Do not put in a drinks bottle
- Not for use by amateur gardeners
- Harmful to animals. Keep all livestock out of treated areas for at least 24 h
- Paraquat may be harmful to hares. Where possible spray stubbles early in the day

Protective clothing/Label precautions
- A, C
- 4c, 5a, 6a, 6b, 17, 18, 21, 28, 29, 36, 37, 40, 47, 54, 60, 64, 65, 70, 79

Harvest interval
- Mint 8 wk [3]

Approval
- Off-label Approval to Aug 1991 for use in grapevines (OLA 0816/89) [1]; (OLA 0815/89) [3]; (OLA 0817/89) [4]; (OLA 0814/89) [6]; to May 1991 for use in mint (OLA 0319/90) [3]

423   penconazole

A protectant conazole fungicide with antisporulant activity
See also captan + penconazole

Products   Topas 100EC   Ciba-Geigy   100 g/l   EC   03231

Uses   Powdery mildew in APPLES, HOPS. Powdery mildew, scab in ORNAMENTAL TREES. Rust in ROSES.

Notes   **Efficacy**
• Apply every 10-14 d (every 7-10 d in warm, humid weather) at first sign of infection or as a protective spray ensuring complete coverage. See label for details of timing
• Increase dose and volume with growth of hops but do not exceed 2000 l/ha
• Antisporulant activity reduces development of secondary mildew in apples

Crop safety/Restrictions
• Spray programme should not exceed 10 sprays per season for apples or 6 for hops
• Check for varietal susceptibility in roses. Some defoliation may occur after repeat applications on Dearest

Special precautions/Environmental safety
• Irritating to eyes and skin
• Harmful to fish. Do not contaminate ponds, waterways or ditches with chemical or used container

Protective clothing/Label precautions
• A, C
• 6a, 6b, 14, 18, 21, 29, 35, 36, 37, 53, 63, 66, 70

Harvest interval
• apples, hops 14 d

424   pencycuron

A non-systemic phenylurea fungicide for use on seed potatoes

Products   Monceren DS   Bayer   12.5% w/w   DP   04160

Uses   Black scurf in SEED POTATOES.

Notes   **Efficacy**
• Provides control of tuber-borne disease and useful, though variable, control of soil-borne disease. Also gives some reduction of stem canker
• Apply to seed tubers in chitting trays or in hopper at planting
• If rain interrupts planting cover tubers in hopper

FOR FULL CONDITIONS OF USE ALWAYS READ THE PRODUCT LABEL

Special precautions/Environmental safety
● Use of suitable dust mask is mandatory when applying dust, filling the hopper or riding on planter

Protective clothing/Label precautions
● F
● 29, 54, 63, 67

425   pendimethalin

A residual dinitroaniline herbicide for cereals and other crops
See also isoproturon + pendimethalin

Products					
	1 Stomp 330	Cyanamid	330 g/l	EC	02028
	2 Stomp 400	Cyanamid	400 g/l	SC	04183
	3 Stomp H	Hortichem	330 g/l	EC	03873

Uses   Annual grasses, annual dicotyledons, annual meadow grass, blackgrass, wild oats, cleavers, speedwells in WINTER WHEAT, DURUM WHEAT, WINTER BARLEY, SPRING BARLEY, WINTER RYE, TRITICALE, COMBINING PEAS, POTATOES [1, 2]. Annual grasses, annual dicotyledons in APPLES, PEARS, PLUMS, CHERRIES, BLACKCURRANTS, GOOSEBERRIES, RASPBERRIES, STRAWBERRIES, HOPS, ONIONS [3]. Annual grasses, annual dicotyledons in EVENING PRIMROSE *(off-label)* [1, 2]. Annual grasses, annual dicotyledons in PARSLEY, SAGE *(off-label)* [2].

Notes   **Efficacy**
● Apply as soon as possible after drilling. Weeds are controlled as they germinate
● For effective blackgrass control apply not more than 2 d before weed seeds germinate
● Tank mixes with approved formulations of isoproturon or chlorotoluron recommended for improved pre- and post-emergence control of blackgrass, with Flexidor for additional broad-leaved weeds [1, 2]
● Best results by application to fine firm, moist, clod-free seedbed when rain follows treatment. Effectiveness reduced by prolonged dry weather after treatment
● Do not use on spring barley after end Mar (mid-Apr in Scotland) because dry conditions likely. Do not apply to dry seedbeds in spring unless rain imminent
● Effectiveness reduced on soils with more than 6% organic matter. Do not use where organic matter exceeds 10%
● Any trash, ash or straw should be incorporated evenly during seedbed preparation
● Do not disturb soil after treatment
● On peas drilled after end Mar (mid-Apr in Scotland) tank-mix with Fortrol [1, 2]
● Apply to potatoes as soon as possible after planting and ridging in tank-mix with Fortrol or Sencorex [1, 2]

Crop safety/Restrictions
● May be applied pre-emergence of cereal crops sown before 30 Nov provided seed covered by at least 32 mm soil, or post-emergence to early tillering stage (GS 23)[1]
● Do not undersow treated crops
● Do not use on crops suffering stress due to disease, drought, waterlogging, poor seedbed conditions or chemical treatment or on soils where water may accumulate
● Apply to combining peas as soon as possible after sowing, not when plumule within 13 mm of surface [1, 2]
● Apply to potatoes up to 7 d before first shoot emerges [1, 2]

475

- In the event of crop failure in autumn, spring wheat, spring barley, maize, potatoes, beans and peas may be sown following ploughing to 15 cm [1, 2]
- After a dry season land must be ploughed to 15 cm before drilling ryegrass
- Apply in top fruit, bush fruit and hops from autumn to early spring when crop dormant. Use on raspberries in Scotland only. See label for details [3]
- Apply in strawberries from autumn to early spring (not before Oct on newly planted beds). Do not apply during flower initiation period (post-harvest to mid-Sep) [3]
- Apply pre-emergence in onions as tank-mixture with propachlor, not on sands, very light, organic or peaty soils [3]

Special precautions/Environmental safety
- Irritating to skin and eyes [1, 3]
- Dangerous to fish. Do not contaminate ponds, waterways or ditches with chemical or used container [1-3]

Protective clothing/Label precautions
- A, C [1]; A [3]
- 6a, 6b [1-3]; 18, 21, 25, 28, 29, 36, 37, 52, 63, 66, 70 [1-3]

Approval
- Off-label Approval to Feb 1993 for use on evening primrose (OLA 0099/90) [1]; to Apr 1993 for use on outdoor parsley and sage (OLA 0304/90) [2]; to Sep 1993 for use on evening primrose (OLA 0502/90) [3]

426   pentachlorophenol

A herbicide, fungicide and insecticide available only in mixtures
See also bromacil + pentachlorophenol

427   pentanochlor

A contact anilide herbicide for various horticultural crops
See also chlorpropham + pentanochlor

Products					
1 Atlas Solan 40	Atlas	400 g/l	EC	03834
2 Croptex Bronze	Hortichem	400 g/l	EC	04087

Uses Annual dicotyledons, annual meadow grass in CARROTS, CELERY, CELERIAC, FENNEL, PARSLEY, PARSNIPS, TOMATOES, APPLES, PEARS, PLUMS, CHERRIES, CURRANTS, GOOSEBERRIES, ANEMONES, ANNUAL FLOWERS, CARNATIONS, CHRYSANTHEMUMS, FREESIAS, SWEET PEAS, NURSERY STOCK, ROSES [1]. Annual dicotyledons, annual meadow grass in CARROTS, PARSNIPS, ANEMONES, ANNUAL FLOWERS, CHRYSANTHEMUMS, FREESIAS, SWEET PEAS, NURSERY STOCK [2].

Notes **Efficacy**
- Best results achieved on young weed seedlings under warm, moist conditions

FOR FULL CONDITIONS OF USE ALWAYS READ THE PRODUCT LABEL

- Weeds most susceptible in cotyledon to 2-leaf stage, up to 3 cm high, redshank, fat-hen, fumitory and some others also controlled at later stages
- Some residual effect in early spring when adequate soil moisture and growing conditions good. Effectiveness reduced by very cold weather or drought

Crop safety/Restrictions
- Apply pre-emergence in anemones and annual flowers
- Apply pre-emergence or after fully expanded cotyledon stage of carrots and related crops. Repeat once if necessary
- Apply as directed spray in fruit crops, nursery stock, perennial flowers and tomatoes
- Any crop may be planted after 4 wk following ploughing and cultivation

Special precautions/Environmental safety
- Irritating to skin and eyes. Harmful if swallowed

Protective clothing/Label precautions
- 6a, 6b, 18, 21, 28, 29, 36, 37, 54, 63, 66

---

428   permethrin

A broad spectrum, contact and ingested pyrethroid insecticide
See also bioallethrin + permethrin
*   fenitrothion + permethrin + resmethrin*

Products					
1 Cooper Coopex Smokes	Wellcome	13.5% w/w	FU	H2661
2 Cooper Coopex WP	Wellcome	25% w/w	WP	H2384
3 Darmycel Agarifume Smoke	Darmycel		FU	00649
4 Fumite Permethrin Smoke	Hunt	2 and 3.5 g a.i.	FU	00940
5 Murfume Permethrin Smoke	DowElanco	13.75% w/w	FU	01427
6 Permasect 10 EC	Mitchell Cotts	10% w/w	EC	03920
7 Permasect 25 EC	Mitchell Cotts	250 g/l	EC	01576
8 Turbair Permethrin	PBI	5 g/l	UL	02246

Uses

Aphids, caterpillars in BRASSICAS [6, 7]. Caterpillars, suckers, sawflies, capsids, fruitlet mining tortrix, codling moth in APPLES, PEARS [6, 7]. Browntail moth in AMENITY TREES AND SHRUBS [8]. Aphids, caterpillars, leaf miners, whitefly in BRASSICAS, CUCUMBERS, TOMATOES, PEPPERS, CHRYSANTHEMUMS, FUCHSIAS, ORNAMENTALS, POT PLANTS [6, 7, 8]. Whitefly in CELERY, CUCUMBERS, PROTECTED LETTUCE, AUBERGINES, PEPPERS, TOMATOES, ORNAMENTALS [4-8]. Mushroom flies in MUSHROOMS [3, 4, 8]. Grain storage pests in GRAIN STORES [1, 2].

Notes

Efficacy
- Sprays give rapid knock-down effect with persistent protection on leaf surfaces
- Apply as soon as pest or damage appears, or as otherwise recommended and repeat as necessary usually at 14 d intervals
- Number and timing of sprays vary with crop and pest. See label for details
- Addition of non-ionic wetter recommended for use on brassicas
- Do not mix Turbair formulation with any other product
- For whitefly or mushroom fly control use at least 3 fumigations at 5-7 d intervals
- Apply in grain stores by fumigation or surface spraying. Foodstuffs should not come into contact with treated surfaces

Crop safety/Restrictions
- Do not use Turbair treatment on Chinese cabbage, coleus, exacum, peperomia, red calceolaria or protected roses
- Cut open blooms before fumigating flower crops
- Do not fumigate young seedlings or plants being hardened off
- Allow 7 d after fumigating before introducing Encarsia or Phytoseiulus
- Check safety of Turbair spray on chrysanthemums, fuchsias, pot plants and ornamentals and of fumigation on plants of unknown sensitivity before treating on a large scale

Special precautions/Environmental safety
- Irritating to eyes, skin and respiratory system [8]; to eyes and respiratory system [3-5]; to eyes and skin [6]; to skin [7]
- Flammable [1-8]
- Dangerous to bees. Do not apply at flowering stage. Keep down flowering weeds
- Extremely dangerous [1-5, 7, 8], dangerous [6], to fish. Do not contaminate ponds, waterways or ditches with chemical or used container

Protective clothing/Label precautions
- F [2]; A, C, [6]; A [7]
- 6a, 6c [3-5]; 6a, 6b [6]; 6b [7]; 6a, 6b, 6c [8]; 12c, 18, 21, 28, 29, 36, 37, 48, 63, 66, 67, 70 [1-8]; 51 [1-5, 7, 8]; 52 [6]

Harvest interval
- zero

Approval
- Approved for ULV application. See label for details. See introductory notes [8]

429 petroleum oil
An insecticide and acaricide oil for horticultural use

Products	Hortichem Spraying Oil	Hortichem	710 g/l	EC	03816

Uses Red spider mites, mealy bugs, scale insects in CUCUMBERS, TOMATOES, GRAPEVINES, POT PLANTS.

Notes

Efficacy
- Spray at 1% (0.5% on tender foliage) to wet plants thoroughly, particularly the underside of leaves, and repeat as necessary
- Apply under quick drying conditions

Crop safety/Restrictions
- Do not apply in bright sun unless glass well shaded
- Treat grapevines before flowering
- On plants of unknown sensitivity test first on a small scale
- Mixtures with certain pesticides may damage crop plants. If mixing, spray a few plants to test for tolerance before treating larger areas

---

FOR FULL CONDITIONS OF USE ALWAYS READ THE PRODUCT LABEL

- Do not mix with sulphur or Plictran or use these materials within 28 d of treatment

Special precautions/Environmental safety
- Harmful if swallowed. Irritating to eyes, skin and respiratory system

Protective clothing/Label precautions
- A, C
- 5c, 6a, 6b, 6c, 18, 21, 28, 36, 37, 63, 65, 70, 78

---

430   phenmedipham
A contact carbamate herbicide for beet crops and strawberries
See also ethofumesate + phenmedipham

Products

1 Atlas Protrum K	Atlas	114 g/l	EC	03089
2 Beetomax	Fine	114 g/l	EC	03129
3 Betanal E	ICI	114 g/l	EC	02674
4 Betanal E	Schering	114 g/l	LI	03862
5 Campbell's Beetup	Campbell	114 g/l	EC	02680
6 Goliath	ABM	114 g/l	EC	03411
7 Gusto	Farm Protection	118 g/l	EC	03110
8 Headland Dephend	WBC Technology	114 g/l	EC	04460
9 Hickson Phenmedipham	Hickson & Welch	118 g/l	EC	02825
10 Pistol 400	Quadrangle	114 g/l	EC	03444
11 Portman Betalion	Portman	110 g/l	EC	04677
12 Power Phenmedipham II	Power	114 g/l	EC	04167
13 Star Phenmedipham	Star	114 g/l	EC	03128
14 Suplex	Unicrop	114 g/l	EC	04072
15 Tripart Beta	Tripart	118 g/l	EC	03111
16 Tripart Beta 2	Tripart	114 g/l	EC	04831
17 Vangard	FCC	114 g/l	EC	02743

Uses

Annual dicotyledons in SUGAR BEET, FODDER BEET, MANGOLDS, RED BEET, STRAWBERRIES [1, 3, 4, 5, 8, 17]. Annual dicotyledons in SUGAR BEET, FODDER BEET, MANGOLDS, RED BEET [2, 7, 9, 11, 12, 15]. Annual dicotyledons in SUGAR BEET, FODDER BEET, MANGOLDS [6, 10, 14, 16]. Annual dicotyledons in SUGAR BEET [13].

Notes

Efficacy
- Best results achieved by application to young seedling weeds, preferably cotyledon stage, under good growing conditions when low doses are effective
- 2-3 repeat applications at 7-10 d intervals using a low dose are recommended on mineral soils, 3-5 applications may be needed on organic soils
- Do not spray wet foliage or if rain imminent
- Addition of adjuvant oil may improve effectiveness on some weeds
- Spray may be applied at conventional or low volume. See label for details
- Various tank-mixtures with other beet herbicides recommended. See label for details
- Use of certain pre-emergence herbicides is recommended in combination with post-emergence treatment. See label for details

Crop safety/Restrictions
- Apply to beet crops at any stage as low dose/low volume spray or from fully developed cotyledon stage with full rate. Apply to red beet after fully developed cotyledon stage

- Use with adjuvant oil on sugar beet only. Crop must have 4 true leaves and not be under stress [5], have at least 2 leaves [7]
- At high temperatures (above 21 °C) reduce rate or spray after 5 pm
- Do not apply immediately after frost or if frost expected
- Do not spray crops stressed by wind damage, nutrient deficiency, pest or disease attack etc. Do not roll or harrow for 7 d before or after treatment
- Apply to strawberries at any time when weeds in susceptible stage, except in period from start of flowering to picking
- Do not use on strawberries under cloches or polythene tunnels

Special precautions/Environmental safety
- Harmful if swallowed [1, 2, 7, 13, 14]; in contact with skin, by inhalation and if swallowed [5, 8]
- Irritating to eyes, skin and respiratory system [1-11, 13-17], to respiratory system [12]
- Harmful to fish. Do not contaminate ponds, waterways or ditches with chemical or used container

Protective clothing/Label precautions
- A, C [1, 6, 8]; A, C, H [2]; A, C, P [7, 9, 15]
- 5c [1, 2, 7, 13, 14]; 5a, 5b, 5c [5, 8]; 6a, 6b, 6c [1-11, 13-17]; 6c [12]; 18, 21, 28, 29, 36, 37, 53, 60, 63, 66, 70, 78 [1-17]

Approval
- Approved for aerial application on beet crops, strawberries. See introductory notes [3, 4]

431   phenothrin

A non-systemic contact and ingested pyrethroid insecticide

| **Products** | Sumithrin 10 SEC | Sumitomo | 103 g/l | EC | 03987 |

Uses   Flies in LIVESTOCK HOUSES.

Notes   **Efficacy**
- Spray walls and other structural surfaces of milking parlours, dairies, cow byres and other animal housing

Crop safety/Restrictions
- Do not apply directly to livestock
- Remove exposed milk and collect eggs before application. Protect milk machinery and containers from contamination

Special precautions/Environmental safety
- Wear approved respiratory equipment and eye protection (goggles) when applying with Microgen equipment
- Not to be sold or supplied to amateur users
- Irritating to eyes

FOR FULL CONDITIONS OF USE ALWAYS READ THE PRODUCT LABEL

- Dangerous to fish. Do not contaminate ponds, waterways or ditches with chemical or used container

Protective clothing/Label precautions
- A, C, D, E, F, H
- 6a, 18, 21, 29, 36, 37, 52, 63, 67

Approval
- May be applied with Microgen ULV equipment. See introductory notes

---

432 phenothrin + tetramethrin
A pyrethroid insecticide mixture for control of flying insects

Products	1 Killgerm ULV 500	Killgerm	48:24 g/l	UL	03540
	2 Sorex Super Fly Spray	Sorex		AE	03144

Uses   Flies, wasps, mosquitoes in AGRICULTURAL PREMISES.

Notes   **Efficacy**
- Close doors and windows and spray in all directions for 3-5 sec. Keep room closed for at least 10 min
- May be used in the presence of poultry and livestock

Special precautions/Environmental safety
- Do not spray directly on food, livestock or poultry
- Remove exposed milk before application. Protect milk machinery and containers from contamination
- Flammable
- Dangerous to fish. Do not contaminate ponds, waterways and ditches with chemical or used container

Protective clothing/Label precautions
- 28, 29, 36, 37, 52

---

433 phenylmercury acetate
An organomercury fungicide seed treatment for cereals and fodder beet
See also gamma-HCH + phenylmercury acetate

Products	1 Agrosan D	ICI	2% w/w	DS	00061
	2 Ceresol	ICI	20 g/l	LS	00458
	3 Ceresol 50	ICI	10 g/l	LS	00459
	4 Single Purpose	DowElanco	20 g/l	LS	03960

Uses   Covered smut, leaf stripe, net blotch in BARLEY [1-4]. Covered smut, pyrenophora leaf spot, loose smut in OATS [1-4]. Bunt in WHEAT [1-4]. Bunt, stripe smut in RYE [1-4].

Efficacy

- Mix dust formulation thoroughly with seed in automatic seed treater or use rotary drum or shake in closed container. Do not mix with shovel on open floor [1]
- Apply liquid formulations with recommended seed treaters. See label for details
- Treat seed at least one day, preferably several days, before sowing [1, 2]

Crop safety/Restrictions

- Do not treat seed with moisture content above 16%
- Do not use on seed already treated with another organomercury seed dressing
- Store treated seed in cool, draught-free place and keep moisture content below 16%
- Do not store treated seed for more than 12 mth [1], 8 wk [2], 6 mth [4]
- Stock carried over to following season should be tested for germination
- Treatment may lower germination capacity, especially if seed not of good quality

Special precautions/Environmental safety

- Harmful in contact with skin, by inhalation or if swallowed [1], in contact with skin or if swallowed [2-4]
- Organomercury compounds can cause rashes or blisters on skin
- Treated seed not to be used as food or feed. Do not re-use sack
- Special precautions needed in event of spillage or leakage. See label for details
- Do not sow treated seed from aircraft
- Harmful to fish. Do not contaminate ponds, waterways or ditches with chemical or used container

Protective clothing/Label precautions

- A
- 5a, 5b, 5c [1]; 5a, 5c [2-4]; 14, 18, 21, 23, 28, 29, 36, 37, 53, 63, 65, 70, 72, 73, 74, 76, 78 [1-4]

Stop-press

- Approvals for sale, supply, storage, advertisement and use of products containing phenylmercury acetate for treatment of beet and sugar beet seed were revoked with effect from 31 Dec 1990 (Pesticides Register, Issue 9 1990)

434   pheromones

Fruit moth pheromone baits

Products

1 Pherocon CM Codelemone	DowElanco	XX	01585
2 Pherocon Archemone	DowElanco	XX	01585
3 Pherocon Adox-O Adoxamone	DowElanco	XX	01585
4 Pherocon GFun Funemone	DowElanco	XX	01585

Uses

Codling moth in APPLES [1]. Fruit tree tortrix moth in TREE FRUIT [2, 3]. Plum fruit moth in PLUMS [4].

FOR FULL CONDITIONS OF USE ALWAYS READ THE PRODUCT LABEL

Efficacy

- Use in conjunction with Pherocon insect traps to monitor level of risk from pest and aid accurate spray timing, effectiveness and economy
- Traps are set on outside of tree canopy and pheromone caps set in centre at appropriate time. See label for trap densities and timings
- Caps are highly specific. Do not place different types in the same trap
- Renew traps after 6 wk and do not allow to become contaminated or soiled
- It is important to avoid cross-contamination of different traps and caps. Handle caps with gloves, keeping one pair for each type
- Store in a cool place, preferably in a refrigerator

Protective clothing/Label precautions

- 29

435   phorate

A systemic organophosphorus insecticide with vapour-phase activity

Products

1 BASF Phorate	BASF	10% w/w	GR	00210
2 Campbell's Phorate	MTM Agrochem.	10% w/w	GR	00418
3 Terrathion Granules	FCC	10% w/w	GR	02106

Uses

Aphids in BEET CROPS, BRASSICAS, CARROTS, POTATOES [1-3]. Aphids, lettuce root aphid in LETTUCE [2]. Aphids in STRAWBERRIES [1, 2]. Aphids, capsids, pea and bean weevils in PEAS [2]. Aphids, capsids, pea and bean weevils in BROAD BEANS, FIELD BEANS [1-3]. Aphids, carrot fly in PARSNIPS [1, 4]. Capsids in BEET CROPS [1-3]. Capsids, leafhoppers, wireworms in POTATOES [1-3]. Carrot fly in CARROTS, CELERY [1-3]. Cabbage root fly in BRASSICAS [1-3]. Cabbage root fly, cabbage stem flea beetle, rape winter stem weevil in OILSEED RAPE [1-3]. Frit fly in MAIZE, SWEETCORN [1-3]. Lettuce root aphid in ICEBERG LETTUCE *(off-label, higher rate)* [2].

Notes

Efficacy

- May be applied as a soil or foliar treatment. Application method, rate and timing vary with crop, pest and soil type. See label for details
- May be applied broadcast, with row crop applicators or in combine drills
- Effectiveness of soil application reduced by hot, dry weather, excessive rainfall or on soils with more than 10% organic matter content
- Foliar application recommended for aphid or capsid control in beet crops, peas, beans and strawberries
- Effectiveness of foliar application reduced on plants suffering from drought stress

Crop safety/Restrictions

- When applied at time of sowing granules must not be in direct contact with seed

Special precautions/Environmental safety

- This product contains an organophosphorus compound. Do not use if under medical advice not to work with such compounds
- Toxic in contact with skin, by inhalation or if swallowed
- Dangerous to game, wild birds and animals
- Dangerous to fish. Do not contaminate ponds, waterways or ditches with chemical or used container

Protective clothing/Label precautions
- A, B, C, C or D + E, H, J, K, M
- 1, 4a, 4b, 4c, 14, 16, 18, 23, 24, 25, 28, 29, 35, 36, 37, 40, 46, 52, 64, 67, 70, 79 [1-3]

Withholding period
- Harmful to livestock. Keep all livestock out of treated areas for at least 6 wk. Bury or remove spillages

Harvest interval
- 6 wk

Approval
- Approved for aerial application on beet crops, spring sown field beans, broad beans, [1]; field beans, broad beans [2, 3]. See introductory notes
- Off-label Approval to Sep 1992 for root aphid control on Iceberg lettuce (OLA 0744/89) [2]

436   phosalone

A contact and ingested organophosphorus insecticide and acaricide

Products					
1 Zolone Liquid	Hortichem	350 g/l	EC	03776
2 Zolone Liquid	RP	350 g/l	EC	02385

Uses

Grain aphid in CEREALS [2]. Seed weevil, brassica pod midge in OILSEED RAPE, FODDER RAPE SEED CROPS, CABBAGE SEED CROPS, KALE SEED CROPS, MUSTARD [1, 2]. Pollen beetles in BRASSICA SEED CROPS, MUSTARD [1, 2]. Aphids, codling moth, tortrix moths, winter moth, bryobia mites, red spider mites in APPLES, PEARS [1]. Red spider mites in PLUMS [1].

Notes

Efficacy
- Spray winter or spring cereals in late spring or early summer
- Spray oilseed rape and brassica seed crops during flowering period. Number and timing of sprays vary with pest and crop. See label for details
- Spray apples and pears pre-blossom and up to 4 times post-blossom at intervals of about 14 d. Spray plums pre-blossom and up to twice post-blossom
- Chemical is most effective during warm weather but should not be applied at low volume during periods of high light intensity or high temperature
- Toxicity hazard to bees and ladybirds is low at recommended rates. To reduce hazard further spray in late evening or early morning
- Do not mix with highly alkaline materials such as lime sulphur

Crop safety/Restrictions
- Apply on cereals up to inflorescence emerging stage (GS 51)
- Apply up to 2 sprays per season on oilseed rape and brassica seed crops

Special precautions/Environmental safety
- This product contains an organophosphorus compound. Do not use if under medical advice not to work with such compounds

---

FOR FULL CONDITIONS OF USE ALWAYS READ THE PRODUCT LABEL

- Harmful in contact with skin or if swallowed. Irritating to eyes
- Harmful to fish. Do not contaminate ponds, waterways or ditches with chemical or used container

Protective clothing/Label precautions
- A, C
- 1, 5a, 5c, 6a, 14, 16, 18, 23, 24, 25, 28, 29, 35, 36, 37, 41, 53, 63, 66, 70, 78

Withholding period
- Harmful to livestock. Keep all livestock out of treated areas for at least 4 wk

Harvest interval
- 3 wk

Approval
- Approved for aerial application on cereals, brassica seed crops

437   picloram

A persistent, translocated picolinic herbicide for non-crop areas

See also bromacil + picloram
2,4-D + picloram

Products					
	Tordon 22K	Chipman	240 g/l	SL	02152

Uses

Annual dicotyledons, perennial dicotyledons, woody weeds, Japanese knotweed, bracken in NON-CROP AREAS, NON-CROP GRASS.

Notes

Efficacy
- May be applied at any time of year. Best results achieved by application as foliage spray in late winter to early spring
- For bracken control apply 2-4 wk before frond emergence
- Clovers are highly sensitive and eliminated at very low doses
- Persists in soil for up to 2 yr

Crop safety/Restrictions
- Do not apply around desirable trees or shrubs where roots may absorb chemical
- Do not apply on slopes where chemical may be leached onto areas of desirable plants

Special precautions/Environmental safety
- Irritating to eyes
- Keep livestock out of treated areas until foliage of any poisonous weeds such as ragwort has died and become unpalatable
- Harmful to fish. Do not contaminate ponds, waterways or ditches with chemical or used container

Protective clothing/Label precautions
- 6a, 21, 29, 43, 53, 63, 66

438   pirimicarb

A carbamate insecticide for aphid control

Products

1 Aphox	ICI	50% w/w	SG	00106
2 Phantom	Bayer	50% w/w	SG	04519
3 Pirimicarb 50 DG	Schering	50% w/w	SP	04063
4 Pirimor	ICI	50% w/w	WP	01594
5 Power Demo	Power	50% w/w	SG	05035

Uses

Aphids in CEREALS, PEAS, POTATOES, SUGAR BEET, BROAD BEANS, FIELD BEANS, RUNNER BEANS, DWARF BEANS, OILSEED RAPE, LEAF BRASSICAS, SWEDES, TURNIPS, CARROTS, PARSNIPS, CELERY, MAIZE, SWEETCORN, GRASSLAND, APPLES, PEARS, CHERRIES, STRAWBERRIES, BLACKCURRANTS, REDCURRANTS, GOOSEBERRIES, RASPBERRIES [1, 2]. Aphids in CEREALS, PEAS, POTATOES, SUGAR BEET, BROAD BEANS, FIELD BEANS, DWARF BEANS, APPLES, PEARS, CHERRIES, STRAWBERRIES [3]. Aphids in STRAWBERRIES, LETTUCE, PROTECTED LETTUCE, ORNAMENTALS, FOREST NURSERIES, CUCUMBERS, TOMATOES, PEPPERS, GLASSHOUSE CUT FLOWERS, POT PLANTS [4]. Aphids in CEREALS, PEAS, POTATOES, SUGAR BEET, BROAD BEANS, FIELD BEANS, LEAF BRASSICAS, APPLES, PEARS, STRAWBERRIES [5]. Aphids in SWEETCORN, RADISHES, COURGETTES, GHERKINS, MARROWS, MELONS, PUMPKINS, SQUASHES *(off-label)* [1, 3, 4]. Aphids in FENNEL, SPINACH, SPINACH BEET *(off-label)* [1].

Notes

Efficacy
- Chemical has contact, fumigant and translaminar activity
- Best results achieved under warm, calm conditions when plants not wilting and spray does not dry too rapidly. Little vapour activity at temperatures below 15°C
- Apply as soon as aphids seen or warning issued and repeat as necessary
- Addition of non-ionic wetter recommended for use on brassicas
- Chemical has little effect on bees, ladybirds and other insects and is suitable for use in integrated control programmes on apples and pears
- On cucumbers and tomatoes a root drench is preferable to spraying when using predators in an integrated control programme
- Aphids resistant to carbamate insecticides have developed in certain areas. Only use aphicides when necessary, not as routine

Special precautions/Environmental safety
- This product contains an anticholinesterase compound. Do not use if under medical advice not to work with such compounds
- Harmful if swallowed

Protective clothing/Label precautions
- A [1-3, 5]; A, F, H [4]
- 2, 5c, 18, 21, 29, 35, 36, 37, 41, 54, 63, 67, 70, 78 [1-5]

Withholding period
- Harmful to livestock. Keep all livestock out of treated areas for at least 7 d

---

FOR FULL CONDITIONS OF USE ALWAYS READ THE PRODUCT LABEL

Harvest interval
- oilseed rape, cereals, sweetcorn, lettuce under glass 14 d; grassland 7d; cucumbers, tomatoes and peppers under glass 2 d; other edible crops 3 d; flowers and ornamentals zero

Approval
- Approved for aerial application on cereals, maize, oilseed rape, swedes, turnips, brassicas, beans, sugar beet, carrots, peas, potatoes [1]; cereals, field beans, sugar beet, peas, potatoes [3]. See introductory notes
- Off-label approval to Feb 1991 for use on fennel, spinach, spinach beet (OLA 0201/88) [1]; radishes (OLA 0202, 0203, 0204/88) [1, 3, 4]; sweetcorn (OLA 0205, 0206, 0207/88) [1, 3, 4]; courgettes, pumpkins, gherkins, marrows, melons, squashes (OLA 0208, 0209, 0210/88) [1, 3, 4]

439   pirimiphos-ethyl

An organophosphorus insecticide for mushrooms and pot plants

Products					
Fernex	Fargro	10% w/w	GR	00862

Uses   Phorid flies, sciarid flies in MUSHROOMS, BEDDING PLANTS, POT PLANTS.

Notes   **Efficacy**
- Incorporate thoroughly into mushroom compost before pasteurization or into casing material. It may be necessary to treat both casing and compost for first cropping cycle if fly population large
- Incorporate thoroughly into compost for pot and bedding plants
- Do not store treated compost for more than 1 mth

Crop safety/Restrictions
- Cropping pattern of new mushroom strains may be upset. Test treat small areas
- In absence of direct fly damage a slight reduction in yield may take place. Advisable to treat one house and observe results before using on large scale
- Do not use in seed compost for bedding or pot plants
- Slight growth retardation may occur in petunias, dahlias and marigolds

Special precautions/Environmental safety
- A chemical subject to the Poisons Rules 1982 and Poisons Act 1972
- This product contains an organophosphorus compound. Do not use if under medical advice not to work with such compounds
- Harmful if swallowed
- Dangerous to game, wild birds and animals
- Dangerous to fish. Do not contaminate ponds, waterways or ditches with chemical or used container

Protective clothing/Label precautions
- A, C, H
- 1, 5a, 5b, 5c, 14, 18, 21, 25, 28, 29, 36, 37, 45, 52, 64, 65, 70, 78

440 pirimiphos-methyl

A contact, fumigant and translaminar organophosphorus insecticide

Products					
1 Actellic D	ICI	250 g/l	EC	H1069
2 Actellic Dust	ICI	2% w/w	DP	H1222
3 Actellic Smoke Generator No 20	ICI	20 g a.i.	FU	00018
4 Actellic 40WP	ICI	40% w/w	WP	H1297
5 Actellifog	ICI	100 g/l	HN	00019
6 Blex	ICI	500 g/l	EC	00284
7 Fumite Pirimiphos Methyl Smoke	Hunt	10 g a.i.	FU	00941

Uses

Wheat bulb fly, frit fly in CEREALS [6]. Flour beetles, flour moths, grain beetles, grain weevils, grain storage mites, warehouse moth, grain storage pests in GRAIN STORES, STORED GRAIN [1-4]. Cabbage stem flea beetle in OILSEED RAPE [6]. Storage pests in STORED OILSEED RAPE, STORED LINSEED [1, 2]. Caterpillars, whitefly in BRASSICAS [6]. Carrot fly in CARROTS [6]. Leaf miners in SUGAR BEET [6]. Aphids, apple sawfly, apple sucker, capsids, codling moth, rust mite, tortrix moths in APPLES [6]. Aphids, caterpillars, rust mite in PEARS [6]. Rust mite in PLUMS [6]. Ants, aphids, capsids, caterpillars, earwigs, leafhoppers, leaf miners, mealy bugs, red spider mites, sawflies, thrips, whitefly, woodlice in GLASSHOUSE CROPS [7]. Aphids, red spider mites, thrips, whitefly, tarsonemid mites in GLASSHOUSE CROPS [5]. French fly in CUCUMBERS [6]. Mushroom flies in MUSHROOM HOUSES [5].

Notes

Efficacy

• Chemical acts rapidly and has short persistence in plants, though spray or dust persists for long periods on inert surfaces
• Apply spray when pest first seen or at time of egg hatch and repeat as necessary
• For wheat bulb fly control apply on receipt of ADAS warning
• Rate, number and timing of sprays vary with crop and pest. See label for details
• For use on Brussels sprouts use of drop-legged sprayer is recommended
• For protection of stored grain disinfect empty stores by spraying surfaces and/or fumigation and treat grain by full or surface admixture or spray bagged grain. See label for details of treatment and suitable application machinery
• Best results obtained when grain stored at 15% moisture or less. If grain moisture higher period of protection can be reduced. Dry and cool moist grain coming into store before treatment
• For control of whitefly and other glasshouse pests apply as smoke or by thermal fogging. See label for details of techniques and suitable machines
• Actellifog may be used to clean up houses before introducing whitefly parasites but do not use when Encarsia present

Crop safety/Restrictions

• Grain treated by admixture as specified may be consumed by humans and livestock
• Do not fumigate glasshouses in bright sunshine or when foliage is wet or roots dry
• Do not fumigate young seedlings or plants being hardened off
• Do not fog open flowers of ornamentals without first consulting firm's representative

FOR FULL CONDITIONS OF USE ALWAYS READ THE PRODUCT LABEL

- Do not fog mushrooms when wet as slight spotting may occur
- Consult processors before using on crops for processing [6]

Special precautions/Environmental safety
- This product contains an organophosphorus compound. Do not use if under medical advice not to work with such compounds
- Irritating to eyes and skin
- Ventilate fumigated or fogged spaces thoroughly before re-entry

Protective clothing/Label precautions
- A [1,6]; A, D, J, L, M [5]
- 1, 16, 18, 21, 28, 29, 35, 36, 37, 39, 53, 63, 65, 70 [1, 2, 3]; 1, 18, 21, 28, 29, 35, 36, 37, 53, 63, 67 [4]; 1, 6a, 6c, 18, 21, 28, 29, 35, 36, 37, 53, 63, 66, 67, 70 [5, 6, 7]

Harvest interval
- zero [3, 5, 7]; brassicas 3 d [6]; winter wheat, sugar beet, carrots, apples, pears 7 d [6]; cucumbers 3 wk [6]

Approval
- One product formulated for application by thermal fogging. See label for details. See introductory notes on ULV application [5]

---

441 potassium sorbate + sodium metabisulphite + sodium propionate
A broad-spectrum fungicide for use in field crops

Products Brimstone Plus Mandops 30 g/l (B):100% w/w(A):200 g/l (B) KK 03695

Uses Powdery mildew, brown rust, yellow rust, septoria, botrytis, sooty moulds, fusarium ear blight in WHEAT. Net blotch, powdery mildew, brown rust, yellow rust, rhynchosporium in BARLEY. Botrytis, light leaf spot in OILSEED RAPE. Blight in POTATOES. Powdery mildew in SUGAR BEET, SWEDES.

Notes
Efficacy
- Apply at first signs of disease and repeat if necessary. See label for details of timing
- Do not use with acidic water of pH below 5.5
- Do not mix with chlormequat other than Mandops Chlormequat or with organophosphorus insecticides

Crop safety/Restrictions
- Do not apply to oilseed rape during flowering

Special precautions/Environmental safety
- Part A of product liberates a toxic gas when in contact with acids

Protective clothing/Label precautions
- 22, 28, 29, 54, 63, 65

● Approval for aerial application on wheat, barley, oilseed rape, potatoes, sugar beet and swedes. See introductory notes

442　prochloraz

A broad-spectrum protectant and eradicant imidazole fungicide
See also carbendazim + prochloraz
　　　　　fenpropimorph + prochloraz

Products

1 Fisons Octave	Fisons	50% w/w	WP	03416
2 Power Prochloraz 4000	Power	400 g/l	EC	04732
3 Prelude 20LF	Agrichem	200 g/l	LS	04371
4 Sporgon 50 WP	Darmycel	50% w/w	WP	02564
5 Sportak	Schering	400 g/l	EC	03871
6 Sportak 45	Schering	450 g/l	EC	03815

Uses　　Eyespot, sharp eyespot in WINTER WHEAT, WINTER BARLEY, WINTER RYE [5, 6]. Glume blotch in WINTER WHEAT [5, 6]. Leaf spot in WINTER WHEAT, WINTER RYE [5, 6]. Net blotch in BARLEY [5, 6]. Powdery mildew in WHEAT, BARLEY, WINTER RYE [2, 5]. Rhynchosporium in BARLEY, WINTER RYE [5]. Alternaria, light leaf spot, canker, white leaf spot, grey mould, sclerotinia stem rot in OILSEED RAPE [2, 5, 6]. Seed-borne diseases in LINSEED, FLAX [3]. Fungus diseases in HARDY ORNAMENTAL NURSERY STOCK [1]. Cobweb, dry bubble, wet bubble in MUSHROOMS [4]. Fungus diseases in NEWLY SOWN LEYS, GRASS SEED CROPS [5].

Notes　　**Efficacy**
● Spray cereals at first signs of disease. Protection of winter crops through season usually requires at least 2 treatments. See label for details of rates and timing. Product active against strains of eyespot resistant to benzimidazole fungicides [5, 6]
● Tank mixes with other fungicides recommended to improve control of rusts in wheat and barley. See label for details [5, 6]
● A period of at least 3 hr without rain should follow spraying
● Apply as drench against soil diseases, as spray against foliar/stem diseases or as dip at propagation [1]
● Apply seed treatment through automated seed treatment machinery, preferably with inert Seed Flow Aid [3]
● Apply to mushrooms as casing treatment or spray between flushes. Timing determined by anticipated disease occurrence [4]

Crop safety/Restrictions
● Do not apply more than twice to any cereal, grass or oilseed rape crop
● Do not exceed stated doses [6]
● See label for details of ornamentals which can be treated [1]

Special precautions/Environmental safety
● Harmful if swallowed [2, 5], in contact with skin [3, 6]. Irritating to eyes and skin [2, 3, 5, 6]

FOR FULL CONDITIONS OF USE ALWAYS READ THE PRODUCT LABEL

- Flammable [2, 3, 5, 6]
- Dangerous to fish. do not contaminate ponds, waterways or ditches with chemical or used container [2, 3, 5, 6]

Protective clothing/Label precautions
- A, C [2, 3, 5, 6]
- 5c [2,5]; 5a [3, 6]; 6a, 6b, 12c [2, 3, 5, 6]; 22, 29, 54, 63, 67 [1,4]; 35, 36, 37, 52, 60, 63, 66, 78 [2, 3, 5, 6]; 72, 73, 74, 75, 76, 77 [3]

Harvest interval
- cereals 6 wk, not after milky ripe stage (GS 75) [2, 3, 5, 6]; oilseed rape 4 wk [5], 6 wk [6]; mushrooms 2 d [4]; grassland harvest/grazing 60 d [5]

Approval
- Approved for aerial application on cereals and oilseed rape. See introductory notes [5]

443   prometryn

A contact and residual triazine herbicide for various field crops

Products					
1 Atlas Prometryne 50 WP	Atlas	50% w/w	WP	03502
2 Gesagard 50 WP	Ciba-Geigy	50% w/w	WP	00981

Uses

Annual dicotyledons, annual grasses in PEAS, POTATOES, CARROTS, CELERY, PARSLEY, TRANSPLANTED LEEKS [1, 2]. Annual dicotyledons, annual grasses in PARSNIPS, CORIANDER [2]. Annual dicotyledons, annual grasses in DRILLED LEEKS *(off-label)* [2].

Notes

Efficacy
- Best results achieved by application to young seedling weeds up to 5 cm high (cotyledon stage for knotgrass, mayweeds and corn marigold) on fine, moist seedbed when rain falls afterwards. Do not use on very cloddy soils
- On organic soils only contact action effective and repeat application may be needed (certain crops only)

Crop safety/Restrictions
- Apply to peas pre-emergence up to 3 d before crop expected to emerge
- Spring sown vining and drying peas may be treated. Damage may occur with Vedette or Printana, especially if emerging under adverse conditions. Do not treat forage peas
- Do not use on peas on very light soils, sands, gravelly or stony soils
- Apply to early potatoes up to 10% emergence
- Apply to carrots, celery, parsley, parsnips or coriander post-emergence after 2-rough leaf stage or after transplants established
- Apply to leeks after transplants established. Do not use on drilled leeks
- Do not use more than one application on peas, potatoes or transplanted leeks
- Excessive rain after treatment may check crop
- In the event of crop failure only plant recommended crops within 8 wk

Protective clothing/Label precautions
- 29, 35, 37, 54, 63, 67, 70

Harvest interval
- 6 wk

• Approved for aerial application on peas, potatoes, carrots, celery, leeks. See introductory notes [2]
• Off-label approval to Feb 1991 for use on drilled leeks (OLA 0212/88) [2]

444 prometryn + terbutryn

A residual, pre-emergence herbicide for use in peas, beans and potatoes

Products					
1 Peaweed	FCC	152:304 g/l	SC	03248
2 Peaweed	PBI	152:304 g/l	SC	03248
3 Spudweed	PBI	152:304 g/l	SC	04965

Uses Annual dicotyledons, annual grasses in DRYING PEAS, VINING PEAS, SPRING FIELD BEANS, SPRING BROAD BEANS [1-3]. Annual dicotyledons, annual grasses in POTATOES [1-3].

Notes

Efficacy
• Weeds are controlled before or shortly after emergence
• Best results on fine, firm, moist seedbeds. Do not use on cloddy or very stony soils
• Residual effects may be reduced on soils with 5-10% organic matter. Do not use where organic matter exceeds 10%
• Effectiveness may be reduced if heavy rain, below average soil temperatures or dry conditions follow application
• Do not cultivate or ridge up potatoes after treatment

Crop safety/Restrictions
• On peas and beans apply within 4 d of drilling, at least 3 d before crop emergence
• Do not use on Vedette or forage peas, seed potatoes or any crop under plastic
• Crop seed should be covered by at least 25 mm of settled soil at time of treatment
• On potatoes apply within 4 d after final ridging and before 10% emergence. See label for tank mixtures
• Do not use on sands, loamy sand or loamy fine sand soils
• Heavy rain after application may cause crop damage on light soils
• Any crop may be drilled or planted after harvesting treated crop following cultivation to 15 cm

Special precautions/Environmental safety
• Harmful if swallowed
• Harmful to fish. Do not contaminate ponds, waterways or ditches with chemical or used container

Protective clothing/Label precautions
• 5c, 28, 29, 36, 37, 53, 60, 63, 66, 67, 70, 78 [1-3]

FOR FULL CONDITIONS OF USE ALWAYS READ THE PRODUCT LABEL

445   propachlor

A pre-emergence amide herbicide for various horticultural crops
See also *chloridazon + propachlor*
               chlorthal-dimethyl + propachlor

Products

1	Albrass	ICI	480 g/l	SC	00069
2	Atlas Orange	Atlas	500 g/l	LI	03096
3	Croptex Amber	Hortichem	500 g/l	LI	03078
4	Portman Propachlor 50 FL	Portman	500 g/l	LI	02784
5	Ramrod Flowable	Monsanto	480 g/l	SC	01688
6	Ramrod Granular	Monsanto	20% w/w	GR	01687
7	Tripart Sentinel	Tripart	500 g/l	LI	03250

Uses

Annual dicotyledons, annual grasses in BROCCOLI, BRUSSELS SPROUTS, CABBAGES, CALABRESE, CAULIFLOWERS, KALE, SWEDES, TURNIPS, ONIONS, LEEKS [1-7]. Annual dicotyledons, annual grasses in OILSEED RAPE [1-5, 7]. Annual dicotyledons, annual grasses in BROWN MUSTARD, WHITE MUSTARD, SAGE [1, 5]. Annual dicotyledons, annual grasses in BLACKCURRANTS, GOOSEBERRIES, RASPBERRIES [1]. Annual dicotyledons, annual grasses in STRAWBERRIES [2-4, 7]. Annual dicotyledons, annual grasses in WOODY ORNAMENTALS, PERENNIALS [1, 6]. Annual dicotyledons, annual grasses in LETTUCE *(off-label)* [5]. Annual dicotyledons, annual grasses in CHINESE CABBAGE *(off-label)* [1].

Notes

Efficacy
- Controls germinating (not emerged) weeds for 6-8 wk
- Best results achieved by application to fine, firm, moist seedbed free of established weeds in spring, summer and early autumn
- Effective results with granules depend on rainfall soon after application [6]
- Use higher rate on soils with more than 10% organic matter
- Recommended in tank-mixture with chlorthal-dimethyl on nursery stock and newly planted bush and cane fruit. See label for details [1]
- See label for details of other recommended tank-mixes

Crop safety/Restrictions
- Apply in brassicas from drilling to time seed chits or after 3-4 true leaf stage but before weed emergence, in swedes and turnips pre-emergence only
- Apply in transplanted brassicas within 48 h of planting in warm weather. Plants must be hardened off and special care needed with block sown or modular propagated plants
- Apply in onions and leeks pre-emergence or from post-crook to young plant stage
- Apply to newly planted strawberries soon after transplanting, to weed-free soil in established crops in early spring before new weeds emerge
- Granules may be applied to onion, leek and brassica nurseries and to most flower crops after bedding out and hardening off
- Do not use pre-emergence of drilled wallflower seed
- Do not use on crops under glass or polythene
- Do not use under extremely wet, dry or other adverse growth conditions
- In the event of crop failure only replant recommended crops in treated soil

Special precautions/Environmental safety
- Irritating to eyes, skin and respiratory system [2, 3, 4], to eyes and skin [1, 5, 7], to skin [6]

- A, C [1-5, 7]; A [6]
- 6a, 6b, 6c, 14, 16, 18, 21, 25, 28, 29, 36, 37, 54, 60, 63, 66, 70 [1-5, 7]; 6a, 14, 18, 29, 36, 37, 54, 63, 67, 70 [6]

Approval
- Off-label Approval to May 1991 for use on outdoor lettuce (OLA 0518/88) [5]; to June 1993 for use on chinese cabbage (OLA 0344/90) [1]

446 propamocarb hydrochloride
A translocated soil-applied protectant carbamate fungicide

Products	Filex	Fisons	722 g/l	SL	00869

Uses

Damping off, downy mildew in BRASSICAS. Damping off in VEGETABLE SEEDLINGS, CELERY, CUCUMBERS, MARROWS, MELONS, PEPPERS, TOMATOES. Downy mildew, phytophthora in ORNAMENTALS. Downy mildew in ONIONS, SPINACH. Red core in STRAWBERRIES. Phytophthora, pythium in BULBS. Phytophthora in CONTAINER-GROWN STOCK, NURSERY STOCK. Root rot, damping off in ROCKWOOL TOMATOES, NFT TOMATOES *(off-label)*. Root rot, damping off in ROCKWOOL CUCUMBERS *(off-label)*. Downy mildew in OUTDOOR RADISHES *(off-label)*.

Notes

Efficacy
- Chemical is absorbed through roots and translocated throughout plant
- Incorporate in compost before use or drench moist compost or soil before sowing, pricking out, striking cuttings or potting up. Use treated compost within 2 wk
- Drench treatment can be repeated at 3-6 wk intervals
- Concentrated solution is corrosive to all metals other than stainless steel
- May also be applied in trickle irrigation systems or feed solution
- To prevent root rot in bulbs apply as dip for 20 min or as a pre-planting drench

Crop safety/Restrictions
- When applied over established seedlings rinse off foliage with water and do not apply under hot, dry conditions
- Do not treat young seedlings with overhead drench
- On plants of unknown tolerance test first on a small scale
- Up to 4 applications per crop may be made at any time of year

Protective clothing/Label precautions
- 21, 29, 35, 54, 60, 63, 66

Harvest interval
- 2 wk

FOR FULL CONDITIONS OF USE ALWAYS READ THE PRODUCT LABEL

• Off-label Approval to Feb 1991 for use on rockwool and NFT tomatoes (OLA 0213/88), rockwool cucumbers (OLA 0214/88); to June 1993 for use on outdoor radishes (OLA 0349/90)

---

447 propham

A pre-sowing incorporated herbicide for beet crops and peas
See also chloridazon + chlorpropham + fenuron + propham
            chloridazon + fenuron + propham
            chlorpropham + diuron + propham
            chlorpropham + fenuron + propham
            chlorpropham + propham

Products	MSS IPC 50 WP	Mirfield	50% w/w	WP	01410

Uses Annual grasses, wild oats, chickweed, polygonums, spurrey in SUGAR BEET, FODDER BEET, MANGOLDS, PEAS.

Notes

Efficacy
• Best control of wild oats achieved by application when seeds germinating, usually not before Mar. Up to 75% reduction possible
• Product is not effective on highly organic (fen) soils

Crop safety/Restrictions
• Apply in beet crops before drilling, incorporate immediately to 10 cm with rotary cultivator and drill within 5 d
• Apply in peas just before final seedbed preparation and incorporate to 10 cm immediately after spraying. Do not drill peas earlier than 5 d after spraying

Protective clothing/Label precautions
• 22, 28, 29, 54, 63, 65

---

448 propiconazole

A systemic, curative and protectant triazole fungicide
See also carbendazim + propiconazole
            chlorothalonil + propiconazole

Products					
1 Power Propiconazole	Power	250 g/l	EC	04489
2 Power Spire	Power	250 g/l	EC	04942
3 Radar	Farm Protection	250 g/l	EC	03000
4 Radar	ICI	250 g/l	EC	01683
5 Tilt 250EC	Ciba-Geigy	250 g/l	EC	02138

Uses Powdery mildew in WHEAT, BARLEY, OATS [4, 5]. Powdery mildew in WHEAT, BARLEY [1-3]. Powdery mildew, rhynchosporium, brown rust in RYE [4, 5]. Brown rust, yellow rust, eyespot in WHEAT, BARLEY [1-5]. Powdery mildew, brown rust, yellow rust, eyespot in DURUM WHEAT [4]. Brown rust, yellow rust in TRITICALE [4, 5]. Rhynchosporium, net blotch in BARLEY [1-5]. Septoria, sooty moulds in

WHEAT [1-5]. Septoria in RYE [4, 5]. Crown rust, mildew, rhynchosporium, drechslera leaf spot in GRASS SEED CROPS, GRASS FOR ENSILING [4, 5]. Mildew, ramularia leaf spot, rust in SUGAR BEET [-4, 5]. Light leaf spot, alternaria in OILSEED RAPE [4, 5]. White rust in CHRYSANTHEMUMS *(off-label)* [3-5].

Notes

Efficacy
- Best results achieved when application made at early stage of disease. Recommended spray programmes vary with crop, disease, season, soil type and product. See label for details

Crop safety/Restrictions
- Maximum number of sprays permitted on cereals varies with product, see label for details and maximum dose rates
- May be applied up to and including grain watery ripe stage of cereals (GS 71)
- Do not apply more than 2 sprays per crop on oilseed rape and grass seed crops
- On oilseed rape do not apply during flowering
- Avoid spraying crops under stress, e.g. during cold weather or periods of frost [3, 4]

Special precautions/Environmental safety
- Irritating to eyes, skin and respiratory system [3], to eyes and skin [4, 5]
- Dangerous to fish. Do not contaminate ponds, waterways or ditches with chemical or used container

Protective clothing/Label precautions
- A, C
- 6a, 6b, 6c [3]; 6a, 6b [4, 5]; 14, 18, 21, 23, 24, 28, 29, 35, 36, 37, 52, 63, 66, 70 [1-5]

Harvest interval
- Cereals 5 wk; oilseed rape 4 wk

Approval
- Approved for aerial application on wheat and barley. See introductory notes [3-5]
- Off-label approval to May 1991 for use only on *confirmed outbreaks* of white rust on chrysanthemums (OLA 0503, 0502, 0501/88) [3-5]

449 propiconazole + tridemorph
A systemic, curative and protectant fungicide for cereals

Products Tilt Turbo 475EC Ciba-Geigy 125:350 g/l EC 03476

Uses Brown rust, yellow rust in WINTER BARLEY, WHEAT. Net blotch, powdery mildew, rhynchosporium in BARLEY. Powdery mildew, leaf spot, sooty moulds in WHEAT.

Notes **Efficacy**
- Apply at start of mildew development and repeat at or just before ear emergence. Treatment of particular benefit where established mildew present at time of spraying

FOR FULL CONDITIONS OF USE ALWAYS READ THE PRODUCT LABEL

- Sprays may be applied in autumn or spring. Optimum timing varies with crop and disease. See label for details
- Autumn treatment in barley suppresses typhula snow rot and eyespot

Crop safety/Restrictions
- Do not apply more than 3 sprays per crop, one of which must be in autumn. If used with other products containing propiconazole do not apply a total of more than 500 g propiconazole/ha in wheat or 375 g/ha in barley in any one season
- Only apply up to grain watery ripe stage (GS 71)

Special precautions/Environmental safety
- Irritating to eyes and skin
- Dangerous to fish. Do not contaminate ponds, waterways or ditches with chemical or used container

Protective clothing/Label precautions
- A, C
- 6a, 6b, 14, 16, 18, 21, 28, 29, 35, 36, 37, 52, 63, 66, 70

Harvest interval
- 5 wk

Approval
- Approved for aerial application on wheat, barley. See introductory notes

450 propoxur

A carbamate insecticide for glasshouse and general use

Products					
	1 Blattanex 20	Bayer	216 g/l	EC	00281
	2 Fumite Propoxur Smoke	Hunt	3.4, 6.8, 11 and 16.5 g a.i.	FU	00942
	3 Murfume Propoxur Smoke	DowElanco	50% w/w	FU	01428

Uses

Whitefly, aphids in TOMATOES, CUCUMBERS, CARNATIONS, CHRYSANTHEMUMS [2, 3]. Ants, earwigs, woodlice in TOMATOES, CUCUMBERS, CARNATIONS, CHRYSANTHEMUMS [2]. General insect control in AGRICULTURAL PREMISES [1]. Aphids, ants, earwigs in CALCEOLARIAS, GERANIUMS, FREESIAS, PRIMULAS, POINSETTIAS, GLASSHOUSE TULIPS, GARDENIAS, CISSUS [3].

Notes

Efficacy
- Fumigate glasshouse at first sign of whitefly infestation and, where serious, repeat after 7 d. Provides partial control of aphids
- Make house smoke tight, water plants several hours previously so that leaves are dry and damp down paths. Fumigate in late afternoon or evening in calm weather when temperature is 16°C or over. Do not fumigate in bright sunshine
- Keep house closed overnight and ventilate well next morning
- For aphid control in chrysanthemums treat when flower colour first appears in bud
- For general control of crawling and flying insects spray areas known to harbour pests. Do not disturb spray deposit for 2-3 d to kill insects emerging after spraying

Crop safety/Restrictions
- Do not treat tomatoes until after third truss has set
- Do not fumigate Chenopodium foliosum, hydrangeas, cyclamen, ferns, young seedlings or plants being hardened off
- Allow 7 d after fumigation before introducing Encarsia or Phytoseiulus
- Open flowers of carnations and chrysanthemums should not be cut before fumigating

Special precautions/Environmental safety
- This product contains an anticholinesterase compound. Do not use if under medical advice not to work with such compounds
- Harmful if swallowed [1], irritating to eyes and respiratory system [2, 3]
- Harmful to fish. Do not contaminate poinds, waterways or ditches with chemical or used container

Protective clothing/Label precautions
- A, C [1]
- 5c, 12, c, 23, 24, 66, 70, 78 [1]; 6a, 6c, 35 [2, 3]; 18, 28, 29, 36, 37, 53, 63 [1-3]

Harvest interval
- 2 d [2, 3]

---

451 propyzamide

A residual amide herbicide for use in a wide range of crops
See also *clopyralid + propyzamide*

Products

1 Campbell's Rapier	MTM Agrochem.	450 g/l	SC	03985
2 Kerb Flo	Rohm & Haas	400 g/l	SC	02759
3 Kerb Flowable	PBI	400 g/l	SC	02759
4 Kerb Granules	PBI	4% w/w	GR	01135
5 Kerb Granules	Rohm & Haas	4% w/w	GR	01136
6 Kerb 50 W	PBI	50% w/w	WP	01133
7 Kerb 50 W	Rohm & Haas	50% w/w	WP	02986
8 Power Propyzamide 50	Power	50% w/w	WP	04816
9 Rapier	FCC	450 g/l	SC	04192

Uses

Annual dicotyledons, annual grasses, perennial grasses in WINTER FIELD BEANS, LUCERNE, WINTER OILSEED RAPE, CLOVER SEED CROPS, SEED BRASSICAS, SUGAR BEET SEED CROPS, LETTUCE, APPLES, PEARS, PLUMS, BLACKCURRANTS, GOOSEBERRIES, REDCURRANTS, BLACKBERRIES, LOGANBERRIES, RASPBERRIES, STRAWBERRIES, RHUBARB, ROSES, WOODY ORNAMENTALS [2, 3, 6, 7]. Annual grasses, wild oats, volunteer cereals, annual dicotyledons in WINTER OILSEED RAPE [1, 8, 9]. Annual grasses, wild oats, volunteer cereals, annual dicotyledons in WINTER FIELD BEANS [8]. Couch, barren brome in FIELD BOUNDARIES [2, 3, 6]. Annual dicotyledons, annual grasses, perennial grasses in CHICORY, SPRING CABBAGE [3, 6]. Annual dicotyledons, annual grasses, perennial grasses in WOODY ORNAMENTALS, FORESTRY [2-7]. Annual grasses, perennial grasses, annual dicotyledons in FARM WOODLAND [3, 4, 6]. Annual dicotyledons, annual grasses, perennial grasses in GRAPEVINES *(off-label)* [6, 7]. Annual dicotyledons, annual grasses, perennial grasses in HOPS *(off-label)* [7].

FOR FULL CONDITIONS OF USE ALWAYS READ THE PRODUCT LABEL

Annual dicotyledons, annual grasses in PROTECTED LETTUCE *(off-label)* [6, 7].
Annual grasses, perennial grasses, annual dicotyledons in CHAMOMILE, EVENING
PRIMROSE, FENUGREEK, SAGE, TARRAGON, RADICCHIO, HERBS, WINTER
VETCHES *(off-label)* [6, 7]. Annual grasses, perennial grasses, annual dicotyledons in
SAGE *(off-label)* [8]. Annual grasses, perennial grasses, annual dicotyledons in HARDY
ORNAMENTAL NURSERY STOCK, NURSERY FRUIT TREES AND BUSHES
(off-label) [4, 5, 6, 7].

Notes

Efficacy
• Active via root uptake. Weeds controlled from germination to young seedling stage,
some species (including many grasses) also when established
• Best results achieved by winter application to fine, firm, moist soil. Rain is required
after application if soil dry
• Do not use on soils with more than 10% organic matter except in forestry
• For heavy couch infestations a repeat application may be needed in following winter
• In lettuce lightly incorporate in top 25 mm pre-drilling or irrigate on dry soil

Crop safety/Restrictions
• Apply to most crops from 1 Oct to 31 Jan (1 Oct to 31 Dec for strawberries, rhubarb,
field beans and forests in south or on peat soils), to chicory in spring/summer, to
lettuce at any time
• Apply as soon as possible after 3-true leaf stage of oilseed rape (GS 1,3), seed brassicas
and spring cabbage, after 4-leaf stage of sugar beet for seed, within 7 d after sowing
but before emergence for field beans, after perennial crops established for at least 1
season, strawberries after 1 yr
• Only apply to strawberries on heavy soils, to field beans on medium and heavy soils, to
established lucerne not less than 7 d after last cut
• Do not treat protected crops or matted row strawberries
• Do not apply more than once within 9 mth
• See label for lists of ornamental and forest species which may be treated
• Lettuce may be sown or planted immediately after treatment, with other crops period
varies from 5 to 40 wk. See label for details

Protective clothing/Label precautions
• 29, 37, 54, 63, 67, 70 [1-9]; 35 [1-5, 9]

Harvest interval
• 6 wk [1-5, 9]

Approval
• May be applied through CDA equipment [2, 3], through CDA equipment in forestry
[6]. See label for details. See introductory notes [2, 3]
• Off-label approval to Feb 1991 for use on grapevines (OLA 0215, 0216/88) [6, 7]; to
Mar 1991 for use on established hops (OLA 0280/88) [7]; to Nov 1991 for use on
protected lettuce (OLA 1356, 1357/88) [7, 6]; to May 1992 for use on tarragon (0474,
0475/89) [7, 6], chamomile (0476, 0477/89) [6, 7], evening primrose (0485, 0486/89)
[6, 7], sage (0487, 0488, 0489/89) [7, 6, 8], fenugreek (OLA 0490, 0491/89) [7, 6]; to
June 1992 for use in hardy ornamental nursery stock, nursery fruit trees and bushes
(OLA 0526, 0527, 0528, 0529/89) [4, 5, 7, 6]l to Aug 1992 for use in radicchio and
herbs (OLA 0680, 0681/89) [7, 6], winter vetches (OLA 0682, 0683/89) [7, 6]

452   pyrazophos

A systemic organophosphorus fungicide with some insecticidal activity

Products

1 Afugan	Promark	300 g/l	EC	00037
2 Missile	Hoechst	300 g/l	EC	03811

Uses

Powdery mildew, net blotch, rhynchosporium in WINTER BARLEY, SPRING BARLEY [2]. Powdery mildew in BRUSSELS SPROUTS [2]. Powdery mildew in APPLES, HOPS, ROSES, POT PLANTS [1]. Powdery mildew in CUCUMBERS *(off-label)* [1]. American serpentine leaf miner in NON-EDIBLE ORNAMENTALS, BEETROOT, LEAF BRASSICAS, ROOT BRASSICAS, CARROTS, CELERIAC, CUCUMBERS, CUCURBITS, FENNEL, PARSNIPS, SALSIFY *(off-label)* [1].

Notes

Efficacy
* Spray apples at pink bud and repeat every 10-14 d until extension growth ceased
* Spray hops at first sign of disease or when shoots 10-12 cm long and repeat every 10-14 d. Treatment suppresses damson-hop aphid
* Spray pot plants and roses at first sign of infection and repeat every 10-14 d
* Spray Brussels sprouts from young plant stage to 2 wk before harvest. Use of pendant lances recommended to give good cover
* Do not spray if rain imminent, on wet crops or within 7 d of late hormone herbicide application [2]
* Temperatures above 30°C may reduce efficacy

Crop safety/Restrictions
* Spray winter or spring barley up to and including complete ear emergence (GS 59)
* Do not tank-mix with sulphur, dinocap or other organophosphorus compounds
* Treatment may cause some yellowing of hop foliage but effect normally outgrown
* Do not spray roses under glass, aquilegia or scorzonera. Some outdoor roses may be slightly sensitive
* Apply a maximum of 3 sprays per season to barley or Brussels sprouts

Special precautions/Environmental safety
* Harmful if swallowed. Irritating to eyes and skin
* Flammable
* This product contains an organophosphorous compound. Do not use if under medical advice not to work with such compounds
* Harmful to game, wild birds and animals
* Dangerous to bees. Do not apply at flowering stage. Keep down flowering weeds
* Dangerous to fish. Do not contaminate ponds, waterways and ditches with chemical or used container

Protective clothing/Label precautions
* A, C
* 1, 5c, 6a, 6b, 12c, 14, 16, 18, 21, 24, 25, 28, 29, 35, 36, 37, 41, 46, 48, 52, 63, 66, 70, 78

Withholding period
* Harmful to livestock. Keep all livestock out of treated areas for at least 2 wk

---

FOR FULL CONDITIONS OF USE ALWAYS READ THE PRODUCT LABEL

Harvest interval
• 2 wk

Approval
• Off-label approval to Mar 1991 for use on cucumbers (OLA 0217/88) [1]; to May 1992 for use on non-edible ornamentals and range of vegetable crops for control of American serpentine leaf miner (OLA 0398/89) [1]

453 pyrethrins
A non-persistent, contact insecticide

See also carbaryl + pyrethrins

Products					
1 Alfadex	Ciba-Geigy	0.75 g/l	RH	00074
2 HC 200 Concentrate	Certified		KN	02703
3 Killgerm ULV 400	Killgerm		UL	03539
4 Pyra-Fog 100	Chemsearch		KN	01659

Uses

Weevils in GRAIN STORES [2, 4]. Flies in DAIRIES, FARM BUILDINGS [1, 2, 3].

Notes

Efficacy
• For fly control close doors and windows and spray or apply fog as appropriate
• For weevil control spray thoroughly in cracks, crevices, around sacks and other places likely to harbour insects, then fog entire storage area

Crop safety/Restrictions
• Do not treat bulk grain products
• Do not allow spray to contact open food products
• Remove exposed milk and collect eggs before application

Special precautions/Environmental safety
• Irritating to eyes, skin and respiratory system
• Extinguish all naked flames, including pilot lights, when applying [4]
• Do not apply directly to livestock
• Harmful to fish. Do not contaminate ponds, waterways or ditches with chemical or used container

Protective clothing/Label precautions
• A, C, D, H [1]; E, F [2]
• 6a, 6b, 6c, 53, [1-4]; 18, 28, 29, 36, 37, 63, 65 [1]; 37, 63, 67 [2]; 28, 29, 37, 63 [4]

Approval
• Product formulated for application by fogging machine. See label for details. See introductory notes on ULV application [1-4]

454   pyrethrins + resmethrin
A contact insecticide for many glasshouse and horticultural crops

Products

1 Pynosect 30 Fogging Solution	Mitchell Cotts	1.6:8.8 g/l	HN	01650
2 Pynosect 30 Water Miscible	Mitchell Cotts	13:84 g/l	EC	01653

Uses

Whitefly, greenfly, blackfly, aphids in CUCUMBERS, TOMATOES, ORNAMENTALS [1, 2]. Mushroom flies in MUSHROOMS [2]. Mushroom flies in MUSHROOMS *(off-label)* [1].

Notes

Efficacy
- Apply when pest first seen and repeat as necessary, for whitefly every 3-4 d until pest controlled
- Apply as high volume spray [2] or with fogging machine [1]. See label for details of suitable equipment
- Fog in late afternoon or early evening, not below 15°C or in bright sunshine

Crop safety/Restrictions
- Do not spray when temperature above 24°C [2] or fog above 32°C [1]
- Care should be taken in spraying 'soft' cucumber plants grown during winter [2]
- Do not treat crops suffering from drought or stress
- On ornamentals a small test spraying is recommended before treating whole crop
- Do not treat red calceolaria [2]

Special precautions/Environmental safety
- Harmful by inhalation or if swallowed. Irritating to eyes, skin and respiratory system [1]
- Harmful to bees. Do not apply at flowering stage. Keep down flowering weeds
- Dangerous to fish. Do not contaminate ponds, waterways or ditches with chemical or used container

Protective clothing/Label precautions
- A, C OR E, G, J. L, M [1]
- 5b, 5c, 6a, 6b, 6c [1]; 14, 18, 23, 24, 29, 36, 37, 50, 52, 63, 65, 78, [1]; 23, 28, 29, 50, 52, 60, 63, 65 [2]

Harvest interval
- zero

Approval
- Product formulated for application by thermal fogging machines [1]. See label for details. See introductory notes
- Off-label Approval to Mar 1991 for treatment of mushrooms using a fogging machine (OLA 0202/90) [1]

FOR FULL CONDITIONS OF USE ALWAYS READ THE PRODUCT LABEL

455   pyridate

A contact herbicide for cereals, oilseed rape and maize

Products   Lentagran WP   Ciba-Geigy   45% w/w   WP   02767

Uses   Annual dicotyledons, cleavers, dead nettle, speedwells in CEREALS, OILSEED RAPE, MAIZE, SWEETCORN.

Notes   **Efficacy**
- Best results achieved by application to actively growing weeds at 6-8 leaf stage when temperatures are above 8°C before crop foliage forms canopy

Crop safety/Restrictions
- Apply to oilseed rape in winter after 6-true leaf stage (GS 1,6) but before mid-Dec, in spring after crop growth started but before 15 cm of stem extension
- Do not apply in mixture with or within 14 d of any other product which may result in dewaxing of crop foliage
- Apply to winter or spring cereals from first-tiller stage to flag leaf just visible (GS 21-37)
- Apply to maize and sweetcorn (in tank-mixture with atrazine) after first-leaf stage. Do not use on cv. Meritosa
- Do not use on crops suffering stress from frost, drought, disease or pest attack
- Do not apply to oilseed rape or cereals when night temperature consistently below 2°C or to oilseed rape when day temperature exceeds 16°C

Special precautions/Environmental safety
- Irritating to eyes and skin. May cause sensitization by skin contact
- Harmful to fish. Do not contaminate ponds, waterways or ditches with chemical or used container

Protective clothing/Label precautions
- A, C
- 6a, 6b, 9, 18, 21, 29, 36, 37, 53, 63, 67

Harvest interval
- maize, sweetcorn 2 mth

456   quassia

An animal repellent of vegetable origin

Products   1 Dog Off   Fieldspray   LI   04397
                2 Hoppit   Fieldspray   LI   04396

Uses   Dogs, rabbits, deer, birds in OUTDOOR CROPS.

Notes   **Efficacy**
- Spray dog sniffing and marking points and repeat treatment as necessary [1]
- Spray areas and plants to be protected and repeat treatment as necessary [2]
- Product requires 6 h drying period [1]

Crop safety/Restrictions
- Food plants may be eaten 24 h after spraying

Protective clothing/Label precautions
- 29, 36, 37, 54, 63, 67

457 quinalphos

A broad spectrum organophosphorus contact and stomach insecticide
See also disulfoton + quinalphos

Products Savall   Farm Protection   25% w/w   EC   03949

Uses Leatherjackets in CEREALS. Caterpillars in BRASSICAS. Carrot fly, cutworms in CARROTS. Celery fly in CELERY. Celery fly in CELERIAC, FENNEL *(off-label)*.

Notes **Efficacy**
- On brassicas apply as soon as caterpillars seen. The addition of an approved non-ionic wetter may improve control
- For control of late generation carrot fly apply in early Aug or following Ministry warning, directing spray towards plant crowns. Repeat in early Sep and Oct if crop to be lifted after end of Oct. Rainfall after application can improve control
- For cutworms in carrots or leatherjackets in cereals apply at first signs of damage or following Ministry warning
- Can be used on mineral or organic soils for protection against cutworms or carrot fly

Special precautions/Environmental safety
- A chemical subject to the Poisons Rules 1982 and Poisons Act 1972
- This product contains an organophosphorus compound. Do not use if under medical advice not to work with such compounds
- Harmful by inhalation and if swallowed. Irritating to eyes
- Flammable
- Dangerous to game, wild birds and animals
- Dangerous to bees. Do not apply at flowering stage. Keep down flowering weeds
- Extremely dangerous to fish. Do not contaminate ponds, waterways or ditches with chemical or used container

Protective clothing/Label precautions
- A, C, H
- 1, 5b, 5c, 6a, 12c, 14, 16, 18, 21, 25, 28, 29, 35, 36, 37, 41, 45, 48, 51, 64, 66, 78

Withholding period
- Harmful to livestock. Keep all livestock out of treated areas for at least 7 d

Harvest interval
- brassicas, celery, fennel 7 d; carrots, cereals, celeriac 3 wk

---

FOR FULL CONDITIONS OF USE ALWAYS READ THE PRODUCT LABEL

• Off-label Approval to Feb 1992 for outdoor use on celeriac and fennel (OLA 0116/89)

458 quinalphos + thiometon
A systemic and contact organophosphorus insecticide for leaf brassicas

Products   Tombel   Hortichem   16:16% w/w   EC   04124

Uses   Aphids, caterpillars in BROCCOLI, BRUSSELS SPROUTS, CABBAGES, CALABRESE, CAULIFLOWERS, SPRING GREENS.

Notes   **Efficacy**
• Spray as soon as pests seen. Addition of non-ionic wetter may improve control

Crop safety/Restrictions
• Avoid spraying in hot weather or when leaves are wilting
• Do not spray cauliflower curds

Special precautions/Environmental safety
• A chemical subject to the Poisons Rules 1982 and Poisons Act 1972
• This product contains an organophosphorus compound. Do not use if under medical advice not to work with such compounds
• Harmful by inhalation or if swallowed. Irritating to eyes
• Flammable
• Dangerous to game, wild birds and animals
• Dangerous to bees. Do not apply at flowering stage. Keep down flowering weeds
• Dangerous to fish. Do not contaminate ponds, waterways or ditches with chemical or used container

Protective clothing/Label precautions
• A, C, H
• 1, 5b, 5c, 6a, 12c, 14, 16, 18, 21, 25, 28, 29, 35, 36, 37, 41, 45, 48, 52, 66, 70, 78

Withholding period
• Harmful to livestock. Keep all livestock out of treated areas for at least 2 wk

Harvest interval
• 7 d

459 quinomethionate
A non-systemic fungicide and acaricide for some horticultural crops

Products   Morestan   Hortichem   25% w/w   WP   01376

Uses   Red spider mites in STRAWBERRIES, GOOSEBERRIES, MARROWS. Powdery mildew, currant leaf spot in BLACKCURRANTS, GOOSEBERRIES. Powdery mildew in MARROWS.

Efficacy
- On recommended varieties of strawberry apply when mites appear in spring/early summer and repeat after harvest if necessary
- For control of mildew and leaf spot apply as soon as disease appears and repeat every 10-14 d if necessary (2 sprays at 2 wk intervals on marrows). See label for details
- Control of leafspot on gooseberries is partial

Crop safety/Restrictions
- Do not apply near hop gardens, pears, alder or Templar strawberry as drift may cause damage
- Establish crop tolerance on a few plants before spraying large areas

Protective clothing/Label precautions
- 21, 29, 54, 63, 67

Harvest interval
- Blackcurrants, gooseberries, strawberries 3 d; marrows (outdoor) 2 wk

460   quintozene
A protectant soil applied fungicide for various horticultural crops

Products					
1 Brabant PCNB 20%	Bos	20% w/w	DP	00313
2 Quintozene WP	RP Environ.	50% w/w	WP	04574

Uses

Dollar spot, fusarium patch, red thread in TURF [2]. Sclerotinia in ANEMONES, IRISES, NARCISSI, TULIPS, BULBS, CHICORY [1]. Damping off and wirestem in BRASSICAS [1]. Botrytis, damping off, sclerotinia in LETTUCE, CHRYSANTHEMUMS, BEDDING PLANTS, POT PLANTS [1]. Rhizoctonia in CUCUMBERS [1]. Botrytis, rhizoctonia in TOMATOES, CARNATIONS, PELARGONIUMS [1]. Botrytis, sclerotinia in DAHLIAS [1].

Notes

Efficacy
- Apply to soil or compost and incorporate before planting or sowing [1]
- For use in turf apply drenching spray when fungal growth active [2]
- Do not apply more than 6 treatments per annum on turf [2]

Crop safety/Restrictions
- In general leave treated soil or compost for 2-3 wk before planting unrooted cuttings, 4 wk before sowing tomatoes or cucumbers and 2-3 d before planting other crops

Special precautions/Environmental safety
- Irritating to skin, eyes and respiratory system

Protective clothing/Label precautions
- A [1]
- 6a, 6b, 6c, 18, 21, 28, 29, 36, 37, 54, 63, 67

FOR FULL CONDITIONS OF USE ALWAYS READ THE PRODUCT LABEL

quizalofop-ethyl

A phenoxypropionic post-emergence grass herbicide

| Products | Pilot | Schering | 500 g/l | SC | 03837 |

Uses Annual grasses, perennial grasses, volunteer cereals, couch in SUGAR BEET, FODDER BEET, MANGOLDS, RED BEET, OILSEED RAPE, MUSTARD.

Notes

Efficacy
- Apply to emerged weeds in mixture with wuitable adjuvant oil, see label for details. Weeds emerging after treatment are not controlled. Effective on annual grasses from 2-leaf stage to fully tillered, on perennials from 4-6 leaf stage to before jointing
- Best results achieved by application to weeds growing actively in warm conditions with adequate soil moisture. Use split treatment to extend period of control
- Annual meadow-grass is not controlled
- Various spray programmes and tank-mixtures recommended to control mixed dicotyledon/grass weed populations. See label for details
- Leave at least 3 d before applying another herbicide
- For effective couch control do not hoe within 21 d after spraying
- At least 4 h rain free period required for effective results

Crop safety/Restrictions
- Apply to beet crops when weeds at appropriate stage and growing actively and not later than 31 Jul. Seed crops may be treated in autumn but consult seed house before use
- Apply to oilseed rape and mustard from expanded cotyledon stage (GS 1,0), before crop covers larger weeds and not later than 31 Jan
- Do not spray crops under stress from any cause or in frosty weather
- In the event of crop failure recommended crops may be resown at any time, broadleaved crops may be sown after 4 wk, cereals after 2 to 6 wk depending on dose applied

Protective clothing/Label precautions
- A
- 22, 29, 35, 54, 60, 63, 66

Harvest interval
- Sugar beet, fodder beet, red beet, mangolds 4 mth; oilseed rape 6 mth

---

462   resmethrin

A contact acting pyrethroid insecticide
See also fenitrothion + permethrin + resmethrin
*   gamma-HCH + resmethrin + tetramethrin*
*   pyrethrins + resmethrin*

| Products | Turbair Resmethrin Extra | PBI | 6 g/l | UL | 02247 |

Uses Aphids, caterpillars in BRASSICAS. Aphids in BEANS, SOFT FRUIT. Pine looper in PINE TREES. Aphids, sciarid flies, whitefly in CUCUMBERS, TOMATOES,

LETTUCE, AUBERGINES, PEPPERS, CARNATIONS, CHRYSANTHEMUMS, POT PLANTS. Sciarid flies in MUSHROOMS.

Notes

Efficacy
- Chemical has rapid contact effect and controls both adult and scale stages of whitefly
- Apply as ULV spray with suitable Turbair sprayer
- Spray when pest appears and repeat as needed, with whitefly every 2-3 d
- Spray when light intensity low. Chemical loses activity when exposed to light
- Do not mix with other products

Crop safety/Restrictions
- If biological control of whitefly being practised only apply to top third of plants to control newly emerged adult whitefly
- Do not treat Chinese cabbage or red calceolaria

Special precautions/Environmental safety
- Irritating to eyes, skin and respiratory system
- Flammable
- Harmful to bees. Do not apply at flowering stage. Keep down flowering weeds
- Harmful to fish. Do not contaminate ponds waterways or ditches with chemical or used container

Protective clothing/Label precautions
- 6a, 6b, 6c, 12c, 18, 28, 29, 35, 36, 37, 50, 53, 63, 65, 70

Harvest interval
- zero

Approval
- Product formulated for application through ULV spraying equipment. See label for details. See introductory notes

463   resmethrin + tetramethrin
A contact acting pyrethroid insecticide mixture

Products   Sorex Wasp Nest Destroyer   Sorex   0.10:0.05% w/w   AE   02344

Uses   Wasps in MISCELLANEOUS SITUATIONS.

Notes

Efficacy
- Product acts by lowering temperature in and around nest and producing a stupefying vapour in addition to its contact insecticidal effect. Spray only on surfaces
- Apply by directing jet at nest from up to 3 m away and approach nest gradually until jet can be directed into entrance hole. Continue spraying until nest saturated

Special precautions/Environmental safety
- For use only by trained operators and owners of commercial and agricultural premises

FOR FULL CONDITIONS OF USE ALWAYS READ THE PRODUCT LABEL

- Harmful by inhalation. Contains perchloroethylene
- Keep off skin. Ensure adequate ventilation
- Harmful to caged birds and pets
- Dangerous to bees. Do not apply at flowering stage. Keep down flowering weeds
- Extremely dangerous to fish. Remove fish tanks before spraying

Protective clothing/Label precautions
- A, C, H
- 5b, 26, 28, 29, 36, 37, 46, 48, 51, 54, 63

---

464   rotenone

A natural, contact insecticide of low persistence

Products					
1 FS Derris Dust	Ford Smith	0.5% w/w	DP	04643
2 FS Liquid Derris	Ford Smith	50 g/l	EC	01213

Uses

Aphids in TOP FRUIT, SOFT FRUIT, VEGETABLES, FLOWERS, ORNAMENTALS, GLASSHOUSE CROPS. Raspberry beetle in CANE FRUIT. Sawflies in GOOSEBERRIES. Slug sawflies in PEARS, ROSES.

Notes

Efficacy
- Apply as high volume spray when pest first seen and repeat as necessary
- Spray to obtain thorough coverage, especially on undersurfaces of leaves

Special precautions/Environmental safety
- Dangerous to fish. Do not contaminate ponds, waterways and ditches with chemical or used container
- Highly flammable [2]

Protective clothing/Label precautions
- 12b [2]; 29, 35, 52, 63, 65, 70 [1, 2]

Harvest interval
- 1 d

---

465   sethoxydim

A cyclohexene-oxime post-emergence grass herbicide

Products					
Checkmate	Embetec	193 g/l	EC	04337

Uses

Annual grasses, volunteer cereals, blackgrass, wild oats in SUGAR BEET, FODDER BEET, MANGOLDS, LINSEED, FLAX, WINTER OILSEED RAPE, SPRING OILSEED RAPE, FIELD BEANS, MUSTARD, COMBINING PEAS, VINING PEAS, POTATOES, BULB ONIONS, SWEDES, TURNIPS, RASPBERRIES, STRAWBERRIES. Couch, bent grasses in SUGAR BEET, FODDER BEET, MANGOLDS, POTATOES, BULB ONIONS, SWEDES, TURNIPS, FIELD BEANS, RASPBERRIES, STRAWBERRIES. Annual grasses, volunteer cereals in PARSNIPS *(off-label).* Blackgrass, wild oats, volunteer cereals in RED FESCUE SEED CROPS, CLOVERS, SAINFOIN, LUCERNE, TREFOIL *(off-label).*

Notes

Efficacy
- Apply to emerged weeds in combination with Adder or Actipron adjuvant oil (except on peas). Weeds emerging after treatment are not controlled
- Best results when weeds growing actively. Action is faster under warm conditions
- Apply to annual grasses from 2-leaf stage to end of tillering, to couch when largest shoots at least 30 cm long. For effective couch control do not cultivate for 2 wk
- Annual meadow-grass is not controlled
- Apply when rain not expected for at least 2 h
- Various tank mixes recommended for grass/dicotyledon weed populations. See label for details of mixtures and sequential treatments
- Leave at least 7 d before applying another herbicide

Crop safety/Restrictions
- See label for details of crop growth stage and timing
- Do not mix with adjuvant oil on peas. Check pea leaf-wax by Crystal Violet test and do not spray where wax insufficient or damaged
- Do not spray crops suffering damage from other herbicides
- In raspberries apply to inter-row or to base of canes only
- Brassicas, field beans and onions may be sown 3 d after treatment, grass and cereal crops after 4 d

Special precautions/Environmental safety
- Irritating to eyes and skin
- Harmful to fish. Do not contaminate ponds, waterways or ditches with chemical or used container

Protective clothing/Label precautions
- A, C
- 6a, 6b, 14, 18, 21, 29, 35, 36, 37, 53, 63, 66, 70

Harvest interval
- vining peas, early potatoes, bulb onions, raspberries, strawberries 4 wk; seed potatoes 7 wk; maincrop potatoes, sugarbeet, parsnips 8 wk; swedes, turnips, fodder beet, mangolds 9 wk; combining peas 11 wk; oilseed rape, linseed, mustard 12 wk

Approval
- Off-label approval to Mar 1991 for use on parsnips (OLA 0220/88), to July 1991 for use on red fescue, clovers, lucerne, sainfoin, trefoil (OLA 0623/88)

FOR FULL CONDITIONS OF USE ALWAYS READ THE PRODUCT LABEL

466   simazine

A soil-acting triazine herbicide with selective and non-selective uses
See also amitrole + 2,4-D + diuron + simazine
*   amitrole + diquat + paraquat + simazine*
*   amitrole + simazine*
*   diquat + paraquat + simazine*
*   diuron + simazine*
*   glyphosate + simazine*
*   metoxuron + simazine*

Products

1 Ashlade Simazine 50 FL	Ashlade	500 g/l	SC	02885
2 Atlas Simazine	Atlas	500 g/l	SC	03099
3 Boroflow S	ABM	500 g/l	SC	04289
4 Gesatop 500FW	Ciba-Geigy	500 g/l	SC	00984
5 Gesatop 50WP	Ciba-Geigy	50% w/w	WP	00983
6 Mascot Simazine 2% Granular	Rigby Taylor	2% w/w	GR	02440
7 Mascot Simazine Residual	Rigby Taylor	500 g/l	SC	02441
8 MSS Simazine 50 FL	Mirfield	500 g/l	SC	01418
9 MSS Simazine 2G	Mirfield	2% w/w	GR	04363
10 MSS Simazine 80 WP	Mirfield	80% w/w	WP	04362
11 Simapron	BP	150 g/l	EC	04162
12 Simflow	RP Environ.	500 g/l	SC	01954
13 Syngran	Synchemicals	2% w/w	GR	02079
14 Truchem Simazine 500L	Truchem	500 g/l	SC	02469
15 Truchem Simazine WP	Truchem	50% w/w	WP	02468
16 Unicrop Flowable Simazine	Unicrop	500 g/l	SC	02271
17 Weedex S2 FG	Hortichem	2% w/w	GR	04223

Uses

Annual dicotyledons, annual grasses in FIELD BEANS, BROAD BEANS, APPLES, PEARS, CURRANTS, GOOSEBERRIES, CANE FRUIT, STRAWBERRIES, HOPS, ASPARAGUS, RHUBARB, MAIZE, TREES AND SHRUBS, ROSES [1, 2, 4, 5, 8, 10, 14-16]. Annual dicotyledons, annual grasses in SWEETCORN [1, 4, 5, 8, 10, 14-16]. Annual dicotyledons, annual grasses in TREES AND SHRUBS, ROSES [6, 7, 11-13, 17]. Annual dicotyledons, annual grasses in FOREST NURSERY BEDS, FORESTRY TRANSPLANT LINES [1, 4, 5, 8, 10, 14-16]. Total vegetation control in NON-CROP AREAS [1-5, 7-12, 14-16]. Annual dicotyledons, annual grasses in GRAPEVINES *(off-label)* [1, 2, 4, 7, 10, 12, 16]. Annual dicotyledons, annual grasses in SAGE *(off-label)* [4]. Annual dicotyledons, annual grasses in MINT *(off-label)* [5].

Notes

Efficacy
- Active via root uptake. Best results achieved by application to fine, firm, moist soil, free of established weeds, when rain falls after treatment
- Do not use on highly organic soils
- Use at higher rates in Oct-Mar for total vegetation control
- Following repeated use of simazine or other triazine herbicides resistant strains of groundsel and some other annual weeds may develop

Crop safety/Restrictions
- Apply to winter beans as soon as possible after sowing but before end Feb, to spring beans and maize within 7 d of sowing. May be applied to spring beans up to 2-true leaf stage but control may be poor if weeds already germinated [4, 5]

- Do not spray beans on sandy or gravelly soils or cultivate after treatment. Do not treat varieties Beryl, Feligreen or Rowena
- Apply in top, bush and cane fruit and woody ornamentals established at least 12 mth in Feb-Mar. Roses may be sprayed immediately after planting. See label for lists of resistant and susceptible species
- Apply to strawberries in Jul-Dec, not in spring. Do not treat spring-planted crops established less than 6 mth, or winter-planted crops less than 9 mth. Do not spray varieties Huxley Giant, Madame Moutot or Regina
- Apply to hops overall in Feb-Apr before weeds emerge
- Apply to forest nursery seedbeds in second year or to transplant lines after plants 5 cm tall. Do not treat Norway spruce
- On sands, stony or gravelly soils there is risk of crop damage, especially with heavy rain
- Allow at least 7 mth before drilling or planting other crops, longer if weather dry
- Do not sow oats in autumn following spring application in maize

Special precautions/Environmental safety
- Irritating to eyes, respiratory system and skin [11]
- Harmful to fish. Do not contaminate ponds, waterways or ditches with chemical or used container [8-11, 16]

Protective clothing/Label precautions
- A, C, E, H, M [11]
- 6a, 6b, 6c [11]; 29, 54, 60, 63, 65, 66, 67, 70 [1, 2, 5-7, 12-16]; 14, 21, 28, 29, 54, 60, 63, 66 [3, 4]; 29, 53, 63, 66, 67 [8-10]; 14, 18, 21, 28, 29, 36, 37, 53, 63, 65 [11]

Approval
- Approved for aerial application on beans [4, 5]. See introductory notes
- Off-label Approval to Aug 1991 for use on grapevines (OLA 0801/88) [1], (OLA 0807/88) [2], (OLA 0805/88) [4], (OLA 0803/88) [7], (OLA 0802/88) [10], (OLA 0800/88) [12], (OLA 0799/88) [16]; to May 1993 for use on sage (OLA 0339/90) [4]; to June 1993 for use on mint (OLA 0357/90) [5]

467   simazine + trietazine

A pre-emergence residual herbicide for peas, field and broad beans

Products					
1 Aventox SC	DowElanco	57.5:402.5 g/l	SC	04229
2 Remtal SC	Schering	57.5:402.5 g/l	SC	03827

Uses   Annual dicotyledons in PEAS, EDIBLE PODDED PEAS, FIELD BEANS, BROAD BEANS [1, 2]. Annual dicotyledons in WHITE LUPINS [2].

Notes   **Efficacy**
- Chemical acts mainly via roots but some contact effect on cotyledon stage weeds. Gives residual control of germinating weeds for season
- Effectiveness increased by rain after spraying
- Weeds of intermediate susceptibility, including annual meadow grass and blackgrass, may be controlled if adequate rainfall occurs soon after spraying

FOR FULL CONDITIONS OF USE ALWAYS READ THE PRODUCT LABEL

- Control of deep germinating weeds on heavier soils may be incomplete
- Do not use on soils with more than 10% organic matter
- With repeated use of simazine or other triazine herbicides resistant strains of groundsel and some other annual weeds may develop

Crop safety/Restrictions
- Apply to spring peas, winter or spring field and broad beans [1, 2], and white lupins [2], between drilling and 5% crop emergence
- On early-drilled peas apply when seeds chitting, on later crops as soon as possible after drilling. Do not spray winter sown peas. Vedette peas may be checked by treatment
- Do not spray any forage pea varieties [2]
- Crop seed should be covered by at least 25 mm of settled soil
- Do not use on sands or on gravelly or cloddy soils
- Other crops may be sown or planted 10 wk or more after spraying. With drilled brassicas allow a minimum of 14 wk
- In the event of crop failure only drill peas, field or broad beans without ploughing but do not respray

Special precautions/Environmental safety
- Harmful if swallowed [1, 2]
- Harmful to fish. Do not contaminate ponds, waterways or ditches with chemical or used container

Protective clothing/Label precautions
- 5c, 18, 29, 36, 37, 53, 63, 66, 70, 78 [1, 2]

Approval
- May be applied through CDA equipment. See label for details. See introductory notes on ULV application [2]

---

468   sodium chlorate

A non-selective inorganic herbicide for total vegetation control
See also atrazine + 2,4-D + sodium chlorate
atrazine + sodium chlorate

Products

1 Arpal Non Selex Powder	Adams	58.2% w/w	SP	04565
2 Atlacide Soluble Powder	Chipman	58.2% w/w	SP	00125
3 Centex	Chemsearch	6.4% w/w	SL	00456
4 Granular Weedkiller	Dimex	30% w/w	GR	03653
5 Lever Sodium Chlorate	Lever	53% w/w	SP	02819
6 Sodium Chlorate (Fire Suppressed) Weedkiller	Ace	50% w/w	SP	04449

Uses

Total vegetation control in NON-CROP AREAS, PATHS AND DRIVES.

Notes

Efficacy
- Active through foliar and root uptake
- Apply as overall spray at any time during growing season. Best results obtained from application in spring or early summer
- Do not apply before heavy rain

Crop safety/Restrictions
- Clothing, paper, plant debris etc become highly inflammable when dry if contaminated with sodium chlorate
- Fire risk has been reduced by inclusion of fire depressant but product should not be used in areas of exceptionally high fire risk e.g. oil installations, timber yards

Special precautions/Environmental safety
- Harmful if swallowed
- Oxidizing - contact with combustible material may cause fire
- Wash clothing thoroughly after use
- If clothes become contaminated do not stand near an open fire
- Keep livestock out of treated areas until foliage of any poisonous weeds such as ragwort has died and become unpalatable

Protective clothing/Label precautions
- A, H, M
- 5c, 14, 18, 29, 36, 37, 43, 54, 63, 67, 70, 78

---

469   sodium cyanide
A poisonous gassing compound for control of rabbits and rats

Products	Cymag	ICI	40% w/w	GE	00623

Uses   Rabbits, rats in FARMLAND.

Notes

Efficacy
- Hydrogen cyanide gas is produced when chemical is placed on moist earth
- Use only in rabbit and rat holes out of doors and well away from farm or domestic buildings
- Place powder in burrows with a spoon or blow it in with a pump and seal openings with sods of grass. See label for details
- Do not use in wet or windy weather

Special precautions/Environmental safety
- A chemical subject to the Poisons Rules 1982 and Poisons Act 1972
- Very toxic in contact with skin, by inhalation or if swallowed
- Cymag and the gas it evolves are deadly poisons. Use only in the presence of another person aware of the symptoms and first aid treatment for hydrogen cyanide poisoning and provided with amyl nitrite for use in an emergency
- Maximum exposure limits apply to this chemical. See HSE Approved Code of Practice for the Control of Susbstances Hazardous to Health
- Dangerous to people and livestock. Keep them out of treated areas during gassing operations
- Dangerous to fish. Do not contaminate ponds, waterways or ditches with chemical or used container

---

FOR FULL CONDITIONS OF USE ALWAYS READ THE PRODUCT LABEL

- 3a, 3b, 3c, 18, 25, 26, 27, 29, 40, 52, 64, 68, 70, 79; because of its highly toxic nature a special set of precautions applies to this product and must be followed carefully. See label for details

470   sodium monochloroacetate
A contact herbicide for various horticultural crops

Products	1 Atlas Somon	Atlas	96% w/w	SP	03045
	2 Croptex Steel	Hortichem	95% w/w	SP	02418

Uses   Annual dicotyledons in BRUSSELS SPROUTS, CABBAGES, KALE, ONIONS, LEEKS, APPLES, PEARS, PLUMS, BLACKCURRANTS, REDCURRANTS, GOOSEBERRIES, HOPS [1, 2]. Annual dicotyledons in BROCCOLI, CALABRESE [2].

Notes   **Efficacy**
- Best results achieved by application to emerged weed seedlings up to young plant stage under good growing conditions
- Effectiveness reduced by rain within 12 h
- May be used for hop defoliation in tank-mixture with tar-oil

Crop safety/Restrictions
- Apply to brassicas from 2-4 leaf stage or after recovery from transplanting
- Do not spray cabbage that has begun to heart
- Apply to onions and leeks after crook stage but before 4-leaf stage
- Safety on brassicas, onions and leeks depends on presence of adequate leaf wax, check by crystal violet wax test. Do not add wetters, pesticides or nutrients to spray
- Apply in fruit crops established for at least 1 yr as a directed spray
- Do not spray if frost likely or if temperature likely to exceed 27°C

Special precautions/Environmental safety
- Irritating to eyes, skin and respiratory system
- Keep livestock, especially poultry, out of treated areas for at least 2 wk
- Harmful to bees. Do not apply at flowering stage. Keep down flowering weeds

Protective clothing/Label precautions
- A
- 6a, 6b, 6c, 18, 21, 28, 29, 36, 37, 42, 50, 54, 63, 65

471   sodium pentachlorophenoxide
A persistent fungicide and wood preservative for mushroom houses

Products	Cryptogil NA	Darmycel		SG	H1341

Uses   Fungus diseases in MUSHROOM HOUSES. Mosses in MUSHROOM HOUSES.

Efficacy
- Sterilise timber or boxes by dipping and spray walls floors etc to prevent mould growth for 3-5 mth
- Spray outside walls and paths to prevent growth of mould, mosses etc

Crop safety/Restrictions
- Do not apply to crop or allow product to become mixed with compost or casing. Stack boxes with at least 2.5 cm between compost surface and bottom of box above
- Do not allow to drain off onto cultivated land

Special precautions/Environmental safety
- Harmful in contact with skin. Fatalities can occur after excessive exposure of skin to concentrated solutions. Take special care in hot weather
- The material should not be handled by anyone suffering from renal deficiencies. If in doubt seek medical advice
- Wear an approved respirator or mask covering nose and mouth in conjunction with an eye shield if exposure to dust anticipated
- Wear an approved respirator if prolonged exposure to spray mist anticipated
- Harmful to animals, birds and bees
- Dangerous to fish. Do not contaminate ponds, waterways or ditches with chemical or used container

Protective clothing/Label precautions
- A, C, D or E + F
- 5a, 14, 16, 21, 25, 28, 29, 46, 52, 64, 67

472   sodium silver thiosulphate
A plant-growth regulator used to extend life of flowers

Products

Argylene	Fargro	8% w/w	SP	03386

Uses

Prolonging flower life in POT PLANTS, GLASSHOUSE FLOWERS.

Notes
Efficacy
- Acts by inhibiting production of ethylene
- Spray flowering pot plants to run off 8-14 d before shipment from glasshouse or at 3 wk intervals. Best time for spraying is late afternoon
- Vase life of cut flowers extended by dip treatment immediately after cutting

Crop safety/Restrictions
- Spraying pot plants just before shipping may cause damage

Special precautions/Environmental safety
- Not to be used on food crops

Protective clothing/Label precautions
- 21, 29, 34, 54, 63, 67

FOR FULL CONDITIONS OF USE ALWAYS READ THE PRODUCT LABEL

73   sodium tetraborate

An inorganic herbicide available only in mixtures
See also aluminium sulphate + copper sulphate + sodium tetraborate
        copper sulphate + sodium tetraborate

74   sulphonated cod liver oil

An animal repellent

Products					
	Scuttle	Fine	900 g/l	EC	04559

Uses

Rabbits, deer in WHEAT, BARLEY, OATS, OILSEED RAPE, CABBAGES, CAULIFLOWERS, LETTUCE, FORESTRY.

Notes

Efficacy
- Apply diluted as a spray before crop is being grazed or attacked or undiluted as a dip for forestry seedlings
- A rain-free period of about 5 h is required after application

Crop safety/Restrictions
- Before using on vegetables users should consider risk of taint and consult processor before using on crops for processing

Special precautions/Environmental safety
- Dangerous to bees. Do not apply at flowering stage. Keep down flowering weeds
- Dangerous to fish. Do not contaminate pounds, waterways or ditches with chemical or used container

Protective clothing/Label precautions
- A, C
- 18, 29, 36, 37, 48, 52, 63, 66

475   sulphur

A broad-spectrum inorganic protectant fungicide and foliar feed
See also carbendazim + maneb + sulphur
        copper oxychloride + maneb + sulphur
        copper sulphate + sulphur

Products

1	Ashlade Sulphur FL	Ashlade	720 g/l	SC	02812
2	Atlas Sulphur 80 FL	Atlas	800 g/l	SC	03802
3	Campbell's Sulphur Flowable	MTM Agrochem.	800 g/l	SC	00424
4	Clifton Sulphur	Clifton	80% w/w	WP	03031
5	Headland Sulphur	WBC Technology	800 g/l	SC	03714
6	Hortag Aquasulf	Avon	900 g/l	SC	03667
7	Intrachem Sulfur W80	Intracrop	80% w/w	WP	02603
8	Kumulus S	BASF	80% w/w	WP	01170
9	Microsul Flowable Sulphur	Stoller	960 g/l	SC	03907
10	MSS Sulphur 80 WP	Mirfield	80% w/w	WP	03225

11 Power Micro-Sulphur	Power	80% w/w	MG	0272
12 Power Sulphur-Flo	Power	850 g/l	SC	0271
13 Solfa	Farm Protection	80% w/w	MG	0362
14 Stoller Flowable Sulphur	Stoller	720 g/l	SC	0376
15 Thiovit	PBI	80% w/w	MG	0212
16 Tripart Imber	Tripart	720 g/l	SC	0405
17 Vassgro Flowable Sulphur	Vass	720 g/l	SC	0340

Uses

Disease control/foliar feed in WHEAT, BARLEY [1-17]. Disease control/foliar feed in WINTER OATS [16]. Powdery mildew in SUGAR BEET [1-17]. Disease control/foliar feed in OILSEED RAPE [1, 2, 4-13, 15-17]. Powdery mildew in SWEDES, TURNIPS [1, 2, 6, 10, 13, 16]. Powdery mildew in SWEDES [1-3, 5, 9, 11-13, 15-17]. Disease control/foliar feed in GRASSLAND [2, 6, 10, 13, 15, 17]. Powdery mildew, scab in APPLES [1-5, 8-10, 13, 14, 16, 17]. Scab in PEARS [3, 5, 9, 14]. Powdery mildew in GOOSEBERRIES [1, 2, 4, 8, 10, 13, 16]. Gall mite in BLACKCURRANTS [1, 4, 8, 10, 13, 16]. Powdery mildew in STRAWBERRIES [1, 3, 4, 5, 8-10, 13, 14]. Powdery mildew in GRAPEVINES [1, 2, 4, 6, 8-10, 13, 14, 16]. Powdery mildew in HOPS [1-6, 8-11, 13-17]. Powdery mildew in ORNAMENTALS [13]. Powdery mildew in PROTECTED TOMATOES *(off-label)* [15].

Notes

Efficacy

- Apply when disease first appears and repeat 2-3 wk later. Details of application rates and timing vary with crop, disease and product. See label for information
- Sulphur acts as foliar feed as well as fungicide and with some crops product labels vary in whether treatment recommended for disease control or growth promotion
- In grassland best at least 2 wk before cutting for hay or silage, 3 wk before grazing
- Treatment unlikely to be effective if disease already established in crop

Crop safety/Restrictions

- May be applied to cereals up to grain watery-ripe stage (GS 71) [1-14, 16, 17], up to milky-ripe stage (GS 75) [15]
- Do not use on sulphur-shy apples (Beauty of Bath, Belle de Boskoop, Cox's Orange Pippin, Lanes Prince Albert, Lord Derby, Newton Wonder, Rival, Stirling Castle) or pears (Doyenne du Comice)
- Do not use on gooseberry cultivars Careless, Early Sulphur, Golden Drop, Leveller, Lord Derby, Roaring Lion, or Yellow Rough
- Do not use on apples or gooseberries when young, under stress or if frost imminent
- Do not use on fruit for processing, on grapevines during flowering or near harvest on grapes for wine-making
- Do not use on hops at or after burr stage
- Do not spray with oil or within 30 d of an oil-containing spray

Protective clothing/Label precautions

- 29, 37, 54, 60, 63, 66, 67, 70 [1-17]

Approval

- Approved for aerial application on wheat, barley, oilseed rape, sugar beet [4]; cereals, oilseed rape, sugar beet, swedes, turnips, hops [7]; wheat, barley, oilseed rape, sugar beet [8]; cereals, oilseed rape, sugar beet, swedes [15]. See introductory notes

---

FOR FULL CONDITIONS OF USE ALWAYS READ THE PRODUCT LABEL

76 tar oils

An insecticidal and fungicidal winter wash

Products

1 Mortegg Emulsion	DowElanco	600 g/l	EC	04233
2 Sterilite Tar Oil Winter Wash 60% Stock Emulsion	Coventry Chemicals	600 g/l	EC	05061
3 Sterilite Tar Oil Winter Wash 80% Stock Emulsion	Coventry Chemicals	800 g/l	EC	05062

Uses

Aphids, scale insects, winter moth in APPLES, PEARS, PLUMS, CHERRIES, CURRANTS, GOOSEBERRIES, CANE FRUIT [1-3]. Aphids, scale insects, winter moth in NECTARINES, PEACHES [2, 3]. Suckers in APPLES [1-3]. Cherry fruit moth in CHERRIES [1]. Raspberry moth in RASPBERRIES [1]. Red spider mites in CURRANTS [1]. Scale insects in GRAPEVINES [1-3]. Chemical stripping in HOPS [1]. Blossom wilt in APPLES, PLUMS [1].

Notes

Efficacy
• Spray kills hibernating insects and eggs as well as moss and lichens on trunk
• Spray winter wash over dormant branches and twigs. See labels for detailed timings
• Ensure that all parts of trees or bushes are completely wetted, especially cracks and crevices. Use higher concentration for cleaning up neglected orchards
• Treatment also partially controls winter and raspberry moths and reduces botrytis on currants and powdery mildew on currants and gooseberries
• Combine with spring insecticide treatment for control of caterpillars
• When used in cold weather always add product to water and stir thoroughly before use
• Apply hop defoliant by spraying when bines 1.5-3 m high and direct spray downwards at 45°. Repeat as necessary until cones are formed

Crop safety/Restrictions
• Only spray fruit trees or bushes when fully dormant before any sign of buds swelling
• Do not spray hops if temperature is above 21°C or after cones have formed. Do not drench rootstocks
• Do not spray on windy, wet or frosty days

Special precautions/Environmental safety
• Harmful if swallowed. Irritating to eyes, skin and respiratory system
• Dangerous to fish. Do not contaminate ponds, waterways or ditches with chemical or used container

Protective clothing/Label precautions
• A, C, H
• 5c, 6a, 6b, 6c, 18, 21, 28, 29, 36, 37, 52, 60, 61, 63, 65, 70, 78

477   TCA
A soil-applied herbicide for grass weed control

Products

1 Farmon TCA	Farm Protection	98% w/w	SP	0084
2 MSS TCA	Mirfield	95% w/w	SP	0142
3 NaTA	Hoechst	95% w/w	SP	0146

Uses

Annual grasses, volunteer cereals, wild oats in OILSEED RAPE, SUGAR BEET, PEAS, KALE, FODDER RAPE [1-3]. Annual grasses, perennial grasses, couch in CEREALS, POTATOES, SUGAR BEET, PEAS, SPRING FIELD BEANS, CARROTS, BRUSSELS SPROUTS, CABBAGES, FODDER RAPE, KALE, MUSTARD, SWEDES, TURNIPS *(delayed sowing)* [1-3]. Annual grasses, perennial grasses, couch LUCERNE, RHUBARB [1-3].

Notes

Efficacy
- Acts by absorption through roots and normally requires soil incorporation
- Best results achieved on warm, moist soil when grasses growing actively
- Effectiveness may be reduced if leached out of soil by heavy rain
- For control of annual grasses incorporate to 12-15 cm with final seedbed preparation, i winter oilseed rape not deeper than 2.5 cm or applied pre-emergence
- For control of perennial grasses in spring or autumn incorporate to 12-15 cm, preferably with a rotary cultivator, to fragment rhizomes

Crop safety/Restrictions
- For annual grass control in sugar beet, peas, oilseed rape, kale and fodder rape apply prior to sowing
- For couch control in lucerne and rhubarb apply when crop dormant
- Time interval required between spraying and sowing or planting may be 1-15 wk and varies with season, rate applied, crop (short with brassicas, long with cereals), weather and product. See label for details
- If used immediately before precision drilled sugar beet increase seed rate
- Do not plant following crop within recommended minimum period
- Do not drill spring cereals after spring application

Special precautions/Environmental safety
- Harmful if swallowed. Irritating to skin, eyes and respiratory system

Protective clothing/Label precautions
- A
- 5c, 6a, 6b, 6c, 16, 18, 21, 26, 27, 28, 29, 36, 37, 54, 60, 63, 65, 70, 78

478   tebutam
A pre-emergence amide herbicide used in brassica crops

Products

1 Comodor	Farm Protection	720 g/l	EC	00565
2 Comodor 600	ICI	600 g/l	EC	04453

FOR FULL CONDITIONS OF USE ALWAYS READ THE PRODUCT LABEL

Uses	Annual dicotyledons, annual grasses, volunteer cereals in WINTER OILSEED RAPE, BROCCOLI, BRUSSELS SPROUTS, CABBAGES, CALABRESE, CAULIFLOWERS, KALE, FODDER RAPE, SWEDES, TURNIPS [1, 2].

Notes

Efficacy
- Acts via roots and inhibits germination of weeds. Emerged weeds are not controlled
- Best results achieved on fine, firm seedbeds, free of clods or trash and in good tilth
- Effectiveness depends on adequate soil moisture. Under very dry conditions 8 mm rain or light irrigation is needed but heavy rain on light soils may reduce effectiveness
- Under dry conditions soil incorporation improves weed control otherwise control best by application after drilling or planting
- Recommended for use in tank-mixture with propachlor and as tank-mix or in sequence with trifluralin. Apply trifluralin tank-mix as pre-plant incorporated treatment
- May be tank-mixed with TCA for improved control of grass weeds in oilseed rape [2]
- On soils with more than 10% organic matter weed control may be reduced

Crop safety/Restrictions
- Apply to oilseed rape or sown or transplanted brassicas shortly after drilling or planting but before weeds or crop germinate
- Transplants must be hardened off before treatment. Do not use on block or modular propagated brassicas. Do not use on brassica seedbeds
- In the event of crop failure only brassicas, oilseed rape, dwarf beans, maize, peas or potatoes may be grown in same season

Special precautions/Environmental safety
- Irritating to eyes and skin [1], to eyes [2]
- Harmful to fish. Do not contaminate ponds or ditches with chemical or used container

Protective clothing/Label precautions
- A, C
- 6a, 6b, 18, 21, 28, 29, 36, 37, 53, 60, 63, 66, 70

---

479   tebuthiuron

A soil-acting urea herbicide for total vegetation control

Products	1 Bushwacker	RP Environ.	80% w/w	WP	03139
	2 Bushwacker Granules	RP Environ.	5% w/w	GR	03140

Uses	Total vegetation control, woody weeds in NON-CROP AREAS, WASTE GROUND.

Notes

Efficacy
- Effective via root uptake. May be applied at any time of year but best control of perennials by application in spring-early summer
- A minimum of 2 yr total weed control can be expected, deeper rooted perennial and woody weeds normally controlled for 3 yr
- To control woody plants apply round base. Application in 30 cm wide bands, 1-3 m apart, controls woody weeds with minimum damage to herbaceous vegetation
- Rainfall after treatment will speed up effects. Do not apply in drought or frost
- Soils with more than 20% clay or 3% organic matter will reduce effectiveness

Crop safety/Restrictions
- Take care to avoid drift onto desirable plants
- Do not apply on slopes where chemical may be washed onto desirable vegetation
- Do not use where desirable plants occur within 2 drip-line lengths of target plants or on areas into which roots of desirable plants may extend
- Treated areas will remain unsuitable for cropping for at least 4-5 yr

Special precautions/Environmental safety
- Harmful if swallowed
- Not to be used on or near field crops

Protective clothing/Label precautions
- A, C
- 5c, 18, 21, 25, 29, 36, 37, 54, 63, 65, 70, 78

---

480 tecnazene

A protectant fungicide and potato sprout suppressant
See also carbendazim + tecnazene
*   gamma-HCH + tecnazene*
*   nonylphenoxypoly(ethyleneoxy)ethanol-iodine complex + tecnazene*

Products

1 Ashlade TCNB3	Ashlade	3% w/w	DP	03168
2 Ashlade TCNB5	Ashlade	5% w/w	GR	03169
3 Ashlade TCNB6	Ashlade	6% w/w	DP	03477
4 Ashlade TCNB 10	Ashlade	10% w/w	GR	03479
5 Ashlade TCNB 12	Ashlade	12% w/w	DP	03478
6 Atlas Tecgran 100	Atlas	10% w/w	GR	03676
7 Atlas Tecnazene 6% Dust	Atlas	6% w/w	DP	03093
8 Bygran S	Dean	10% w/w	GR	00366
9 Fumite TCNB Smoke	Hunt	10.2 or 31 g	FU	04140
10 Fusarex 6% Dust	ICI	6% w/w	DP	00954
11 Fusarex 10% Granules	ICI	10% w/w	GR	00955
12 Hickstor 3	Hickson & Welch	3% w/w	DP	03105
13 Hickstor 5 Granules	Hickson & Welch	5% w/w	GR	03180
14 Hickstor 10 Granules	Hickson & Welch	10% w/w	GR	03121
15 Hortag Tecnazene 9% Dust	Avon	9% w/w	DP	02951
16 Hortag Tecnazene Double Dust	Avon	6% w/w	DP	01072
17 Hortag Tecnazene Dust	Avon	3% w/w	DP	01074
18 Hortag Tecnazene 10% Granules	Avon	10% w/w	GR	03966
19 Hortag Tecnazene Potato Granules	Avon	5% w/w	GR	01075
20 Hystore 10	Agrichem	10% w/w	GR	03581
21 Hytec	Agrichem	3% w/w	DP	01099
22 Hytec 6	Agrichem	6% w/w	DP	03580
23 Nebulin	Dean	300 g/l	HN	01469
24 New Hickstor 6	Hickson & Welch	6% w/w	DP	04221
25 New Hystore	Agrichem	5% w/w	GR	01485

FOR FULL CONDITIONS OF USE ALWAYS READ THE PRODUCT LABEL

26	Power Tecnazene 3D	Power	3% w/w	DP	03382
27	Power Tecnazene 10G	Power	10% w/w	GR	03381
28	Quad-Keep	Quadrangle	3% w/w	DP	03170
29	Quad-Store	Quadrangle	5% w/w	GR	03171
30	Tripart Arena 3	Tripart	3% w/w	DP	03818
31	Tripart Arena 6	Tripart	6% w/w	DP	02574
32	Tripart Arena 5G	Tripart	5% w/w	GR	04441
33	Tripart Arena 10G	Tripart	10% w/w	GR	03122
34	Tubodust	FCC	6% w/w	DP	03876
35	Tubostore Granules	FCC	5% w/w	GR	03875

Uses

Sprout suppression in STORED POTATOES [1-8, 10-35]. Dry rot in STORED POTATOES [1-8, 10-35]. Botrytis in TOMATOES, PROTECTED LETTUCE, CHRYSANTHEMUMS, ORNAMENTALS [9]. Leaf mould in TOMATOES [9].

Notes

Efficacy
- Apply dust or granules evenly to potatoes with appropriate equipment (see label for details) during loading. Works by vapour phase action so clamps or bins should be covered to avoid loss of vapour
- Best results achieved on dry, dirt-free potatoes. Fungicidal activity best at 12-14°C, sprouting control best at 9-12°C. For best sprouting control maintain temperature at 10-14°C for first 14 d, then reduce to 7°C [10, 11]
- Under suitable storage conditions sprouting prevented for 4-6 mth. Effectiveness may be decreased with excessive ventilation or inadequate covering
- Sprouting not controlled in tubers which have already broken dormancy
- Treatment can be used in sequence with chlorpropham treatment [10, 11]
- Treatment gives some control of skin spot, gangrene and silver scurf
- For glasshouse botrytis control, fumigate at first sign of disease and repeat at 2-3 wk intervals (not less than 3 wk for lettuce) [9]

Crop safety/Restrictions
- Air tubers for 6 wk before planting; chitting should have commenced
- Do not remove potatoes for sale or processing for at least 6 wk after treatment. Ventilate before marketing
- To prevent contamination of other crops remove all traces of treated potatoes and soil after the store has been emptied
- Do not apply fumigation treatment in bright sunshine or to plants which have wet foliage or are dry at roots [9]
- Apply to tomatoes only after 4-5 trusses thrown, to lettuce 4 wk after planting, to ornamentals when well established and growing strongly [9]
- Do not use on cucurbits, schizanthus and some orchids [9]

Special precautions/Environmental safety
- Irritating to eyes and respiratory system [9]
- See Pesticides Register Issue No 7, 1989 for details of Advisory MRL and guidance on methods for achieving requirements

Protective clothing/Label precautions
- A [23]
- 6a, 6c [9]; 28, 29, 35, 54, 63, 65, 67, 70 [1-35]; 21 [23]

Harvest interval
- lettuce, tomatoes 2 d [9]

• One product formulated for application by thermal fogging [23]. See label for details. See introductory notes on ULV application

481 tecnazene + thiabendazole
A fungicide and potato sprout suppressant

Products

1 Hickstor TBZ 3	Hickson & Welch	3:0.9% w/w	DP	04208
2 Hickstor TBZ 6	Hickson & Welch	6:1.8% w/w	DP	04209
3 Hortag Tecnazene Plus	Avon	6:1.8% w/w	DP	01073
4 Hytec Super	Agrichem	6:1.8% w/w	DP	01100
5 Storaid Dust	MSD Agvet	6:1.8% w/w	DP	02031
6 Storite SS	MSD Agvet	300:100 g/l	SS	02034
7 Tripart Arena 6 + TBZ	Tripart	6:1.8% w/w	DP	04442

Uses

Dry rot, gangrene, silver scurf, skin spot in STORED POTATOES [1-7]. Sprout suppression in STORED POTATOES [1-7].

Notes

Efficacy
• Apply dust evenly with suitable dusting machine (not by hand) as tubers enter store and cover as soon as possible 1-5, 7
• Apply liquid with suitable low-volume mist applicator or spinning disc [6]
• Treat tubers as soon as possible after lifting, no longer than 14 d
• Under suitable storage conditions sprouting prevented for 3-6 mth. Effectiveness may be decreased with excessive ventilation or inadequate covering
• Best results achieved on dry, dirt-free tubers
• Sprouting not controlled on tubers that have broken dormancy

Crop safety/Restrictions
• Air tubers for 6 wk before planting, chitting should have commenced
• Do not mix Storite SS with thiabendazole (Storite Clear Liquid)

Special precautions/Environmental safety
• Harmful to fish. Do not contaminate ponds, waterways or ditches with chemical or used container
• See Pesticides Register Issue No 7, 1989 for details of Advisory MRL for tecnazene and guidance for achieving requirements

Protective clothing/Label precautions
• A
• 28, 29, 37, 53, 63 [1-5, 7]; 21, 29, 53, 63, 65 [6]

Harvest interval
• ware potatoes, 6 wk

FOR FULL CONDITIONS OF USE ALWAYS READ THE PRODUCT LABEL

482 terbacil

A residual uracil herbicide for apples, asparagus and strawberries

Products Sinbar   Du Pont   80% w/w   WP   01956

Uses Annual dicotyledons, annual grasses, couch, bent grasses in APPLES, ASPARAGUS, STRAWBERRIES. Annual dicotyledons, annual grasses, couch, bent grasses in MINT *(off-label)*.

Notes **Efficacy**
- Has some foliar activity but uptake mainly through roots
- Adequate rainfall needed to ensure penetration to weed root zone
- Best results achieved by application in early spring, before mid-Apr

Crop safety/Restrictions
- Apply round base of apple trees established at least 4 yr and keep off foliage
- Apply to asparagus established at least 2 yr
- Apply as directed spot-treatment in strawberries to prevent spread of perennial grasses. Sprayed plants will be severely damaged and will not fruit in following season
- Do not use on sand, loamy sand or gravelly soils or soils with less than 1% organic matter
- Do not use for at least 2 yr prior to grubbing and replanting

Protective clothing/Label precautions
- 21, 28, 29, 54, 63, 65

Approval
- Off-label Approval to May 1993 for use in mint (OLA 0320/90)

483 terbuthylazine

A triazine herbicide available only in mixtures
See also atrazine + terbuthylazine

484 terbuthylazine + terbutryn

A pre-emergence herbicide for peas, beans, lupins and potatoes

Products Opogard 500FW   Ciba-Geigy   150:350 g/l   SC   01514

Uses Annual dicotyledons, annual grasses in PEAS, POTATOES, FIELD BEANS, BROAD BEANS, LUPINS.

Notes **Efficacy**
- Active via root uptake and with foliar activity on cotyledon stage weeds
- Best results achieved by application to fine, firm, moist seedbed, preferably at weed emergence, when rain falls after spraying
- Effectiveness may be reduced by excessive rain, drought or cold

- Residual control lasts for up to 8 wk on mineral soils. Effectiveness reduced on highly organic soils and subsequent use of post-emergence treatment recommended
- Do not cultivate after treatment
- Better control of black nightshade achieved in vining than in dried peas because soil moisture levels generally higher
- Do not use on soils which are very cloddy or have more than 10% organic matter

Crop safety/Restrictions
- Apply to spring sown peas, field and broad beans and lupins as soon as possible after drilling and at least 3 d before crop emergence
- Crop seeds must be covered by at least 25 mm of settled soil
- Vedette and Printana peas may be damaged by treatment. Do not use on forage peas
- Heavy rain after application may cause damage to peas on light soils
- Apply to potatoes before 10% emergence and before weeds past cotyledon stage
- Do not treat peas, beans or lupins on soils lighter than loamy fine sand, potatoes on soils lighter than loamy sand or on very stony soils or lupins on silty clay soil
- Subsequent crops may be sown or planted after 12 wk (14 wk if prolonged drought)

Special precautions/Environmental safety
- Harmful if swallowed
- Harmful to fish. Do not contaminate ponds, waterways or ditches with chemical or used container

Protective clothing/Label precautions
- A, C
- 5c, 18, 21, 28, 29, 36, 37, 53, 63, 66, 70

Approval
- May be applied through CDA equipment for use on peas, beans and lupins. See label for details. See introductory notes on ULV spraying

---

485   terbutryn

A residual triazine herbicide for cereals and aquatic weed control
See also *linuron + terbutryn*
         prometryn + terbutryn
         terbuthylazine + terbutryn

Products					
1 Clarosan 1FG	Ciba-Geigy	1% w/w	GR	03859
2 Prebane 500FW	Ciba-Geigy	500 g/l	SC	01627

Uses   Annual dicotyledons, annual meadow grass, rough meadow grass in WINTER WHEAT, WINTER BARLEY, WINTER OATS, WINTER RYE, TRITICALE, DURUM WHEAT [2]. Aquatic weeds in AREAS OF WATER [1].

Notes   **Efficacy**
- Apply pre-weed emergence in autumn. Best results given on fine, firm seedbed in which any trash or ash dispersed. Do not use on soils with more than 10% organic matter

---

FOR FULL CONDITIONS OF USE ALWAYS READ THE PRODUCT LABEL

- Effectiveness reduced by long, dry period after spraying or by waterlogging
- Do not harrow after treatment
- Various tank mixes recommended for control of emerged weeds
- For aquatic weed control apply granules to water surface with suitable equipment, when growth active but before heavy infestation develops, usually Apr-May, sometimes to Aug. Do not use if maximum water flow exceeds 1 m/3 min
- Effectiveness reduced in water with peaty bottom
- See label for list of susceptible species

Crop safety/Restrictions
- Apply to cereal crops after drilling but before emergence and before 30 Nov
- Only use lower rate on durum wheat, winter oats, triticale or winter rye. Higher rate used to control blackgrass and ryegrass should only be used on winter wheat and barley (not on barley on sands or very light soils)
- Do not use on spring barley cultivars sown in autumn
- Before spraying direct-drilled crops ensure seed well covered and drill slots closed
- Risk of crop damage on stony or gravelly soils especially if heavy rain falls soon after treatment

Special precautions/Environmental safety
- Do not use treated water for irrigation purposes within 7 d of treatment [1]
- Concentration in water must not exceed 10 ppm [1]
- Harmful to fish. Do not contaminate ponds, waterways or ditches with chemical or used container [2]

Protective clothing/Label precautions
- 29, 55, 56, 63, 67 [1]; 29, 53, 63, 66, 70 [2]

Approval
- Approved for aquatic weed control. See introductory notes on use of herbicides in or near water [1]
- Approved for aerial application on cereals, rye, triticale [2]. See introductory notes
- May be applied through CDA equipment. See label for details. See introductory notes on ULV application [2]

---

486   terbutryn + trietazine

A residual triazine herbicide for peas, potatoes and field beans

Products	Senate	Schering	250:250 g/l	SC	04275

Uses   Annual dicotyledons, annual meadow grass in PEAS, POTATOES, FIELD BEANS.

Notes   **Efficacy**
- Product acts mainly through roots of germinating seedlings but also has some contact effect on cotyledon stage weeds
- Weeds germinating from deeper than 2.5 cm may not be completely controlled
- Weed control is improved by light rain after spraying but reduced by excessive rain, drought or cold
- Persistence may be reduced on soils with high organic matter content

Crop safety/Restrictions
- Apply between drilling and 5% crop emergence (peas), before emergence (field beans) between sowing and 10% emergence (potatoes), provided weeds not beyond cotyledon stage
- May be used on spring sown vining, combining or forage peas and field beans covered by not less than 3 cm soil
- Do not spray any winter sown peas or beans sown before Jan
- May be used on early and maincrop potaotes, including seed crops
- On stony or gravelly soils there is risk of crop damage, especially if heavy rain falls soon after treatment. Frost after application may also check crop
- Do not use on soils classed as sands
- Plough or cultivate to 15 cm before sowing or planting another crop. Any crop may be sown or planted after 12 wk (14 wk with drilled brassicas)
- If sprayed pea crop fails redrill without ploughing but do not respray

Special precautions/Environmental safety
- Harmful if swallowed
- Harmful to fish. Do not contaminate ponds, waterways or ditches with chemical or used container

Protective clothing/Label precautions
- A, C
- 5c, 18, 21, 29, 36, 37, 53, 63, 66, 70, 78

487   terbutryn + trifluralin
A residual, pre-emergence herbicide for winter cereals

Products					
1 Ashlade Summit	Ashlade	150:200 g/l	SC	04496
2 Tripart Opera	Tripart	150:200 g/l	SC	04259

Uses

Annual dicotyledons, annual grasses, chickweed, mayweeds, speedwells, annual meadow grass, blackgrass in WINTER WHEAT, WINTER BARLEY.

Notes

Efficacy
- Apply as soon as possible after drilling, before weed or crop emergence when soil moist
- Best results achieved by application to fine, firm, moist seedbed when adequate rain falls after application. Do not use on crops drilled after 30 Nov
- Do not use on uncultivated stubbles or soils with trash or more than 10% organic matter. Do not harrow after treatment
- Effectiveness reduced in mild, wet winters, especially in south-west

Crop safety/Restrictions
- Crops should only be treated if drilled 25-35 mm deep
- Do not use on sandy soils (coarse sandy loam-coarse sand) or on spring-drilled crops
- Do not treat undersown crops or crops to be undersown
- Crops suffering stress due to waterlogging, deficiency, poor seedbed preparation or pest problems may be damaged

FOR FULL CONDITIONS OF USE ALWAYS READ THE PRODUCT LABEL

- Any crop may be sown following harvest of treated crop. In the event of crop failure only resow with wheat or barley

Special precautions/Environmental safety
- Irritating to skin, eyes and respiratory system
- Harmful to fish. Do not contaminate ponds, waterways or ditches with chemical or used container

Protective clothing/Label precautions
- 6a, 6b, 6c, 18, 21, 25, 29, 36, 37, 53, 63, 66, 70

---

488   tetradifon

A bridged diphenyl acaricide for use in horticultural crops
See also dicofol + tetradifon

Products	Tedion V-18 EC	Hortichem	80 g/l	EC	03820

Uses   Red spider mites in FRENCH BEANS, APPLES, PEARS, PLUMS, CHERRIES, APRICOTS, PEACHES, NECTARINES, BLACKCURRANTS, GOOSEBERRIES, RASPBERRIES, BLACKBERRIES, LOGANBERRIES, GRAPEVINES, HOPS, FLOWERS, ORNAMENTALS, TOMATOES, PEPPERS, CUCUMBERS, MELONS.

Notes

Efficacy
- Chemical active against summer eggs and all stages of red spider larvae but does not kill adult mites
- Apply as soon as first mites seen and repeat as required. See label for details of timing
- May be used in conjunction with biological control
- Resistant strains of red spider mite have developed in some areas

Crop safety/Restrictions
- Some rose cultivars are slightly susceptible. Do not treat cissus, dahlia, ficus, kalanchoe or primula

Special precautions/Environmental safety
- Harmful in contact with skin and if swallowed. Irritating to eyes and skin
- Flammable

Protective clothing/Label precautions
- A, C
- 5a, 5c, 6a, 6b, 18, 21, 29, 36, 37, 54, 63 ,65, 70, 78

---

489   tetramethrin

A contact acting pyrethroid insecticide
See also gamma-HCH + resmethrin + tetramethrin
*   phenothrin + tetramethrin*
*   resmethrin + tetramethrin*

Products	Killgerm Py-Kill W	Killgerm	22 g/l	EC	01148

Flies in LIVESTOCK HOUSES, AGRICULTURAL PREMISES.

Notes

Efficacy
- Dilute in accordance with directions and apply as space or surface spray

Crop safety/Restrictions
- Do not apply directly on food or livestock
- Remove exposed milk before application. Protect milk machinery and containers from contamination

Special precautions/Environmental safety
- Harmful to fish. Do not contaminate ponds, waterways or ditches with chemical or used container

Protective clothing/Label precautions
- 12c, 21, 28, 29, 53, 63, 65

490   thiabendazole

A systemic, curative and protectant MBC fungicide
See also captan + fosetyl-aluminium + thiabendazole
carboxin + imazalil + thiabendazole
carboxin + thiabendazole
ethirimol + flutriafol + thiabendazole
gamma-HCH + thiabendazole + thiram
imazalil + thiabendazole
iprodione + metalaxyl + thiabendazole
metalaxyl + thiabendazole
metalaxyl + thiabendazole + thiram
nonylphenoxypoly(ethyleneoxy)ethanol-iodine complex + thiabendazole
tecnazene + thiabendazole

Products					
1 Ceratotect	MSD Agvet	266 g/l	SL	03554
2 Hymush	Agrichem	60% w/w	WP	01092
3 Storite Clear Liquid	MSD Agvet	266 g/l	LS	02032
4 Storite Flowable	MSD Agvet	450 g/l	FS	02033
5 Tecto Dust	MSD Agvet	2% w/w	DS	02092
6 Tecto Systemic Turf Fungicide	Synchemicals	450 g/l	SC	02094
7 Tecto 60% WP	MSD Agvet	60% w/w	WP	02095

Uses

Dry rot, gangrene, silver scurf, skin spot in POTATOES [3, 4, 5]. Canker in OILSEED RAPE *(seed treatment)* [7]. Dollar spot, fusarium patch, red thread in TURF, AMENITY GRASS [6]. Fusarium basal rot in NARCISSI [3, 7]. Dutch elm disease in ELM TREES *(injection)* [1]. Dry bubble, wet bubble in MUSHROOMS [2, 7]. Fungus diseases in ASPARAGUS *(off-label)* [3]. Fungus diseases in ORNAMENTAL BULBS *(off-label)* [3].

FOR FULL CONDITIONS OF USE ALWAYS READ THE PRODUCT LABEL

Efficacy
- Apply to potatoes using suitable equipment within 24 h of lifting. See label for details. Best results achieved on clean, dry tubers. Even distribution essential
- Apply to rapeseed as dry powder or slurry. Slurry must be dry before storing seed [7]
- Apply to turf as spray after mowing and do not mow for at least 48 h. Best results on red thread obtained in combination with fertilizer [6]
- Apply to bulbs as dip. Ensure bulbs are clean [3, 7]
- Use on elms as curative treatment when infection seen or earlier as protective treatment. Inject at a minimum of 3-4 points round base of trunk. Only treat trees with less than 5% crown symptoms [1]
- Apply to mushroom casing as drench, by mechanical incorporation or by spraying between flushes [2, 7]

Crop safety/Restrictions
- Some elms may show phytotoxic damage soon after injection [1]

Special precautions/Environmental safety
- Harmful to fish. Do not contaminate ponds, waterways or ditches with chemical or used container

Protective clothing/Label precautions
- 28, 29, 37, 53, 63, 65, 67

Harvest interval
- potatoes, 21 d [3, 4, 5]

Approval
- One product formulated for application with ULV equipment. See label for details. See introductory notes [4]
- Off-label Approval to Feb 1991 for use on field-grown asparagus (OLA 0221/88) [3]; to May 1992 for use on ornamental bulbs (OLA 0472/89) [3]

---

491   thiabendazole + thiram

A fungicide seed dressing mixture for field and vegetable crops

Products

1 Ascot 480FS	Ciba-Geigy	180:300 g/l	FS	04047
2 Hy-TL	Agrichem	225:300 g/l	FS	03352
3 Hy-Vic	Agrichem	255:255 g/l	FS	02858

Uses

Ascochyta, damping off in FIELD BEANS, PEAS [1]. Ascochyta, damping off in PEAS [2]. Canker in BRASSICAS [3]. Chocolate spot in BROAD BEANS, FIELD BEANS [3]. Neck rot in ONIONS [3]. Seed-borne diseases, damping off in BROAD BEANS, FIELD BEANS, CEREALS, GRASS SEED CROPS, LUPINS, VETCHES, FLAX, VEGETABLES, ORNAMENTALS, CUCURBITS, CARROTS, LEEKS, PARSNIPS [3]. Damping off and wirestem in BRASSICAS [3]. Ascochyta, damping off in EDIBLE PODDED PEAS *(off-label)* [2].

Notes

Efficacy
- Apply through continuous flow seed treaters [1]
- Dress seed as near to sowing as possible
- Dilution may be needed with particularly absorbent types of seed. If diluted material used, seed may require drying before storage

Special precautions/Environmental safety
- Harmful if swallowed. Irritating to eyes, skin and respiratory system
- Treated seed not to be used as food or feed. Do not re-use sack
- Harmful to fish. Do not contaminate ponds, waterways or ditches with chemical or used container

Protective clothing/Label precautions
- A, C
- 5c, 6a, 6b, 6c, 18, 21, 28, 29, 36, 37, 53, 60, 63, 65, 67, 70, 73, 74, 75, 76, 77, 78

Environmental safety
- Off-label Approval to Apr 1992 for use as seed treatment on edible-podded peas (OLA 0336/89) [2]

492 thifensulfuron-methyl

A sulphonyl urea herbicide available only in mixtures
See also *metsulfuron-methyl*
          thifensulfuron-methyl

493 thiometon

A systemic organophosphorus insecticide available only in mixtures
See also *quinalphos + thiometon*

494 thiophanate-methyl

A systemic MBC fungicide with protectant and curative activity
See also *gamma-HCH + thiophanate-methyl*
          iprodione + thiophanate methyl

Products

1 CDA Mildothane	RP Environ.	500 g/l	UL	04154
2 Cercobin Liquid	RP	500 g/l	SC	00457
3 Mildothane Liquid	DowElanco	500 g/l	SC	04345
4 Mildothane Liquid	Hortichem	500 g/l	SC	04552
5 Mildothane Turf Liquid	RP Environ.	500 g/l	SC	01356

Uses

Botrytis, eyespot, fusarium ear blight, sooty moulds in WINTER WHEAT [2]. Eyespot, powdery mildew, rhynchosporium in BARLEY [2]. Clubroot in CABBAGES, CAULIFLOWERS, BRUSSELS SPROUTS [3, 4]. Botrytis, light leaf spot, sclerotinia stem rot in OILSEED RAPE [2]. Chocolate spot in FIELD BEANS [2]. Dollar spot, fusarium patch, red thread in TURF [1, 5]. Scab, powdery mildew in APPLES, PEARS [3, 4]. Canker, storage rots in APPLES [3, 4]. Powdery mildew in CUCUMBERS [4].

Notes   **Efficacy**
- For eyespot control spray when 10-20% tillers attacked between 5-leaf stage and first node detectable (GS 15-31)

FOR FULL CONDITIONS OF USE ALWAYS READ THE PRODUCT LABEL

- For rhynchosporium spray at first signs of disease, not before 3-leaf (GS 13) and not later than boot stage (GS 45)
- To protect against ear disease apply at full ear emergence (GS 59)
- Number and timing of sprays varies with crop and disease, see label for details. On tree fruit sprays should be repeated at 14 d intervals
- Spray treatments can reduce wood canker on apples
- Apply a post-blossom spray or dip fruit to reduce storage rot diseases
- Apply as dip to brassica transplants to control clubroot
- Apply spray or drench treatment on cucumbers. High volume sprays are not recommended where Phytoseiulus is used to control red spider mites
- Apply turf liquid during period of active growth and do not mow for 48 h
- Do not mix with MCPB herbicides or copper. See label for compatible mixtures

Crop safety/Restrictions
- A maximum of 3 applications may be made per crop of cereals
- Do not use after grain watery-ripe stage of cereals (GS 71)
- Spraying apples from blossom to fruitlet stage may increase russeting on prone cultivars
- Do not use on fruit trees at petal fall or early fruitlet stage when preceded by a protectant such as captan

Protective clothing/Label precautions
- A, C, H, M [1]
- 29, 54, 60, 63, 67, 70

195 thiram

A protectant, dithiocarbamate fungicide and animal repellant
See also *carboxin + gamma-HCH + thiram*
fenpropimorph + gamma-HCH + thiram
gamma-HCH + thiabendazole + thiram
gamma-HCH + thiram
metalaxyl + thiabendazole + thiram
metalaxyl + thiram
thiabendazole + thiram

Products

1	Agrichem Flowable Thiram	Agrichem	600 g/l	FS	02857
2	FS Thiram 15% Dust	Ford Smith	15% w/w	DP	04728
3	Hortag Thiram 80	Avon	80% w/w	WP	03753
4	Hortag Thiram Dust	Avon	15% w/w	DP	01076
5	Hortag Thiram Flowable	Avon	600 g/l	SC	02856
6	Unicrop Thianosan	Unicrop	80% w/w	WG	04516

Uses

Damping off in MAIZE, FIELD BEANS, DWARF BEANS, PEAS, VEGETABLES, FLOWERS *(seed treatment)* [1]. Scab in PEARS [6]. Rust in BLACKCURRANTS, CHRYSANTHEMUMS, CARNATIONS [5, 6]. Cane spot, spur blight, botrytis in RASPBERRIES [3, 6]. Botrytis in STRAWBERRIES, OUTDOOR LETTUCE, PROTECTED LETTUCE, POT PLANTS, CHRYSANTHEMUMS, CARNATIONS, FREESIAS, GLASSHOUSE FLOWERS, TOMATOES [2-6]. Fire in TULIPS [3, 6].

Notes

Efficacy
- Spray before onset of disease and repeat every 7-14 d. Spray interval varies with crop and disease. See label for details

- Apply as seed treatment for protection against damping off [1]
- Do not spray when rain imminent
- For use on tulips, chrysanthemums and carnations add non-ionic wetter

Crop safety/Restrictions
- On protected lettuce only 2 post-planting applications of thiram or any combination of products containing thiram or other EBDC fungicides (mancozeb, maneb, zineb), either as a spray or a dust, are permitted within 2 wk of planting out and none thereafter. On winter lettuce if only thiram based products are being used 3 applications may be made within 3 wk of planting out and none thereafter. Do not apply to Worcester Pearmain apples or hydrangeas
- If fruit is to be used for processing consult processors before applying
- Do not dip roots of forestry transplants

Special precautions/Environmental safety
- Irritating to eyes, skin and respiratory system

Protective clothing/Label precautions
- A, C [1]; A [2-6]
- 5a, 5c [1,5]; 6a, 6b, 6c, 18, 21, 28, 29, 36, 37, 54, 63, 65, 70 [1-6]; 78 [1,5]

Harvest interval
- outdoor lettuce 2 wk; protected lettuce 3 wk; other edible crops 1 wk

496   tolclofos-methyl
A protectant organophosphorus fungicide for soil-borne diseases

Products					
	1 Basilex	Fisons	50% w/w	WP	02847
	2 Rizolex	Schering	10% w/w	DS	03826
	3 Rizolex Flowable	Schering	500 g/l	FS	03719

Uses   Damping off and wirestem in LEAF BRASSICAS [1]. Damping off in LETTUCE, PROTECTED LETTUCE, ORNAMENTALS, SEEDLINGS OF ORNAMENTALS [1]. Black scurf and stem canker in SEED POTATOES [2, 3].

Notes   **Efficacy**
- Dust seed potatoes during planting with automatic potato planter (see label for details of suitable applicator) or apply normally during hopper loading [2]
- Apply flowable formulation to clean tubers with suitable misting equipment over a roller table. Spray as potatoes taken into store (first earlies) or as taken out of store (second earlies, maincrop, crops for seed) pre-chitting [3]
- Do not mix flowable formulation with any other product
- To control rhizoctonia in vegetables and ornamentals apply as drench before sowing, pricking out or planting or incorporate into compost
- On established seedlings and pot plants apply as drench and rinse off foliage

---

FOR FULL CONDITIONS OF USE ALWAYS READ THE PRODUCT LABEL

Crop safety/Restrictions
- Apply to seed potatoes before chitting. Not recommended where hot water treatment used or to be used
- Do not apply as overhead drench to vegetables or ornamentals when hot and sunny
- Apply only once before transplanting lettuce or brassicas
- Apply only one treatment to ornamentals at each stage of growth (i.e. sowing, pricking out and potting). Do not use on heathers [1]

Special precautions/Environmental safety
- This product contains an anticholinesterase organophosphorous compound. Do not use if under medical advice not to work with such compounds
- Dangerous to fish. Do not contaminate ponds, waterways and ditches with chemical or used container
- Treated tubers to be used as seed only, not for food or feed

Protective clothing/Label precautions
- 1, 28, 29, 52, 63, 67 [1-3]; 60, 66 [3]

---

97   triadimefon

A systemic triazole fungicide with curative and protectant action

Products					
1 Bayleton	Bayer	25% w/w	WP	00221
2 Bayleton 5	Bayer	5% w/w	WP	00222
3 Portman Triadimefon 25	Portman	25% w/w	WP	03500

Uses

Rhynchosporium, powdery mildew, rust in CEREALS, GRASSLAND [1, 3]. Powdery mildew in FODDER BEET, SUGAR BEET, BRUSSELS SPROUTS, CABBAGES, TURNIPS, SWEDES, PARSNIPS, APPLES, HOPS [1]. Powdery mildew in SUGAR BEET [3]. Rust in LEEKS [1]. Powdery mildew in APPLES, HOPS, RASPBERRIES, STRAWBERRIES, CANE FRUIT, GRAPEVINES [2]. American gooseberry mildew in BLACKCURRANTS, GOOSEBERRIES [2]. Rust in CHIVES, MARJORAM, MINT, SORREL, TARRAGON *(off-label)* [2].

Notes

Efficacy
- Apply at first sign of disease and repeat as necessary, applications to established infections are less effective. Spray programme varies with crop and disease, see label
- Applications to control mildew will also reduce crown rust in oats, snow rot in winter barley and rust in beet [1]
- Can be used effectively in integrated control programmes [2]
- Apply to sugar beet at first signs of mildew [3]

Crop safety/Restrictions
- Up to 3 applications may be made on winter barley, 2 on wheat, oats, rye and spring barley [1]. Up to 6 applications may be made on grapevines [2]
- Continued use of fungicides from the same group in cereals may lead to reduced effectiveness against mildew

Special precautions/Environmental safety
- Harmful to fish. Do not contaminate ponds, waterways or ditches with chemical or used container

Protective clothing/Label precautions
• 29, 35, 41, 53, 63, 67

Withholding period
• Harmful to livestock. Keep all livestock out of treated areas for 21 d

Harvest interval
• Sugar beet, Brussels sprouts, cabbages, turnips, swedes, leeks, apples, hops, cane fruit, strawberries, blackcurrants, gooseberries 14 d; grassland 3 wk; grapes 6 wk

Approval
• May be applied through CDA equipment. See label for details. See introductory notes on ULV application [1]
• Approved for aerial application on sugar beet, cereals, brassicas [1]; blackcurrants [2]. See introductory notes
• Off-label Approval to July 1993 for use on chives, marjoram, mint, sorrel, tarragon (OLA 0425/90) [2]

498   triadimenol

A systemic triazole fungicide for cereals, beet and brassicas
See also fuberidazole + imazalil + triadimenol
fuberidazole + triadimenol

Products					
1 Bayfidan	Bayer	250 g/l	EC	02672
2 Spinnaker	Shell	250 g/l	EC	03941

Uses

Powdery mildew, rust, rhynchosporium, snow rot in WHEAT, BARLEY, OATS, RYE [1, 2]. Powdery mildew in SUGAR BEET, FODDER BEET, TURNIPS, SWEDES, CABBAGES, BRUSSELS SPROUTS [1]. Mildew, rust in SUGAR BEET [2]. Mildew in SWEDES [2].

Notes

Efficacy
• Apply sprays at first signs of disease and ensure good cover
• Treatment also reduces septoria in wheat, crown rust in oats [1, 2] and rust in beet [1]
• Continued use of fungicides from same group can result in reduced effectiveness against powdery mildew. See label for recommended tank mixtures

Crop safety/Restrictions
• Up to 3 applications may be made on winter barley (only 2 in spring), 2 applications may be made on wheat, oats, rye and spring barley
• Do not apply after kernel milky ripe stage of cereals (GS 75)

Special precautions/Environmental safety
• Harmful if swallowed. Irritating to eyes
• Harmful to fish. Do not contaminate ponds, waterways or ditches with chemical or used container

Protective clothing/Label precautions
• A, C

FOR FULL CONDITIONS OF USE ALWAYS READ THE PRODUCT LABEL

• 5c, 6a, 18, 21, 23, 24, 25, 29, 36, 37, 53, 63, 66, 70, 78

Harvest interval
• Beet crops, brassicas 14 d

499   triadimenol + tridemorph
A systemic fungicide mixture with protectant and curative action

Products	Dorin	Bayer	125:375 g/l	EC	03292

Uses   Powdery mildew, rust in WINTER WHEAT, SPRING WHEAT, BARLEY, OATS. Septoria leaf spot in WHEAT. Rhynchosporium, snow rot in BARLEY. Crown rust in OATS.

Notes

Efficacy
• Apply at first signs of disease or as 2 or 3-spray protectant programme
• Treatment for mildew control will reduce crown rust infection on oats

Crop safety/Restrictions
• Up to 3 applications may be used per crop
• Do not apply after kernel milky ripe stage (GS 75)
• Crop scorch may occur if sprayed in frosty weather
• Scorch may occur on wheat if applied under very warm or drought conditions

Special precautions/Environmental safety
• Harmful if swallowed. Irritating to skin
• Grazing livestock must be kept out of treated areas for at least 14 d
• Dangerous to fish. Do not contaminate ponds, waterways or ditches with chemical or used container

Protective clothing/Label precautions
• A, C
• 5c, 6b, 18, 23, 24, 25, 29, 36, 37, 52, 63, 66, 70, 78

500   tri-allate
A soil-acting thiocarbamate herbicide for grass weed control

Products					
1 Avadex BW	Monsanto	400 g/l	EC	00173
2 Avadex BW Granular	Monsanto	10% w/w	GR	00174

Uses   Wild oats, blackgrass, annual meadow grass in WINTER WHEAT, WINTER BARLEY, SPRING BARLEY, SPRING RYE, DURUM WHEAT, TRITICALE, BEET CROPS, BRASSICAS, WINTER OILSEED RAPE, CARROTS, FIELD BEANS, BROAD BEANS, FRENCH BEANS, RUNNER BEANS, FLAX, FODDER BRASSICAS, FODDER LEGUMES, LINSEED, MAIZE, ONIONS, PARSLEY, PARSNIPS, PEAS, SEED BRASSICAS [1]. Wild oats, blackgrass, annual meadow grass in WINTER WHEAT, SPRING WHEAT, WINTER BARLEY, SPRING BARLEY, DURUM WHEAT, TRITICALE, FIELD BEANS, PEAS, SUGAR BEET

[2]. Wild oats, blackgrass, annual meadow grass in MANGE-TOUT PEAS *(off-label)* [1, 2].

Notes

Efficacy
- Mix thoroughly into top 1.5-2.5 cm of soil immediately after spraying and drill up to 21 days later (incorporation after drilling but pre-emergence possible with some crops) [1]. Incorporate or apply to surface pre-emergence (post-emergence application possible in winter cereals up to 2-leaf stage of wild oats) [2]
- On soils with more than 10% organic matter use higher rates [1] or do not use [2]
- Wild oats only controlled before emergence [1], up to 2-leaf stage [2]
- If applied to dry soil, rainfall needed for full effectiveness, especially with granules on surface. Do not use if top 5-8 cm bone dry
- Do not apply to cloddy seedbeds
- Tank mixture with various other pre-emergence herbicides recommended to increase control of dicotyledon weeds. Mixture with Betanal E recommended early post-emergence in sugar and fodder beet (see label for details) [1]
- Use sequential treatments to improve control of barren brome and annual dicotyledons (see label for details)

Crop safety/Restrictions
- Do not apply Avadex BW to spring wheat. Apply Avadex BW Granular to spring wheat post-emergence only to crops drilled in winter
- Consolidate loose, puffy seedbeds before drilling to avoid chemical contact with seed
- Drill cereals well below treated layer of soil (see label for safe drilling depths)
- Do not use on direct-drilled crops or undersow grasses into treated crops
- Do not sow oats or grasses within one year of treatment

Special precautions/Environmental safety
- Harmful in contact with skin and if swallowed [1]. Irritating to eyes and skin [1, 2]
- Harmful to fish. Do not contaminate ponds, waterways or ditches with chemical or used container

Protective clothing/Label precautions
- A, C, H [1]; A [2]
- 5a, 6a, 6b, 14, 16, 18, 21, 28, 29, 36, 37, 53, 63, 66, 70, 78 [1]; 6a, 6b, 18, 21, 36, 37, 53, 63, 67 [2]

Approval
- Approved for aerial application on wheat, barley, winter beans, peas. See introductory notes [2]
- Off-label approval to Feb 1991 for use in mange-tout peas (OLA 0226, 0227/88) [1, 2]

---

501   triazophos

A contact and stomach acting organophosphorus insecticide

Products					
Hostathion	Hoechst	420 g/l	EC	01080

FOR FULL CONDITIONS OF USE ALWAYS READ THE PRODUCT LABEL

Frit fly, leatherjackets in CEREALS. Wheat-blossom midges in WHEAT. Frit fly in MAIZE, SWEETCORN. Frit fly, leatherjackets in GRASSLAND. Pod midge, seed weevil in WINTER OILSEED RAPE. Caterpillars in BRASSICAS. Cabbage root fly in BRUSSELS SPROUTS. Bean beetle in BROAD BEANS. Pea and bean weevils in FIELD BEANS. Pea moth, pea and bean weevils, pea midge, aphids in PEAS. Carrot fly in CARROTS. Cutworms in LEEKS, POTATOES, SUGAR BEET, PARSNIPS, CARROTS, ONIONS, BRASSICAS, CELERY. Capsids, codling moth, tortrix moths, red spider mites in APPLES. Red spider mites, tortrix moths in STRAWBERRIES. American serpentine leaf miner in ARTICHOKES, ASPARAGUS, BEANS, CELERY, CUCURBITS, FENNEL, MANGE-TOUT PEAS, ONIONS, LEEKS, GARLIC, SHALLOTS, POTATOES, RHUBARB, SUGAR BEET, ORNAMENTALS *(off-label)*. Colorado beetle in POTATOES *(off-label)*. Carrot fly in PARSNIPS *(off-label)*. Seed weevil, pod midge in BROWN MUSTARD *(off-label)*. Leaf miners in OUTSIDE ORNAMENTALS *(off-label)*. Caterpillars, cutworms, capsids, codling moth, tortrix moths, red spider mites in NURSERY FRUIT TREES AND BUSHES *(off-label)*. American serpentine leaf miner, non-indigenous leaf miners in ORNAMENTALS *(off-label)*.

Efficacy

- Chemical gives rapid kill of insects and persists for 3-4 wk in plants or soil. It is especially active against caterpillar and fly larvae
- Persistence may be reduced under hot, dry conditions
- Rate, timing and number of sprays vary with pest and crop. See label for details
- For use on brassicas addition of non-ionic wetter is recommended
- For cabbage root fly control in Brussels sprout buttons use of pendant lances advised to obtain maximum coverage of lower buttons
- For control of second generation carrot fly direct spray towards carrot crowns. Rain soon after application may help to distribute chemical in upper layers of soil
- Do not apply to wet crops if run-off likely or heavy rain imminent

Crop safety/Restrictions

- Do not spray cereals from Jan to end of Mar
- To minimize hazard to bees spray oilseed rape when petal-fall complete and apply in early morning or late evening
- Do not use to control pollen beetle
- Do not spray spring oilseed rape
- Do not spray crops suffering from stress (especially Desiree potatoes and fruit trees) or at temperatures above 30°C
- Do not spray Laxton's Fortune or June Wealthy apples or Malus species grown as pollinators

Special precautions/Environmental safety

- A chemical subject to the Poisons Rules 1982 and Poisons Act 1972
- This product contains an organophosphorus compound. Do not use if under medical advice not to work with such compounds
- Toxic if swallowed, harmful in contact with skin
- Harmful to livestock. Keep all livestock out of treated areas for at least 7 d
- Harmful to game, wild birds and animals. Do not spray cereals in areas of known wild-fowl migration
- Dangerous to bees. Do not apply at flowering stage except as directed on peas. Keep down flowering weeds in all crops
- Harmful to fish. Do not contaminate ponds, waterways or ditches with chemical or used container

Protective clothing/Label precautions
- A, C, H
- 1, 4c, 5a, 12c, 16, 18, 23, 24, 28, 29, 35, 36, 37, 41, 46, 48, 53, 60, 64, 66, 70, 79

Withholding period
- Livestock must not be allowed to graze treated cereals or grassland until 4 wk after application

Harvest interval
- oilseed rape, peas 3 wk; other crops 4 wk

Approval
- Approved for aerial application on peas. See introductory notes
- Off-label approval to May 1992 for use on various field, vegetable and glasshouse crops and ornamentals. Specific restrictions apply (OLA 0425/89). Off-label approval to Nov 1991 for aerial application on potatoes for Colorado beetle control (OLA 1366/88); to Mar 1992 for use on parsnips (OLA 0226/89), brown mustard (OLA 0227/89), outdoor ornamentals (OLA 0228/89); to Apr 1992 for use on nursery fruit trees and bushes (OLA 0375/89); to May 1992 for use on ornamentals (OLA 1065/89)

502 trichlorfon

A contact and ingested organophosphorus insecticide

Products Dipterex 80 Bayer 80% w/w SP 00711

Uses Flea beetles, mangold fly in SUGAR BEET, RED BEET, FODDER BEET, MANGOLDS, SPINACH. Caterpillars, leaf miners in BRASSICAS. Cabbage root fly in BRUSSELS SPROUTS. Cutworms in VEGETABLES. Celery fly in CELERY. Cherry bark tortrix in APPLES. Earwigs in BLACKCURRANTS. Strawberry tortrix in STRAWBERRIES. Caterpillars in ROSES. Browntail moth, small ermine moth in HEDGES. Flies in MANURE HEAPS. Leaf miners, caterpillars in TOMATOES *(off-label)*. Leaf miners in SPINACH, SPINACH BEET *(off-label)*. Cabbage root fly in RADISHES, MOOLI *(off-label)*.

Notes **Efficacy**
- Apply at appearance of caterpillars or larvae for most uses and repeat at 7-10 d intervals if necessary. See label for full details
- To control cabbage root fly in Brussels sprouts buttons spray 1 mth before expected harvest and repeat twice at 7 d intervals using pendant lances to obtain good cover
- Addition of non-ionic wetter recommended for use on brassicas
- For cutworm control apply as drenching spray
- For cherry bark tortrix control spray trunks and main branches in mid-May, taking care to avoid leaves

Special precautions/Environmental safety
- This product contains an organophosphorus compound. Do not use if under medical advice not to work with such compounds

---

FOR FULL CONDITIONS OF USE ALWAYS READ THE PRODUCT LABEL

- Harmful to fish. Do not contaminate ponds, waterways or ditches with chemical or used container

Protective clothing/Label precautions
- 1, 21, 28, 29, 35, 53, 63, 67

Harvest interval
- Outdoor tomatoes 14 d; glasshouse tomatoes 3 d; other crops 2 d

Approval
- Approved for aerial application on beet crops, brassicas, spinach. See introductory notes
- Off-label approval to Feb 1991 for use on spinach and spinach beet (OLA 0223/88), on radishes and mooli (OLA 0228/88)

---

503 Trichoderma viride
A biological fungicide for certain tree diseases

Products	Binab T	HDRA	WP, PT 00264

Uses Silver leaf in FRUIT TREES, ORNAMENTAL TREES.

Notes

Efficacy
- Product contains a fungus antagonistic to growth of disease-causing fungi
- Apply wettable powder formulation as a paste onto pruning wounds to protect against silver leaf and other wood decaying fungi
- To treat silver leaf apply pellets into holes drilled every 10 cm round trunk and seal
- Store below 10°C

Protective clothing/Label precautions
- 29, 54, 67

---

504 triclopyr
A pyridine herbicide for perennial and woody weed control
See also clopyralid + triclopyr
2,4-D + dicamba + triclopyr
dicamba + mecoprop + triclopyr

Products					
1 Garlon 2	ICI	240 g/l	EC	03767
2 Garlon 4	Chipman	480 g/l	EC	03768
3 Timbrel	DowElanco	480 g/l	EC	04108

Uses Perennial dicotyledons, woody weeds, stinging nettle, docks, brambles, broom, gorse, hard rush in ESTABLISHED GRASSLAND, ROUGH GRAZING, NON-CROP AREAS, APPLE ORCHARDS [1]. Perennial dicotyledons, woody weeds, stinging nettle, docks, brambles, broom, gorse in FORESTRY, NON-CROP AREAS, ROADSIDE GRASS, NON-CROP GRASS [2]. Broom, gorse, rhododendrons, woody weeds in FORESTRY [3].

Efficacy
- Apply in grassland as spot treatment or overall foliage spray when weeds in active growth in spring or summer. Details of dose and timing vary with species - see label
- Apply to woody weeds as summer foliage, winter shoot, basal bark, cut stump or tree injection treatment
- Apply foliage spray in water when leaves fully expanded but not senescent
- Apply winter shoot, basal bark or cut stump sprays in paraffin or diesel oil. Dose and timing vary with species. See label for details
- Inject undiluted or 1:1 dilution into cuts spaced every 7.5 cm round trunk
- Do not spray in drought, in very hot or cold conditions
- Effectiveness of foliar sprays may be reduced if rain falls within 2 h

Crop safety/Restrictions
- Clover will be killed or severely checked by application in grassland. Do not apply to grass leys less than 1 yr old [1]
- In apples established for at least 4 yr apply as carefully directed spray in Mar- early Apr avoiding leaves and branches [1]
- Avoid spray drift onto edible crops, ornamental plants, Douglas fir, larches or pines. Vapour drift may occur under hot conditions
- Do not drill kale, swedes, turnips, grass or mixtures containing clover within 6 wk of treatment. Allow at least 6 wk before planting trees

Special precautions/Environmental safety
- Irritating to skin [1-3]. Harmful if swallowed [2]
- Flammable [1, 3]
- Not to be used on food crops
- keep livestock out of treated areas for at least 7 d and until foliage of any poisonous weeds such as buttercups or ragwort has died and become unpalatable
- Dangerous to fish. Do not contaminate ponds, waterways or ditches with chemical or used container
- Not to be applied in or near water [1]. Do not apply from tractor-mounted sprayer within 250 m of ponds, lakes or watercourses [1, 3]

Protective clothing/Label precautions
- A, C, H, M
- 5c [2]; 6b [1-3]; 12c [1,3]; 14, 18, 21, 28, 29, 34, 36, 37, 43, 52, 63, 66, 70 [1-3]; 78 [2]

---

505 tridemorph

A systemic, eradicant and protectant morpholine fungicide
See also *carbendazim + maneb + tridemorph*
   propiconazole + tridemorph
   triadimenol + tridemorph

Products

1	Calixin	BASF	750 g/l	EC	00369
2	Power Tridemorph 750	Power	750 g/l	EC	04703

Uses   Powdery mildew in BARLEY, OATS, WINTER WHEAT, SWEDES, TURNIPS.

FOR FULL CONDITIONS OF USE ALWAYS READ THE PRODUCT LABEL

Efficacy
- Apply to cereals when mildew starts to build up, normally May-early Jun and repeat once if necessary. Treatment may also be made in autumn if required
- Reduced rate recommended in tank mix with triazole fungicides (flutriafol, propiconazole, triadimefon, triadimenol) for control of established mildew in barley - see label for details [1]
- Tank mix with carbendazim recommended for rhynchosporium control in barley, with other products for eyespot and yellow rust, see label for details
- Apply to swedes and turnips at first signs of disease, normally Jul-Aug and repeat at 2 wk intervals if required up to a maximum of 3 sprays
- Product is rainfast after 2 h

Crop safety/Restrictions
- Permitted crop growth stages vary with crop, level of infection and product. See label for details
- Do not spray winter wheat in high temperatures or drought
- Do not apply if frost expected
- Do not mix with other chemicals on swedes or turnips

Special precautions/Environmental safety
- Harmful in contact with skin and if swallowed. Irritating to eyes and skin
- Harmful to fish. Do not contaminate ponds, waterways or ditches with chemical or used container

Protective clothing/Label precautions
- A, C
- 5a, 5c, 6a, 6b, 16, 18, 21, 29, 36, 37, 53, 63, 66, 70, 78

Withholding period
- Keep all livestock out of treated areas for at least 14 d

Harvest interval
- swedes, turnips 2 wk; barley, oats 4 wk; wheat 6 wk

Approval
- Approved for aerial application on barley, oats, winter wheat, swedes, turnips [1]. See introductory notes

506   trietazine

A triazine herbicide available only in mixtures

See also linuron + trietazine
linuron + trietazine + trifluralin
simazine + trietazine
terbutryn + trietazine

507   trifluralin

A soil-incorporated dinitroaniline herbicide for use in various crops
See also diflufenican + trifluralin
isoproturon + trifluralin
linuron + trietazine + trifluralin
linuron + trifluralin
napropamide + trifluralin
terbutryn + trifluralin

Products

1 Ashlade Trimaran	Ashlade	480 g/l	EC	04485
2 Atlas Trifluralin	Atlas	480 g/l	EC	03051
3 Campbell's Trifluralin	MTM Agrochem.	480 g/l	EC	00425
4 Farmon Trifluralin	Farm Protection	480 g/l	EC	03396
5 Portman Trifluralin	Portman	480 g/l	EC	02957
6 Treflan	DowElanco	480 g/l	EC	03174
7 Trigard	FCC	480 g/l	EC	02178
8 Tripart Trifluralin 48 EC	Tripart	480 g/l	EC	02215
9 Tristar	PBI	480 g/l	EC	02219

Uses

Annual grasses, annual dicotyledons in OILSEED RAPE, MUSTARD, TURNIPS, BROCCOLI, BRUSSELS SPROUTS, CABBAGES, CAULIFLOWERS, KALE, SUGAR BEET, BROAD BEANS, FRENCH BEANS, RUNNER BEANS, CARROTS, PARSNIPS, LETTUCE, RASPBERRIES, STRAWBERRIES [1-9]. Annual grasses, annual dicotyledons in WINTER WHEAT, WINTER BARLEY [2, 4, 5, 6, 9]. Annual grasses, annual dicotyledons in WINTER CEREALS [1]. Annual grasses, annual dicotyledons in SWEDES [1, 3-9]. Annual grasses, annual dicotyledons in CALABRESE [1-4, 6, 9]. Annual grasses, annual dicotyledons in FIELD BEANS [3, 5, 7, 8]. Annual grasses, annual dicotyledons in NAVY BEANS [1, 2, 4, 6, 9]. Annual grasses, annual dicotyledons in PARSLEY [1, 4, 6, 9]. Annual grasses, annual dicotyledons in ORNAMENTALS [6]. Annual grasses, annual dicotyledons in HERBS *(off-label)* [2-9]. Annual grasses, annual dicotyledons in FENUGREEK, SOYA BEANS, EVENING PRIMROSE *(off-label)* [6, 7, 8]. Annual grasses, annual dicotyledons in PEAS, SOYA BEANS, CELERIAC, SUNFLOWERS, LINSEED, EVENING PRIMROSE, ORNAMENTALS *(off-label)* [9]. Annual grasses, annual dicotyledons in PEAS, RADISH SEED CROPS, SUNFLOWERS, GOLD-OF-PLEASURE, NURSERY FRUIT TREES AND BUSHES, POT PLANTS, BEDDING PLANTS *(off-label)* [6].

Notes

Efficacy
- Acts on germinating weeds and requires soil incorporation to 5 cm (10 cm for crops to be grown on ridges) within 30 min of spraying (2 h [9], except in mixtures on cereals). See label for details of suitable application equipment
- Best results achieved by application to fine, firm seedbed, free of clods, crop residues and established weeds
- Do not use on sands, fen soil or soils with more than 10% organic matter
- In winter cereals normally applied as surface treatment without incorporation in tank-mixture with other cereal herbicides to increase spectrum of weed control. See label for details and for recommended spray programmes in other crops

---

FOR FULL CONDITIONS OF USE ALWAYS READ THE PRODUCT LABEL

Crop safety/Restrictions

- Apply and incorporate at any time during 2 wk before sowing or planting
- Do not apply to brassica plant raising beds
- Transplants should be hardened off prior to transplanting
- Apply in sugar beet after plants have 4-10 leaves and are 10 cm high and harrow into soil
- Has been used successfully on a wide variety of herbaceous and woody ornamentals. Details available from firm [6]
- Only plant recommended crops within 5 mth of treatment following ploughing to at least 15 cm (12 mth with sugarbeet)

Special precautions/Environmental safety

- Irritating to eyes, skin and respiratory system [2, 4], to eyes and skin [3, 8], to eyes [5]
- Flammable
- Harmful to fish. Do not contaminate ponds, waterways or ditches with chemical or used container

Protective clothing/Label precautions

- A [2, 4, 5]
- 6a, 6b, 6c, 12c, 18, 21, 25, 29, 36, 37, 53, 60, 63, 66

Approval

- Off-label Approval to June 1992 for use on peas, celeriac, sunflower (OLA 0567/89) [9], linseed and ornamentals (OLA 0581/89) [9]; to July 1992 for use on herbs (OLA 0558/89) [2], (OLA 0557/89) [3], (OLA 0554/89) [4], (OLA 0561/89) [5], (OLA 0560/89) [7], (OLA 0563/89) [8]; to July 1992 for use on evening primrose (OLA 0549/89) [7], (OLA 0553/89) [8], fenugreek and soya beans (OLA 0550/89) [7], (OLA 0551/89) [8]; to Jan 1993 for use on peas, sunflower, nursery fruit trees and bushes (OLA 0057/90) [6], pot plants, bedding plants, radish seed crops, herbs, gold-of-pleasure (OLA 0058/90) [6], fenugreek, soya bean (0059/90) [6], herbs (OLA 0072/90) [9]; to Feb 1993 for use on evening primrose (OLA 0060/90, 0062/90) [6, 9], soya bean (OLA 0061/90) [9]

508   triforine

A locally systemic fungicide with protectant and curative activity
See also bupirimate + triforine

Products	1 Fairy Ring Destroyer	Synchemicals	190 g/l	EC	04774
	2 Funginex	Synchemicals	66 g/l	EC	00949
	3 Saprol	Promark	190 g/l	EC	01863

Uses   Net blotch, powdery mildew in BARLEY [3]. Fairy rings in TURF [1]. Powdery mildew, scab in APPLES [3]. American gooseberry mildew, currant leaf spot in BLACKCURRANTS [3]. American gooseberry mildew in GOOSEBERRIES [3]. Powdery mildew in HOPS [3]. Powdery mildew, rust in ORNAMENTALS, ROSES, CHRYSANTHEMUMS, CARNATIONS, BEGONIAS, FUCHSIAS [2, 3]. Black spot in ROSES [2, 3]. Powdery mildew in CUCUMBERS *(off-label)* [3].

Notes   **Efficacy**
- Apply to spring barley for mildew control as soon as disease becomes apparent
- For control of net blotch, yellow and brown rust use mixture with mancozeb

- For fairy ring control apply as high volume spray or drench as soon as infection noted and repeat twice at 14 d intervals (3 times if necessary)
- Apply to apples at green cluster and repeat at 10-14 d intervals. Precede by specific scab treatment where high level of scab infection present
- Spray programme on apples also suppresses red spider mite
- On ornamentals apply at 10 d intervals for protective programme, 7 d intervals as curative programme or to control rust
- Number and timing of sprays varies with crop and disease. See label for details
- Apply seed treatment with suitable seed treatment machinery. See label for details

Crop safety/Restrictions
- Do not spray apples following frost injury or when trees are under drought or other stress. Do not use on Golden Delicious or Ingrid Marie or apply before 90% petal fall on Cox's Orange Pippin
- Apply up to 4 sprays on hops in season, not recommended as full programme
- Do not spray apples or hops when temperature exceeds 20°C. In hot weather spray ornamentals in cool part of the day [2]
- On ornamentals of doubtful tolerance treat a small number of plants in first instance. Avoid spraying Loraine begonias, cyclamen and primula in flower

Special precautions/Environmental safety
- Harmful in contact with skin, irritating to eyes [1-2]

Protective clothing/Label precautions
- A, C
- 5a, 6a, 18, 21, 28, 29, 36, 37, 54, 60, 63, 66, 70, 78 [1, 2]

Harvest interval
- apples 7 d; blackcurrants, gooseberries, hops 14 d

Approval
- Approved for aerial application on spring barley. See introductory notes. Consult firm for details [2]
- Off-label Approval to Mar 1991 for use on cucumbers (OLA 0176/88) [2]

509 Verticillium lecanii

A fungal parasite of aphids and whitefly

Products					
1 Mycotal	Koppert	16.1% w/w	WP	04782
2 Vertalec	Koppert	16.1% w/w	WP	04781

Uses

Aphids, whitefly in PROTECTED BEANS, PROTECTED LETTUCE, PROTECTED ORNAMENTALS, PROTECTED CUT FLOWERS, CUCUMBERS, PEPPERS, AUBERGINES.

Notes

Efficacy
- Apply spore powder as spray as part of biological control programme

FOR FULL CONDITIONS OF USE ALWAYS READ THE PRODUCT LABEL

Crop safety/Restrictions
- At high dose up to 3 treatments may be made per crop [1] (up to 2 treatments per crop on chrysanthemums), 1 per crop on all other crops [2]
- At low dose and low volume up to 12 treatments may be made per crop [2]

Protective clothing/Label precautions
- No information

---

510   vinclozolin
A protectant dicarboximide fungicide for a wide range of crops

Products

1 Fumite Ronilan Smoke	Hunt		FU	03572
2 Mascot Contact Turf Fungicide	Rigby Taylor	500 g/l	SC	02711
3 Power Drive	Power	500 g/l	SC	04950
4 Ronilan	BASF	50% w/w	WP	01821
5 Ronilan FL	BASF	500 g/l	SC	02960

Uses

Alternaria, botrytis, sclerotinia stem rot, light leaf spot in OILSEED RAPE [3, 4, 5]. Chocolate spot in BROAD BEANS, FIELD BEANS [3, 4, 5]. Botrytis in DWARF BEANS, RUNNER BEANS, COMBINING PEAS, VINING PEAS [4, 5]. Botrytis in DWARF BEANS, PEAS [3]. Botrytis, sclerotinia in CELERY [4, 5]. White rot in SALAD ONIONS [4, 5]. Botrytis in LETTUCE, PROTECTED LETTUCE, TOMATOES, ONIONS, RASPBERRIES, BLACKBERRIES, GOOSEBERRIES, CURRANTS, STRAWBERRIES, GRAPEVINES, ORNAMENTALS, AUBERGINES, PEPPERS, CUCUMBERS [4, 5]. Dollar spot, fusarium patch, red thread, melting out in AMENITY TURF, SPORTS TURF [2]. Blossom wilt in APPLES, CHERRIES, PLUMS, PEACHES [4, 5]. Storage rots in APPLES, PEARS [4, 5]. Botrytis in ORNAMENTALS [1]. Botrytis, penicillium rot in BULBS *(pre-storage dip)* [4, 5]. Didymella stem rot in TOMATOES [4, 5]. Botrytis, sclerotinia in PEAS-ALL TYPES *(off-label)* [4].

Notes

Efficacy
- Number and timing of sprays varies with crop and disease. See label for details
- Where dicarboximide resistance developed use in spray programme with fungicide of different mode of action. See label for recommendations [4, 5]
- Do not apply sprays if crop wet or if rain or frost expected
- Apply as pre-storage dip to apples and pears for protection against storage disease, to bulbs for botrytis control and suppression of penicillium rot
- Apply as seed dressing and basal stem spray to control white rot in salad onions
- When using as fumigant ensure house smoke-tight, water several hours beforehand, ensure foliage dry and treat in late afternoon/early evening, in calm weather when temperature above 16°C. Keep house closed overnight and ventilate well next morning

Crop safety/Restrictions
- Do not treat lettuce in conditions of low temperature and/or high humidity
- Do not treat celery under hot or humid conditions
- Do not treat mange-tout peas [4, 5]
- Do not treat pot plant seedlings before 3-4 leaf stage
- Do not use on seedlings in blocks for at least 7 d after transplanting
- Do not treat rooted cuttings during first 3-4 d post-planting or treat unrooted cuttings

- Do not spray pot plants during flowering or fumigate plants dry at roots
- Do not fumigate seedlings not hardened off, or rare/unusual plants without small scale test first [1]

Special precautions/Environmental safety
- Irritating to skin [2-5], to eyes and respiratory system [1]
- Remove domestic animals and birds from vicinity of buildings before fumigating [1]
- Harmful to fish. Do not contaminate ponds, waterways or ditches with chemical or used container

Protective clothing/Label precautions
- A, C
- 6a, 6b, 18, 21, 25, 28, 29, 36, 37, 53, 63, 65, 70, 78 [2, 3, 5]; 6a, 6b, 6c, 18, 21, 28, 29, 36, 37, 53, 63, 67 [1, 4]

Harvest interval
- apples, pears (dip) zero;cucumbers, tomatoes 1 d; strawberries 3 d; aubergines, bush fruit, cane fruit, peppers 7 d; apples (foliar spray), beans, celery, cherries, onions (foliar spray), peas, peaches, plums, grapevines, lettuce (Mar-Sep) 14 d; oilseed rape 3 wk; onions (basal spray), lettuce (Oct-Feb) 4 wk

Approval
- Approved for aerial application on oilseed rape, dwarf and field beans [3], on oilseed rape, dwarf and field beans, ornamental bulbs [4, 5]. See introductory notes
- Off-label Approval to Feb 1991 for use on peas (OLA 0231/88) [4]

511　warfarin

A hydroxycoumarin rodenticide

Products					
1	RCR Grey Squirrel Killer Concentrate	Layson	0.5%	CB	01009
2	Sakarat Concentrate	Killgerm	0.5% w/w	CB	01849
3	Sakarat Ready-to-Use (cut wheat)	Killgerm	0.025% w/w	RB	04340
4	Sakarat Ready-to-Use (whole wheat)	Killgerm	0.025% w/w	RB	01850
5	Sakarat X Ready-to-Use Warfarin Rat Bait	Killgerm	0.05% w/w	RB	01851
6	Sakarat X Warfarin Concentrate	Killgerm	1% w/w	CB	01852
7	Sewarin Extra	Killgerm	0.05% w/w	RB	03426
8	Sewarin P	Killgerm	0.025% w/w	RB	01930
9	Sewercide Cut Wheat Rat Bait	Killgerm	0.05% w/w	RB	03761
10	Sewercide Whole Wheat Rat Bait	Killgerm	0.05% w/w	RB	03759
11	Sorexa Plus Rat Bait	Sorex	0.025%	RB	01986
12	Warfarin 0.5% Concentrate	BH&B	0.5%	CB	02325
13	Warfarin Ready Mixed Bait	BH&B	0.025%	RB	02333

FOR FULL CONDITIONS OF USE ALWAYS READ THE PRODUCT LABEL

Grey squirrels in FORESTRY [1]. Rats, mice in AGRICULTURAL PREMISES [2-13].

Efficacy
- For rodent control place ready-to-use or prepared baits at many points wherever rats and mice active. Out of doors shelter bait from weather
- Inspect baits frequently and replace or top up as long as evidence of feeding. Do not underbait
- For grey squirrel control mix with whole wheat and leave to stand for 2-3 h before use
- Use bait in specially constructed hoppers and inspect every 2-3 d. Replace as necessary

Special precautions/Environmental safety
- For use only by local authorities, professional operators providing a pest control service and persons occupying industrial, agricultural or horticultural premises
- Prevent access to baits by children and animals, especially cats, dogs and pigs
- Rodent bodies must be searched for and burned or buried, not placed in refuse bins or rubbish tips. Remains of bait and containers must be removed after treatment and burned or buried
- Bait must not be used where food, feed or water could become contaminated
- The use of warfarin to control grey squirrels is illegal unless the provisions of the Grey Squirrels Order 1973 are observed. See label for list of counties in which bait may not be used

Protective clothing/Label precautions
- 25, 29, 63, 67, 81, 82, 83, 84, 85 [1-13]

512 zineb

A protectant dithiocarbamate fungicide for many horticultural crops
See also ferbam + maneb + zineb

Products

1 Hortag Zineb 7% Dust	Avon	7% w/w	DP	01077
2 Hortag Zineb Wettable	Avon	70% w/w	WP	03752
3 Unicrop Zineb	Unicrop	70% w/w	WP	02279

Uses

Blight in POTATOES, TOMATOES [2, 3]. Downy mildew in LETTUCE, PROTECTED LETTUCE [1, 3]. Downy mildew in ONIONS, LETTUCE, SPINACH [2]. Celery leaf spot in CELERY [2, 3]. Currant leaf spot in BLACKCURRANTS [2, 3]. Downy mildew in HOPS [2, 3]. Botrytis in ANEMONES [2, 3]. Rust in CARNATIONS [1, 3]. Brown rust, white rust, blotch, petal blight, ray blight in CHRYSANTHEMUMS [1]. Rust in CHRYSANTHEMUMS [1, 2]. Fire in TULIPS [3]. Blight, leaf mould in TOMATOES [1-3]. Buck-eye rot, damping off, root rot in TOMATOES [3]. Bubble, cobweb in MUSHROOMS [1, 2]. Fungus diseases in CABBAGES, CAULIFLOWERS, CALABRESE, BROCCOLI, CHINESE CABBAGE, KALE, LEEKS, MARROWS, SALAD ONIONS *(off-label)* [1-3]. Fungus diseases in KOHLRABI *(off-label)* [2, 3].

Notes

Efficacy
- Spray at early stage of disease or as preventive treatment and repeat every 7-14 d
- Number and timing of sprays varies with crop and disease. See label for details
- Increase spray volume on hops as bine develops

- Best results obtained by spraying in settled weather. Do not spray if rain imminent
- Apply as drench to tomatoes at about time of set of 5th truss for control of root rots

Crop safety/Restrictions
- Pre-picking sprays should not be applied to blackcurrants intended for canning
- Apply to hops only before burr stage
- On protected lettuce only 2 post-emergence applications of zineb or of any combination of products containing EBDC fungicides (mancozeb, maneb, thiram, zineb) either as a spray or a dust are permitted within 2 wk of planting out and none thereafter
- Growth of lettuce may be checked under low temperature conditions [1]

Special precautions/Environmental safety
- Irritating to eyes, skin and respiratory system

Protective clothing/Label precautions
- A
- 6a, 6b, 6c, 21, 27, 29, 35, 36, 37, 54, 63, 67, 70 [1-3]

Harvest interval
- blackcurrants, gooseberries 4 wk; protected lettuce 3 wk; outdoor lettuce 2 wk; other edible outdoor crops 1 wk; other edible glasshouse crops 2 d; hops to burr stage only

Approval
- Approved for aerial application on potatoes. See introductory notes [3]
- Off-label Approval to Feb 1991 for use on leafy brassicas, leeks, salad onions and marrows (OLA 0233/88) [1], (OLA 0232/88) [2], (OLA 0238/88) [3]

513   zineb-ethylenethiuram disulphide complex
A protectant dithiocarbamate fungicide available only in mixtures
See also nitrothal-isopropyl + zineb-ethylenethiuram disulphide complex

514   ziram
A dithiocarbamate bird and animal repellent

Products   AAprotect   Unicrop   32% w/w   PA   03784

Uses   Deer, hares, rabbits, birds in FORESTRY, AMENITY TREES AND SHRUBS, TOP FRUIT, FIELD CROPS.

Notes   **Efficacy**
- Apply undiluted to main stems up to knee height to protect against browsing animals at any time of year or spray 1:1 dilution on stems and branches in dormant season
- Use dilute spray on fully dormant fruit buds to protect against bullfinches
- Only apply to dry stems, branches or buds
- Use of diluted spray can give limited protection to field crops in areas of high risk during establishment period

FOR FULL CONDITIONS OF USE ALWAYS READ THE PRODUCT LABEL

Crop safety/Restrictions
• Do not spray elongating shoots or buds about to open

Special precautions/Environmental safety
• Irritating to eyes, skin and respiratory system

Protective clothing/Label precautions
• A, C
• 6a, 6b, 6c, 18, 21, 28, 29, 35, 36, 37

Harvest interval
• 8 wk

INDEX OF PROPRIETARY NAMES
OF PESTICIDES

The references are to entry numbers, not pages.

The references are to entry numbers, not pages.

Appendix 1
SUPPLIERS OF PESTICIDES AND ADJUVANTS

ABM: **ABM Chemicals Ltd.**
Poleacre Lane,
Woodley, Stockport,
Cheshire SK6 1PQ
Tel: (061-430) 4391
FAX: (061-430) 4364

Ace: **Ace Chemicals Ltd.**
Loanwath Road, Gretna,
Dumfriesshire
Tel: (0461) 37572

Adams: **R.P. Adams Ltd.**
Arpal Works, Riverside Road,
Selkirk, Scotland TD7 5DU
Tel: (0750) 21586
Fax: (0750) 21506

Agrichem: **Agrichem (International) Ltd.**
Industrial Estate,
Station Road, Whittlesey,
Peterborough PE7 2EY
Tel: (0733) 204019
Fax: (0733) 204162

Agri-Technics: **Agri-Technics Ltd.**
Muston Gorse, Redmile,
Nottingham NG13 0GN
Tel: (0949) 42255
Fax: (0949) 43407

Agstock: **Agstock Chemicals Ltd.**
PO Box 46, Wokingham,
Berks. RG11 1PF
Tel: (0734) 781163

Aitken: **R. Aitken,**
123 Harmony Road,
Govan, Glasgow GS1 3NB
Tel: (041-440) 0033
Fax: (041-440) 2744

ANP: **ANP Developments,**
40 East Street,
Havant, Hants. PO9 1AJ

Antec: **Antec (AH) International**
Windham Road,
Chilton Industrial Estate,
Sudbury, Suffolk CO10 6XD
Tel: (0787) 77305
Fax: (0787) 310846

Aquaspersions: **Aquaspersions Ltd.**
Charlestown Works,
Charlestown,
Hebden Bridge,
West Yorks. HX7 6PL
Tel: (0422) 843715
Fax: (0422) 845067

Armillatox: **Armillatox Ltd.**
The Old Malt House,
Brassington, Derby DE4 4HJ
Tel: (0629) 85205

Ashlade: **Ashlade Formulations Ltd.**
Ness Road, Slade Green,
Erith, Kent. DA8 2LD
Tel: (0322) 331671
Fax: (0322) 332279

Atlas: **Atlas Interlates Ltd.** Green book.
P.O. Box 38, Low Moor,
Bradford, W. Yorks. BD12 0JZ
Tel: (0274) 671267
Fax: (0274) 606499

Avon: **Avon Packers Ltd.**
Salisbury Road,
Downton, Wilts. SP5 3JJ
Tel: (0725) 22822
Fax: (0725) 22840

Barclay: **Barclay Chemicals (UK),**
28 Howard Street,
Glossop,
Derbyshire SK13 9DD
Tel: (0457) 853386
Fax: (0457) 853557

BASF: **BASF plc.**
Agricultural Division,
Lady Lane, Hadleigh,
Ipswich,   १९९१ ×१
Suffolk IP7 6BQ
Tel: (0473) 822531
Fax: (0473) 827450

Bayer: **Bayer UK Ltd.**
Agrochem Business Group,
Eastern Way,
Bury St Edmunds,
Suffolk IP32 7AH   १९९१ ×१
Tel: (0284) 763200
Fax: (0284) 702810

BDH: **BDH Chemicals Ltd.**
Broom Road,
Poole BH12 4NN
Tel: (0202) 745520
Fax: (0202) 738299

B H & B: **Battle, Hayward & Bower Ltd.**
Victoria Chemical Works,
Crofton Drive,
Allenby Road Industrial Estate,
Lincoln LN3 4NP
Tel: (0522) 29206/41241
Fax: (0522) 38960

Bos: **Bos Chemicals Ltd.**
Paget Hall, Tydd St Giles,
Wisbech, Cambs. PE13 5FL
Tel: (0945) 870118
Fax: (0945) 870264

BP: **BP Oil Ltd.**
BP House, Breakspear Way,
Hemel Hempstead,
Herts HP2 4UL
Tel: (0442) 232323
Fax: (0442) 225225

BritAg: **BritAg Industries Ltd.**
Waterfront House,
Skeldergate Bridge,
York Y01 1DR
Tel: (0904) 611800
Fax: (0904) 627473

Bromine & Chem: **Bromine and Chemicals Ltd.**
6 Arlington Street,
St James's,
London SW1A 1RE
Tel: 071-493 9711
Fax: 071-493 9714

Brown Butlin: **Brown Butlin Ltd.**
Brook House, Ruskington,
Sleaford,
Lincs. NG34 9EP
Tel: (0526) 832771
Fax: (0526) 832967

CDA Chem: **CDA Chemicals Ltd.**
The Granary,
Billingborough, Sleaford,
Lincs. NG34 0QS
Tel: (0529) 240456
Fax: (0529) 240053

Certified: **Certified Laboratories Ltd.**
PO Box 70, Oldbury,
Warley,
W. Midlands B69 4AD
Tel: (021-525) 6678
Fax: (021-500) 5386

Chemsearch: **Chemsearch (UK) Ltd.**
Landchard House,
Victoria Street,
West Bromwich,
W. Midlands B70 8ER
Tel: (021-525) 1666
Fax: (021-500) 5386

Chemtech: **Chemtech (Crop Protection) Ltd.**
The Arable Centre,
Winterbourne Monkton,
Swindon,
Wilts. SN4 9NW
Tel: (06723) 591

Chiltern: **Chiltern Farm Chemicals Ltd.**
11 High Street,
Thornborough,
Buckingham MK18 2DF
Tel: (0280) 822400
Fax: (0280) 822082

Chipman: **Chipman Ltd.**
Portland Building,
Portland Street,
Staple Hill,
Bristol BS16 4PS
Tel: (0272) 574574
Fax: (0272) 563461

Ciba-Geigy: **Ciba-Geigy Agrochemicals**
Whittlesford,
Cambridge CB2 4QT
Tel: (0223) 833621
Fax: (0223) 835211

1990 x 2

Cleanacres: **Cleanacres Ltd.**
Andoversford, Cheltenham,
Glos. GL54 4LZ
Tel: (0242) 820481
Fax: (0242) 820807

Clifton: **Clifton Chemicals Ltd.**
119 Grenville Street,
Edgeley, Stockport,
Cheshire SK3 9EU
Tel: (061-476) 1128
Fax: (061-476) 1280

Coventry Chemicals: **Coventry Chemicals Ltd.**
Herald Way, Binley,
Coventry CV3 2NY
Tel: (0203) 449425
Fax: (0203) 635463

CSC: **Chemical Spraying Co. Ltd.**
Chemical House,
Glenearn Road,
Perth PH2 0NL
Tel: (0738) 23201
Fax: (0738) 30360

Cyanamid: **Cyanamid of Great Britain Ltd.**
Crop Protection Division,
Cyanamid House,
Fareham Road, PO Box 7,
Gosport,
Hants. PO13 0AS
Tel: (0329) 224000
Fax: (0329) 230213

1991 x 1

Darmycel: **Darmycel UK.**
Station Road, Rustington,
Littlehampton,
West Sussex BN16 3RF
Tel: (0903) 775111
Fax: (0903) 787022

Dax: **Dax Products Ltd.**
76 Cyprus Road,
Nottingham NG3 5ED
Tel: (0602) 609996

Dean: **Dean Agrochemicals Ltd.**
19 Whitehouse Gardens,
York YO2 2DZ
Tel: (0904) 629657

Denoon: **Denoon CDS,**
The Campbell House,
Bilbrough,
York YO2 3PN
Tel: (0937) 835515

Deosan: **Deosan Ltd.**
Weston Favell Centre,
Northampton NN3 4PD
Tel: (0604) 414000
Fax: (0604) 406809

Dermaglen: **Dermaglen Ltd.**
PO Box 60,
Northampton NN4 9JH
Tel: (0604) 766361
Fax: (0604) 701238

Dimex: **Dimex Ltd,**
Dimex House,
116 High Street,
Solihull,
West Midlands
Tel: (021-704) 3551
Fax: (021-704) 4024

Doff Portland: **Doff Portland Ltd.**
Bolsover Street,
Hucknall,
Nottingham NG15 7TY
Tel: (0602) 632842
Fax: (0602) 638657

DowElanco: **DowElanco Ltd.**
Latchmore Court,
Brand Street,
Hitchen,
Herts. SG5 1HZ
Tel: (0462) 457272
Fax: (0462) 453906

Du Pont: **Du Pont (UK) Ltd.**
Agricultural Products Department,
Wedgwood Way,
Stevenage,
Herts. SG1 4QN
Tel: (0438) 734000
Fax: (0438) 734154

Elliott: **Thomas Elliott Ltd.**
Hast Hill, Hayes,
Bromley, Kent
Tel: (081-462) 2271

Embetec: **Embetec Crop Protection**
Springfield House,
Kings Road, Harrogate,
N. Yorks. HG1 5JJ
Tel: (0423) 509731
Fax: (0423) 509736

English Woodland: **English Woodlands Ltd.**
Graffham, Petworth,
Sussex GU28 0LR
Tel: (07986) 574
Fax: (042879) 4711

Fargro: **Fargro Ltd.**
Toddington Lane,
Littlehampton,
Sussex BN17 7PP
Tel: (0903) 721591
Fax: (0903) 730737

Farm Protection: **Farm Protection**
Woolmead House West
Bear Lane, Farnham,
Surrey GU9 7UB
Tel: (0252) 734201
Fax: (0252) 736333

FCC: **Farmers Crop Chemicals Ltd.**
County Mills,
Worcester WR1 3NU
Tel: (0905) 27733
Fax: (0905) 723412

Fermenta: **Fermenta ASC Europe Ltd.**
Bayheath House,
4 Fairway, Petts Wood,
Orpington,
Kent BR5 1EG
Tel: (0689) 74011
Fax: (0689) 74085

Fieldspray: **Fieldspray Division, Nilco Chemical Co, Ltd.**
Stewart Road, Kingsland Industrial Park,
Basingstoke,
Hants. RG24 0GX
Tel: (0256) 474661
Fax: (0256) 450603

Fine: **Fine Agrochemicals Ltd.**
3 The Bull Ring,
Worcester WR2 5AA
Tel: (0905) 748444
Fax: (0905) 748440

Fisons: **Fisons plc.**
Horticulture Division,
Levington Research Station,
Levington, Ipswich,
Suffolk IP10 0NG
Tel: (0473) 717811
Fax: (0473) 715549

Fletcher: **Sam Fletcher Agricultural Specialists, Division of Banks Odam Dennick Ltd.**
Fleet Road Industrial Estate,
Holbeach, Spalding,
Lincs. PE12 7EG
Tel: (0406) 22207
Fax: (0406) 26525

Flowering Plants: **Flowering Plants Ltd.**
55 Well Street, Buckingham,
Bucks. MK18 1EP
Tel: (0280) 813764
Fax: (0280) 822238

Ford Smith: **Ford Smith & Co. Ltd.**
Lyndean Industrial Estate,
Felixstowe Road,
Abbey Wood, London SE2 9SG
Tel: (081-310) 8127
Fax: (081-310) 9563

Geeco: **Geeco Division of McKechnie**
Consumer Products Ltd.
Gore Road Industrial Estate,
New Milton,
Hants. BH25 6SE
Tel: (0425) 614600
Fax: (0425) 619463

Gordon: **Gordon International Corp.**
Bossington Farms Ltd.
Houghton, Stockbridge,
Hants. SO20 6LY
Tel: (0794) 388265

HDRA: **Henry Doubleday Research**
Association
Ryton-on-Dunsmore,
Coventry CV8 3LG
Tel: (0203) 303517
Fax: (0203) 639229

Hickson & Welch: **Hickson & Welch Ltd.**
Wheldon Road,
Castleford,
W. Yorks. WF10 2JT
Tel: (0977) 556565
Fax: (0977) 518058

Hoechst: **Hoechst UK Ltd.**
Agriculture Division,
East Winch Hall,
East Winch, King's Lynn,
Norfolk PE32 1HN
Tel: (0553) 841581
Fax: (0553) 841090

Hortichem: **Hortichem Ltd.**
1 Edison Road,
Churchfields Industrial Estate,
Salisbury,
Wilts. SP2 7NU
Tel: (0722) 20133
Fax: (0722) 26799

Hunt: **Octavius Hunt Ltd.**
5 Dove Lane, Redfield,
Bristol BS5 9NQ
Tel: (0272) 556107
Fax: (0272) 557875

ICI: **ICI Agrochemicals**
Woolmead House West,
Bear Lane, Farnham,
Surrey GU9 7UB
Tel: (0252) 733888
Fax: (0252) 736333

ICI Professional: **ICI Agrochemicals**
Professional Products,
Woolmead House East,
Woolmead Walk,
Farnham,
Surrey GU9 7UB
Tel: (0252) 733919
Fax: (0252) 736222

Ideal: **Ideal Manufacturing Ltd.**
Atlas House, Burton Road,
Finedon, Wellingborough,
Northants. NN9 5HX
Tel: (0933) 681616
Fax: (0933) 681042

Intracrop: **Intracrop Ltd.**
The Crop Centre,
Waterstock,
Oxford OX9 1LJ
Tel: (08447) 377
Fax: (08447) 267

Janssen: **Janssen Pharmaceutical Ltd.**
Grove, Wantage,
Oxon. OX12 0DQ
Tel: (0235) 772966
Fax: (0235) 772901

Jones: **Brian Jones and Associates Ltd.**
Fluorocarbon Building
Caxton Hill,
Hertford,
Herts. SG13 7NH
Tel: (0992) 550731
Fax: (0992) 584697

Killgerm: **Killgerm Chemicals Ltd.**
PO Box 2, Wakefield Road,
Flushdyke, Ossett,
W. Yorks. WF5 9BW
Tel: (0924) 277631/3
Fax: (0924) 264757

Koppert: **Koppert (UK) Ltd.**
1 Wadhurst Business Park,
Faircrouch Lane, Wadhurst,
E. Sussex TN5 6PT
Tel: (0892) 884411
Fax: (0892) 882469

Layson: **Leo Fay Ltd.** (trading as Layson Ltd)
35 Tatton Court,
Kingsland Grange,
Warrington,
Cheshire WA1 4RR
Tel: (0925) 814025
Fax: (0925) 837814

Lever: **Lever Industrial Ltd.**
PO Box 100, Runcorn,
Cheshire WA7 3JZ
Tel: (0928) 719000
Fax: (0928) 714628

Mandops: **Mandops (UK) Ltd.**
165 King Street (Unit 2),
Norwich,
Norfolk NR1 1QH
Tel: (0603) 623307
Fax: (0603) 765816

Maxicrop: **Maxicrop International Ltd.**
Bridge House, 97–101 High Street,
Tonbridge,
Kent TN9 1DR
Tel: (0732) 366710
Fax: (0732) 365939

Maxwell Hart: **Maxwell Hart Ltd.**
17 Adlington Court, Birchwood,
Warrington WA3 6PL
Tel: (0925) 825501
Fax: (0925) 812712

McKechnie: **McKechnie Chemicals Ltd.**
PO Box 4, Tanhouse Lane,
Widnes,
Cheshire WA8 0PG
Tel: (051-424) 2611
Fax: (051-424) 4221

Micro-Bio: **Micro-Biologicals Ltd.**
Salisbury Street,
Fordingbridge,
Hants. SP6 1AE
Tel: (0425) 52205
Fax: (0425) 54325

Microcide: **Microcide Ltd.**
Shepherds Grove,
Stanton,
Bury St. Edmunds,
Suffolk IP31 2AR
Tel: (0359) 51077
Fax: (0284) 706410

Midkem: **Midkem Agrochemicals,**
20 Rothersthorpe Avenue,
Northampton NN4 9JH
Tel: (0604) 764027
Fax: (0604) 701238

Mirfield: **Mirfield Sales Services Ltd.**
Moorend House,
Moorend Lane, Dewsbury,
W. Yorks. WF13 4QQ
Tel: (0484) 842451
Fax: (0484) 847066

Mitchell Cotts: **Mitchell Cotts Chemicals Ltd.**
PO Box 6, Steanard Lane,
Mirfield,
W. Yorks. WF14 8QB
Tel: (0924) 493861
Fax: (0924) 490972

Monro: **Monro Horticulture Ltd.**
Goodwood, Chichester,
Sussex PO18 0PJ
Tel: (0243) 533700
Fax: (0243) 784512

Monsanto: **Monsanto plc.**
Thames Tower, Burleys Way,
Leicester LE1 3TP
Tel: (0533) 620864 |990 × 1
Fax: (0533) 530320

MSD Agvet: **MSD Agvet,**
Hertford Road,
Hoddesdon,
Herts. EN11 9BU
Tel: (0992) 467272
Fax: (0992) 467270

MTM Agrochem: **MTM Agrochemicals Ltd.**
18 Liverpool Road,
Great Sankey, Warrington,
Cheshire WA5 1QR
Tel: (0925) 33232
Fax: (0925) 52679

Newman: **Newman Agrochemicals Ltd.**
Cavendish House, Barton,
Cambs. CB3 7AR
Tel: (0223) 263777
Fax: (0223) 264038

Nickerson: **Nickerson Seeds Ltd.**
JNRC, Rothwell,
Lincs. LN7 6DT
Tel: (0472) 89471
Fax: (0472) 89602

PBI: **Pan Britannica Industries Ltd.**
Britannica House,
Waltham Cross,
Herts. EN8 7DY
Tel: (0992) 23691
Fax: (0992) 26452

Pennwalt: **Pennwalt Chemicals Ltd.**
see Bos Chemicals Ltd.
distributors in UK.

Portman: **Portman Agrochemicals Ltd**
Apex House, Grand Arcade,
Tally-Ho Corner,
North Finchley,
London N12 0EH
Tel: (081-446) 8383
Fax: (081-445) 6045

Powaspray: **Powaspray (CDA) Ltd.**
Middlefield Farm,
Great Shelford,
Cambs. CB2 5AN
Tel: (0223) 841995
Fax: (0223) 845576

Power: contact **Kommer-Brookwick**
88 Westlaw Place,
Whitehill Estate,
Glenrothes,
Fife KY6 2RZ
Tel: (0592) 630052
Fax: (0592) 630109

Promark: **Hoechst (UK) Ltd.**
Promark Division,
East Winch Hall,
East Winch, King's Lynn,
Norfolk PE32 1HN
Tel: (0553) 841581
Fax: (0553) 841090

Quadrangle: **Quadrangle Agrochemicals**
Langthorpe, Boroughbridge,
N. Yorks. YO5 9BZ
Tel: (0423) 322429
Fax: (0845) 537668

Rentokil: **Rentokil Ltd.**
Felcourt, East Grinstead,
West Sussex RH19 2JY
Tel: (0342) 833022
Fax: (0342) 326229

Rigby Taylor: **Rigby Taylor Ltd.**
The Riverway Estate,
Portsmouth Road,
Peasmarsh, Guildford,
Surrey GU3 1LZ
Tel: (0483) 35657
Fax: (0483) 34058

Rohm & Haas: **Rohm & Haas (UK) Ltd.**
Lennig House,
2 Masons Avenue,
Croydon,
Surrey CR9 3NB
Tel: (081-686) 8844
Fax: (081-686) 8329

R.P: **Rhone Poulenc Crop Protection,**
Regent House, Hubert Road,
Brentwood,
Essex CM14 4TZ *1990 x 2*
Tel: (0277) 261414
Fax: (0277) 260621

R.P. Environ: **Rhone Poulenc Environmental**
Products,
Address as above

Schering: **Schering Agriculture**
Nottingham Road,
Stapleford, *1990 x 91*
Nottingham NG9 8AJ *x 1*
Tel: (0602) 390202
Fax: (0602) 398031

Seedcote: **Seedcote Systems Ltd.**
Telford Way, Thetford,
Norfolk IP24 1HU
Tel: (0842) 66261
Fax: (0842) 66263

Selectokil: **Selectokil Ltd.**
Abbey Gate Place, Tovil,
Maidstone,
Kent ME15 0PP
Tel: (0622) 755471
Fax: (0622) 692113

Service Chemicals: **Service Chemicals Ltd.**
Lanchester Way,
Royal Oak Industrial Estate,
Daventry,
Northants. NN11 5PH
Tel: (0327) 704444
Fax: (0327) 71154

Shell: **Shell Chemicals UK Ltd.**
Agriculture Division,
Heronshaw House, *1991 x 1*
Ermine Business Park,
Huntingdon,
Cambs. PE18 6YA
Tel: (0480) 414140
Fax: (0480) 444444

Sinclair: **Sinclair Horticulture & Leisure Ltd.**
Firth Road,
Lincoln LN6 7AH
Tel: (0522) 537561
Fax: (0522) 513609

Sipcam: **Sipcam UK Ltd.**
100 Chalk Farm Road,
London NW1 8EH

Smyth-Morris: **Smyth-Morris Chemicals Ltd.**
Bassington Industrial Estate,
Cramlington,
Northumberland ME23 8AD
Tel: (0670) 713411
Fax: (0670) 590235

Sorex: **Sorex Ltd.**
St Michael's Industrial Estate,
Hale Road, Widnes,
Cheshire WA8 8TJ
Tel: (051-420) 7151
Fax: (051-495) 1163

Sphere: **Sphere Laboratories (London) Ltd.**
The Yews, Main Street,
Chilton,
Oxon. OX11 0RZ
Tel: (0235) 833896

Spraydex: **Spraydex Ltd.**
Moreton Avenue,
Wallingford,
Oxon OX10 9DE
Tel: (0491) 25251
Fax: (0491) 25014

Star: **Star Agrochem Ltd.**
Odder Farm,
Saxilby Road, Burton,
Lincoln LN1 2BB
Tel: (0522) 703777

Staveley: **Staveley Chemicals Ltd.**
Staveley Works,
Chesterfield,
Derbyshire S43 2PB
Tel: (0246) 277251
Fax: (0246) 280090

Stoller: **Stoller Chemical Ltd.**
Unit 23, Marathon Place,
Moss Side Trading Centre,
Leyland,
Lancs. PR5 3QN
Tel: (0772) 454443
Fax: (0772) 622320

Sumito: **Sumito Chemical (UK) plc.**
107 Cheapside,
London EC2V 6DQ
Tel: (071-220) 9284
Fax: (071-796) 3533

Synchemicals: **Synchemicals Ltd**
Owen Street, Coalville,
Leics. LE6 2DE
Tel: (0530) 510060
Fax: (0530) 510299

Syntex: **Syntex Manufacturing Ltd.**
Mid Road, Blailinn Industrial Estate,
Cumbernauld GR7 2TL
Tel: (0236) 739696

Tripart: **Tripart Farm Chemicals Ltd.**
Swan House, Beulah Street,
Gaywood, King's Lynn,
Norfolk PE30 4DN
Tel: (0553) 674303
Fax: (0553) 674422

Truchem: **Truchem Ltd.**
Brook House, 30 Larwood Grove,
Sherwood,
Nottingham NG5 3JD
Tel: (0602) 260762
Fax: (0602) 671153

Unicrop: **Universal Crop Protection Ltd.**
Park House, Cookham,
Maidenhead,
Berks. SL6 9DS
Tel: (0628) 526083
Fax: (0628) 810457

Uniroyal: **Uniroyal Chemical Ltd.**
Kennet House,
4 Langley Quay,
Slough,
Berks. SL3 6EH
Tel: (0753) 580888
Fax: (0753) 591350

Vass: **L.W. Vass (Agricultural) Ltd.**
Springfield Farm,
Silsoe Road, Maulden,
Bedford MK45 2AX
Tel: (0525) 403041
Fax: (0525) 402282

Vitax: **Vitax Ltd.**
Selby Place, Stanley Industrial Estate,
Skelmersdale,
Lancs. WN8 8EF
Tel: (0695) 51834
Fax: (0695) 51823

Walkover: **Walkover Sprayers Ltd.**
21 London Road,
Great Shelford,
Cambridge CB2 5DF
Tel: (0223) 844024

WBC Technology: **WBC Technology Ltd.**
Norfolk House,
Gt. Chesterford Court,
Gt. Chesterford,
Saffron Walden,
Essex CB10 1PF
Tel: (0799) 30146
Fax: (0799) 30229

Wellcome: **The Wellcome Foundation Ltd.**
Crewe Hall,
Cheshire CW1 1UB
Tel: (0270) 583151
Fax: (0270) 589305

Appendix 2
A. KEY TO LABEL REQUIREMENTS FOR PROTECTIVE CLOTHING

Engineering control of operator exposure must be used where practicable in addition to the items of protective clothing specified on the product label, but may replace the protective clothing if it provides an equal or higher standard of protection. The series of letters entered in the profiles under the heading Protective clothing / label precautions refers to the protective items listed below :

- A. Suitable protective gloves (the product label should be consulted for any specific requirements about the material of which the gloves should be made)
- B. Rubber gauntlet gloves
- C. Face-shield
- D. Approved respiratory protective equipment
- E. Goggles
- F. Dust mask
- G. Full face-piece respirator
- H. Coverall
- J. Hood
- K. Rubber apron
- L. Waterproof coat
- M. Rubber boots
- N. Waterproof jacket and trousers
- P. Suitable protective clothing

Appendix 2
B. KEY TO NUMBERED LABEL PRECAUTIONS

The series of numbers listed in this section refers to the numbered precautions below. Where the generalised wording includes a phrase such as "...for xx days" the specific requirement for each pesticide is given in full in the "Special Precautions" section.

Medical Restrictions on Use
1. This product contains an organophosphorus compound. DO NOT USE if under medical advice NOT to work with such compounds
2. This product contains an anticholinesterase compound. DO NOT USE if under medical advice NOT to work with such compounds

Hazard Classification
3. Very toxic
a.   In contact with skin
b.   By inhalation
c.   If swallowed

VERY TOXIC

4. Toxic
a.   In contact with skin
b.   By inhalation
c.   If swallowed

TOXIC

5. Harmful
a.   In contact with skin
b.   By inhalation
c.   If swallowed

HARMFUL

6. Irritant
a.   Irritating to eyes
b.   Irritating to skin
c.   Irritating to respiratory system

IRRITANT.

7. Corrosive
8. Causes burns
9. Risk of serious damage to eyes

CORROSIVE

10a. May cause sensitization by skin contact
10b. May cause sensitization by inhalation
11. Danger of serious damage to health by prolonged exposure
12a. Extremely flammable
12b. Highly flammable
12c. Flammable
13. Explosive when mixed with oxidizing substances

Protection of User
14. Wash all protective clothing thoroughly after use, especially the inside of gloves
15. Wash splashes off gloves immediately
16. Take off immediately all contaminated clothing
17. Take off immediately all contaminated clothing and wash underlying skin. Wash clothes before re-use
18. When using do not eat, drink or smoke

19. When using do not eat, drink, smoke or use naked lights
20. Handle with care and mix only in a closed container
21. Wash concentrate from skin or eyes immediately
23. Wash any contamination from skin or eyes immediately
23. After contact with skin or eyes wash immediately with plenty of water
24. In case of contact with eyes rinse immediately with plenty of water and seek medical advice
25. Avoid all contact by mouth
26. Avoid all contact with skin
27. Avoid all contact with eyes
28. Do not breathe gas/fumes/vapour/spray mist/dust
29. Wash hands and exposed skin before eating, drinking or smoking, before meals and after work
30. Wash hands before meals and after work
31. Before entering treated crops, cover exposed skin areas, particularly arms and legs
32. Do not touch cachet with wet hands or gloves
33. Ventilate treated areas thoroughly when smoke has cleared

Protection of Consumers
34. Not to be used on food crops
35. Do not harvest for human or animal consumption for at least xx days/weeks after last application
36. Keep away from food, drink and animal feeding-stuffs

Protection of Public, Livestock, Wildlife etc.
37. Keep out of reach of children
38. Keep unprotected persons out of treated areas for at least 24 hours/other intervals
39. Cover water storage tanks before application
40. Dangerous to livestock. Keep all livestock out of treated areas for at least xx days/weeks. Bury or remove spillages
41. Harmful to livestock. Keep all livestock out of treated areas for at least xx days/weeks. Bury or remove spillages
42. Keep poultry out of treated areas for at least xx days/weeks
43. Keep livestock out of treated areas for at least xx weeks and until foliage of any poisonous weeds such as ragwort has died and become unpalatable
44. Do not feed treated straw or haulm to livestock within 4 days of spraying
45. Dangerous to game, wild birds and animals
46. Harmful to game, wild birds and animals
47. Harmful to animals. Paraquat may be harmful to hares, where possible spray stubbles early in the day
48. Dangerous to bees. Do not apply at flowering stage. Keep down flowering weeds
49. Dangerous to bees. Do not apply at flowering stage except as directed on cereals and peas. Keep down flowering weeds in all crops.
50. Harmful to bees. Do not apply at flowering stage. Keep down flowering weeds
51. Extremely dangerous to fish. Do not contaminate ponds, waterways or ditches with chemical or used container
52. Dangerous to fish. Do not contaminate ponds, waterways or ditches with chemical or used container
53. Harmful to fish. Do not contaminate ponds, waterways or ditches with chemical or used container
54. Do not contaminate ponds, waterways or ditches with chemical or used container

55. Do not dump surplus herbicide in water or ditch bottoms
56. Do not use treated water for irrigation purposes within xx d/wk of treatment
57. Avoid damage by drift onto susceptible crops or water courses
58. Store away from seeds, fertilizers, fungicides and insecticides
59. Store well away from corms, bulbs, tubers and seeds
60. Store away from frost
61. Store away from heat
62. Store under cool, dry conditions
63. Keep in original container, tightly closed, in a safe place
64. Keep in original container, tightly closed, in a safe place, under lock and key
65. Wash out container thoroughly and dispose of safely
66. Wash out container thoroughly, empty washings into spray tank and dispose of safely
67. Empty container completely and dispose of safely
68. Empty container completely and dispose of it in the specified manner
69. Return empty container as instructed by supplier
70. Do not re-use this container for any other purpose
71. Do not burn this container

Sack Labels for Treated Seed
72. Do not handle seed unnecessarily
73. Treated seed not to be used as food or feed
74. Do not re-use sack for food or feed
75. Harmful to game and wild life. Bury spillages
76. Wash hands and exposed skin before meals and after work
77. Do not apply from the air

Medical Advice
78. If you feel unwell seek medical advice (show label where possible) – for 'harmful' pesticides
79. In case of accident or if you feel unwell, seek medical advice immediately (show label where possible) – for 'toxic' and 'very toxic' pesticides
80. If swallowed, seek medical advice immediately and show container or label

Precautions when using Rodenticides
81. Prevent access to baits/powder by children or domestic animals, particularly cats, dogs, pigs and poultry
82. Do not lay baits/powder where food, animal feed or water could become contaminated
83. Remove all remains of bait, tracking powder or bait containers after use and burn or bury
84. Search for and burn or bury all rodent bodies. Do not place in refuse bins or on rubbish
85. Use bait containers clearly marked POISON at all surface baiting points

Appendix 3
KEY TO CROP GROWTH STAGES

Decimal code for the growth stages of cereals

0 Germination
00 Dry seed
03 Imbibition complete
05 Radicle emerged from caryopsis
07 Coleoptile emerged from caryopsis
09 Leaf at coleoptile tip

1 Seedling growth
10 First leaf through coleoptile
11 First leaf unfolded
12 2 leaves unfolded
13 3 leaves unfolded
14 4 leaves unfolded
15 5 leaves unfolded
16 6 leaves unfolded
17 7 leaves unfolded
18 8 leaves unfolded
19 9 or more leaves unfolded

2 Tillering
20 Main shoot only
21 Main shoot and 1 tiller
22 Main shoot and 2 tillers
23 Main shoot and 3 tillers
24 Main shoot and 4 tillers
25 Main shoot and 5 tillers
26 Main shoot and 6 tillers
27 Main shoot and 7 tillers
28 Main shoot and 8 tillers
29 Main shoot and 9 or more tillers

3 Stem elongation
30 ear at 1 cm
31 1st node detectable
32 2nd node detectable
33 3rd node detectable
34 4th node detectable
35 5th node detectable
36 6th node detectable
37 Flag leaf just visible
39 Flag leaf ligule/collar just visible

4 Booting
41 Flag leaf sheath extending
43 Roots just visibly swollen
45 Roots swollen
47 Flag leaf sheath opening
49 First awns visible

5 Inflorescence
51 First spikelet of inflorescence just visible
52 $1/4$ of inflorescence emerged
55 $1/2$ of inflorescence emerged
57 $3/4$ of inflorescence emerged
59 inflorescence completed

6 Anthesis
60
61 } Beginning of anthesis
64
65 } Anthesis half way
68
69 } Anthesis complete

7 Milk development
71 Caryopsis watery ripe
73 Early milk
75 Medium milk
77 Late milk

8 Dough development
83 Early dough
85 Soft dough
87 Hard dough

9 Ripening
91 Caryopsis hard (difficult to divide by thumb-nail)
92 Caryopsis hard (can no longer be dented by thumb-nail)
93 Caryopsis loosening in daytime

Stages of development of oilseed rape

0 Germination and emergence

1 Leaf production
- 1,0 Both cotyledons unfolded and green
- 1,1 First true leaf
- 1,2 Second true leaf
- 1,3 Third true leaf
- 1,4 Fourth true leaf
- 1,5 Fifth true leaf
- 1,10 About tenth true leaf
- 1,15 About fifteenth true leaf

2 Stem extension
- 2,0 No internodes ('rosette')
- 2,5 About five internodes

3 Flower bud development
- 3,0 Only leaf buds present
- 3,1 Flower buds present but enclosed by leaves
- 3,3 Flower buds visible from above ('green bud')
- 3,5 Flower buds raised above leaves
- 3,6 First flower stalks extending
- 3,7 First flower buds yellow ('yellow bud')

4 Flowering
- 4,0 First flower opened
- 4,1 10% all buds opened
- 4,3 30% all buds opened
- 4,5 50% all buds opened

5 Pod development
- 5,3 30% potential pods
- 5,5 50% potential pods
- 5,7 70% potential pods
- 5,9 All potential pods

6 Seed development
- 6,1 Seeds expanding
- 6,2 Most seeds translucent but full size
- 6,3 Most seeds green
- 6,4 Most seeds green-brown mottled
- 6,5 Most seeds brown
- 6,6 Most seeds dark brown
- 6,7 Most seeds black but soft
- 6,8 Most seeds black and hard
- 6,9 All seeds black and hard

7 Leaf senescence

8 Stem senescence
- 8,1 Most stem green
- 8,5 Half stem green
- 8,9 Little stem green

9 Pod senescence
- 9,1 Most pods green
- 9,5 Half pods green
- 9,9 Few pods green

THE UK PESTICIDE GUIDE 1991
RE-ORDER FORM

Surname _____ Initials _____ Dr/Mr/Mrs/Ms

Address _____

_____

Post/Zip Code _____ Country _____

Date _____ Signed _____

Please return your order to any of the addresses overleaf

Please send me _____ more copies of *The UK Pesticide Guide 1991*

☐ Payment enclosed, cheques made payable to CAB International or BCPC Publications Sales

☐ Please invoice (please attach your official company order)

☐ I wish to pay by credit card

Please debit my Access/Master Card/Eurocard/American Express/Visa Card account

by the sum of _____ Name of cardholder _____

Card No _____ Expiry Date _____

Address of cardholder (if different from above) _____

_____

STANDING ORDERS

☐ Yes, I wish to take out a standing order for _____ copy/copies of each new
edition of *The UK Pesticide Guide*

OR TELEPHONE YOUR ORDER NOW AND ASK FOR THE SALES DEPARTMENT

Post-Publication Bulk Discount Prices	
Number of copies	Price per copy
100+	£9.70
50–99	£10.15
20–49	£11.55
10–19	£13.00
1–9	£14.45

1992 EDITION

Substantial discounts will be available for orders placed **in advance of publication**

☐ Please send me advance price details for the 1992 Edition of *The UK Pesticide Guide* as
soon as they are available

Please detach and return to:

CAB International
Wallingford
Oxon OX10 8DE
UK

Telephone: (0491) 32111
Telex: 847964 (COMAGG G)
Fax: (0491) 33508

BCPC Publications Sales
Bear Farm
Binfield
Bracknell
Berkshire RG12 5QE
UK

Telephone: (0734) 341998
Fax: (0252) 727194